No. 2612
$25.95

SOLUTIONS of partial differential equations

Dean G. Duffy

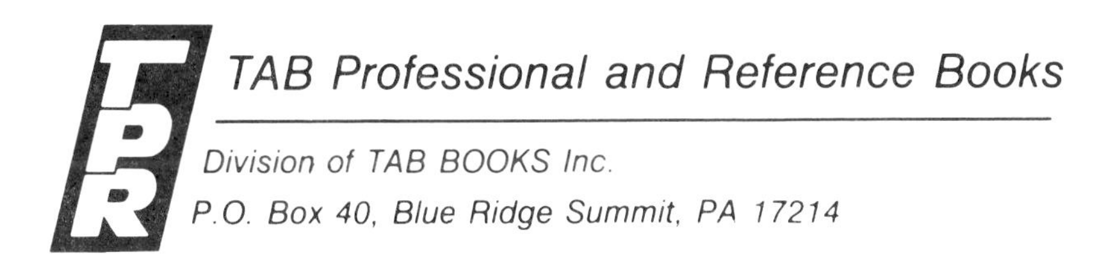

TAB Professional and Reference Books

Division of TAB BOOKS Inc.
P.O. Box 40, Blue Ridge Summit, PA 17214

Dedicated to
the Brigade of Midshipmen
of the United States Naval Academy

FIRST EDITION

FIRST PRINTING

Printed in the United States of America

Library of Congress Cataloging in Publication Data

Duffy, Dean G.
Solutions of partial differential equations.

Includes index.
1. Differential equations, Partial—Numerical solutions. I. Title.
QA374.D84 1986 515.3′53 85-27877
ISBN 0-8306-0412-X

Contents

2 Fourier Series and Integral 63

3 Laplace transforms 132

Acknowledgments

No book of this magnitude can be undertaken without the encouragement and inspiration of many individuals. For this author, it has been the many midshipmen who have honored him by taking Engineering Mathematics SM311 and SM312. Of these many fine students, the author must single out Ensigns Karl Diederich, Eric Krebs, and Dave Magnoni of the Class of 1985 for special thanks. They have made this task especially worthwhile.

Introduction

This book evolved from a two semester sequence of advanced engineering mathematics given to the midshipmen at the United States Naval Academy. This material is usually presented in their second or third years (sophomore or junior years) after they have had the typical sequence of differential, integral and multivariable calculus along with differential equations. Because these students are typically science and engineering majors, it was designed for the student who needs to solve a partial differential equation in the course of his undergraduate studies.

The emphasis of this book is on the mechanics of how to solve a given partial differential equation rather than on the fundamental theory underlying these equations. For undergraduates these techniques usually fall into the two areas of orthogonal expansions and Laplace transforms, and these techniques constitute the majority of the topics presented. Chapter 1 is devoted to the development of the concept of eigenfunction expansions. First, the general Sturm-Liouville theory is presented in its simplest form. Eventually the more difficult Fourier-Bessel and Fourier-Legendre series are treated.

Chapter 2 is devoted to the most common use of eigenfunction expansions—the Fourier series. The Fourier series is first presented, re-expressed in terms of complex notation and finally generalized to the Fourier integral. Along with many illustrations of its use, the finite Fourier series is introduced and illustrated.

In Chapter 3, Laplace transforms are introduced. In addition to the fundamental theorems, convolution and simple inversion methods, numerous examples of its use in solving ordinary differential equations are presented. Of particular importance for future reference is the discussion of the inversion integral. Because this integral involves integration in the complex plane, it is desirable for the reader to have had a course in complex variables before tackling this material. However, the relevant material is presented in Appendix B.

After these preliminaries the remaining portions of the book are devoted to partial differential equations. Chapter 4 contains the standard techniques for solving first-order partial differential equations. Although these equations do not play as large of a role as the classical equations of mathematical physics in engineering and science, they do provide a natural introduction to the concept of characteristics.

The wave equation is introduced in Chapter 5. In the first few sections, the simple wave equation is slowly modified to introduce the concepts of dispersion and damping (Sections 5.3-5.4). Then the use of eigenfunction expansions in Bessel functions and Legendre polynomials are explored (Sections 5.7-5.8).

The use of Laplace transforms in solving the wave equation is shown in Sections 5.14-5.16. First, simple examples are worked and then the equation of telegraphy (Section 5.15) is worked out in detail. The especially difficult use of Laplace transforms with branch cuts is illustrated in Section 5.16.

Because the wave equation often applies to regions of infinite domain, the use of Fourier transforms, the method of characteristics and d'Alembert's formula has been included in Sections 5.11-5.13 and 5.17.

Chapter 6 is devoted to the heat equation. For various boundary conditions, the heat equation is solved first using sines and cosines (Sections 6.3, 6.8) and then with Bessel functions (Sections 6.4, 6.9). The use of Laplace transforms in solving the heat equation is presented in Sections 6.14-6.16.

The book concludes with a study of Laplace's equation. First, the equation is solved by separation of variables in rectangular (Section 7.3), cylindrical (Section 7.4) and spherical coordinates (Section 7.5-7.7). The book closes with the classic application of conformal mapping in solving Laplace's equation.

It is hoped that by keeping the most commonly used techniques in solving linear partial differential equations rooted in the applications that gave them birth, the student will find this a refreshing technique of learning and a valuable reference for future work. Certainly it will answer that age-old question: "What's this good for, anyway?"

Chapter 1

Sturm-Liouville Problem

THE SUBJECT OF MOST OF THIS BOOK IS PARTIAL DIFFERENtial equations: their physical meaning, problems in which they appear, and their solution. As we shall see, there are a number of techniques that have been developed for their solution. Many of these techniques require separating a partial differential equation into ordinary differential equations.

1.1 INTRODUCTION

An important class of these ordinary differential equations in one dimension involve those that satisfy certain conditions at two end points. Such problems are called *boundary-value problems*, in distinction with *initial-value problems*, where all conditions are specified at one point. The solution of these boundary-value problems is often easily and elegantly constructed as a series of orthogonal functions. Exactly what an eigenfunction is and how you construct a series of them is the subject of this chapter.

The study of eigenfunctions is intimately connected with the second-order linear differential equation

$$\frac{d}{dx}\left(p(x)\frac{dy}{dx}\right) + [q(x) + \lambda r(x)]y = 0 \qquad \textbf{(1.1.1)}$$

together with the boundary conditions

$$\alpha y(a) + \beta y'(a) = 0 \text{ and } \gamma y(b) + \delta y'(b) = 0. \qquad \textbf{(1.1.2)}$$

The important thing to notice about the boundary conditions is that they are specified at *each* end of the range. These equations include most of the boundary-value problems occurring naturally in physics and engineering. In Eq. (1.1.1), $p(x)$, $q(x)$ and $r(x)$ are real functions of x; λ is a parameter; and $p(x)$ and $r(x)$ are functions that are assumed to be continuous and positive on the interval $a \leq x < b$. Taken together, Eq. (1.1.1) and (1.1.2) are known as the regular *Sturm-Liouville problem* because it was first intensively studied by the French mathematicians Sturm and Liouville in the 1830s. In the case when $p(x)$ or $r(x)$ vanishes at some point in the interval $[a,b]$ or when the interval is of infinite length, the problem is called a singular Sturm-Liouville problem.

Another important boundary condition that often arises with the differential equation (1.1.1) is

$$y(a) = y(b) \text{ and } y'(\text{a}) = y'(\text{b}). \qquad \textbf{(1.1.3)}$$

These relationships are called *periodic boundary conditions*, because both the solution and its derivatives are required to have the same value at the end points of the interval $[a,b]$. Periodic boundary conditions are not associated with a regular Sturm-Liouville problem.

Although Eq. (1.1.1) appears to be rather restrictive, many boundary-value problems may be written in this canonical form. Often this can be done by inspection. However, it can be shown that any equation of the form

$$a_0(x)\frac{d^2y}{dx^2} + a_1(x)\frac{dy}{dx} + (a_2(x) + \lambda a_3(x))\,y = 0 \qquad \textbf{(1.1.4)}$$

can be written in the form of Eq. (1.1.1) by setting

$$p(x) = \exp\left(\int \frac{a_1(x)}{a_0(x)}\,dx\right), \quad q(x) = \frac{a_2(x)p(x)}{a_0(x)} \quad \text{and}$$

$$r(x) = \frac{a_3(x)p(x)}{a_0(x)}. \qquad \textbf{(1.1.5)}$$

In the case of regular Sturm-Liouville problems, we can solve the problem in either closed form or as a power series solution (the method of Frobenius) or numerically. From a thorough study of these solutions, it is found that unless the λ's are restricted to particular values, the only possible solution is $y = 0$. These λ's which allow a regular Sturm-Liouville problem to have a nontrival solution are called the *eigenvalues* of the problem (1.1.1) and (1.1.2) and the corresponding nontrival solutions are called *eigenfunction.* If there is only one independent eigenfunction for each eigenvalue, that eigenvalue is said to be simple. When more than one eigenfunction belongs to a single eigenvalue, the problem is referred to as *degenerate*. It is possible to show (See Morse, P.M. and H. Feshbach, 1953: *Methods of Theoretical Physics*. McGraw Hill Book Co., Inc., 722-724 for the proof.) that a regular Sturm-Liouville problem has infinitely many real and simple eigenvalues λ_n, $n = 1,2,3,$ . . . which can be arranged in a monotonically increasing sequence $\lambda_1, \lambda_2, \lambda_3, \ldots, \lambda_n$ such that λ_n tends to infinity as n tends to infinity.

1.2 EXAMPLES OF THE STURM-LIOUVILLE PROBLEM: THE BUCKLING OF A TAPERED COLUMN

In this section we shall consider several examples of the Sturm-Liouville problem. First, let us find the eigenvalues and eigenfunctions of

$$u'' + \lambda u = 0 \tag{1.2.1}$$

subject to the boundary conditions that

$$u(0) = 0 \text{ and } u(\pi) - u'(\pi) = 0. \tag{1.2.2}$$

Our first task is to check to see whether the problem is indeed a regular Sturm-Liouville problem. A comparison between Eq. (1.2.1) and Eq. (1.1.1) shows that

$$\frac{d}{dx}\left(p(x)\frac{dy}{dx}\right) + \left(q(x) + \lambda r(x)\right) y = 0 \tag{1.1.1}$$

and

$$\frac{d}{dx}\left(1 \ \frac{du}{dx}\right) + (0 + \lambda\, 1)\, u = 0. \tag{1.2.1}$$

By allowing $p(x) = 1$, $q(x) = 0$ and $r(x) = 1$, Eq. (1.1.1) becomes equal to Eq. (1.2.1). A comparison of the boundary conditions shows that

$$\alpha y(a) + \beta y'(a) = 0 \text{ and } \gamma y(b) + \delta y'(b) = 0 \tag{1.1.2}$$

$$1u(0) + 0u'(0) = 0 \text{ and } 1u(\pi) + (-1)u'(\pi) = 0. \tag{1.2.2}$$

By allowing $\alpha = \gamma = 1$, $\delta = -1$, $\beta = 0$, $a = 0$ and $b = \pi$, Eq. (1.1.2) and Eq. (1.2.2) become identical.

Having shown that the posed problem is a regular Sturm-Liouville problem, we next solve the differential equation. The general solution of the differential equation is

$$u(x) = A \cosh(mx) + B \sinh(mx) \text{ if } \lambda < 0 \tag{1.2.3}$$

$$u(x) = C + Dx \qquad \text{if } \lambda = 0 \tag{1.2.4}$$

and

$$u(x) = E \cos(kx) + F \sin(kx) \qquad \text{if } \lambda > 0 \tag{1.2.5}$$

where $\lambda = -m^2 < 0$ in Eq. (1.2.3) and $\lambda = k^2 > 0$ in Eq. (1.2.5). Both k and m are real and positive by these definitions. Turning to the condition that $u(0) = 0$, we find that $A = C = E = 0$. The other boundary condition $u(\pi) - u'(\pi) = 0$ gives

$$B(\sinh(m\pi) - m \cosh(m\pi)) = 0 \tag{1.2.6}$$

$$D = 0 \tag{1.2.7}$$

and

$$F\,[\sin(k\pi) - k \cos(k\pi)] = 0. \tag{1.2.8}$$

If $\sinh(m\pi) - m \cosh(m\pi)$ is graphed for all positive m, it is always found to be negative. Consequently B must be taken equal to zero. However, in Eq. (1.2.8), a nontrival solution (i.e., $F \neq 0$) may be found if

$$\tan(k\pi) = k. \tag{1.2.9}$$

The eigenvalues of the problem are consequently the square of the roots of the transcendental equation (1.2.9). The roots may be found either graphically or through the use of a numerical algorithm. The graphical solution to the problem is shown in Fig. 1.2.1. The root $k = 0$ has been excluded because $\lambda = 0$ is a trival solution. The corresponding eigenfunction, which is traditionally written without the arbitrary amplitude constant, is given by

$$u_n(x) = \sin(k_n x) \tag{1.2.10}$$

and the first four are shown in Fig. 1.2.2.

For a second example let us solve the Sturm-Liouville problem

$$u'' + \lambda u = 0 \tag{1.2.11}$$

with the boundary conditions

$$u(0) - u'(0) = 0 \tag{1.2.12}$$

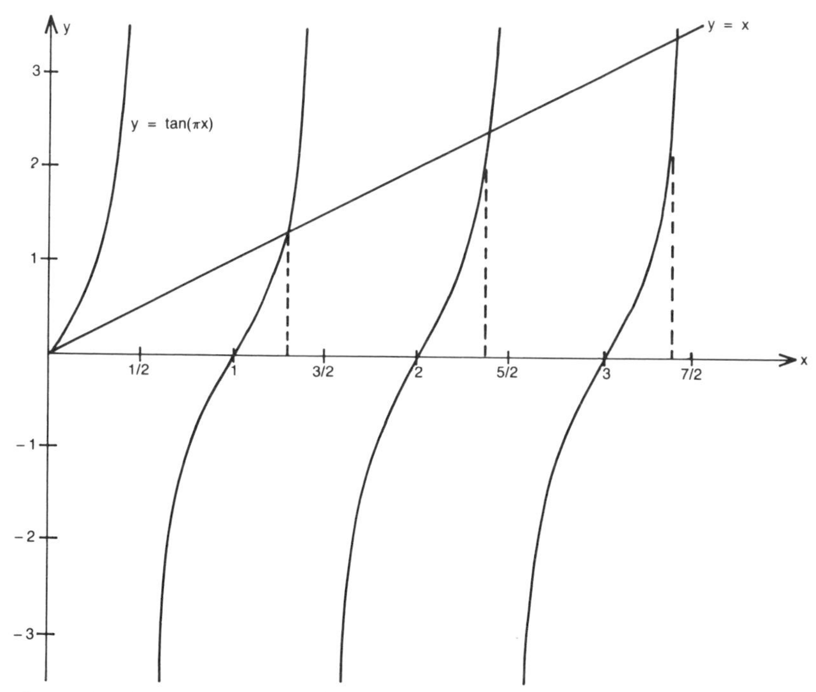

Fig. 1.2.1: Graphical solution of tan(π x) = x.

$$u(\pi) - u'(\pi) = 0. \qquad \textbf{(1.2.13)}$$

Once again the three possible solutions to Eq. (1.2.11) are

$$u(x) = A\cosh(mx) + B\sinh(mx) \text{ if } \lambda < 0 \qquad \textbf{(1.2.14)}$$

$$u(x) = C + Dx \qquad \lambda = 0 \qquad \textbf{(1.2.15)}$$

$$u(x) = E\cos(k\,x) + F\sin(kx) \qquad \lambda > 0 \qquad \textbf{(1.2.16)}$$

Two simultaneous equations result from the substitution of Eq. (1.2.14) into Eqs. (1.2.12) and (1.2.13):

$$A - mB = 0 \qquad \textbf{(1.2.17)}$$

$$[\cosh(m\pi) - m\sinh(m\pi)]\,A + [\sinh(m\pi) - m\cosh(m\pi)]\,B = 0 \qquad \textbf{(1.2.18)}$$

The elimination of A between the two equations yields

$$\sinh(m\pi)\,(1 - m^2)\,B = 0. \qquad \textbf{(1.2.19)}$$

If Eq.(1.2.14) is to be a nontrival solution, $B \neq 0$ and

$$\sinh(m\pi) = 0 \qquad \textbf{(1.2.20)}$$

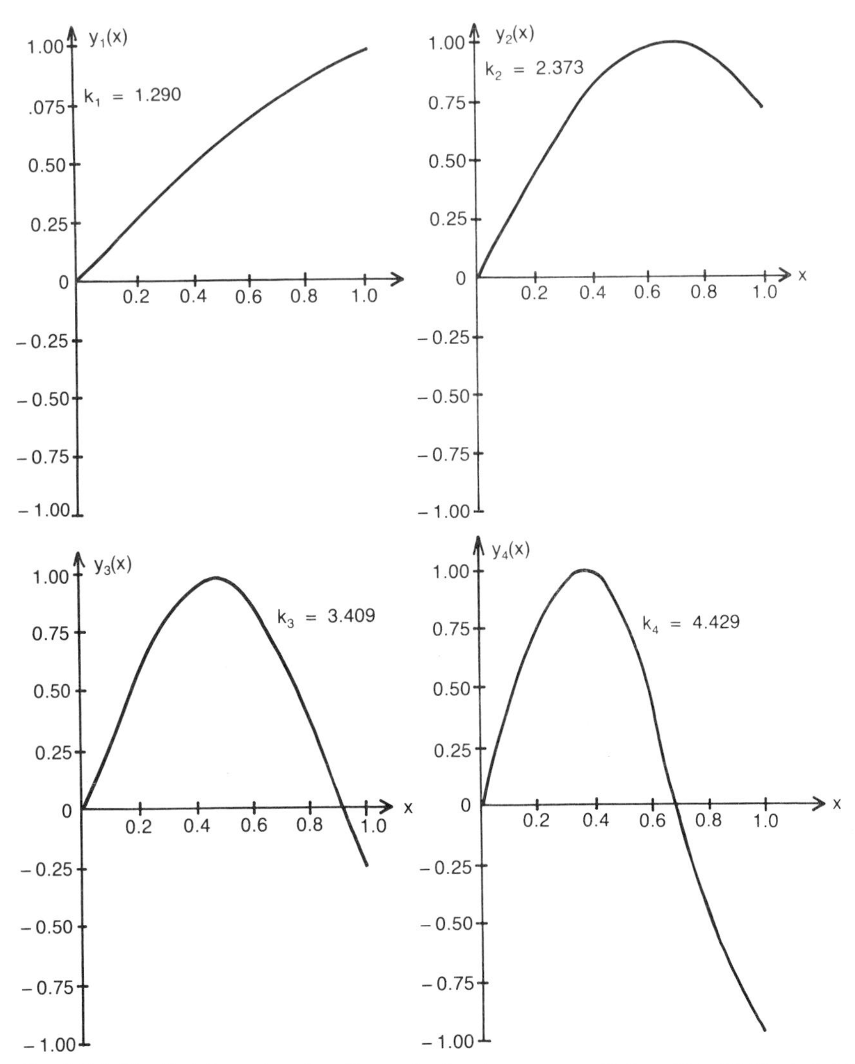

Fig. 1.2.2: The first four eigenfunctions sin $(k_n x)$ corresponding the eigenvalue problem tan $(k\pi) = k$.

or

$$m^2 = 1. \qquad \textbf{(1.2.21)}$$

Condition (1.2.20) cannot hold because it implies $m = \lambda = 0$ which contradicts the assumption used in deriving Eq. (1.2.14). On the other hand, condition (1.2.21) is quite acceptable. It corresponds to $\lambda = -1$ and the eigenfunction is

$$u_0(x) = \cosh(x) + \sinh(x) = e^x \qquad \textbf{(1.2.22)}$$

because it satisfies the differential equation

$$u_0'' - u_0 = 0 \qquad \textbf{(1.2.23)}$$

and the boundary conditions

$$u_0(0) - u_0'(0) = 0 \quad \textbf{(1.2.24)}$$

$$u_0(\pi) - u_0'(\pi) = 0. \quad \textbf{(1.2.25)}$$

An alternative method of finding m, which is quite popular because it can be used in more difficult problems, follows from viewing Eqs. (1.2.17) and (1.2.18) as a system of homogeneous linear equations where A and B are the unknowns. It is well known (see Jeffreys, H. and B. Jeffreys, 1972: *Methods of Mathematical Physics.* Cambridge: At the University Press, Section 4.05) that in order for Eqs. (1.2.17)–(1.2.18) to have a nontrival solution (i.e., $A \neq 0$ and/or $B \neq 0$) the determinant of the coefficients must vanish:

$$\begin{vmatrix} 1 & -m \\ \cosh(m\pi) - m\sinh(m\pi) & \sinh(m\pi) - m\cosh(m\pi) \end{vmatrix} = 0.$$

Expanding the determinant, we have

$$\sinh(m\pi)\,(1 - m^2) = 0$$

which leads directly to Eqs. (1.2.20) and (1.2.21).

Considering Eq. (1.2.15) next, we have

$$C - D = 0 \quad \textbf{(1.2.26)}$$

and

$$C + D\pi - D = 0. \quad \textbf{(1.2.27)}$$

This set of simultaneous equations yields $C = D = 0$ and we have only trival solutions for $\lambda = 0$.

Finally, we examine solution (1.2.16) and find that

$$E - kF = 0 \quad \textbf{(1.2.28)}$$

and

$$(\cos(k\pi) + k\sin(k\pi))E + (\sin(k\pi) - k\cos(k\pi))F = 0 \quad \textbf{(1.2.29)}$$

The elimination of E from Eqs. (1.2.28) and (1.2.29) gives

$$F(1 + k^2)\sin(k\pi) = 0. \quad \textbf{(1.2.30)}$$

In order that Eq. (1.2.16) is nontrival, $F \neq 0$ and

$$k^2 = -1 \tag{1.2.31}$$

or

$$\sin(k\pi) = 0 \tag{1.2.32}$$

Condition (1.2.31) violates our assumption that k is real. However, the condition (1.2.32) can be satisfied if

$$k = n, \; n = 1, 2, 3, \ldots \tag{1.2.33}$$

because k must be positive. Consequently, we have the additional eigenvalues of

$$\lambda_n = n^2 \tag{1.2.34}$$

and the corresponding eigenfunctions

$$u_n(x) = \sin(nx) + n\cos(nx). \tag{1.2.35}$$

because $u(x) = F\sin(kx) + F\,k\cos(kx)$ since $E = kF$. Figure 1.2.3 illustrates some of the eigenfunctions given by Eq. (1.2.22) and (1.2.35).

Consider now the problem

$$u'' + \lambda u = 0 \tag{1.2.36}$$

with

$$u(\pi) = u(-\pi) \tag{1.2.37}$$

and

$$u'(\pi) = u'(-\pi). \tag{1.2.38}$$

This is *not* a regular Sturm-Liouville problem because the boundary conditions are periodic and do not conform to the canonical boundary condition (1.1.2).

The general solution to Eq. (1.2.36) is given by Eqs. (1.2.3) – (1.2.5). Substituting these solutions into the boundary condition (1.2.37), we obtain

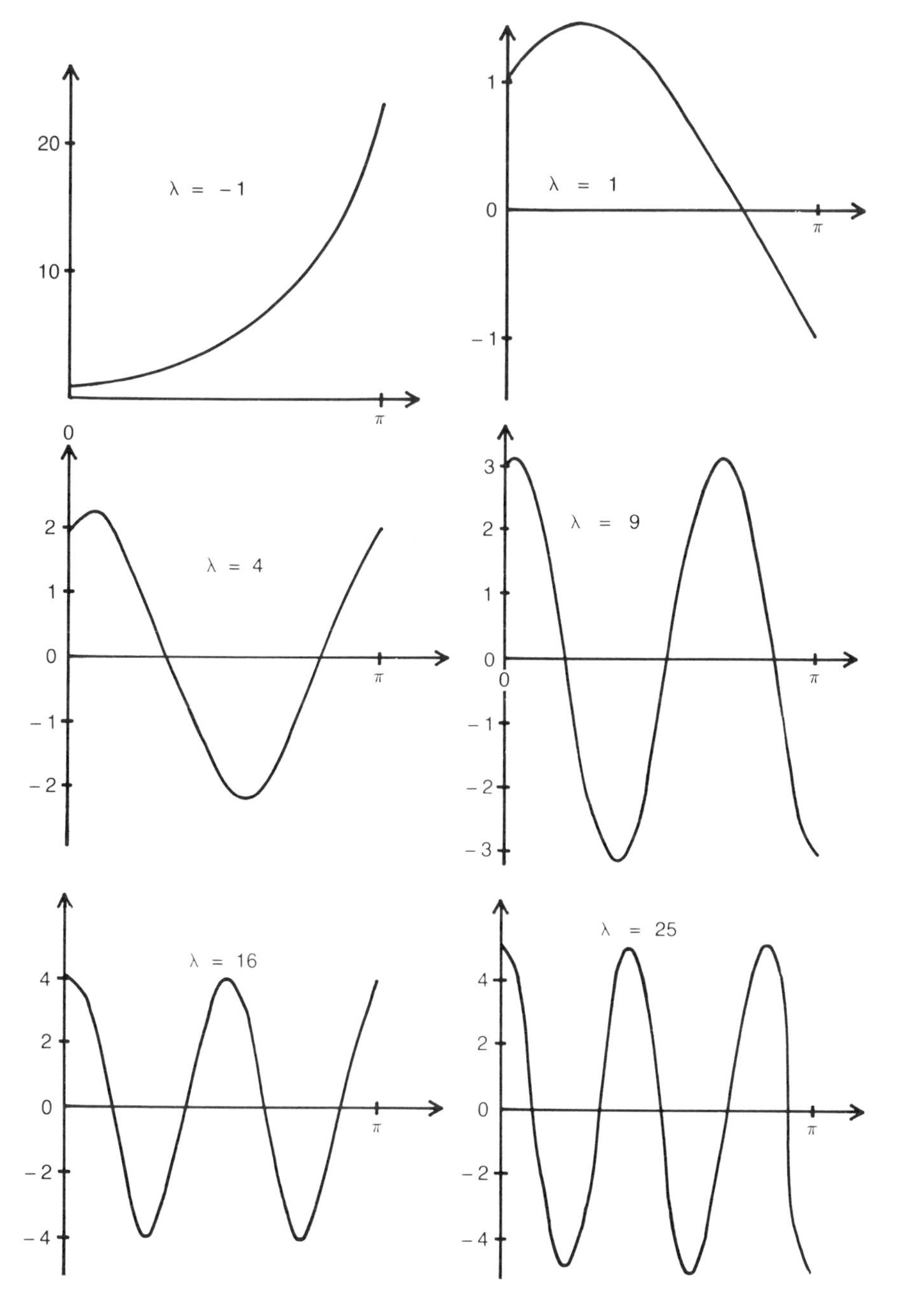

Fig. 1.2.3: The first six eigenfunctions for the Sturm-Liouville problem (1.2.11).

$$A\cosh(m\pi) + B\sinh(m\pi) = A\cosh(-m\pi) + B\sinh(-m\pi) \quad \textbf{(1.2.39)}$$

$$C + D\pi = C - D\pi \quad \textbf{(1.2.40)}$$

$$E\cos(kx) + F\sin(k\pi) = E\cos(-k\pi) + F\sin(-k\pi) \quad \textbf{(1.2.41)}$$

or

$$B \sinh(m\pi) = 0 \tag{1.2.42}$$

$$D = 0 \tag{1.2.43}$$

$$F \sin(k\pi) = 0 \tag{1.2.44}$$

because hyperbolic cosine and cosine are even functions and hyperbolic sine and sine, odd functions. Because m must be positive, $\sinh(m\pi)$ cannot equal zero and $B = 0$. On the other hand, if

$\sin(k\pi) = 0$ or $k = n$,

$n = 1, 2, 3, 4, \ldots$, we have a nontrival solution for positive λ and $\lambda_n = n^2$. Note that we still have A, C, E and F free.

From the boundary condition (1.2.38), we have

$$A \sinh(m\pi) = A \sinh(-m\pi) \tag{1.2.45}$$

and

$$-E \sin(k\pi) + F \cos(k\pi) = -E \sin(-k\pi) + F \cos(-k\pi). \tag{1.2.46}$$

The solution $u_0(x) = C$ identically satisfies the boundary condition (1.2.37) for all C. Because m and $\sinh(m\pi)$ must be positive, $A = 0$. From Eq. (1.2.46), we once again have

$$\sin(k\pi) = 0 \text{ and } k = n.$$

Consequently, the eigenfunction solutions to Eq. (1.2.36) – (1.2.38) are

$$\lambda = 0, \; u_0(x) = 1 \tag{1.2.47}$$

$$\lambda_n = n^2, \; u_n(x) = \begin{cases} \sin(nx) \\ \cos(nx) \end{cases} \tag{1.2.48}$$

and we have a degenerate set of eigenfunctions to the Sturm-Liouville problem (1.2.36) with the periodic boundary conditions (1.2.37) – (1.2.38).

Another example of a Sturm-Liouville problem arises in the study of a tapered column subjected to an axial load P where the origin of the abscissa x occurs at the fictitious vertex S of the cone and the column is clamped at its base $x = b$. See Fig. 1.2.4. The equation describing this problem is derived from the laws of elasticity and is

$$\frac{d^2w}{dx^2} + \frac{Pb^4}{EI_b} \frac{w}{x^4} = 0 \tag{1.2.49}$$

where I_b is the moment of inertia of the cross section at the base

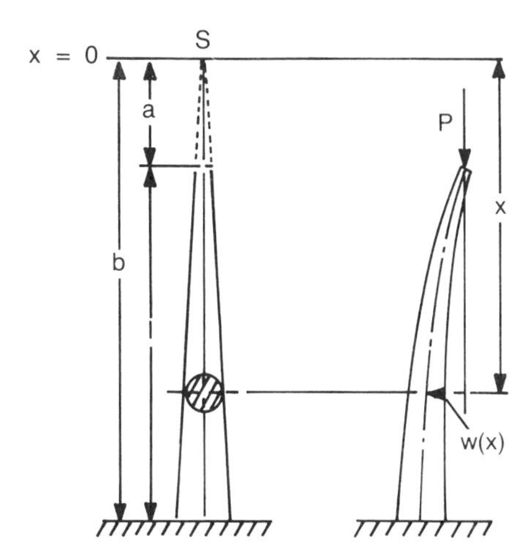

Fig. 1.2.4: Buckling of a tapered column. Taken with permission from von Karman, Th. and M. A. Biot, 1940: *Mathematical Methods in Engineering*. New York: McGraw-Hill Book Co., Inc. p. 299.

$x = b$, E the Young's modulus of elasticity and w the displacement away from equilibrium (see von Karman, Th. and M.A. Biot, 1940: *Mathematical Methods in Engineering*. New York: McGraw Hill Book Co., pp. 299-304 for details). This equation (1.2.49) corresponds to a regular Sturm-Liouville problem with $p(x) = 1$, $q(x) = 0$, $r(x) = 1/x^4$ and $\lambda = Pb^4/EI_b$.

Equation (1.2.49) may be solved by introducing the new dependent variable $y(x) = w(x)/x$ and the new independent variable $s = 1/x$. Then, because

$$\frac{d^2w}{dx^2} = x\frac{d^2y}{dx^2} + 2\frac{dy}{dx}, \tag{1.2.50}$$

$$w = xy, \tag{1.2.51}$$

$$\frac{d}{dx} = -\frac{1}{x^2}\frac{d}{ds}, \tag{1.2.52}$$

and

$$\frac{d^2}{dx^2} = \frac{2}{x^3}\frac{d}{ds} + \frac{1}{x^4}\frac{d^2}{ds^2}, \tag{1.2.53}$$

Equation (1.2.49) becomes

$$\frac{d^2y}{ds^2} + k^2\,y = 0 \tag{1.2.54}$$

which has the solution

$$y(s) = A\cos(ks) + B\sin(ks) \qquad (1.2.55)$$

where $k^2 = Pb^4/EI_b$. Equation (1.2.55) can be written in terms of w and x to give

$$w(x) = x\left(A\cos\left(\frac{k}{x}\right) + B\sin\left(\frac{k}{x}\right)\right) \qquad (1.2.56)$$

The boundary conditions are $w = 0$ at $x = b$ and $dw/dx = 0$ at $x = a$. Once again the boundary conditions correspond to those required for a regular Sturm-Liouville problem. Hence, we have

$$A\cos\left(\frac{k}{b}\right) + B\sin\left(\frac{k}{b}\right) = 0 \qquad (1.2.57)$$

$$A\left(\cos\left(\frac{k}{a}\right) + \frac{k}{a}\sin\left(\frac{k}{a}\right)\right) + B\left(\sin\left(\frac{k}{a}\right) - \frac{k}{a}\cos\left(\frac{k}{a}\right)\right) = 0. \qquad (1.2.58)$$

Elimination of A and B between these two equations yields the following condition for k:

$$\tan\left(\frac{k}{b} - \frac{k}{a}\right) = -\frac{k}{a} \qquad (1.2.59)$$

With the notation $a - b = \ell$ and $k\ell/ab = \alpha$, we obtain the following transcendental equation for α:

$$\tan(\alpha) = -\frac{b\alpha}{\ell}. \qquad (1.2.60)$$

This equation has an infinite number of roots α_k (see Fig. 1.2.5) given by the intersection of the curves

$$\eta_1 = \tan(\alpha) \text{ and } \eta_2 = -b\alpha/\ell.$$

Because the parameters b, E and I_b are fixed for a given column, each different α_k corresponds to a particular critical load which will cause buckling:

$$P_k = \alpha_k^2 \left(\frac{a}{b}\right)^2 \frac{EI_b}{\ell^2}. \qquad (1.2.61)$$

Consequently buckling will *only* occur if the load equals one of these critical loads. When a critical load is applied, the displacement $w_k(x)$ will be given by

$$A_k\, x \sin\left[\alpha_k \frac{a}{\ell}\left(1 - \frac{b}{x}\right)\right]. \qquad (1.2.62)$$

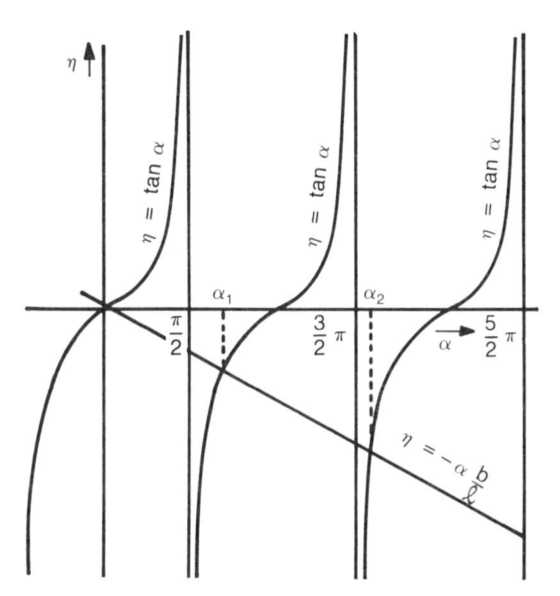

Fig. 1.2.5: Graphical construction for determining the critical buckling loads of a tapered column.

Our analysis has shown that as more and more pressure is loaded onto a tapered column, buckling will not occur until the load equals

$$P_1 = \alpha_1^2 \left(\frac{a}{b}\right)^2 \frac{EI_b}{\ell^2} .$$

Then the column will buckle and its shape is given in Fig. 1.2.6 with $\alpha_1 = 2.4557$. Had we applied a load greater than P_1, then the next load which would result in buckling is

$$P_2 = \alpha_2^2 \left(\frac{a}{b}\right)^2 \frac{EI_b}{\ell^2}$$

and the displacement is shown in Fig. 1.2.6 with $\alpha_2 = 5.2329$. Similar considerations hold for α_3 and α_4. As the figure shows, the largest displacement occurs near the top, which is an agreement with experience.

Exercises

For problems 1-5, find the eigenvalues and eigenfunctions for each of the following:

1. $u'' + \lambda u = 0 \qquad u'(0) = 0 \qquad u(L) = 0$
2. $u'' + \lambda u = 0 \qquad u'(0) = 0 \qquad u'(\pi) = 0$
3. $u'' + \lambda u = 0 \qquad u(0) + u'(0) = 0 \qquad u(\pi) + u'(\pi) = 0$
4. $u'' + \lambda u = 0 \qquad u'(0) = 0 \qquad u(\pi) - u'(\pi) = 0$
5. $u'''' - \lambda u = 0 \qquad u(0) = u''(0) = 0 \qquad u(L) = u''(L) = 0$

For each of problems 6 through 10, find an equation from which you could find λ and give the form of the eigenfunction.

6. $u'' + \lambda u = 0 \qquad u(0) + u'(0) = 0 \qquad u(1) = 0$
7. $u'' + \lambda u = 0 \qquad u(0) = 0 \qquad u(\pi) + u'(\pi) = 0$
8. $u'' + \lambda u = 0 \qquad u'(0) = 0 \qquad u(1) - u'(1) = 0$

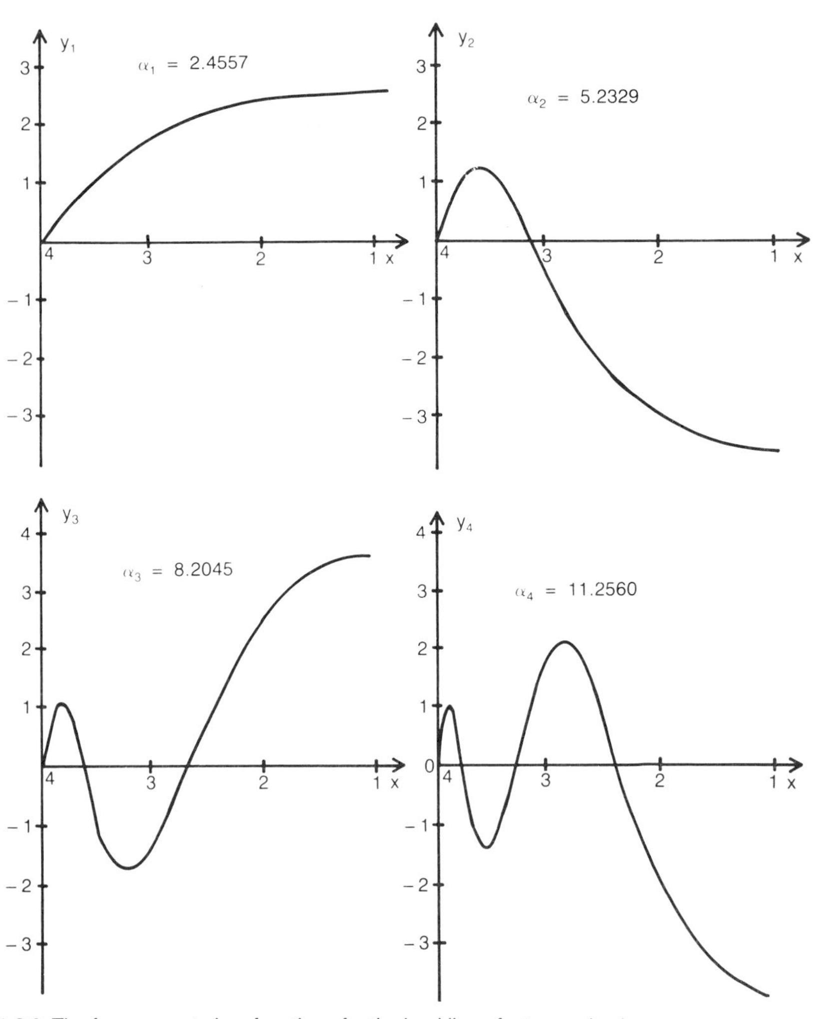

Fig. 1.2.6: The four gravest eigenfunctions for the buckling of a tapered column.

9. $u'' + \lambda u = 0 \qquad u(0) + u'(0) = 0 \qquad u'(\pi) = 0$

10. $u'' + \lambda u = 0 \qquad u(0) + u'(0) = 0 \qquad u(\pi) - u'(\pi) = 0$

11. Find the eigenvalues and eigenfunctions of the Sturm-Liouville problem

$$\frac{d}{dx}\left(x\frac{du}{dx}\right) + \frac{\lambda}{x}u = 0 \qquad 1 \leq x \leq e$$

for each of the following boundary conditions:

(a) $u(1) = u(e) = 0$; *(b)* $u(1) = u'(e) = 0$; *(c)* $u'(1) = u'(e) = 0$.

12. Find the eigenvalues and eigenfunctions of the Sturm-Liouville problem

$$x^2 u'' + 2x u' + \lambda u = 0,\ u(1) = u(e) = 0,\ 1 \leq x < e.$$

13. Find the eigenvalues and eigenfunctions of the Sturm-Liouville problem

$$\frac{d}{dx}(x^3 u') + \lambda x u = 0,\ u(1) = u(e^\pi) = 0,\ 1 \leq x \leq e^\pi.$$

14. Find the eigenvalues and eigenfunctions of the Sturm-Liouville problem

$$\frac{d}{dx}\left(\frac{1}{x}u'\right) + \frac{\lambda}{x^3}u = 0,\ u(1) = u(e) = 0, 1 \leq x \leq e.$$

15. Consider the boundary value problem

$$u'' + \lambda u = 0,\ u(-L) - u(L) = u'(-L) - u'(L) = 0,\ -L \leq x \leq L$$

with periodic boundary conditions. Find the eigenvalues of the problem and show that to each eigenvalue $\lambda \neq 0$, there corresponds two linearly independent eigenfunctions.

16. Find the eigenvalues and eigenfunctions of the problem

$$u'' + \lambda u = 0,\ u(0) - u(L) = 0,\ u'(0) - u'(L) = 0,\ 0 \leq x \leq L$$

with periodic boundary conditions.

1.3 ORTHOGONALITY OF EIGENFUNCTIONS

In the previous sections we have seen how solutions to the regular Sturm-Liouville problem may be characterized by eigenvalues and eigenfunctions. Our next task is to relate one eigenfunction to another.

Suppose that y_n and y_m are eigenfunctions corresponding to two different eigenvalues λ_n and λ_m. Then

$$\frac{d}{dx}\left(p(x)\frac{dy_n}{dx}\right) + (q(x) + \lambda_n r(x))\, y_n = 0 \qquad \textbf{(1.3.1)}$$

$$\frac{d}{dx}\left(p(x)\frac{dy_m}{dx}\right) + (q(x) + \lambda_m r(x))\, y_m = 0 \qquad \textbf{(1.3.2)}$$

and both functions satisfy the boundary conditions. Let us multiply the first differential equation by y_m; the second, by y_n. Next, we subtract these two equations and move the terms containing $y_n y_m$ to the right-hand side. The resulting equation is

$$y_n\,(p(x)y_m')' - y_m\,(p(x)y_n)' = (\lambda_n - \lambda_m)r(x)y_n y_m \qquad \textbf{(1.3.3)}$$

Integrating Eq. (1.3.3) from a to b yields

$$\int_a^b y_n\,(p(x)y_m')' - y_m\,(p(x)y_n')'\,dx = \int_a^b (\lambda_n - \lambda_m)r(x)y_n y_m\,dx. \qquad \textbf{(1.3.4)}$$

The left-hand side may be simplified by integrating by parts to give

$$\int_a^b \Big[[p(x)y_m']'\,y_n - [p(x)y_n']\,y_m'\Big]\,dx = p(x)y_m'\,y_n - p(x)y_n'\,y_m\Big|_a^b - \int_a^b p(x)\,(y_n'y_m' - y_n'y_m')\,dx \qquad \textbf{(1.3.5)}$$

The second integral is identically zero. Because y_n and y_m satisfy the boundary condition at $x = a$,

$$\alpha y_n(a) + \beta y_n'(a) = 0 \qquad \textbf{(1.3.6)}$$
$$\alpha y_m(a) + \beta y_m'(a) = 0 \qquad \textbf{(1.3.7)}$$

These two equations may be considered simultaneous equations in α and β. At least one of the numbers α and β is nonzero; otherwise, there would be no boundary conditions at $x = a$. Hence, the determinant of the equation must be zero:

$$y_n'(a)\,y_m(a) - y_m'(a)\,y_n(a) = 0. \qquad \textbf{(1.3.8)}$$

Similarly, at the other end:

$$y_n'(b)\, y_m(b) - y_m'(b)\, y_n(b) = 0. \qquad \textbf{(1.3.9)}$$

Consequently, the right-hand side of Eq. (1.3.5) vanishes which reduces Eq. (1.3.4) to

$$\int_a^b r(x)\, y_n(x)\, y_m(x)\, dx = 0 \qquad \textbf{(1.3.10)}$$

if $\lambda_n \neq \lambda_m$. Equation (1.3.10) states that the eigenfunctions y_n and y_m are mutually *orthogonal* to each other with respect to the weighting function $r(x)$ over the interval $[a,b]$. The name "weight" appears to be borrowed from statistics: The function $r(x)$ gives more weight to some x's and less to others, just as if the y's were experimental data measured in greater abundance for some x's and more sparsely for others. If $r(x) = 1$, then the system is simply said to be orthogonal on $[a,b]$.

For example, the orthogonality condition for the first set of eigenfunctions found in the previous section is

$$\int_0^\pi \sin(k_n x) \sin(k_m x)\, dx = 0 \qquad \text{if } m \neq n \qquad \textbf{(1.3.11)}$$

because $r(x) = 1$. Note, however, that if $n = m$, we have

$$\begin{aligned}
\int_0^\pi \sin(k_n x) \sin(k_n x)\, dx &= 1/2 \int_0^\pi 1 - \cos(2k_n x)\, dx \\
&= \frac{\pi}{2} - \frac{1}{4k_n} \sin(2k_n \pi) \\
&= 1/2\, (\pi - \cos^2(k_n \pi)) > 0
\end{aligned} \qquad \textbf{(1.3.12)}$$

because $\sin(2A) = 2 \sin(A) \cos(A)$ and $k_n = \tan(k_n \pi)$. That is, any eigenfunction *cannot* be orthogonal to itself.

This name "orthogonality" has its origins in linear algebra. Two N-dimensional vectors are said to be mutually perpendicular or orthogonal if their scalar (or inner) product is zero, i.e.,

$$\mathbf{x} \bullet \mathbf{y} = \sum_{n=0}^{N} x_n y_n = 0 \qquad \text{if } x \neq y. \qquad \textbf{(1.3.13)}$$

In the limit of N going to infinity, we obtain Eq. (1.3.10) if $r(x) = 1$.

In closing we note that had we defined the eigenfunction as

$$\frac{\sin (k_n x)}{\{1/2[\pi - \cos^2 (k_n \pi)]\}^{1/2}}$$

rather than

$$\sin (k_n x),$$

the orthogonality condition would have been

$$\int_0^\pi y_n(x)\, y_m(x)\, dx = \begin{cases} 0 & \text{if } m \neq n \\ 1 & \text{if } m = n. \end{cases} \tag{1.3.14}$$

This process of *normalizing* an eigenfunction so that the orthogonality condition becomes

$$\int_a^b r(x)\, y_n(x)\, y_m(x)\, dx = \begin{cases} 0 & \text{if } m \neq n \\ 1 & \text{if } m = n \end{cases} \tag{1.3.15}$$

is done to generate *orthonormal* eigenfunctions. The convenience of doing this will be seen later.

Exercises

1. Given $u'' + \lambda^2 u = 0$, $u(0) = u(\pi) - u'(\pi) = 0$, show that $\int_0^\pi \sin (\lambda_1 x) \sin (\lambda_2 x)\, dx = 0$ where λ_1 and λ_2 are the roots of the equation $\tan (\lambda \pi) = \lambda$.

2. Given $(xu')' + \lambda u/x = 0$, $u(1) = u(e) = 0$ show that $\int_1^e \frac{1}{x} \sin [n\pi \ell n(x)] \sin [m\pi \ell n(x)]\, dx = 0$ where m and n are distinct integers.

3. (a) Given $u'' + \lambda u = 0$, $u(-L) = u(L) = 0$, show that

$$\int_{-L}^{L} \sin \left(\frac{n\pi x}{L}\right) \sin \left(\frac{m\pi x}{L}\right) dx = 0 \qquad \text{if } m \neq n.$$

(b) Given $u'' + \lambda u = 0$, $u'(-L) = u'(L) = 0$, show that

$$\int_{-L}^{L} \cos \left(\frac{n\pi x}{L}\right) \cos \left(\frac{m\pi x}{L}\right) dx = 0 \qquad \text{if } m \neq n.$$

1.4 EXPANSION IN SERIES OF EIGENFUNCTIONS

The most important practical use of a system of eigenfunctions is the approximation of an arbitrary function by an infinite series of them. Generally the eigenfunctions used in the expansion are chosen because they enable the series to satisfy some imposed boundary condition.

Suppose that a function $f(x)$ is given in the interval $a<x<b$ and that we wish to express $f(x)$ in terms of the eigenfunctions $y_n(x)$ given by a regular Sturm-Liouville problem. That is, we wish to have

$$f(x) = \sum_{n=1}^{\infty} c_n y_n(x). \tag{1.4.1}$$

The orthogonality relation Eq. (1.3.10) gives us the method for computing the coefficients c_n. We multiply both sides of Eq. (1.4.1) by $r(x)\, y_m(x)$ (where m is a fixed integer) and integrate it from a to b. If we assume that the series can be integrated term by term, we get

$$\int_a^b r(x)\, f(x)\, y_m(x)\, dx = \sum_{n=1}^{\infty} c_n \int_a^b r(x)\, y_n(x)\, y_m(x)\, dx. \tag{1.4.2}$$

The orthogonality relationship says that all of the terms on the right-hand side of (1.4.2), except for the one in which $n = m$, must disappear. Thus, we are left with

$$\int_a^b r(x)\, f(x)\, y_m(x)\, dx = c_m \int_a^b r(x)\, y_m^2(x)\, dx \tag{1.4.3}$$

$$c_n = \frac{\int_a^b f(x)\, r(x)\, y_n(x)\, dx}{\int_a^b r(x)\, y_n^2(x)\, dx} \tag{1.4.4}$$

if we replace m by n in Eq. (1.4.3).

The series given by Eq. (1.4.1) with the coefficients found by Eq. (1.4.4) is called a *generalized Fourier series* of the function $f(x)$ with respect to the eigenfunction system $y_n(x)$. Note that if we had used an orthonormal set of eigenfunctions, then the denominator of Eq. (1.4.4) would be one and our work is reduced in half. The coefficients c_n are called the Fourier coefficients.

One of the most remarkable facts about generalized Fourier series is their applicability even when the function has a finite number of bounded discontinuities in the range $[a,b]$. Because continuous eigenfunctions are used in the series representation of discontinuous functions, we are not merely rewriting $f(x)$ in a new form. The c_n's are actually chosen so that the eigenfunctions are fitting $f(x)$ in the "least squares" sense of

$$\int_a^b r(x) \mid f(x) - \sum_{n=1}^{\infty} c_n y_n(x) \mid^2 dx = 0. \qquad \textbf{(1.4.5)}$$

(See Morse, P.M. and F. Feshbach, 1953: *Methods of Theoretical Physics*. McGraw Hill Book Co., p. 736 for further discussion.) Consequently we should expect peculiar things to occur at the points of discontinuity. Spurious oscillations will occur in the neighborhood of the discontinuity. This is known as *Gibbs phenomena* and this phenomena will be illustrated later.

1.5 AN EXAMPLE OF AN EIGENFUNCTION EXPANSION: CHEBYSHEV POLYNOMIALS

To illustrate the concept of an eigenfunction expansion, let us find the eigenfunction expansion for $f(x) = x$ over the interval $0 \leq x \leq \pi$ using the solutions to the regular Sturm-Liouville problem of $y'' + \lambda y = 0$, $y(0) = y(\pi) = 0$. Because the eigenfunctions are $y_n(x) = \sin(nx)$ $(n = 1, 2, 3, \ldots)$, $r(x) = 1$, $a = 0$ and $b = \pi$, Eq. (1.4.4) gives

$$c_n = \frac{\int_0^{\pi} x \sin(nx)\, dx}{\int_0^{\pi} \sin^2(nx)\, dx} = \frac{-\dfrac{x\cos(nx)}{n} + \dfrac{\sin(nx)}{n^2} \Big|_0^{\pi}}{\dfrac{x}{2} - \dfrac{1}{4n}\sin(2nx) \Big|_0^{\pi}} \qquad \textbf{(1.5.1)}$$

$$= -\frac{2}{n}\cos(n\pi) = \frac{2}{n}(-1)^n \qquad \textbf{(1.5.2)}$$

Eq. (1.4.1) then gives that

$$f(x) = -2 \sum_{n=1}^{\infty} \frac{(-1)^n}{n} \sin(nx). \qquad \textbf{(1.5.3)}$$

This is a particular example of what we shall later call a half-range sine expansion in Section 2.6.

At $x = \pi$, we notice that the series converges to zero while $f(\pi) = \pi$. At $x = 0$, both the series and the function converge to zero. Hence the series expansion (1.5.3) is valid for $0 \leq x < \pi$.

An important use of eigenfunction expansions is in the approximating of a function over a given interval. Taylor series expansions represent a function with great precision near the point of expansion, but the error increases rapidly (proportional to a power) as we employ them at points farther away. Because we often want an expansion which is fairly accurate over a larger region than that given by a Taylor series, orthogonal functions provide us with a method where we trade some of the excessive precision of Taylor expansions near the center of the interval for a reduction in the error farther away.

Although there are many orthogonal functions from which to choose, a popular set is the *Chebyshev polynomials*. Their popularity stems partially from the fact that most (but not all) expansions require fewer terms using Chebyshev polynomials to achieve a given accuracy than if any other orthogonal function had been used. Furthermore, Chebyshev polynomials possess several properties that greatly facilitate their use in expansions.

It is readily shown that the set of functions $\cos(n\theta)$ is an orthogonal set over the interval $0 \leq x \leq \pi$. (Consider the regular Sturm-Liouville problem of $u'' + \lambda u = 0$, $u'(0) = u'(\pi) = 0$.) By direct integration we find that

$$\int_0^{\pi} \cos(m\theta)\cos(n\theta)\,d\theta = \begin{cases} 0 \text{ if } m \neq n \\ \pi/2 \text{ if } m = n \neq 0. \\ \pi \text{ if } m = n = 0 \end{cases} \qquad \textbf{(1.5.4)}$$

This set of orthogonal functions can be redefined as

$$T_n(x) = \cos\,[n \arccos(x)] \qquad \textbf{(1.5.5)}$$

by letting

$$\theta = \arccos(x) \qquad \textbf{(1.5.6)}$$

In this special formulation, they are given the special name of *Chebyshev polynomials.*

The definition of Chebyshev polynomial given by Eq. (1.5.5), although exact, is not very convenient for computational purposes. They are usually reexpressed in terms of powers of x:

$T_0(x) = 1$		$1 = T_0$
$T_1(x) = x$		$x = T_1$
$T_2(x) = 2x^2 - 1$		$x^2 = 1/2(T_0 + T_2)$
$T_3(x) = 4x^3 - 3x$	or	$x^3 = 1/4(3T_1 + T_3)$
$T_4(x) = 8x^4 - 8x^2 + 1$		$x^4 = (3T_0 + 4T_2 + T_4)/8$
$T_5(x) = 16x^5 - 20x^3 + 5x$		$x^5 = (10T_1 + 5T_3 + T_5)/16$

These results follow from the three-term recurrence relation

$$T_{n+1}(x) - 2x\,T_n(x) + T_{n-1}(x) = 0 \qquad (1.5.7)$$

which is really the trigonometric identity

$$\cos[(n+1)\phi] + \cos[(n-1)\phi] = 2\cos(\phi)\cos(n\phi) \qquad (1.5.8)$$

in disguise. Note the coefficients of this equation do not depend on n. A quick check confirms the fact that $T_{2n}(x)$ is an even function while $T_{2n+1}(x)$ is an odd function.

We now turn to the question of creating an expansion in Chebyshev polynomials. Clearly we want to reexpress $f(x)$ by the expansion of Chebyshev polynomials:

$$f(x) = a_0/2 + \sum_{n=1}^{\infty} a_n T_n(x). \qquad (1.5.9)$$

Because we can write Eq. (1.5.9) as

$$f(\theta) = a_0/2 + \sum_{n=1}^{\infty} a_n \cos(n\theta), \qquad (1.5.10)$$

Equation (1.4.4) gives a_n as

$$a_n = \frac{2}{\pi}\int_0^{\pi} f\,(\cos(\theta))\cos(n\theta)\,d\theta. \qquad (1.5.11)$$

Because $\theta = \arccos(x)$, $T_n(x) = \cos(n\theta)$ and $d\theta = dx/(1-x^2)^{1/2}$, Eq. (1.5.11) becomes

$$a_n = \frac{2}{\pi}\int_{-1}^{1} f(x)\,\frac{T_n(x)}{\sqrt{1-x^2}}\,dx. \qquad (1.5.12)$$

Consequently, given a $f(x)$, we could perform the integral (1.5.11) or (1.5.12) to find a_n. In those cases when integration of (1.5.11) or (1.5.12) fails by conventional methods, these integrations may be carried out numerically.

For example, let us reexpress the function $f(x) = \sqrt{1-x^2}$ in Chebyshev polynomials. The Taylor expansion for $f(x)$ about $x = 0$ is

$$f(x) = 1 - x^2/2 + \frac{3}{8}x^4 - \frac{5}{16}x^6 + \ldots \quad |x|<1. \qquad (1.5.13)$$

The expansion in Chebyshev polynomials is given by

$$f(x) = a_0/2 + \sum_{n=1}^{\infty} a_n T_n(x) \qquad \textbf{(1.5.14)}$$

where

$$a_n = \frac{2}{\pi}\int_0^{\pi} (1\text{-}\cos^2(\theta))^{1/2} \cos(n\theta)\, d\theta$$

$$= \frac{2}{\pi}\int_0^{\pi} \sin(\theta) \cos(n\theta)\, d\theta$$

$$= \begin{cases} \dfrac{1}{\pi}\dfrac{-4}{n^2-1} & n = 0,2,4,6,\ldots \\ 0 & n = 1,3,5,7,\ldots \end{cases} \qquad \textbf{(1.5.15)}$$

In Fig. 1.5.1 are shown the graphs of the exact function along with values given by the first two terms in the Taylor expansion (1.5.13) and the Chebyshev expansion (1.5.14) over the interval $[-1,1]$. As can be readily seen, the Chebyshev expansion approximates the function more closely over the entire domain than the Taylor expansion.

Exercises

1. Find the formal eigenfunction expansion of the function $f(x) = x^2$, $0 \leq x \leq \pi$, with respect to the system of functions $[(2/\pi)^{1/2} \sin(nx)]$, $r(x) = 1$.
2. Find the formal eigenfunction of $f(x) = x$, $0 \leq x \leq L$, with respect to the eigenfunctions $\cos\left[\dfrac{(2n-1)\pi x}{2L}\right]$, $n = 1,2,3,4,\ldots,$ and $r(x) = 1$.
3. Find the formal eigenfunction expansion of $f(x) = x$, $0 \leq x < \pi$, with respect to the eigenfunctions $\cos(nx)$, $n = 0, 1, 2, 3, \ldots$ (Watch out for the case $n = 0$.) with $r(x) = 1$.
4. Derive the following Chebyshev expansions:

(a) $$|x| = \frac{2}{\pi}\left[1 - 2\sum_{n=1}^{\infty} \frac{(-1)^n}{4n^2-1} T_{2n}(x)\right]$$

(b) $$\text{sgn}(x) = \frac{4}{\pi}\sum_{n=0}^{\infty} \frac{(-1)^n}{2n+1} T_{2n+1}(x)$$

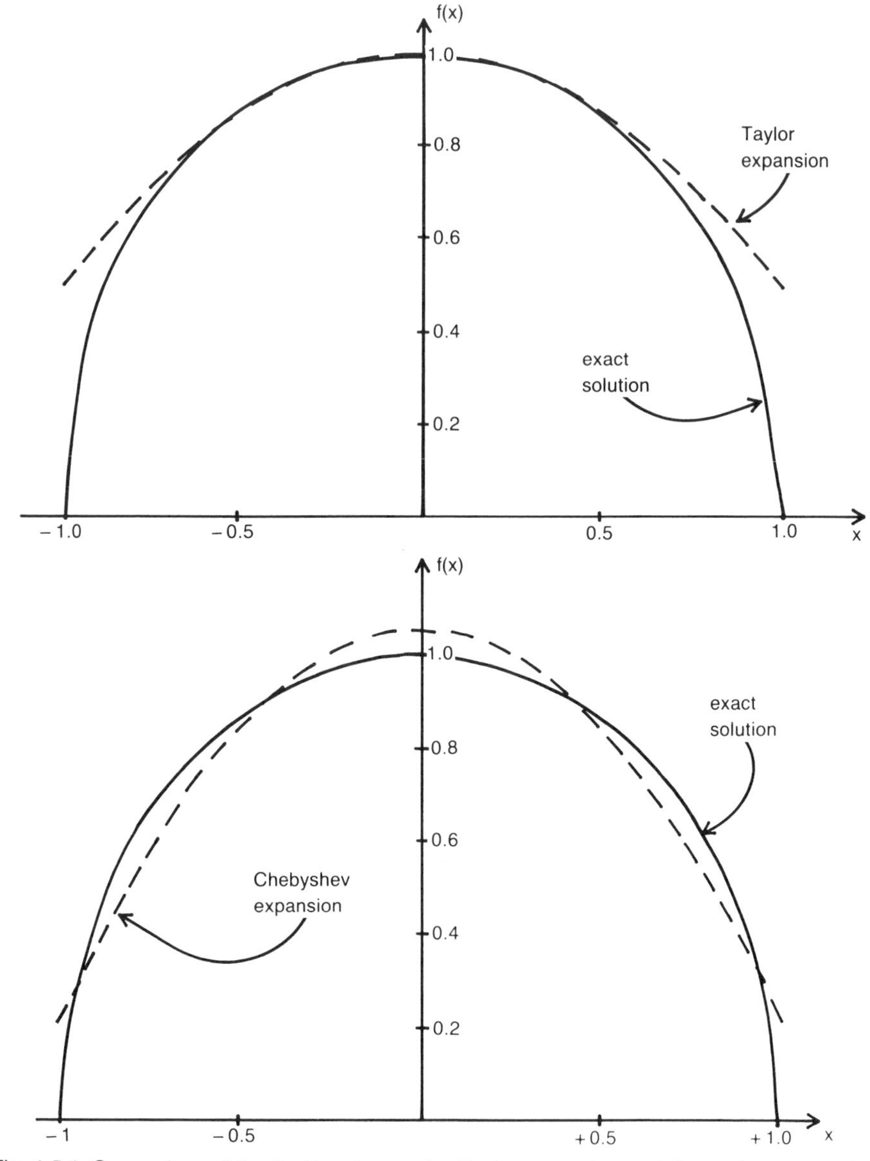

Fig. 1.5.1: Comparison of the first two terms of a Taylor expansion and Chebyshev expansion of $(1 - x^2)^{1/2}$.

$$\text{(c) } \delta(x) = \frac{1}{\pi}\left[1 - 2 \sum_{n=1}^{\infty} (-1)^{n+1} T_{2n}(x)\right]$$

$$\text{(d) } x\sqrt{1-x^2} = \sum_{n=1}^{\infty} \frac{-4}{\pi(2n-3)(2n+1)} T_{2n-1}(x)$$

where

$$\operatorname{sgn}(x) = \begin{cases} 1 & x>0 \\ 0 & x=0 \\ -1 & x<0 \end{cases}$$

and the Dirac delta function is given by

$$\delta(x) = \begin{cases} 1 & |x|<\epsilon \\ 0 & \text{otherwise,} \end{cases}$$

in the limit of $\epsilon \to 0$.

Review of the Regular Sturm-Liouville Problem

1. Sturm-Liouville equation

$$\frac{d}{dx}\left[p(x)\frac{dy}{dx}\right] + \left[q(x) + r(x)\lambda\right] y = 0 \qquad \text{with } p(x), r(x) > 0.$$

2. The boundary conditions

$$\alpha y(a) + \beta y'(a) = 0 \text{ and } \gamma y(b) + \delta y'(b) = 0$$

3. How to find the eigenvalue and eigenfunction:

(a) Obtain the most general solution to the differential equation and write it as $y(x) = A\, y_1(x) + B\, y_2(x)$.

(b) Use the boundary conditions to obtain two homogeneous equations involving A and B:

$$[\alpha y_1(a) + \beta y_1'(a)]\, A + [\alpha y_2(a) + \beta\, y_2'(a)]B = 0 \qquad (1)$$

$$[\gamma y_1(b) + \delta y_1'(b)]\, A + [\gamma y_2(b) + \delta\, y_2'(b)]B = 0 \qquad (2)$$

(c) To obtain a nontrival solution, find the λ's such that

$$\begin{vmatrix} \alpha y_1(a) + \beta y_1(a) & \alpha y_2(a) + \beta y_2'(a) \\ \gamma y_1(b) + \delta y_1'(b) & \gamma y_2(b) + \delta y_2'(b) \end{vmatrix} = 0$$

(d) To find the eigenfunction, use either (1) or (2) to eliminate either A or B from the general solution. The eigenfunction is given by this particular solution after discarding the arbitrary constant (i.e., either B or A).

4. Orthogonality:

$$\int_a^b r(x)\, y_n(x)\, y_m(x)\, dx = \begin{cases} 0 & \text{if } m \neq n \\ \text{positive constant} & \text{if } m = n \end{cases}$$

5. Orthogonal expansions:

If
$$f(x) = \sum_{n=0}^{\infty} c_n y_n(x)$$

then
$$c_n = \frac{\int_a^b r(x)\, f(x)\, y_n(x)\, dx}{\int_a^b r(x)\, y_n^2(x)\, dx}$$

1.6 AN EXAMPLE OF A SINGULAR STURM-LIOUVILLE PROBLEM: LEGENDRE'S EQUATION

In the previous sections we have used solutions to a regular Sturm-Liouville problem in the orthogonal expansion of the function $f(x)$. The fundamental reason why we could form such an expansion was the orthogonality condition (1.3.10). We now want to address the question of whether we can still retain our orthogonality condition (and therefore still be able to construct an orthogonal expansion) even when some of the conditions associated with the regular Sturm-Liouville problem no longer hold.

To answer this question, we combine Eqs. (1.3.4) and (1.3.5) to obtain

$$(\lambda_n - \lambda_m) \int_a^b r(x)\, y_n(x)\, y_m(x)\, dx = \Big\{ p(b) y_m'(b) y_n(b) - p(b) y_n'(b) y_m(b) - p(a) y_m'(a) y_n(a) + p(a) y_n'(a) y_m(a) \Big\} \quad \textbf{(1.6.1)}$$

From Eq. (1.6.1) we see that the right-hand side will vanish (and orthogonality will be preserved) if y_n is finite and $p(x) y_n'(x)$ tends to zero at both end points. An important example of this *singular Sturm-Liouville problem* is Legendre's equation:

$$(1 - x^2) \frac{d^2y}{dx^2} - 2x \frac{dy}{dx} + n(n+1)y = 0$$

or

$$\frac{d}{dx}\left((1 - x^2) \frac{dy}{dx}\right) + n(n+1)y = 0 \quad \textbf{(1.6.2)}$$

where $a = -1$, $b = 1$, $\lambda = n(n+1)$, $p(x) = 1 - x^2$, $q(x) = 0$ and

$r(x) = 1$. This equation arises in the solution of partial differential equations on a sphere.

Equation (1.6.2) does not have a simple general solution. (If $n = 0$, $y(x) = 1$ is a solution.) Consequently we try to solve it with the power series:

$$y(x) = \sum_{k=0}^{\infty} A_k x^k \tag{1.6.3}$$

$$y'(x) = \sum_{k=0}^{\infty} k A_k x^{k-1} \tag{1.6.4}$$

$$y''(x) = \sum_{k=0}^{\infty} k(k-1) A_k x^{k-2} \tag{1.6.5}$$

Substituting into Eq. (1.6.2), we obtain

$$\sum_{k=0}^{\infty} k(k-1) A_k x^{k-2} + \sum_{k=0}^{\infty} \left\{ n(n+1) - 2k - k(k-1) \right\} A_k x^k = 0 \tag{1.6.6}$$

which equals

$$\sum_{m=2}^{\infty} m(m-1) A_m x^{m-2} + \sum_{k=0}^{\infty} \left[n(n+1) - k(k+1) \right] A_k x^k = 0 \tag{1.6.7}$$

If in the first summation we define $k = m + 2$, then

$$\sum_{k=0}^{\infty} (k+2)(k+1) A_{k+2} x^k + \sum_{k=0}^{\infty} \left[n(n+1) - k(k+1) \right] A_k x^k = 0 \tag{1.6.8}$$

Because Eq. (1.6.8) must be true for all x, each power of x must vanish separtely from which it follows that

$$(k+2)(k+1) A_{k+2} = [k(k+1) - n(n+1)] A_k \tag{1.6.9}$$

or

$$A_{k+2} = \frac{k(k+1) - n(n+1)}{(k+1)(k+2)} A_k \cdot \; k = 0\ 1,2,3\ ,... \tag{1.6.10}$$

It should be remembered that we still have the constants A_0 and A_1, which represent the two arbitrary constants of the general solution to Eq. (1.6.2).

The first few terms of the solution associated with A_0 are

$$u_p(x) = 1 - \frac{n(n+1)}{2!} x^2 + \frac{n(n-2)(n+1)(n+3)}{4!} x^4 - \frac{n(n-2)(n-4)(n+1)(n+3)(n+5)}{6!} x^6 + \ldots \quad \textbf{(1.6.11)}$$

while the first few terms associated with the A_1 coefficient are

$$v_p(x) = x - \frac{(n-1)(n+2)}{3!} x^3 + \frac{(n-1)(n-3)(n+2)(n+4)}{5!} x^5 - \frac{(n-1)(n-3)(n-5)(n+2)(n+4)(n+6)}{7!} x^7 + \ldots \quad \textbf{(1.6.12)}$$

We notice that if n is an *even* positive integer (including n equal to zero), then the series (1.6.11) terminates with the term involving x^n. The solution is a polynomial of degree n. Similarly if n is an *odd* integer, the series (1.6.12) terminates with the term involving x^n. Otherwise, for n noninteger the expressions are infinite series.

For reasons that will become apparent, we restrict ourselves to positive integers n. (Actually this includes all possible integers because the negative integer $-n-1$ has the same Legendre's equation and solution in common with the positive integer n.) In this case one of the solutions, either Eq. (1.6.11) or Eq. (1.6.12), is a polynomial of degree n while the other solution is an infinite series. These polynomials are called *Legendre polynomials* and may be computed by the power series

$$P_n(x) = \sum_{k=0}^{m} (-1)^k \frac{(2n-2k)!}{2^n\, k!\, (n-k)!\, (n-2k)!} x^{n-2k} \quad \textbf{(1.6.13)}$$

where $m = n/2$ or $(n-1)/2$, depending upon which is an integer. We have chosen to use Eq. (1.6.13) over Eq. (1.6.11) or Eq. (1.6.12) because Eq. (1.6.13) has the advantage that $P_n(1) = 1$. The first ten Legendre polynomials are given in Table 1.6.1.

The other solution, the infinite series, is called the Legendre function of the second kind, $Q_n(x)$. In Fig. 1.6.1, the first four Legendre polynomials $P_n(x)$ and the first four Legendre functions of the second kind $Q_n(x)$ are plotted. From this figure we see that $Q_n(x)$ becomes infinite at the points $x = \pm 1$. This is important

Table 1.6.1. The First Tan Legendre Polynomials.

$P_0(x)$	=	1
$P_1(x)$	=	x
$P_2(x)$	=	$1/2\,(3x^2 - 1)$
$P_3(x)$	=	$1/2\,(5x^3 - 3x)$
$P_4(x)$	=	$1/8\,(35x^4 - 30x^2 + 3)$
$P_5(x)$	=	$1/8\,(63x^5 - 70x^3 + 15x)$
$P_6(x)$	=	$1/16\,(231\,x^6 - 315\,x^4 + 105\,x^2 - 5)$
$P_7(x)$	=	$1/16\,(429\,x^7 - 693\,x^5 + 315\,x^3 - 35x)$
$P_8(x)$	=	$1/128\,(6435\,x^8 - 12012\,x^6 + 6930\,x^4 - 1260\,x^2 + 35)$
$P_9(x)$	=	$1/128\,(12155x^9 - 25740x^7 + 18018x^5 - 4620x^3 + 315x)$
$P_{10}(x)$	=	$1/256\,(46189x^{10} - 109395x^8 + 90090x^6 - 30030x^4 + 3465x^2 - 63)$

because we are interested only in solutions to Legendre's equation that are finite over the entire interval $[-1,1]$. On the other hand, in problems where the points $x = \pm 1$ are excluded, Legendre functions of the second kind will appear in the general solution

$$y(x) = A\,P_n(x) + B\,Q_n(x). \tag{1.6.14}$$

(See Smythe, W. R., 1950: *Static and Dynamic Electricity*. New York: McGraw Hill Book Co., Inc., Section 5.215 for an example.)

In the case when n is not an integer, it can be shown that we can construct a solution where the solution remains finite at $x = 1$ but it does not remain finite at $x = -1$. Furthermore, we can construct a solution which is finite at $x = -1$ but it does not remain finite at $x = 1$. (See Carrier, G. F., M. Krook and C. E. Pearson, 1966: *Functions of the Complex Variable*. New York: McGraw Hill Book Co., Inc., 212-213.) Consequently these solutions must also be rejected from further consideration and we are left only with the Legendre polynomials. From now on, we will only consider the properties and uses of these functions.

Although we have the series (1.6.13) to compute $P_n(x)$, several alternative methods are found to be desirable. The first, known as *Rodrigues' formula*, may be obtained by writing Eq. (1.6.13) in the form

$$P_n(x) = \frac{1}{2^n n!} \sum_{k=0}^{m} (-1)^k \frac{n!}{k!(n-k)!} \frac{(2n-2k)!}{(n-2k)!} x^{n-2k} \tag{1.6.15}$$

$$= \frac{1}{2^n n!} \frac{d^n}{dx^n} \sum_{k=0}^{m} (-1)^k \frac{n!}{k!(n-k)!} x^{2n-2k} \tag{1.6.16}$$

The last summation is the binomial expansion of $(x^2-1)^n$ so that

$$\boxed{P_n(x) = \frac{1}{2^n\,n!} \frac{d^n}{dx^n}, (x^2-1)^n.} \tag{1.6.17}$$

Legendre functions of the first kind

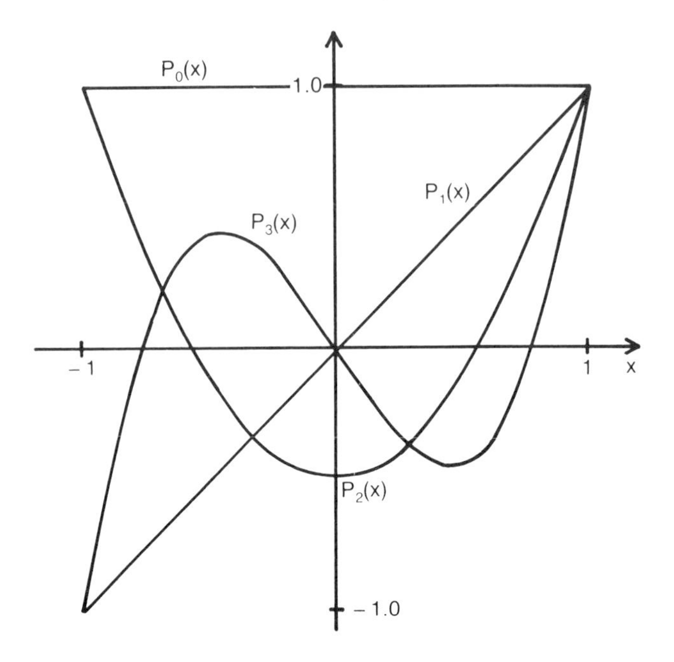

Legendre functions of the second kind

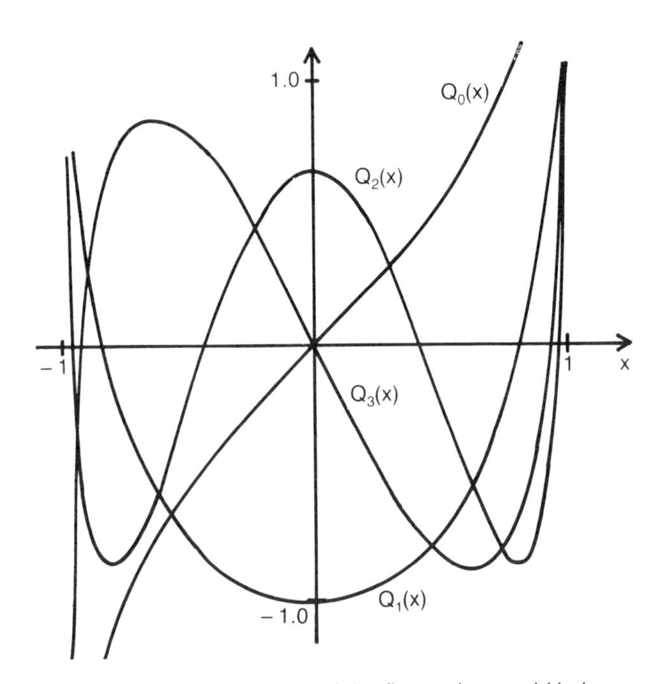

Fig. 1.6.1: The first four Legendre functions of the first and second kind.

Example 1.6.1

Let us use Rodrigues' formula to compute $P_2(x)$. From Eq. (1.6.17) with $n = 2$, we have

$$P_2(x) = \frac{1}{2^2 2!}\frac{d^2}{dx^2}((x^2-1)^2) \tag{1.6.18}$$

$$= \frac{1}{8}\frac{d^2}{dx^2}(x^4 - 2x^2 - 1) \tag{1.6.19}$$

$$= \frac{1}{2}(3x^2 - 1). \tag{1.6.20}$$

Another method involves the use of recurrence formulas. The first step in finding these formuli is to establish the fact that

$$(1 + h^2 - 2xh)^{-1/2} = P_0(x) + h\,P_1(x) + h^2 P_2(x) + \ldots \tag{1.6.21}$$

The function $(1 + h^2 - 2xh)^{-1/2}$ is called the *generating function* for $P_n(x)$.

The expansion may be obtained via the formal binomial expansion

$$(1 + h^2 - 2xh)^{-1/2} = 1 - 1/2(2xh - h^2) + \frac{1}{2}\frac{3}{2}\frac{1}{2!}(2xh - h^2)^2 - \frac{1}{2}\frac{3}{2}\frac{5}{2}\frac{1}{3!}(2xh - h^2)^3 + \ldots \tag{1.6.22}$$

Upon expanding the terms contained in $2xh - h^2$ and grouping like powers of h, we have

$$(1 + h^2 - 2hx)^{-1/2} = 1 + xh + (\frac{3}{2}x^2 - 1/2)h^2 + \ldots \tag{1.6.23}$$

A direct comparison between the coefficients for each power of h and the Legendre polynomial $P_n(x)$ completes the demonstration. Note that these results hold only if $|x|$ and $|h| < 1$.

If we define $W(x,h) = (1 + h^2 - 2xh)^{-1/2}$, a quick check shows that $W(x,h)$ satisfies the first-order partial differential equation

$$(1 - 2xh + h^2)\frac{\partial W}{\partial h} + (h - x)\,W = 0. \tag{1.6.24}$$

The substitution of Eq. (1.6.21) into Eq. (1.6.24) yields

$$(1 - 2xh + h^2)\sum_{n=0}^{\infty} n\,P_n(x)\,h^{n-1} + (h - x)\sum_{n=0}^{\infty} P_n(x)h^n = 0 \tag{1.6.25}$$

Setting the coefficients of h^n equal to zero, we find that

$$(n+1)\,P_{n+1}(x) - 2nx\,P_n(x) + (n-1)\,P_{n-1}(x) + P_{n-1}(x) - xP_n(x) = 0 \tag{1.6.26}$$

or

$$\boxed{(n+1)\,P_{n+1}(x) - (2n+1)\,x\,P_n(x) + n\,P_{n-1}(x) = 0} \tag{1.6.27}$$

with $n = 1,2,3, \ldots$ Similarly, the first-order partial differential equation

$$(1 - 2xh + h^2)\,\frac{\partial W}{\partial x} - hW = 0 \tag{1.6.28}$$

leads to

$$(1 - 2xh + h^2) \sum_{n=0}^{\infty} P_n'(x)\,h^n - \sum_{n=0}^{\infty} P_n(x)\,h^n = 0 \tag{1.6.29}$$

which implies

$$P'_{n+1}(x) - 2x\,P'_n(x) + P'_{n-1}(x) - P_n(x) = 0. \tag{1.6.30}$$

Differentiating Eq. (1.6.27), we first eliminate $P'_{n-1}(x)$ and then $P'_{n+1}(x)$ from the resulting equation and Eq. (1.6.30). This gives two further recurrence relationships:

$$P'_{n+1}(x) - x\,P'_n(x) - (n+1)\,P_n(x) = 0, n = 0,1,2, \ldots \tag{1.6.31}$$

and

$$x\,P'_n(x) - P'_{n-1}(x) - n\,P_n(x) = 0, \quad n = 1,2,3, \ldots \tag{1.6.32}$$

Adding Eqs. (1.6.31) and (1.6.32), we obtain the more symmetric formula

$$\boxed{P'_{n+1}(x) - P'_{n-1}(x) = (2n+1)\,P_n(x), n = 1,2,3, \ldots} \tag{1.6.33}$$

Given any two of the polynomials $P_{n+1}(x)$, $P_n(x)$ and $P_{n-1}(x)$, the third can be found from either Eq. (1.6.27) or Eq. (1.6.33).

Example 1.6.2

Let us compute $P_3(x)$ from a recurrence relation. From Eq. (1.6.27) with $n = 2$, we have

$$3\,P_3(x) - 5x\,P_2(x) + 2\,P_1(x) = 0. \tag{1.6.34}$$

But, $P_2(x) = (3x^2 - 1)/2$ and $P_1(x) = x$, so that

$$3\,P_3(x) = 5x\,P_2(x) - 2\,P_1(x) = 5x\,[(3x^2 - 1)/2] - 2x \tag{1.6.35}$$

$$= \frac{15}{2}\,x^3 - \frac{9}{2}\,x \tag{1.6.36}$$

or

$$P_3(x) = (5x^3 - 3x)/2. \tag{1.6.37}$$

Example 1.6.3

We want to show that

$$\int_{-1}^{1} P_n(x)\,dx = 0. \tag{1.6.38}$$

From Eq. (1.6.33), we have

$$(2n+1)\int_{-1}^{1} P_n(x)\,dx = \int_{-1}^{1} P'_{n+1}(x) - P'_{n-1}(x)\,dx \tag{1.6.39}$$

$$= P_{n+1}(x) - P_{n-1}(x)\Big|_{-1}^{1} \tag{1.6.40}$$

$$= P_{n+1}(1) - P_{n+1}(-1) - P_{n-1}(1) + P_{n-1}(-1) \tag{1.6.41}$$

$$= 0 \tag{1.6.42}$$

because $P_n(1) = 1$ and $P_n(-1) = (-1)^n$.

Having determined several methods for finding the Legendre polynomial $P_n(x)$, we now turn to the actual orthogonality condition. Consider the integral

$$J = \int_{-1}^{1} \frac{dx}{\sqrt{1 + h^2 - 2xh}\,\sqrt{1 + t^2 - 2xt}} \qquad |h|,|t|<1 \tag{1.6.43}$$

$$J = \int_{-1}^{1} (P_0(x) + h\,P_1(x) + \ldots + h^n\,P_n(x) + \ldots)$$

$$(P_0(x) + t\,P_1(x) + \ldots + t^m\,P_m(x) + \ldots)\,dx \tag{1.6.44}$$

$$J = \sum_{m,n=0}^{\infty} h^n t^m \int_{-1}^{1} P_n(x)\, P_m(x)\, dx. \qquad \textbf{(1.6.45)}$$

On the other hand, let $a = (1 + h^2)/2h$ and $b = (1 + t^2)/2t$, we see that the integral J becomes

$$J = \int_{-1}^{1} \frac{dx}{\sqrt{1 + h^2 - 2xh}\,\sqrt{1 + t^2 - 2xt}} \qquad \textbf{(1.6.46)}$$

$$= \frac{1}{2\sqrt{ht}} \int_{-1}^{1} \frac{dx}{\sqrt{a - x}\sqrt{b - x}} \qquad \textbf{(1.6.47)}$$

$$= \frac{-1}{\sqrt{ht}} \int_{-1}^{1} \frac{1/2\left(\frac{-1}{\sqrt{a-x}} + \frac{-1}{\sqrt{b-x}}\right)}{\sqrt{a - x} + \sqrt{b - x}}\, dx \qquad \textbf{(1.6.48)}$$

$$= -\frac{1}{\sqrt{ht}} \ell n\, (\sqrt{a - x} + \sqrt{b - x}) \Big|_{-1}^{1} \qquad \textbf{(1.6.49)}$$

$$= \frac{1}{\sqrt{ht}} \ell n \left\{ \frac{\sqrt{a + 1} + \sqrt{b + 1}}{\sqrt{a - 1} + \sqrt{b - 1}} \right\} \qquad \textbf{(1.6.50)}$$

But $a + 1 = (1 + h^2 + 2h)/2h = (1 + h)^2/2h$ and $a - 1 = (1 - h)^2/2h$. After a little algebra,

$$J = \frac{1}{\sqrt{ht}} \ell n \left[\frac{1 + \sqrt{ht}}{1 - \sqrt{ht}} \right] \qquad \textbf{(1.6.51)}$$

$$= \frac{2}{\sqrt{ht}} \left(\sqrt{ht} + \frac{(ht)^{3/2}}{3} + \frac{(ht)^{5/2}}{5} + \ldots\right) \qquad \textbf{(1.6.52)}$$

$$= 2\left(1 + \frac{ht}{3} + \frac{h^2t^2}{5} + \ldots + \frac{h^n t^n}{2n+1} + \ldots\right) \qquad \textbf{(1.6.53)}$$

As we noted earlier, the coefficients of $h^n t^m$ in this series is $\int_{-1}^{1} P_n(x)P_m(x)dx$. If we match the powers of $h^n t^m$, the orthogonality condition is

$$\boxed{\int_{-1}^{1} P_n(x)P_m(x)\, dx = \begin{cases} 0 & \text{if } m \neq n \\ \dfrac{2}{2n+1} & \text{if } m = n \end{cases}} \qquad \textbf{(1.6.54)}$$

This demonstration was purposed by Symons, B., 1983: Legendre polynomials and their orthogonality. *Mathematical Gazette*. pp. 152-154.

With the orthogonality condition (1.6.42), we are ready to determine how a function $f(x)$, which is piecewise differentiable (i.e., in a finite range it is possible to divide that range into a finite number of intervals so that the derivative of $f(x)$ exists inside each interval) in the interval $(-1,1)$ can be represented by the series.

$$f(x) = \sum_{m=0}^{\infty} A_m P_m(x) \qquad (-1 \leq x \leq 1) \qquad \textbf{(1.6.55)}$$

To find A_m we multiply both sides of Eq. (1.6.55) by $P_n(x)\,dx$ and integrate from -1 to 1:

$$\int_{-1}^{1} f(x)\,P_n(x)\,dx = \sum_{m=0}^{\infty} A_m \int_{-1}^{1} P_n(x)\,P_m(x)\,dx. \qquad \textbf{(1.6.56)}$$

All of the terms on the right-hand side vanish except for $n = m$ because of the orthogonality condition (1.6.55). Consequently the coefficients A_n is given by

$$A_n \int_{-1}^{1} P_n^2(x)\,dx = \int_{-1}^{1} f(x)\,P_n(x)\,dx \qquad \textbf{(1.6.57)}$$

or

$$\boxed{A_n = \frac{2n+1}{2} \int_{-1}^{1} f(x)\,P_n(x)\,dx.} \qquad \textbf{(1.6.58)}$$

In the special case when $f(x)$ and its first n derivatives are continuous throughout the interval $(-1,1)$, we may use Rodrigues' formula to evaluate

$$\int_{-1}^{1} f(x)\,P_n(x)\,dx = \frac{1}{2^n n!} \int_{-1}^{1} f(x)\,\frac{d^n(x^2-1)^n}{dx^n}\,dx \qquad \textbf{(1.6.59)}$$

$$= \frac{(-1)^n}{2^n n!} \int_{-1}^{1} (x^2-1)^n\,\frac{d^n f(x)}{dx^n}\,dx \qquad \textbf{(1.6.60)}$$

by integrating by parts n times. Consequently

$$A_n = \frac{2n+1}{2^{n+1}n!} \int_{-1}^{1} (1 - x^2)^n \frac{d^n f(x)}{dx^n} dx. \qquad \textbf{(1.6.61)}$$

A particularly useful result from Eq. (1.6.61) follows if $f(x)$ is a *polynomial of degree k*. Because all derivatives of $f(x)$ of order n vanish identically when $n>k$, we conclude that $A_n = 0$ if $n>k$. This means that any polynomial of degree k can be expressed as a linear combination of the first $k+1$ Legendre polynomials $[P_0(x), \ldots, P_k(x)]$.

Example 1.6.4

Let us express x^2 in terms of Legendre polynomials, the results from the previous paragraph mean that we need only worry about $P_0(x)$, $P_1(x)$ and $P_2(x)$:

$$x^2 = A_0 P_0(x) + A_1 P_1(x) + A_2 P_2(x). \qquad \textbf{(1.6.62)}$$

Substituting for the Legendre polynomial

$$x^2 = A_0 + A_1 x + 1/2\, A_2 (3x^2 - 1), \qquad \textbf{(1.6.63)}$$

we find

$$A_0 = \frac{1}{3}, A_1 = 0, \text{ and } A_2 \ \frac{2}{3}. \qquad \textbf{(1.6.64)}$$

Example 1.6.5

Let us find the expansion in Legendre polynomials of the function:

$$f(x) = \begin{cases} 0 & -1<x<0 \\ 1 & 0<x<1. \end{cases} \qquad \textbf{(1.6.65)}$$

In this problem, we find that

$$A_n = \frac{2n+1}{2} \int_{0}^{1} P_n(x)\, dx. \qquad \textbf{(1.6.66)}$$

$$A_0 = 1/2 \int_0^1 1\, dx = 1/2 \qquad A_1 = \frac{3}{2} \int_0^1 x\, dx = \frac{3}{4}$$

$$A_2 = \frac{5}{2}\,\frac{1}{2}\int_0^1 3\,x^2 - 1\;dx = 0$$

$$A_3 = \frac{7}{2}\,\frac{1}{2}\int_0^1 5\,x^3 - 3x\;\;dx = -\,\frac{7}{16}$$

so that

$$f(x) = 1/2\;P_0(x) + \frac{3}{4}\;P_1(x) - \frac{7}{16}\;P_3(x) + \frac{11}{32}\;P_5(x) + \ldots \qquad \textbf{(1.6.67)}$$

In Fig. 1.6.2 the expansion (1.6.67) is graphed where we have used only the first five terms. The spurious oscillations arise from trying to represent a discontinuous function by five continuous, oscillatory functions. Even if additional terms are added, the spurious oscillations will persist although located nearer to the discontinuity. This behavior is called *Gibbs phenomena*.

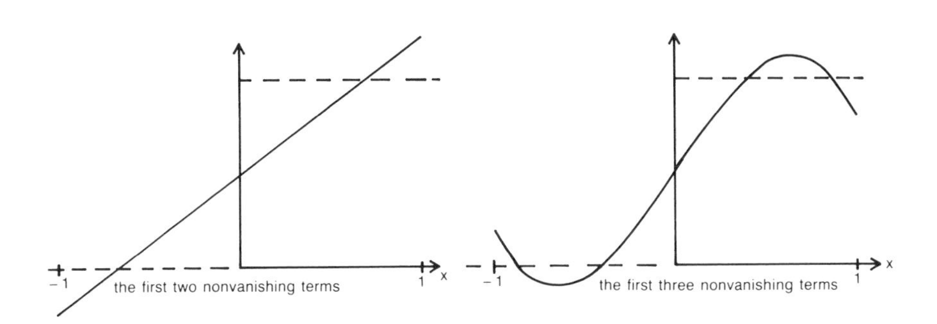

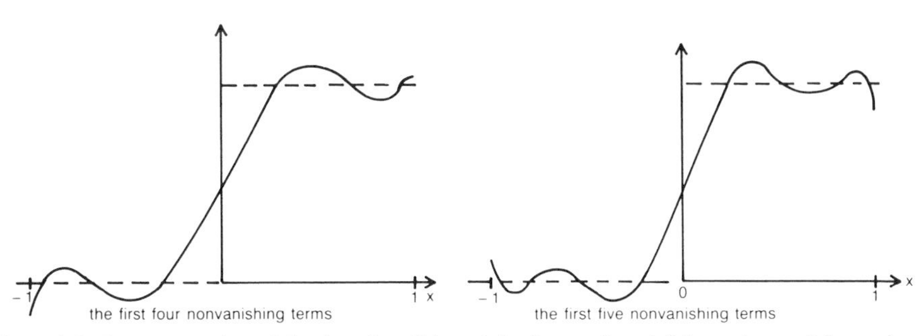

Fig. 1.6.2: Representation of the function f(x) = 1 for 0<x<1 and 0 for −1<x<0 by various partial summations of its Legendre polynomial expansion. The dashed lines denote the exact function.

Summary of Legendre's Equation

1. Legendre's equation:

$$(1-x^2)y'' - 2x\,y' + n(n+1)y = 0$$

2. Over the domain $[-1,1]$, the only finite solution is the Legendre polynomial $P_n(x)$.

3. Three ways of computing $P_n(x)$:

(a) Power series

$$P_n(x) = \sum_{k=0}^{m} (-1)^k \frac{(2n-2k)!}{2^n k!\,(n-k)!\,(n-2k)!} x^{n-2k},$$

$$m = n/2 \text{ or } (n-1)/2$$

(b) Rodrigues' formula

$$P_n(x) = \frac{1}{2^n n!} \frac{d^n}{dx^n} (x^2-1)^n \qquad n = 0,1,2,\ldots$$

(c) Recurrence formulas

$$(n+1)\,P_{n+1}(x) - (2n+1)xP_n(x) + n\,P_{n-1}(x) = 0$$

$$(2n+1)\,P_n(x) = P'_{n+1}(x) - P'_{n-1}(x), \; n = 1,2,3,\ldots$$

4. $P_{2n}(x)$ is an even function; $P_{2n+1}(x)$ is an odd function.

5. $$P_{2n+1}(0) = 0; \quad P_{2n}(0) = (-1)^n \frac{(2n)!}{2^{2n}(n!)^2}$$

6. Orthogonality

$$\int_{-1}^{1} P_n(x)\,P_m(x)\,dx = \begin{cases} 0 & m \neq n \\ \dfrac{2}{2n+1} & m = n \end{cases}$$

7. Generating function

$$(1 + h^2 - 2xh)^{-1/2} = P_0(x) + h\,P_1(x) + h^2\,P_2(x) + \ldots$$

8. Expansions in Legendre polynomials

$$f(x) = \sum_{n=0}^{\infty} A_n\,P_n(x)$$

where

$$A_n = \frac{2n+1}{2}\int_{-1}^{1} f(x)\, P_n(x)\, dx$$

or

$$A_n = \frac{2n+1}{2^{n+1}n!}\int_{-1}^{1} (1-x^2)^n \frac{d^n f(x)}{dx^n}\, dx$$

if $f(x)$ and the first n derivatives are continuous.

9. $$\int_{-1}^{1} x^m P_n(x)\, dx = 0 \qquad n > m$$

10. $$\int_{x}^{1} P_n(t)\, dt = \frac{1}{2n+1}\,[P_{n-1}(x) - P_{n+1}(x)] \qquad n > 1$$

Exercises

1. Find the first three nonvanishing coefficients in the Legendre expansions for the function:

a) $f(x) = \begin{cases} 0 & -1 < x < 0 \\ x & 0 < x < 1 \end{cases}$ b) $f(x) = \begin{cases} 1/(2\epsilon) & |x| < \epsilon \\ 0 & \epsilon < |x| < 1 \end{cases}$

c) $f(x) = |x| \quad |x| < 1$ d) $f(x) = x^3 \quad |x| < 1$

e) $f(x) = \begin{cases} -1 & -1 < x < 0 \\ 1 & 0 < x < 1 \end{cases}$ f) $f(x) = \begin{cases} -1 & -1 < x < 0 \\ x & 0 < x < 1 \end{cases}$

2. Use Rodrigues' formula to show that

$$P_4(x) = \frac{1}{8}\,(35x^4 - 30x^2 + 3).$$

3. Given $P_5(x) = \frac{63}{8}x^5 - \frac{70}{8}x^3 + \frac{15}{8}x$ and $P_4(x)$ from problem 2, use the recurrence formula for $P_{n+1}(x)$ to find $P_6(x)$.

4. Show that

$$P_n(1) = 1 \qquad P_n(-1) = (-1)^n$$

$$P_{2n+1}(0) = 0 \qquad P_{2n}(0) = (-1)^n \frac{(2n)!}{2^{2n}n!n!}$$

5. Prove that

$$\int_x^1 P_n(t)\,dt = \frac{1}{2n+1}\,[P_{n-1}(x) - P_{n+1}(x)]$$

1.7 ANOTHER EXAMPLE OF A SINGULAR STURM-LIOUVILLE PROBLEM: BESSEL'S EQUATION

In the previous section we discussed the solutions to Legendre's equation, especially with regards to their use in orthogonal expansions. In this section we consider another classic equation, Bessel's equation,

$$x^2 y'' + xy' + (\mu^2 x^2 - n^2)y = 0$$

or

$$\frac{d}{dx}\left(x\,\frac{dy}{dx}\right) + \left(-\frac{n^2}{x} + \mu^2 x\right)y = 0 \qquad \textbf{(1.7.1)}$$

which arises in many solutions of partial differential equations in cylindrical coordinates. Once again our ultimate goal will be to use these solutions in orthogonal expansions.

A quick check of Bessel's equation shows that it conforms to the canonical form of the Sturm-Liouville problem: $p(x) = x$, $q(x) = -n^2/x)$, $r(x) = x$ and $\lambda = \mu^2$. Restricting our attention to the interval $[0,L]$, we see that because $p(0) = 0$ the Sturm-Liouville problem involving Eq. (1.7.1) is singular. From Eq. (1.6.1) in the previous section, it is easily shown that eigenfunctions to a singular Sturm-Liouville problem will still be orthogonal over the interval $[0,L]$ if 1) at $x = 0$ $y(x)$ is finite and $xy'(x)$ is zero and 2) at $x = L$ a homogeneous boundary condition (1.1.2) is prescribed. Consequently we will only seek solutions that are finite at the origin and where $x\,y'(x)$ equals zero at $x = 0$.

The solution to Bessel's equation cannot be written down in a simple closed form. As in the case with Legendre's equation, the solution must be found by power series. Because we intend to make the expansion about $x = 0$ and this point is a regular singular point, we must use the method of Frobenius (see Appendix A). Because the quantity n^2 appears in Eq. (1.7.1), we can consider n to be non-negative without any loss of generality.

At present we have placed no restriction on n and it may assume values from 0 to positive infinity. However, in the majority of problems associated with Bessel's equation, n will equal a non-negative integer. This case is much simpler than the case of arbitrary n. For those cases of noninteger n, see Hillebrand, F. B., 1962: *Advanced*

Calculus for Applications. Englewood Cliff, NJ: Prentice-Hall, Inc. Section 4.8.

To simplify matters, we first find the solution when $\mu = 1$. The solution for $\mu \neq 1$ can then be found by substituting μx for x. Consequently we seek solutions of the form:

$$y(x) = \sum_{k=0}^{\infty} B_k x^{2k+s} \quad \textbf{(1.7.2)}$$

$$y'(x) = \sum_{k=0}^{\infty} (2k+s) B_k x^{2k+s-1} \quad \textbf{(1.7.3)}$$

and

$$y''(x) = \sum_{k=0}^{\infty} (2k+s)(2k+s-1) B_k x^{2k+s-2}. \quad \textbf{(1.7.4)}$$

The substitution of Eqs. (1.7.2)-(1.7.4) into Eq. (1.7.1) with $\mu = 1$ yields

$$\sum_{k=0}^{\infty} (2k+s)(2k+s-1)B_k x^{2k+s} + \sum_{k=0}^{\infty} (2k+s)B_k x^{2k+s}$$

$$+ \sum_{k=0}^{\infty} B_k x^{2k+s+2} - n^2 \sum_{k=0}^{\infty} B_k x^{2k+s} = 0 \quad \textbf{(1.7.5)}$$

or

$$\sum_{k=0}^{\infty} \{(2k+s)^2 - n^2\}B_k x^{2k} + \sum_{k=0}^{\infty} B_k x^{2k+2} = 0. \quad \textbf{(1.7.6)}$$

If we explicitly separate the $k = 0$ term from the other terms in the first summation in Eq. (1.7.6), Eq. (1.7.6) becomes

$$(s^2 - n^2)B_0 + \sum_{m=1}^{\infty} \{(2m+s)^2 - n^2\}B_m x^{2m} + \sum_{k=0}^{\infty} B_k x^{2k+2} = 0. \quad \textbf{(1.7.7)}$$

We now change the dummy integer in the first summation of Eq. (1.7.7) by letting $m = k+1$ so that

$$(s^2 - n^2)B_0 + \sum_{k=0}^{\infty} \{ [(2k+s+2)^2 - n^2]B_{k+1} + B_k\}x^{2k+2} = 0. \qquad \textbf{(1.7.8)}$$

Because Eq. (1.7.8) must be true for all x, each power of x must vanish identically:

$$s = \pm n \qquad \textbf{(1.7.9)}$$

and

$$[(2k+s+2)^2 - n^2]B_{k+1} + B_k = 0. \qquad \textbf{(1.7.10)}$$

Because the difference of the larger indicial root from the lower root equals the integer $2n$, we are only guaranteed a power series solution of the form (1.7.2) for $s = n$ (see Section A.3). If we use this indicial root and the recurrence formula (1.7.10), this solution, known as the Bessel function of the first kind of order n and denoted by $J_n(x)$, is

$$J_n(x) = \sum_{k=0}^{\infty} \frac{(-1)^k (x/2)^{n+2k}}{k!\,(n+k)!} \qquad \textbf{(1.7.11)}$$

To find the second general solution to Bessel's equation, the one corresponding to $s = -n$, the most economical method is to express it in terms of partial derivatives of $J_n(x)$ with respect to its order n

$$Y_n(x) = \left[\frac{\partial J_\nu(x)}{\partial x} - (-1)^n \frac{\partial J_\nu(x)}{\partial x} \right]_{\nu = n} \qquad \textbf{(1.7.12)}$$

(See Watson, G. N., 1966: *A Treatise on the Theory of Bessel Functions*. Cambridge: At the University Press. Section 3.5. for the derivation.) Upon substituting the power series representation (1.7.11) into Eq. (1.7.12), we find

$$Y_n(x) = \frac{2}{\pi} J_n(x) \ln(x/2) - \frac{1}{\pi} \sum_{k=0}^{n-1} \frac{(n-k-1)!}{k!} (x/2)^{2k-n}$$

$$- \frac{1}{\pi} \sum_{k=0}^{\infty} \frac{(-1)^k (x/2)^{n+2k}}{k!\,(n+k)!} \{\psi(k+1) + \psi(k+n+1)\} \qquad \textbf{(1.7.13)}$$

where

$$\psi(m+1) = -\gamma + 1 + \frac{1}{2} + \ldots + \frac{1}{m}\,, \tag{1.7.14}$$

$$\psi(1) = -\gamma \tag{1.7.15}$$

and γ is Euler's constant (0.5772157). In the case $n = 0$, the first sum in Eq. (1.7.13) should be set equal to zero. This function $Y_n(x)$ is called Neumann's Bessel function of the second kind of order n. Consequently the general solution to Eq. (1.7.1) is

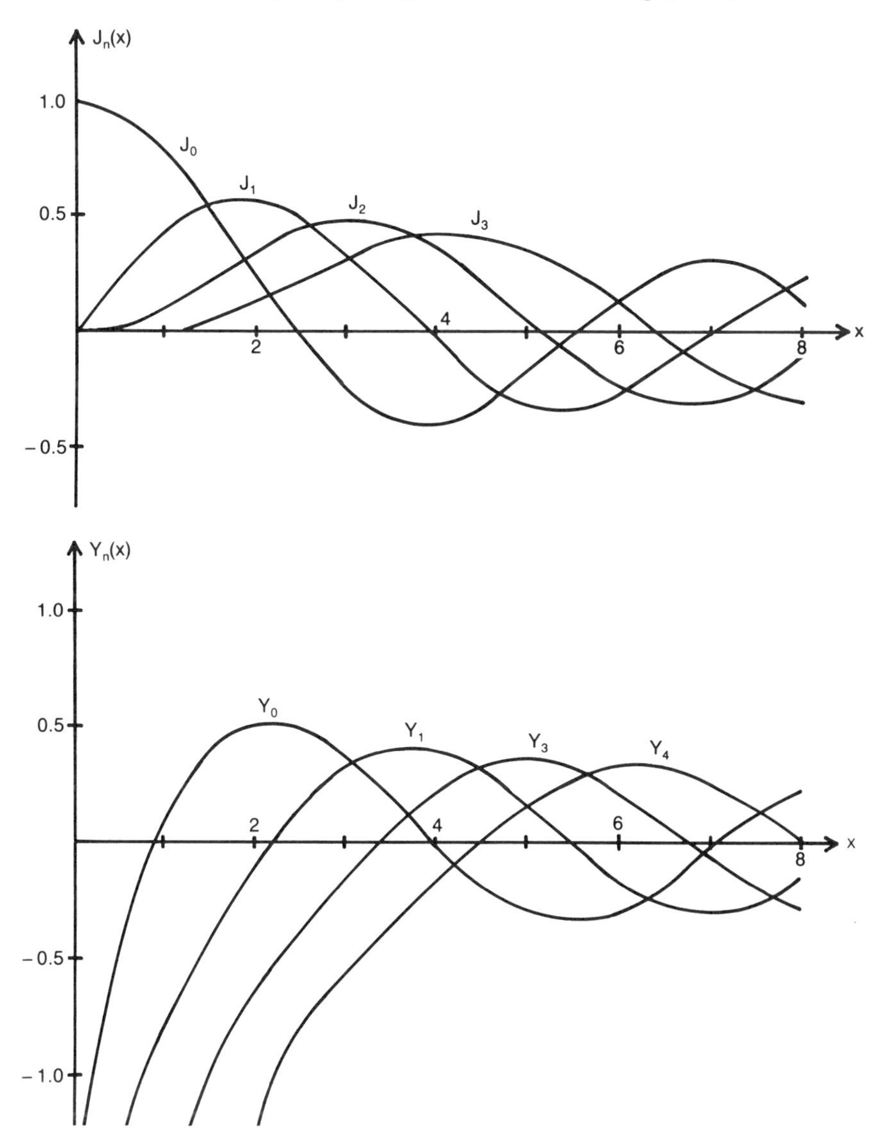

Fig. 1.7.1: The first four Bessel functions of the first kind (top) and second kind (bottom) over the interval from zero to eight.

$$y(x) = A\,J_n(\mu x) + B\,Y_n(\mu x). \quad \textbf{(1.7.16)}$$

In Fig. 1.7.1 the function $J_0(x)$, $J_1(x)$, $J_2(x)$, $J_3(x)$, $Y_0(x)$, $Y_1(x)$, $Y_2(x)$ and $Y_3(x)$ are illustrated.

Example 1.7.1

Starting with Bessel's equation, we want to show that the solution to

$$y'' - \left(\frac{2a-1}{x}\right) y' + \left(b^2c^2x^{2c-2} + \frac{a^2 - n^2c^2}{x^2}\right)y = 0 \quad \textbf{(1.7.17)}$$

is

$$y(x) = A\,x^a J_n(bx^c) + B\,x^a Y_n(bx^c) \quad \textbf{(1.7.18)}$$

provided that $bx^c > 0$ so that $Y_n(bx^c)$ is defined.

The general solution to

$$X^2 \frac{d^2Y}{dX^2} + X\frac{dY}{dX} + (X^2 - n^2)Y = 0 \quad \textbf{(1.7.19)}$$

is

$$Y = A\,J_n(X) + B\,Y_n(X). \quad \textbf{(1.7.20)}$$

If we now let $Y = y(x)/x^a$ and $X = bx^c$, then

$$\frac{d}{dX} = \frac{dx}{d(bx^c)}\frac{d}{dx} = \frac{x^{1-c}}{bc}\frac{d}{dx} \quad \textbf{(1.7.21)}$$

$$\frac{d^2}{dX^2} = \frac{x^{2-2c}}{b^2c^2}\frac{d^2}{dx^2} - \frac{c-1}{b^2c^2}\,x^{1-2c}\frac{d}{dx} \quad \textbf{(1.7.22)}$$

$$\frac{d}{dx}\left(\frac{y}{x^a}\right) = \frac{1}{x^a}\frac{dy}{dx} - \frac{a}{x^{a+1}}\,y \quad \textbf{(1.7.23)}$$

and

$$\frac{d^2}{dx^2}\left(\frac{y}{x^a}\right) = \frac{1}{x^a}\frac{d^2y}{dx^2} - \frac{2a}{x^{a+1}}\frac{dy}{dx} + \frac{a(a+1)}{x^{a+2}}\,y\,. \quad \textbf{(1.7.24)}$$

Substituting Eqs. (1.7.21)-(1.7.24) into Eq. (1.7.19) and simplifying, we find

$$y'' + \left(\frac{1-2a}{x}\right) y' + \left(b^2c^2x^{2c-2} + \frac{a^2 - n^2c^2}{x^2}\right)y = 0. \quad \textbf{(1.7.25)}$$

Therefore, the solution is given by Eq. (1.7.18).

An equation which is very similar to Eq. (1.7.1) is

$$x^2 \frac{d^2y}{dx^2} + x \frac{dy}{dx} - (n^2 + x^2)y = 0. \quad \textbf{(1.7.26)}$$

It arises in the solution of partial differential equations in cylindrical coordinates. If we substitute $ix = t$ (where $i = \sqrt{-1}$) into Eq. (1.7.26), it becomes Bessel's equation:

$$t^2 \frac{d^2y}{dt^2} + t \frac{dy}{dt} + (t^2 - n^2)y = 0. \quad \textbf{(1.7.27)}$$

Consequently, we may immediately write the solution as

$$y(x) = c_1 J_n(ix) + c_2 Y_n(ix) \quad \textbf{(1.7.28)}$$

if n is an integer. Traditionally, the solution to Eq. (1.7.26) has been written

$$y(x) = c_1 I_n(x) + c_2 K_n(x) \quad \textbf{(1.7.29)}$$

rather than in terms of $J_n(ix)$ and $Y_n(ix)$ where

$$I_n(x) = \sum_{k=0}^{\infty} \frac{1}{k!(k+n)!} \left(\frac{x}{2}\right)^{2k+n} \quad \textbf{(1.7.30)}$$

and

$$K_n(x) = \frac{\pi}{2} i^{n+1}[J_n(ix) + i\, Y_n(x)]. \quad \textbf{(1.7.31)}$$

The function $I_n(x)$ is known as the modified Bessel function of the first kind, of order n while $K_n(x)$ is known as the modified Bessel function of the second kind, of order n. In Fig. 1.7.2 I_0, I_1, I_2, I_3, K_0, K_1, K_2 and K_3 have been graphed. Note that $K_n(x)$ has no real zeros while $I_n(x)$ equals zero only at $x = 0$ for $n \geq 1$.

From Fig. 1.7.1 we see that Bessel functions of the first kind remain finite at $x = 0$ while Bessel functions of the second kind do not. Similarly the product $x J_n'(x)$ goes to zero at $x = 0$. Consequently, for the remaining portions of this section, we shall con-

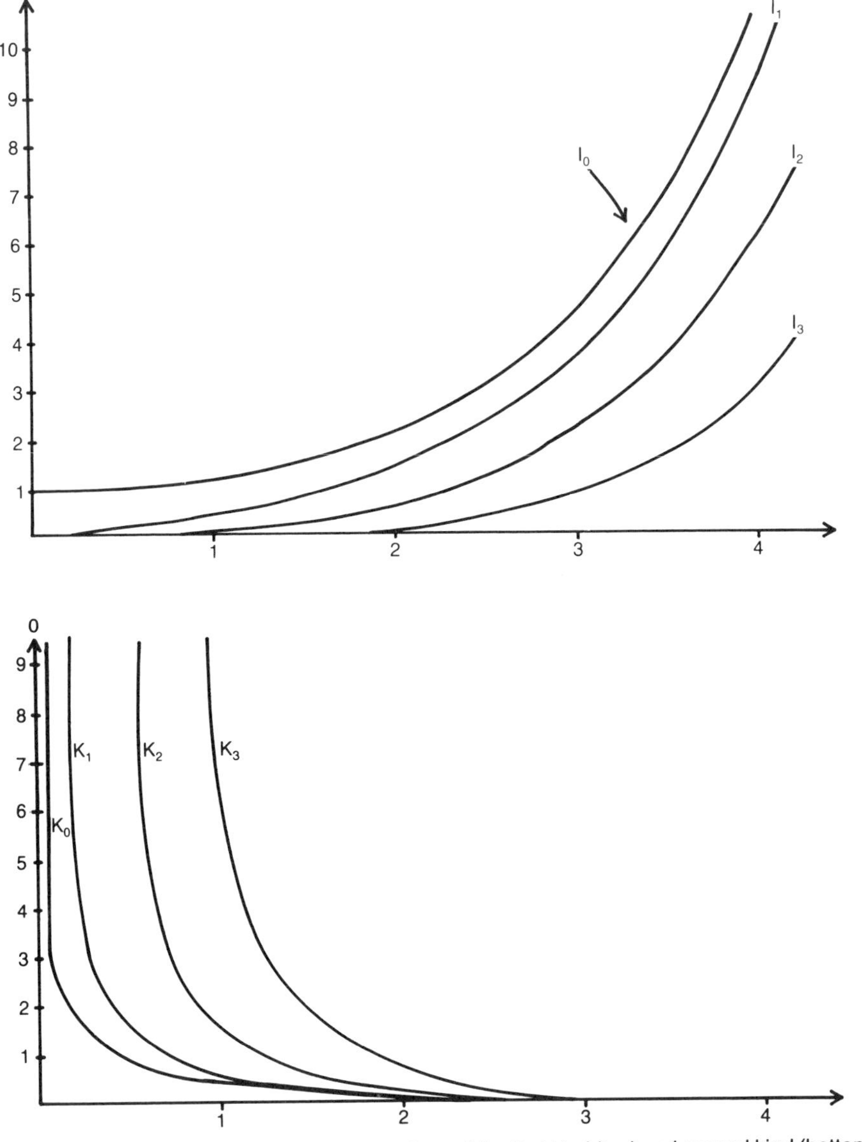

Fig. 1.7.2: The first four modified Bessel functions of the first kind (top) and second kind (bottom) over the interval from zero to four.

cern ourselves only with Bessel functions of the first kind because only they may be used in a Fourier-Bessel expansion over the interval $[0,L]$.

The Bessel function $J_n(x)$ is related to $J_{n+1}(x)$ and $J_{n-1}(x)$ by recurrence formulae. Assuming that n is a positive integer, we multiply the series (1.7.11) by x^n and then differentiate with respect to x. This gives

$$\frac{d}{dx}\left(x^n J_n(x)\right) = \sum_{k=0}^{\infty} \frac{(-1)^k (2n+2k)\, x^{2n+2k-1}}{2^{n+2k}\, k!\, (n+k)!} \qquad (1.7.32)$$

$$= x^n \sum_{k=0}^{\infty} \frac{(-1)^k}{k!(n-1+k)!} \left(\frac{x}{2}\right)^{n-1+2k} = x^n J_{n-1}(x) \qquad (1.7.33)$$

or

$$\boxed{\frac{d}{dx}\left[x^n J_n(x)\right] = x^n J_{n-1}(x)} \qquad (1.7.34)$$

for $n = 1, 2, 3, \ldots$ Similarly, multiplying Eq. (1.7.11) by x^{-n}, we find that

$$\boxed{\frac{d}{dx}\left[x^{-n} J_n(x)\right] = -x^{-n} J_{n+1}(x)} \qquad (1.7.35)$$

for $n = 0, 1, 2, \ldots$ If we now differentiate Eq. (1.7.34) and (1.7.35) and divide by the factors $x^{\pm n}$, we find

$$J_n'(x) + \frac{n}{x} J_n(x) = J_{n-1}(x) \qquad (1.7.36)$$

and

$$J_n'(x) - \frac{n}{x} J_n(x) = -J_{n+1}(x) \qquad (1.7.37)$$

Eqs. (1.7.36)-(1.7.37) immediately yield the recurrence relationship

$$\boxed{J_{n-1}(x) + J_{n+1}(x) = \frac{2n}{x} J_n(x)} \qquad (1.7.38)$$

and

$$\boxed{J_{n-1}(x) - J_{n+1}(x) = 2\, J_n'(x)} \qquad (1.7.39)$$

for $n = 1, 2, 3, \ldots$ For $n = 0$, Eq. (1.7.39) should be replaced by

$$J_0'(x) = -J_1(x). \qquad (1.7.40)$$

Example 1.7.2

We want to show that

$$x^2 J''_n(x) = (n^2 - n - x^2) J_n(x) + x J_{n+1}(x). \quad (1.7.41)$$

From Eq. (1.7.37), we have

$$J_n'(\mathrm{x}) = \frac{n}{x} J_n(x) - J_{n+1}(x) \quad (1.7.42)$$

$$J_n''(x) = -\frac{n}{x^2} J_n(x) + \frac{n}{x} J_n'(x) - J_{n+1}'(x) \quad (1.7.43)$$

and

$$J_n''(x) = -\frac{n}{x^2} J_n(x) + \frac{n}{x} [\frac{n}{x} J_n(x) - J_{n+1}' (x)]$$

$$- [-\frac{n+1}{x} J_{n+1}(x) + J_n(x)] \quad (1.7.44)$$

after using Eqs. (1.7.36)-(1.7.37). Simplifying, we have

$$J_n''(x) = \left(\frac{n^2 - n}{x^2} - 1\right) J_n(x) + \frac{J_{n+1}(x)}{x} \quad (1.7.45)$$

or

$$x^2 J_n''(x) = (n^2 - n - x^2) J_n(x) + x J_{n+1}(x). \quad (1.7.46)$$

Example 1.7.3

Show that

$$\int_0^a x^5 J_2(x)\, dx = a^5 J_3(a) - 2 a^4 J_4(a). \quad (1.7.47)$$

The evaluation of Eq. (1.7.47) is accomplished by integration by parts. If $u = x^2$ and $dv = x^3 J_2(x)\, dx$, then

$$\int_0^a x^5 J_2(x)\, dx = x^5 J_3(x) \Big|_0^a - 2 \int_0^a x^4 J_3(x)\, dx \quad (1.7.48)$$

because $d[x^3 J_3(x)] = x^2 J_2(x)\, dx$ by Eq. (1.7.34). Finally, because $x^4 J_3(x)\, dx = d[x^4 J_4(x)]$ by Eq. (1.7.34), we have

$$\int_0^a x^5 J_2(x)\,dx = a^5 J_3(a) - 2x^4 J_4(x)\Big|_0^a \quad \textbf{(1.7.49)}$$

$$= a^5 J_3(a) - 2a^4 J_4(a). \quad \textbf{(1.7.50)}$$

Before we can use Bessel functions in an orthogonal expansion, we must discuss the boundary conditions at $x = L$. There are three cases. One of them is the requirement that $y(L) = 0$ and results in the condition that $J_n(\mu_k L) = 0$. Another condition is $y'(L) = 0$ and gives $J_n'(\mu_k L) = 0$. Finally, if $h\,y(L) + y'(L) = 0$, then $h\,J_n(\mu_k L) + \mu_k J_n'(\mu_k L) = 0$. In all of these cases the eigenfunction that is used is $J_n(\mu_k x)$ so that the function is given by

$$f(x) = \sum_{k=1}^{\infty} A_k J_n(\mu_k x) \quad \textbf{(1.7.51)}$$

where the numbers μ_k are the positive solution of either

$$J_n(\mu_k L) = 0 \quad \textbf{(1.7.52)}$$

$$J_n'(\mu_k L) = 0 \quad \textbf{(1.7.53)}$$

or

$$h\,J_n(\mu_k L) + \mu_k J_n'(\mu_k L) = 0. \quad \textbf{(1.7.54)}$$

We now multiply Eq. (1.7.51) by $x\,J_n(\mu_m x)\,dx$ and integrate from 0 to L:

$$\sum_{k=1}^{\infty} A_k \int_0^L x\,J_n(\mu_k x)\,J_n(\mu_m)\,dx = \int_0^L x\,f(x)\,J_n(\mu_m x)\,dx \quad \textbf{(1.7.55)}$$

From the general orthogonality condition (1.3.10)

$$\int_0^L x\,J_n(\mu_k x)\,J_n(\mu_m x)\,dx = 0 \quad \text{if } k \neq m, \quad \textbf{(1.7.56)}$$

Equation (1.7.55) simplifies to

$$A_m \int_0^L x\,J_n(\mu_m x)^2\,dx = \int_0^L x\,f(x)\,J_n(\mu_m x)\,dx \quad \textbf{(1.7.57)}$$

or

$$\boxed{A_k = \frac{1}{C_k} \int_0^L x\, f(x)\, J_n(\mu_k x)\, dx} \tag{1.7.58}$$

where

$$C_k = \int_0^L x\, J_n(\mu_k x)^2\, dx \tag{1.7.59}$$

and m has been replaced by k in Eq. (1.7.57).

The factor C_k depends upon the nature of the boundary condition at $x = L$. In all cases, we start from Bessel's equation

$$\frac{d}{dx} \left(x \frac{dJ_n}{dx}\right) + \left(\mu_k^2 x - \frac{n^2}{x}\right) J_n = 0 \tag{1.7.60}$$

If we multiply both sides of Eq. (1.7.60) by $2\, x\, J_n'(\mu_k x)$, the resulting equation can be written

$$(\mu_k^2 x^2 - n^2) \frac{d}{dx} (J_n(\mu_k x)^2) = - \frac{d}{dx} \left(x \frac{dJ_n}{dx}\right)^2. \tag{1.7.61}$$

The integration of Eq. (1.7.61) from 0 to L followed by the subsequent use of integration by parts, results in

$$(\mu_k^2 x^2 - n^2) J_n(\mu_k x)^2 \Big|_0^L - 2\mu_k^2 \int_0^L x\, J_n(\mu_k x)^2\, dx = - \left[x^2 \left(\frac{dJ_n}{dx} x\right)^2\right] \Big|_0^L \tag{1.7.62}$$

Because $J_n(0) = 0$ for $n > 0$, $J_0(0) = 1$ and $x\, J_n'(x) = 0$ at $x = 0$, the integrated terms vanish at the lower limits. Thus we have

$$C_k = \int_0^L x\, J_n(\mu_k x)^2\, dx \tag{1.7.63}$$

$$= \frac{1}{2\mu_k^2} (\mu_k^2 L^2 - n^2) J_n(\mu_k L)^2 + L^2 \left(\frac{dJ_n(\mu_k L)}{dx}\right)^2]. \tag{1.7.64}$$

Because

$$\frac{d}{dx} [J_n(\mu_k x)] = -\mu_k\, J_{n+1}(\mu_k x) + \frac{n}{x} J_n(\mu_k x) \tag{1.7.65}$$

from Eq. (1.7.37), C_k becomes

$$\boxed{C_k = 1/2\, L^2\, J_{n+1}(\mu_k L)^2} \tag{1.7.66}$$

if $J_n(\mu_k L) = 0$. Otherwise, if $J_n'(\mu_k L) = 0$, then

$$\boxed{C_k = \frac{\mu_k^2 L^2 - n^2}{2\mu_k^2} J_n(\mu_k L)^2} \tag{1.7.67}$$

Finally

$$\boxed{C_k = \frac{\mu_k^2 L^2 - n^2 + h^2 L^2}{2\mu_k^2} J_n(\mu_k L)^2} \tag{1.7.68}$$

if $\mu_k J_n'(\mu_k L) = -h J_n(\mu_k L)$.

All of the preceding results must be slightly modified when $n = 0$ and the boundary condition is

$$J_0'(\mu_k L) = 0 \text{ or } \mu_k J_1(\mu_k L) = 0. \tag{1.7.69}$$

This modification results from the additional eigenvalue $\mu_0 = 0$ being present and the additional term A_0 must be added to the expansion. When the series is modified to read

$$f(x) = A_0 + \sum_{k=1}^{\infty} A_k J_0(\mu_k x), \tag{1.7.70}$$

the equation for finding A_0 is found to be

$$A_0 = \frac{2}{L^2} \int_0^L f(x)\, x\, dx \tag{1.7.71}$$

and the remaining coefficients are given by Eq. (1.7.58) and (1.7.61) with $n = 0$.

Example 1.7.4

Let us expand $f(x) = x$, $0 \leq x \leq 1$, in the series

$$f(x) = \sum_{k=1}^{\infty} A_k J_1(\mu_k x) \tag{1.7.72}$$

where μ_k denotes the kth zero of $J_1(x)$. From Eqs. (1.7.58) and

(1.7.66), we have

$$A_k = \frac{2}{J_2(\mu_k)^2} \int_0^1 x^2 J_1(\mu_k x)\, dx. \quad \textbf{(1.7.73)}$$

However, from Eq. (1.7.34),

$$\frac{d}{dx}(x^2 J_2(x)) = x^2 J_1(x) \quad \textbf{(1.7.74)}$$

if $n = 2$. Therefore Eq. (1.7.73) becomes

$$A_k = \frac{2}{J_2(\mu_k)^2} \frac{1}{\mu_k{}^3} x^2 \quad J_2(x) \Big|_0^{\mu_k} \quad \textbf{(1.7.75)}$$

$$= \frac{2}{\mu_k J_2(\mu_k)} \quad \textbf{(1.7.76)}$$

and the resulting expansion is

$$x = 2 \sum_{k=1}^{\infty} \frac{J_1(\mu_k x)}{\mu_k J_2(\mu_k)} \quad 0<x<1. \quad \textbf{(1.7.77)}$$

In Fig. 1.7.3 the Bessel function expansion of $f(x) = x$ is shown in a truncated form when only one, two, three or four terms are used. The value of the coefficients for the first four terms are 1.29615, −0.95001, 0.78736 and −0.68747.

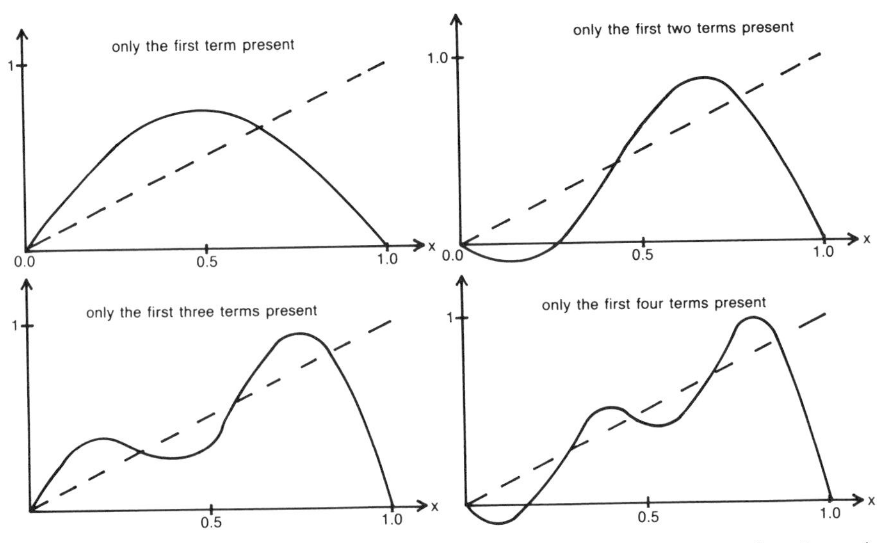

Fig. 1.7.3: The representation (1.7.77) of x over the interval from zero to one when the series is truncated so that only the first, first two, first three and first four terms are used.

EXAMPLE 1.7.5

Let us expand the function $f(x) = x^2$, $0 \le x \le 1$ in the series

$$f(x) = \sum_{k=1}^{\infty} A_k J_0(\mu_k x) \tag{1.7.78}$$

where μ_k denotes the kth positive zero of $J_0(\mu)$. From Eqs. (1.7.58) and (1.7.66), we have

$$A_k = \frac{2}{J_1(\mu_k)^2} \int_0^1 x^3 J_0(\mu_k x)\, dx. \tag{1.7.79}$$

If we let $t = \mu_k x$, the integration (1.7.79) becomes

$$A_k = \frac{2}{\mu_k^4 J_1(\mu_k)^2} \int_0^{\mu_k} t^3 J_0(t)\, dt. \tag{1.7.80}$$

We now let $u = t^2$ and $dv = t J_0(t)\, dt$ so that integration by parts results in

$$A_k = \frac{2}{\mu_k^4 J_1(\mu_k)^2} \left[t^3 J_0(t) \Big|_0^{\mu_k} - 2 \int_0^{\mu_k} t^2 J_1(t)\, dt \right] \tag{1.7.81}$$

$$= \frac{2}{\mu_k^4 J_1(\mu_k)^2} \left[\mu_k^3 J_1(\mu_k) - 2 \int_0^{\mu_k} t^2 J_1(t)\, dt \right] \tag{1.7.82}$$

because $v = t J_1(t)$ from Eq. (1.7.34). If we integrate by parts once more, we find that

$$A_k = \frac{2}{\mu_k^4 J_1(\mu_k)^2} \left[\mu_k^3 J_1(\mu_k) - 2 \mu_k^2 J_2(\mu_k) \right] \tag{1.7.83}$$

$$= \frac{2}{J_1(\mu_k)^2} \left[\frac{J_1(\mu_k)}{\mu_k} - \frac{2 J_2(\mu_k)}{\mu_k^2} \right] \tag{1.7.84}$$

However, from Eq. (1.7.38) with $n = 1$, we have

$$J_1(\mu_k) = 1/2\ \mu_k\ [J_2(\mu_k) + J_0(\mu_k)] \tag{1.7.85}$$

or

$$J_2(\mu_k) = \frac{2 J_1(\mu_k)}{\mu_k} \tag{1.7.86}$$

because $J_0(\mu_k) = 0$. Therefore

$$A_k = \frac{2}{J_1(\mu_k)^2} \frac{2\,(\mu_k^2 - 4)}{\mu_k^3} J_1(\mu_k) \qquad \textbf{(1.7.87)}$$

and

$$x^2 = 2 \sum_{k=1}^{\infty} \frac{(\mu_k{}^2 - 4)\, J_0(\mu_k x)}{\mu^3{}_k\, J_1(\mu_k)} \qquad 0 \leq x \leq 1 \qquad \textbf{(1.7.88)}$$

In Fig. 1.7.4 the representatiuon of x^2 by the Fourier-Bessel series is shown when the series is truncated so that only one, two, three or fours terms are included. The corresponding coefficients are 0.49397, -0.92501, 0.80596 and -0.70869.

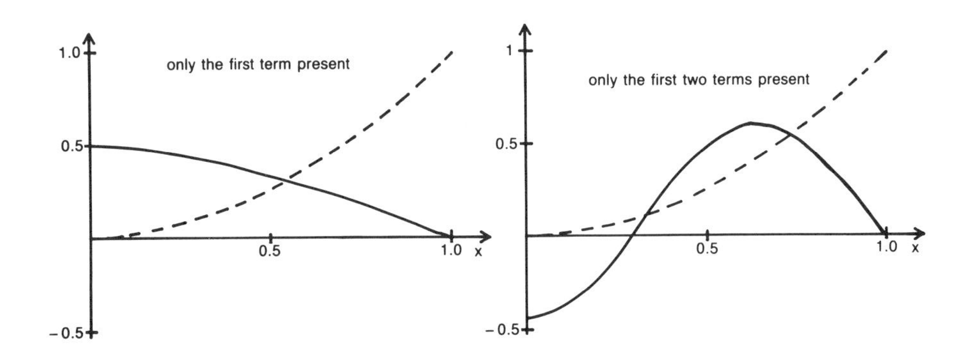

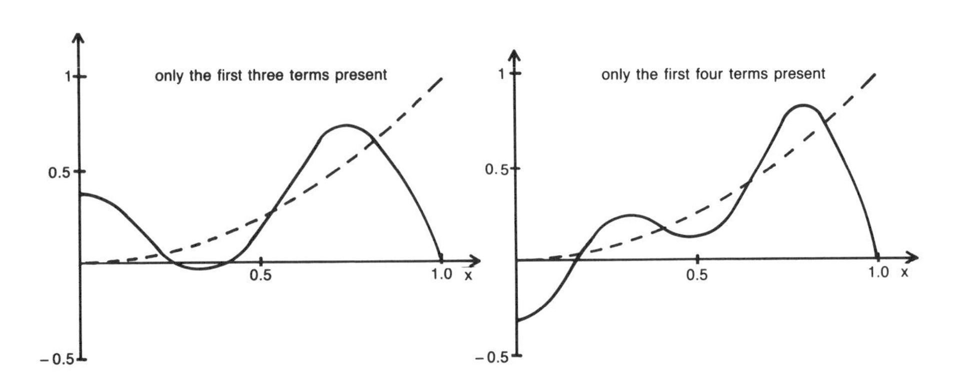

Fig. 1.7.4: The representation (1.7.88) of x over the interval from zero to one when the series is truncated so that only the first, first two, first three and first four terms are used.

Summary of Orthogonal Expansions in $J_n(\mu_k x)$

1. For non-negative integer n,

$$J_n(x) = \sum_{k=0}^{\infty} \frac{(-1)^k (x/2)^{n+2k}}{k!\,(n+k)!}$$

2.

$$\frac{d}{dx}[x^n J_n(x)] = x^n J_{n-1}(x) \qquad n = 1,2,3,\ldots$$

$$\frac{d}{dx}[x^{-n} J_n(x)] = -x^{-n} J_{n+1}(x) \qquad n = 1,2,3,\ldots$$

3. Recurrence relationship

$$J_{n-1}(x) + J_{n+1}(x) = \frac{2n}{x} J_n(x) \qquad n = 1,2,3,\ldots$$

$$J_{n-1}(x) - J_{n+1}(x) = 2\,J_n'(x) \qquad n = 1,2,3,\ldots$$

$$J_0'(x) = -J_1(x)$$

4. If

$$f(x) = \sum_{k=1}^{\infty} A_k J_n(\mu_k x),$$

then

$$A_k = \frac{1}{C_k}\int_0^L x\,f(x)\,J_n(\mu_k x)\,dx$$

where

$$C_k = 1/2\,L^2 J_{n+1}(\mu_k L)^2 \qquad \text{if } J_n(\mu_k L) = 0$$

$$C_k = \frac{\mu_k^2 L^2 - n^2}{2\mu_k^2} J_n(\mu_k L)^2 \qquad \text{if } J_n'(\mu_k L) = 0 \; *$$

$$C_k = \frac{\mu_k^2 L^2 - n^2 + h^2 L^2}{2\mu_k^2} J_n(\mu_k L)^2 \quad \text{if } \mu_k J_n'(\mu_k L) = -h\,J_n(\mu_k L)$$

* For the special case $J_0'(\mu_k L) = 0$, see Eqs. (1.7.70)-(1.7.71).

1.8 THE SWING OF A PENDULUM OF VARIABLE LENGTH

In this section we illustrate the use of Bessel functions in the solution of an ordinary differential equation. This situation arises in the study of a pendulum where the length is increasing at the constant rate a.

The differential equation is obtained by equating the rate of change of angular momentum to the gravitation restoring torque:

$$\frac{d}{dt}\left[L\left(mL\frac{d\theta}{dt}\right)\right] = -L\,mg\sin(\theta) \qquad \textbf{(1.8.1)}$$

where m is the mass of the bob at the end of a massless cord of instantaneous length $L(t)$ making the angle θ from the vertical plane. Adopting the traditional approximation that θ is sufficiently small so that $\sin(\theta)$ may be approximated by θ, we have

$$\frac{d^2\theta}{dt^2} + \frac{2}{L}\frac{dL}{dt}\frac{d\theta}{dt} + \frac{g}{L}\theta = 0 \qquad \textbf{(1.8.2)}$$

or

$$\frac{d^2\theta}{dt^2} + \frac{2a}{L_0+at}\frac{d\theta}{dt} + \frac{g}{L_0+at}\theta = 0 \qquad \textbf{(1.8.3)}$$

if $L = L_0 + at$. Finally, we define $x = (L_0+at)/a$ so that

$$\frac{d^2\theta}{dx^2} + \frac{2}{x}\frac{d\theta}{dx} + \frac{k^2\theta}{x} = 0 \qquad \textbf{(1.8.4)}$$

where $k^2 = g/a$. Comparing Eq. (1.8.4) with Eq. (1.7.17), we can immediately write down the solution as

$$\theta(t) = x^{-1/2}\left[A\,J_1(2kx^{1/2}) + B\,Y_1(2k\,x^{1/2})\right] \qquad \textbf{(1.8.5)}$$

If we assume that the pendulum is initially released in a manner so that $\theta = \theta_0$ and $d\theta/dt = 0$ at $t = 0$, the initial conditions in terms of x are

$$\theta(L_0/a) = \theta_0 \qquad \textbf{(1.8.6)}$$

and

$$\frac{d\theta(L_0/a)}{dt} = 0 \qquad \textbf{(1.8.7)}$$

Applying initial condition (1.8.6) to the general solution, we have

$$A\,J_1(c) + B\,Y_1(c) = \theta_0 \left(\frac{L_0}{a}\right)^{1/2} \tag{1.8.8}$$

where $c = 2k(L_0/a)^{1/2} = 2\,(gL_0)^{1/2}/a$.

To apply initial condition (1.8.7), we first employ Eq. (1.7.29) with $n = 1$ to give

$$\frac{d}{du}\left(u^{-1} J_1(u)\right) = -\,u^{-1} J_2(u) \tag{1.8.9}$$

If we take $u = 2k\,x^{1/2}$, then

$$2k\frac{d}{dx}\left(u^{-1} J_1(u)\right) = -2k\,u^{-1} J_2(u)\,\frac{du}{dx} = -\frac{k}{x} J_2(2kx^{1/2}). \tag{1.8.10}$$

A similar result holds for the Y_1 function. Therefore,

$$\frac{d\theta}{dx} = -\frac{k}{x}\left[A\,J_2(2k\,x^{1/2}) + B\,Y_2(2k\,x^{1/2})\right] \tag{1.8.11}$$

and

$$A\,J_2(c) + B\,Y_2(c) = 0. \tag{1.8.12}$$

Solving for A and B from Eq. (1.8.8) and (1.8.12), we have

$$A = \frac{\theta_0 (L_0/a)^{1/2}\,Y_2(c)}{J_1(c)Y_2(c) - J_2(c)Y_1(c)} \tag{1.8.13}$$

and

$$B = \frac{-\,\theta_0 (L_0/a)^{1/2}\,J_2(c)}{J_1(c)Y_2(c) - J_2(c)Y_1(c)} \tag{1.8.14}$$

The Wronskian relation

$$J_n(x)\,Y_{n+1}(x) - J_{n+1}(x)\,Y_n(x) = -\frac{2}{n\pi} \tag{1.8.15}$$

(Watson, G.N., 1966: *A Treatise on the Theory of Bessel Functions.*

Cambridge: At the University Press. p. 77) may be used to simply Eq. (1.8.13)-(1.8.5) to yield

$$\theta = \frac{\theta_0 \, g^{1/2} L_0}{a(L_0 + at)^{1/2}} \{ J_2(c) Y_1(z) - Y_2(c) J_1(z) \} \qquad \textbf{(1.8.16)}$$

where $z = 2 \, g^{1/2}(L_0 + at)^{1/2}/a$. For typical values of a and L_0 (e.g., $a = 10^{-3}$ m/sec and $L_0 = 1.$ m), c is very large. Consequently, Eq. (1.8.16) can be approximated very well by using the asymptotic expansion for Bessel functions of large arguments (see Exercise 7) and becomes

$$\theta = \frac{\theta_0}{\left(1 + \dfrac{at}{L_0}\right)^{3/4}} \cos\left(\frac{act}{L_0}\right). \qquad \textbf{(1.8.17)}$$

Therefore, over one cycle, the amplitude of the oscillation decreases slightly.

Exercises

1. Find the values of the parameter k for which the following problem has nontrival solutions:

$$\frac{1}{r}\frac{d}{dr}\left(r\frac{du}{dr}\right) + k^2 u = 0, \; 0<r<a$$

with $u(a) = 0$ and $u(0)$ bounded.

2. Show from the series solution that

$$\frac{d}{dr} J_0(kr) = - k \, J_1(kr).$$

3. From the recurrence formulas, show these following relations:

$$2J_0''(x) = J_2(x) - J_0(x)$$

$$J_2(x) = J_0''(x) - J_0'(x)/x$$

$$J_0'''(x) = J_0(x)/x + \left(\frac{2}{x^2} - 1\right) J_0'(x)$$

$$J_2(x)/J_1(x) = 1/x - J_0''(x)/J_0'(x)$$

$$J_2(x)/J_1(x) = 2/x - J_0(x)/J_1(x)$$

$$J_2(x)/J_1(x) = 2/x + J_0(x)/J_0'(x)$$

$$J_4(x) = \left(\frac{48}{x^3} - \frac{8}{x}\right) J_1(x) - \left(\frac{24}{x^2} - 1\right) J_0(x)$$

$$J_{n+2}(x) = \left(2n + 1 - \frac{2n(n^2 - 1)}{x^2}\right) J_n(x) + 2(n+1) J_n''(x)$$

$$J_3(x) = \left(\frac{8}{x^2} - 1\right) J_1(x) - \frac{4}{x} J_0(x)$$

$$4 J_n''(x) = J_{n-2}(x) - 2 J_n(x) + J_{n+2}(x)$$

4. Show that the maximum and minimum values of $J_n(x)$ occur when

$$x = \frac{n J_n(x)}{J_{n+1}(x)} \qquad x = \frac{n J_n(x)}{J_{n-1}(x)} \qquad \text{and } J_{n-1}(x) = J_{n+1}(x).$$

5. Show that

$$\frac{d}{dx}\left[x^2 J_3(2x)\right] = -x J_3(2x) + 2x^2 J_2(2x)$$

$$\frac{d}{dx}\left[x J_0(x^2)\right] = J_0(x^2) - 2x^2 J_1(x^2).$$

6. Show that

$$\int x^3 J_2(3x)\, dx = \frac{1}{3} x^3 J_3(3x) + C$$

$$\int x^{-2} J_3(2x)\, dx = -1/2\, x^{-2} J_2(2x) + C$$

$$\int x \ln(x) J_0(x)\, dx = J_0(x) + x \ln(x) J_1(x) + C$$

$$\int_0^a x J_0(kx)\, dx = \frac{a^2 J_1(ka)}{ka}$$

$$\int_0^1 x(1 - x^2) J_0(kx)\, dx = \frac{4}{k^3} J_1(k) - \frac{2}{k^2} J_0(k)$$

$$\int_0^1 x^3 J_0(kx)\, dx = \frac{k^2 - 4}{k^3} J_1(k) + \frac{2}{k^2} J_0(k)$$

7. Show that by changing the dependent variables to $y = x^{-1/2}u(x)$ that Bessel's equation of order n becomes

$$\frac{d^2u}{dx^2} + (1 + \frac{1 - 4n^2}{4x^2}) u = 0.$$

Hence, show that for large values of x, solutions of Bessel's equation are approximately given by

$$y(x) \sim c_1 \frac{\sin(x)}{\sqrt{x}} + c_2 \frac{\cos(x)}{\sqrt{x}}$$

Note that these solutions are independent of n.

8. Show that

$$1 = \sum_{k=1}^{\infty} \frac{2}{\mu_k J_1(\mu_k)} J_0(\mu_k x) \quad 0<x<1$$

where μ_k are the positive roots of $J_0(\mu) = 0$.

9. Show that

$$\frac{1 - x^2}{8} = \sum_{k=1}^{\infty} \frac{J_0(\mu_k x)}{\mu_k^3 J_1(\mu_k)} \quad 0<x<1$$

where μ_k are the positive roots of $J_0(\mu) = 0$.

10. Show that

$$4x - x^3 = -16 \sum_{k=1}^{\infty} \frac{J_1(\mu_k x)}{\mu_k^3 J_0(2\mu_k)} \quad 0<x<2$$

where μ_k are the positive roots of $J_1(2\mu) = 0$.

11. Show that

$$x^3 = \sum_{k=1}^{\infty} \frac{2 (8 - \mu_k^2) J_1(\mu_k x)}{\mu_k^3 J_1'(\mu_k)} \quad 0<x<1$$

where μ_k are the positive roots of $J_1 (\mu) = 0$.

Answers

P. 23

1. $\lambda_n = \frac{(2n-1)^2\pi^2}{4L^2}$, $u_n(x) = \sin(\lambda_n^{1/2}x)$ $\quad n = 1,2,3, \ldots$

2. $\lambda_n = n^2$, $\quad u_n(x) = \cos(nx)$ $\quad n = 0,1,2, \ldots$

3. $\lambda_0 = -1,\ u_0(x) = e^{-x};\ \lambda_n = n^2,$
$u_n(x) = \sin(nx) - n\cos(nx),\ n = 1,2,3,\ldots$

4. $k_0 \tanh(k_0\pi) = 1,\ u_0(x) = \cosh(k_0x);\ k_n \tan(k_n\pi) = -1,$
$u_n(x) = \cos(k_nx)$

5. $\lambda_n = \dfrac{n^4\pi^4}{L^4},\ u_n(x) = \sin(\dfrac{n\pi x}{L}),\ n = 1,2,3,4,\ldots$

6. $k_n = \tan(k_n),\ u_n(x) = \sin(k_nx) - k_n\cos(k_nx);$
$\lambda_0 = 0,\ u_0(x) = 1 - x$

7. $k_n = -\tan(k_n\pi), u_n(x) = \sin(k_nx)$

8. $k_n = -\cot(k_n),\ u_n(x) = \cos(k_nx)$

9. $k_n \tan(k_n\pi) = -1,\ u_n(x) = \sin(k_nx) - k_n\cos(k_nx)$

10. $(1-k_n^2)\tan(k_n\pi) = 2k_n,\ u_n(x) = \sin(k_nx) - k_n\cos(k_nx)$

11. (a) $\lambda_n = n^2\pi^2,\ u_n(x) = \sin[n\pi\ \ell n(x)],\ n = 1,2,3,\ldots$

(b) $\lambda_n = \dfrac{(2n-1)^2\pi^2}{4},\ u_n(x) = \sin[(2n-1)\pi\ \ell n(x)/2)$

(c) $\lambda_n = n^2\pi^2,\ u_n(x) = \cos[n\pi\ \ell n(x)]$

12. $\lambda_n = (4n^2\pi^2 + 1)/4\ u_n(x) = x^{-1/2}\sin[n\pi\ \ell n(x)]$

13. $\lambda_n = n^2 + 1,\ u_n(x) = x^{-1}\sin[n\pi\ \ell n(x)]$

14. $\lambda_n = n^2\pi^2 + 1,\quad u_n(x) = x\sin[n\pi\ \ell n(x)],\ n = 1,2,3,$

15. $\lambda_n = \dfrac{n^2\pi^2}{L^2}\qquad u_n(x) = A_n\cos\left(\dfrac{n\pi x}{L}\right) + B_n\sin\left(\dfrac{n\pi x}{L}\right)$

A_n and B_n are arbitrary constants.

16. $\lambda_n = \dfrac{4n^2\pi^2}{L^2}\qquad u_n(x) = A_n\cos\left(\dfrac{2n\pi x}{L}\right) + B_n\sin\left(\dfrac{2n\pi x}{L}\right)$

A_n and B_n are arbitrary constants.

P. 33

1. $f(x) = \frac{2}{\pi} \sum_{n=1}^{\infty} \frac{(2 - n^2\pi^2)(-1)^n - 2}{n^3} \sin(nx)$

2. $f(x) = \frac{4L}{\pi} \sum_{n=1}^{\infty} \left[\frac{-2}{(2n-1)^2\pi} + \frac{(-1)^{n+1}}{(2n-1)} \right] \cos\left(\frac{(2n-1)\pi x}{2L}\right)$

3. $f(x) = \pi/2 - \frac{2}{\pi} \sum_{n=1}^{\infty} \frac{\cos((2n-1)x)}{(2n-1)^2}$

P. 48

1. (a) $f(x) = 1/4\, P_0(x) + 1/2\, P_1(x) + \frac{5}{16} P_2(x) + \ldots$

(b) $f(x) = 1/2\, P_0(x) - \frac{5}{4}(1 - \epsilon^2)P_2(x)$

$+ \frac{9}{16}(2 - 10\epsilon^2 + 7\epsilon^4)\, P_4(x) + \ldots$

(c) $f(x) = 1/2\, P_0(x) + \frac{5}{8} P_2(x) - \frac{3}{16} P_4(x) + \ldots$

(d) $f(x) = \frac{3}{5} P_1(x) + \frac{2}{5} P_3(x)$

(e) $f(x) = \frac{3}{2} P_1(x) - \frac{7}{8} P_3(x) + \frac{11}{16} P_5(x) + \ldots$

(f) $f(x) = -1/4\, P_0(x) + \frac{5}{4} P_1(x) + \frac{5}{16} P_2(x) + \ldots$

Chapter 2

Fourier Series and Integral

ONE OF THE FUNDAMENTAL AND MOST COMMON METHODS of solving partial differential equations leads to the Fourier series. A Fourier series is an expansion of any arbitrary function, defined over the interval $a \leq x \leq b$, in terms of sines and/or cosines that have multiples of the independent variable as their argument. It is the most commonly used application of eigenfunction expansions that we discussed in Chapter 1.

2.1 INTRODUCTION TO FOURIER SERIES

Fourier series are quite useful for phenomena that are periodic either in time or space. For example, a musical note consists of a simple sound—the fundamental—compounded with a series of auxiliary vibrations called overtones. The power behind the Fourier series is that it will allow us to resolve any complex sound into a sound made of the fundamental and overtones. Furthermore, because of the periodic nature of the phenomena, Fourier series that are constructed over any one period will also be valid for all time and space.

But we can go further than this. We can also use Fourier's theorem to expand aperiodic phenomena over the time interval $a \leq t \leq b$. True, in this particular case, Fourier's theorem is an artificial way of representing any physical quantity by a series of waves or vibrations. It will give incorrect results outside of the interval (a,b) but, the convenience of dealing with sines and cosines is so appealing that this mathematical fiction often becomes a necessity.

2.2 THE ORTHOGONALITY OF TRIGONOMETRIC FUNCTIONS

Before any function may be considered for use in an orthogonal expansion, it must first be demonstrated that the function satisfies some sort of orthogonality condition.

Consider the Sturm-Liouville problem

$$u'' + \lambda u = 0, \; u(0) = 0, \; u(L) = 0. \qquad \textbf{(2.2.1)}$$

This problem has the eigenfunctions

$$u_n(x) = \sin\left(\frac{n\pi x}{L}\right) \qquad \textbf{(2.2.2)}$$

corresponding to the eigenvalues

$$\lambda_n = \frac{n^2\pi^2}{L^2} \quad (n = 1,2,3,4,5, \ldots). \qquad \textbf{(2.2.3)}$$

If we replace the boundary conditions in Eq. (2.2.1) with $u'(0) = u'(L) = 0$, then the Sturm-Liouville problem has the eigenfunctions

$$u_n(x) = \cos\left(\frac{n\pi x}{L}\right) \qquad \textbf{(2.2.4)}$$

which correspond to the eigenvalues

$$\lambda_n = \frac{n^2\pi^2}{L^2} \quad (n = 1,2,3, \ldots). \qquad \textbf{(2.2.5)}$$

The orthogonality condition (1.3.10) in the previous chapter states that each of the sets of eigenfunctions given by Eq. (2.2.2) and Eq. (2.2.4) forms an orthogonal system on the interval $0 \leq x \leq L$. That is,

$$\int_0^L \sin\left(\frac{m\pi x}{L}\right) \sin\left(\frac{n\pi x}{L}\right) dx = 0 \quad \text{if } m \neq n \qquad \textbf{(2.2.6)}$$

and

$$\int_0^L \cos\left(\frac{m\pi x}{L}\right)\cos\left(\frac{n\pi x}{L}\right) dx = 0 \quad \text{if } m \neq n.$$

Now, let us consider the set of functions

$$1, \cos\left(\frac{n\pi x}{L}\right), \quad \sin\left(\frac{n\pi x}{L}\right) \quad (n = 1,2,3,4, \ldots) \qquad \textbf{(2.2.7)}$$

which is the collection of the eigenfunctions Eq. (2.2.2) and (2.2.4).

These functions are all periodic and have the common period $2L$; that is

$$\cos\left[\frac{n\pi}{L}(x+2L)\right] = \cos\left(\frac{n\pi x}{L}\right)$$

$$\sin\left[\frac{n\pi}{L}(x+2L)\right] = \sin\left(\frac{n\pi x}{L}\right) \qquad \textbf{(2.2.8)}$$

for all n. But, are they orthogonal to each other over the interval $-L$ to L? For each integer $n \geq 1$, it is clear that

$$\int_{-L}^{L} 1 \cos\left(\frac{n\pi x}{L}\right) dx = \int_{-L}^{L} 1 \sin\left(\frac{n\pi x}{L}\right) dx = 0 \qquad \textbf{(2.2.9)}$$

Therefore, the function 1 is orthogonal to the functions $\cos\left(\frac{n\pi x}{L}\right)$ and $\sin\left(\frac{n\pi x}{L}\right)$ for $n \geq 1$ on the interval $[-L, L]$.

Next, let m and n be two distinct integers. From the trigonometric identities

$$\cos\left(\frac{m\pi x}{L}\right)\cos\left(\frac{n\pi x}{L}\right) = 1/2\left[\cos\left(\frac{(m-n)\pi x}{L}\right) + \cos\left(\frac{(m+n)\pi x}{L}\right)\right]$$

and

$$\sin\left(\frac{m\pi x}{L}\right)\sin\left(\frac{n\pi x}{L}\right) = 1/2\left[\cos\left(\frac{(m-n)\pi x}{L}\right) - \cos\left(\frac{(m+n)\pi x}{L}\right)\right]$$

(2.2.10)

we immediately obtain that

$$\int_{-L}^{L} \cos\left(\frac{m\pi x}{L}\right) \cos\left(\frac{n\pi x}{L}\right) dx = \begin{cases} 2L & m = n = 0 \\ L & m = n \neq 0 \\ 0 & m \neq n \end{cases} \quad \textbf{(2.2.11)}$$

$$\int_{-L}^{L} \sin\left(\frac{m\pi x}{L}\right) \sin\left(\frac{n\pi x}{L}\right) dx = \begin{cases} 0 & m = n = 0 \\ L & m = n \neq 0 \\ 0 & m \neq n \end{cases} \quad \textbf{(2.2.12)}$$

Formulas (2.2.11) and (2.2.12) state that the orthogonality of the eigenfunctions (2.2.2) and (2.2.4) remains valid on the extended interval $-L \leq x \leq L$.

Finally, from the identity that

$$\sin\left(\frac{m\pi x}{L}\right) \cos\left(\frac{n\pi x}{L}\right) = 1/2 \left[\sin\left(\frac{(m-n)\pi x}{L}\right) + \sin\left(\frac{(m+n)\pi x}{L}\right)\right] \quad \textbf{(2.2.13)}$$

we see that for all integers m and n

$$\int_{-L}^{L} \sin\left(\frac{m\pi x}{L}\right) \cos\left(\frac{n\pi x}{L}\right) dx = 0 \quad \textbf{(2.2.14)}$$

Each of the functions in Eq. (2.2.2) is orthogonal to each of the functions in Eq. (2.2.4) on the interval $-L \leq x \leq L$, and vice versa. Therefore, by definition, it follows that the functions in Eq. (2.2.7) form an orthogonal system on the interval $-L \leq x \leq L$.

The functions in Eq. (2.2.7) actually constitute a set of eigenfunctions for the eigenvalue problem

$$u'' + \lambda u = 0,\ u(-L) = u(L),\ u'(-L) = u'(L) \quad \textbf{(2.2.15)}$$

This problem is *not* of the regular Sturm-Liouville type because its boundary conditions are periodic. Nevertheless, it has real eigenvalues given by $\lambda_n = n^2\pi^2/L^2$, $n = 0,1,2,3,\ldots$ For each integer $n \geq 1$, the functions $u_n(x) = \cos\left(\frac{n\pi x}{L}\right)$ and $v_n(x) = \sin\left(\frac{n\pi x}{L}\right)$ constitute two linearly independent set of eigenfunctions, corresponding to the same eigenvalue $\lambda_n = \frac{n^2\pi^2}{L^2}$, which are orthogonal on the interval $-L \leq x \leq L$. This is, therefore, an example of a degenerate Sturm-Liouville problem. The function $u_0 = 1$ is the only eigenfunction associated with the smallest eigenvalue $\lambda_0 = 0$.

Despite the fact that the functions (2.2.7) do not truly belong to a regular Sturm-Liouville problem, we have established the fact that they are orthogonal to one another. In the next section, we use that fact to derive an expansion for any arbitrary function in terms of the functions given by (2.2.7).

2.3 FOURIER SERIES

Many functions that occur in engineering and science are periodic in time and space. Consequently it is often desirable to represent them in terms of the very simple periodic functions 1, sin $(\pi x/L)$, cos $(\pi x/L)$, sin $(2\pi x/L)$, cos $(2\pi x/L)$, and so forth. *All* of these functions have the common period of $2L$, although some have other periods as well.

If $f(x)$ is periodic with period $2L$, finite in the interval $[-L,L]$, and has a finite number of discontinuities, then we will attempt to represent it in the form of an infinite series. If the sum of the series exists,

$$f(x) = a_0/2 + \sum_{n=1}^{\infty} a_n \cos\left(\frac{n\pi x}{L}\right) + b_n \sin\left(\frac{n\pi x}{L}\right) . \qquad \textbf{(2.3.1)}$$

We require that the sum of this series converges to the value of the function $f(x)$ at every point in the interval $[-L,L]$ with the possible exceptions of the points at the discontinuities and the endpoints of the interval. Each term of the series has a period of $2L$; therefore, it will also be a function of period $2L$.

We must find some easy method for computing the a_n's and b_n's. As a first attempt, we integrate Eq. (2.3.1) term by term from $-L$ to L. (We must assume that the integration of the series can be carried out term-by-term. This is sometimes difficult to justify, but we do it anyway.) On the right-hand side we find that all of the a_n and b_n terms vanish because the average of cos $(n\pi x/L)$ and sin$(n\pi x/L)$ is zero and we are left with

$$\boxed{a_0 = \frac{1}{L}\int_{-L}^{L} f(x)\, dx.} \qquad \textbf{(2.3.2)}$$

Consequently a_0 is twice the mean value of $f(x)$ over one period.

We next multiply each side of Eq. (2.3.1) by $\sin(\frac{m\pi x}{L})$, where m is a fixed integer. Integrating from $-L$ to L, we find

$$\int_{-L}^{L} f(x) \sin\left(\frac{m\pi x}{L}\right) dx = a_0/2 \int_{-L}^{L} \sin\left(\frac{m\pi x}{L}\right) dx$$

$$+ \sum_{n=1}^{\infty} a_n \int_{-L}^{L} \cos\left(\frac{n\pi x}{L}\right) \sin\left(\frac{m\pi x}{L}\right) dx$$

$$+ \sum_{n=1}^{\infty} b_n \int_{-L}^{L} \sin\left(\frac{n\pi x}{L}\right) \sin\left(\frac{m\pi x}{L}\right) dx \qquad \textbf{(2.3.3)}$$

The a_0 term vanishes by direct integration, the a_n terms by Eq. (2.2.14). Furthermore, of all the b_n terms, only the $n = m$ term is not zero by Eq. (2.2.12). Consequently we are left with

$$\boxed{b_m = \frac{1}{L} \int_{-L}^{L} f(x) \sin\left(\frac{m\pi x}{L}\right) dx} \qquad \textbf{(2.3.4)}$$

Finally, by multiplying both sides of Eq. (2.3.1) by cos $(m\pi x/L)$ (m is again a fixed integer) and integrating, we also find that

$$\boxed{a_m = \frac{1}{L} \int_{-L}^{L} f(x) \cos\left(\frac{m\pi x}{L}\right) dx} \qquad \textbf{(2.3.5)}$$

Although we have derived Eqs. (2.3.2), (2.3.4) and (2.3.5) to obtain a_0, a_n, and b_n for periodic functions over the interval $[-L,L]$, in certain applications it is convenient to use the interval $[\tau, \tau + 2L]$ where τ is any real number. In that case, the Fourier series of $f(x)$ is still given by Eq. (2.3.1) and

$$a_0 = \frac{1}{L} \int_{\tau}^{\tau+2L} f(x)\, dx, \qquad \textbf{(2.3.6)}$$

$$a_n = \frac{1}{L} \int_{\tau}^{\tau+2L} f(x) \cos\left(\frac{n\pi x}{L}\right) dx, \qquad \textbf{(2.3.7)}$$

$$b_n = \frac{1}{L} \int_{\tau}^{\tau+2L} f(x) \sin\left(\frac{n\pi x}{L}\right) dx. \qquad \textbf{(2.3.8)}$$

These results follow when we remember that the function $f(x)$ is assumed to be a periodic function that extends from plus infinity to minus infinity. The results must remain unchanged, therefore, when we shift from the interval $[-L,L]$ to the new interval $[\tau,\tau + 2L]$.

A question of some importance is what type of functions can be developed in convergent Fourier series? Secondly, if a function is discontinuous at a certain point, what value is given by the Fourier series? These questions were first answered in detail by Dirichlet in a series of analyses given in 1829 and 1837 (Dirichlet, P. G. L., 1829: Sur la convergence des series trigonometriques qui servent a representer une function arbitraire entre des limites donnees. *Journal fur die reine und angewandte Mathematik, 4*, 157-169 and Dirichlet, P. G. L., 1837: Sur l 'usage des integrales defines dans la sommation des series finies ou infinies. *Journal fur die reine und angewandte Mathematik, 17*, 57-67). Dirichlet showed that if any arbitrary function is finite in a given interval and has a finite number of maxima and minima, then the Fourier series is convergent. If $f(x)$ is continuous at the point x, the series converges to $f(x)$. If the function $f(x)$ is discontinuous at the point x and has two different values at $f(x-0)$ and $f(x+0)$, the Fourier series converges to the mean value of $1/2[f(x+0) + f(x-0)]$.

Dirichlet also investigated functions that are infinite at a finite number of x's. The portions of the interval outside of an arbitrary close neighborhood around the points of infinity are assumed to be finite with a finite number of maxima and minima. When these conditions are fulfilled, the Fourier series is still convergent if the integral $\int |f(x)|\, dx$ over the whole interval is finite. The series will give the value of $f(x)$ at every point except at the points of infinite discontinuity. Because these conditions are very mild, it is very rare that a function appearing in an engineering or scientific problem cannot be expanded into a convergent Fourier series.

2.4 EXAMPLES OF FOURIER SERIES. THE DESIGN OF SNOW TIRES

In this section we shall work out some detailed examples of Fourier series. For our first example, we find the Fourier series of the function

$$f(x) = \begin{cases} 0 & -\pi \leq x < 0 \\ x & 0 < x \leq \pi \end{cases} \qquad \textbf{(2.4.1)}$$

We compute the Fourier coefficients a_n and b_n according to the formuli (2.3.6) through (2.3.8) noting that $L = \pi$ and $\tau = -\pi$. We find

$$a_0 = \frac{1}{\pi} \int_{-\pi}^{\pi} f(x)\, dx = \frac{1}{\pi} (0 + \int_0^{\pi} x\, dx) = \frac{\pi}{2} \qquad \textbf{(2.4.2)}$$

$$a_n = \frac{1}{\pi}\int_0^{\pi} x\cos(nx)\,dx = \frac{1}{\pi}\left(\frac{x\sin(nx)}{n} + \frac{\cos(nx)}{n^2}\right)\Bigg|_0^{\pi}$$

$$= \frac{\cos(n\pi) - 1}{\pi n^2} = \frac{(-1)^n - 1}{\pi\, n^2} \qquad \textbf{(2.4.3)}$$

because $\cos(n\pi) = (-1)^n$.

$$b_n = \frac{1}{\pi}\int_0^{\pi} x\sin(nx)\,dx$$

$$= \frac{1}{\pi}\left(\frac{-x\cos(nx)}{\mathrm{n}} + \frac{\sin(nx)}{\mathrm{n}^2}\right)\Bigg|_0^{\pi}$$

$$= -\frac{\cos(n\pi)}{n} = \frac{(-1)^{n+1}}{n} \qquad \textbf{(2.4.4)}$$

for $n = 1,2,3, \ldots$ Thus, the Fourier series for $f(x)$ is

$$f(x) = \frac{\pi}{4} + \sum_{n=1}^{\infty} \frac{(-1)^n - 1}{\pi n^2}\cos(nx) + \frac{(-1)^{n+1}}{n}\sin(nx)$$

$$= \frac{\pi}{4} - \sum_{m=1}^{\infty} \frac{2}{\pi(2m-1)^2}\cos((2m-1)x) - \sum_{n=1}^{\infty}\frac{(-1)^n}{n}\sin(nx)$$

(2.4.5)

We note that at the points $x = \pm(2n-1)\pi$, the function jumps from zero to π. To what value does the Fourier series converge to at these points? From Dirichlet's celebrated theorem that we stated at the end of the previous section, the series converges to the average of the values of the function just to the right and left of the point of discontinuity, i.e., $(\pi + 0)/2 = \pi/2$. At the remaining points, the series converges to $f(x)$. In Fig. 2.4.1, how well Eq. (2.4.5) approximates the function is shown by graphing various partial sums of Eq. (2.4.5) as we include more and more terms (harmonics).

Fourier series expansions also provide a method to simplify summations involving integers. For example, if we were to evaluate

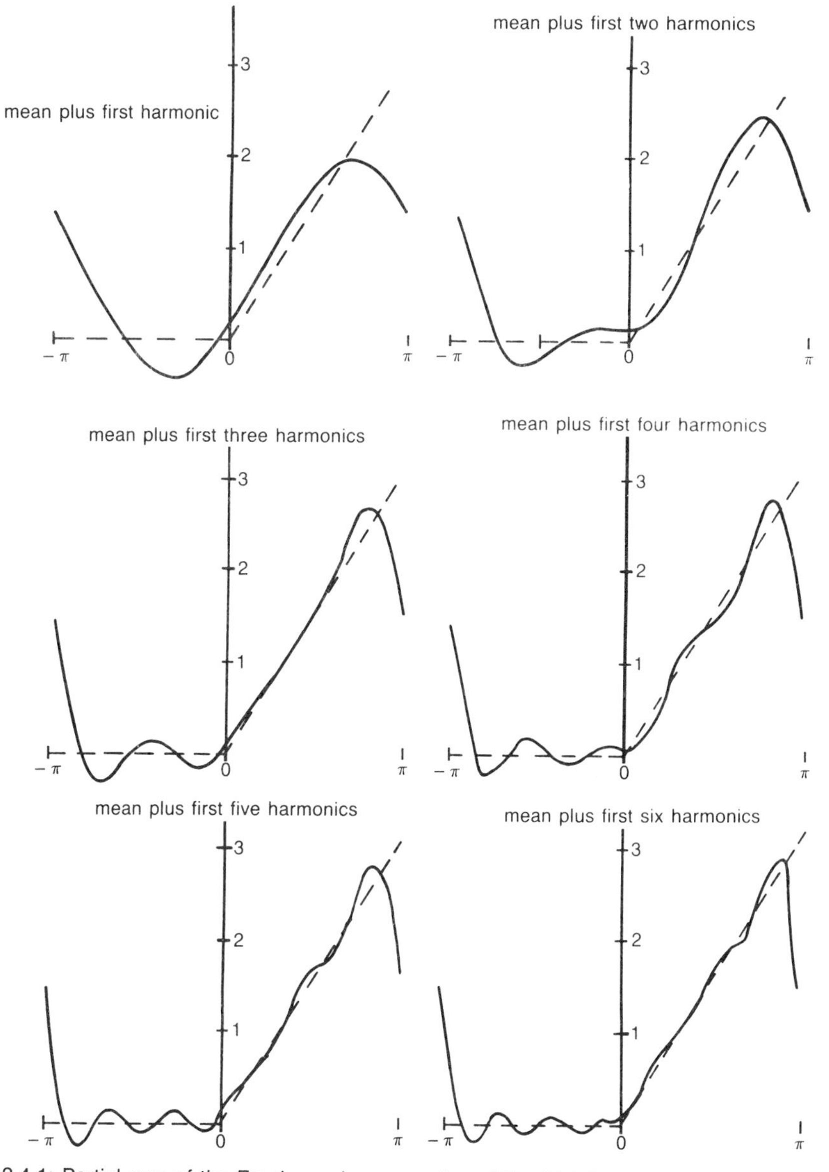

Fig. 2.4.1: Partial sum of the Fourier series expansion of Eq. (2.4.1).

our Fourier series (2.4.5) at $x = \pi$, we would obtain

$$\pi/4 + \frac{2}{\pi} \sum_{n=1}^{\infty} \frac{1}{(2n-1)^2} \tag{2.4.6}$$

because $\sin(n\pi) = 0$ and $\cos[(2m-1)\pi] = -1$. The Fourier series converges to $\pi/2$ because at $x = \pi$ there is a simple discontinuity where the function jumps from π to zero. Consequently

$$\pi/2 = \pi/4 + \frac{2}{\pi} \sum_{n=1}^{\infty} \frac{1}{(2n-1)^2} \tag{2.4.7}$$

or

$$\frac{\pi^2}{8} = 1 + \frac{1}{3^2} + \frac{1}{5^2} + \ldots \tag{2.4.8}$$

As a second example, we calculate the Fourier series of the function $f(x) = |x|$, $-\pi < x < \pi$. The Fourier coefficients of the function are

$$a_0 = \frac{1}{\pi} \left(\int_{-\pi}^{0} -x\, dx + \int_{0}^{\pi} x\, dx \right) = \frac{\pi}{2} + \frac{\pi}{2} = \pi \tag{2.4.9}$$

$$a_n = \frac{1}{\pi} \left(\int_{-\pi}^{0} -x \cos(nx)\, dx + \int_{0}^{\pi} x \cos(nx)\, dx \right)$$

$$= -\left. \frac{nx \sin(nx) + \cos(nx)}{n^2 \pi} \right|_{-\pi}^{0} + \left. \frac{nx \sin(nx) + \cos(nx)}{n^2 \pi} \right|_{0}^{\pi}$$

$$= \frac{2}{n^2 \pi} [(-1)^n - 1] \tag{2.4.10}$$

$$b_n = \frac{1}{\pi} \left(\int_{-\pi}^{0} -x \sin(nx)\, dx + \int_{0}^{\pi} x \sin(nx)\, dx \right)$$

$$= \left. \frac{nx \cos(nx) - \sin(nx)}{n^2 \pi} \right|_{-\pi}^{0} - \left. \frac{nx \cos(nx) - \sin(nx)}{n^2 \pi} \right|_{0}^{\pi}$$

$$= 0 \tag{2.4.11}$$

for $n = 1, 2, 3, \ldots$ Hence,

$$|x| = \frac{\pi}{2} + \frac{2}{\pi} \sum_{n=1}^{\infty} \frac{(-1)^n - 1}{n^2} \cos(nx)$$

$$= \frac{\pi}{2} - \frac{4}{\pi} \sum_{n=1}^{\infty} \frac{\cos[(2n-1)x]}{(2n-1)^2} \quad (-\pi \leq x \leq \pi) \tag{2.4.12}$$

Table 2.4.1. Trigonometric Identities.

	n	n even	n odd	n/2 odd	n/2 even
sin (nπ)	0	0	0	0	0
cos (nπ)	$(-1)^n$	1	−1	1	1
sin (nπ/2)		0	$(-1)^{(n-1)/2}$	0	0
cos (nπ/2)		$(-1)^{n/2}$	0	−1	1
sin (nπ/4)			$\frac{\sqrt{2}}{2}(-1)^{(n^2+4n+11)/8}$	$(-1)^{(n-2)/4}$	0

In Table 2.4.1, some trigonometric identities have been included that may prove useful in simplifying Fourier series expansions.

At this point it is useful to discuss whether we can use the Fourier series representation of a function for the purpose of computing its derivative or integral. Clearly we can differentiate or integrate Eq. (2.4.12) term-by-term to obtain the derivative or integral. However, we know that the derivative of the absolute value function is undefined at the origin. On the other hand, the integral, which is merely the area under the function, is well defined. This illustrates the general result that differentiating a Fourier series to compute the derivative may not be justified but integrating the series to find the integral always is. (See Jeffrey, H. and B.S. Jeffrey, 1972: *Methods of Mathematical Physics*. Cambridge (England): At the University Press, Sections 14.06 and 14.061 for further discussion.)

An application of Fourier series to a problem in industry occurred several years ago, when it was found that snow tires driven on dry pavement produced a loud whine. In those days the 42 treads were evenly spaced apart. As each one of those treads struck the pavement, the resulting force with time looked something like part A of Fig. 2.4.2. If we perform a Fourier analysis of this forcing, we find that

$$a_0 = \frac{1}{\pi}\int_{-\pi}^{\pi} f(t)\,dt$$

$$= \frac{1}{\pi}\left[\int_{-\pi/2-\epsilon}^{-\pi/2+\epsilon} 1\,dt + \int_{\pi/2-\epsilon}^{\pi/2+\epsilon} 1\,dt\right] = \frac{4\epsilon}{\pi} \qquad \textbf{(2.4.13)}$$

$$a_n = \frac{1}{\pi}\int_{-\pi}^{\pi} f(t)\cos(nt)\,dt$$

$$= \frac{1}{n\pi}\left[\sin(nt)\Big|_{-\pi/2-\epsilon}^{-\pi/2+\epsilon} + \sin(nt)\Big|_{\pi/2-\epsilon}^{\pi/2+\epsilon}\right]$$

$$a_n = \frac{1}{n\pi}\left[\sin\left(-\frac{n\pi}{2} + n\epsilon\right) - \sin\left(-\frac{n\pi}{2} - n\epsilon\right) + \sin\left(\frac{n\pi}{2} + n\epsilon\right) - \sin\left(\frac{n\pi}{2} - n\epsilon\right)\right]$$

$$= \frac{2}{n\pi}\left[2\cos\left(-\frac{n\pi}{2}\right) + 2\cos\left(\frac{n\pi}{2}\right)\right]\sin(n\epsilon) = \frac{4}{n\pi}\cos\left(\frac{n\pi}{2}\right)\sin(n\epsilon)$$

$$= \frac{4\epsilon}{\pi}\cos\left(\frac{n\pi}{2}\right) \qquad \textbf{(2.4.14)}$$

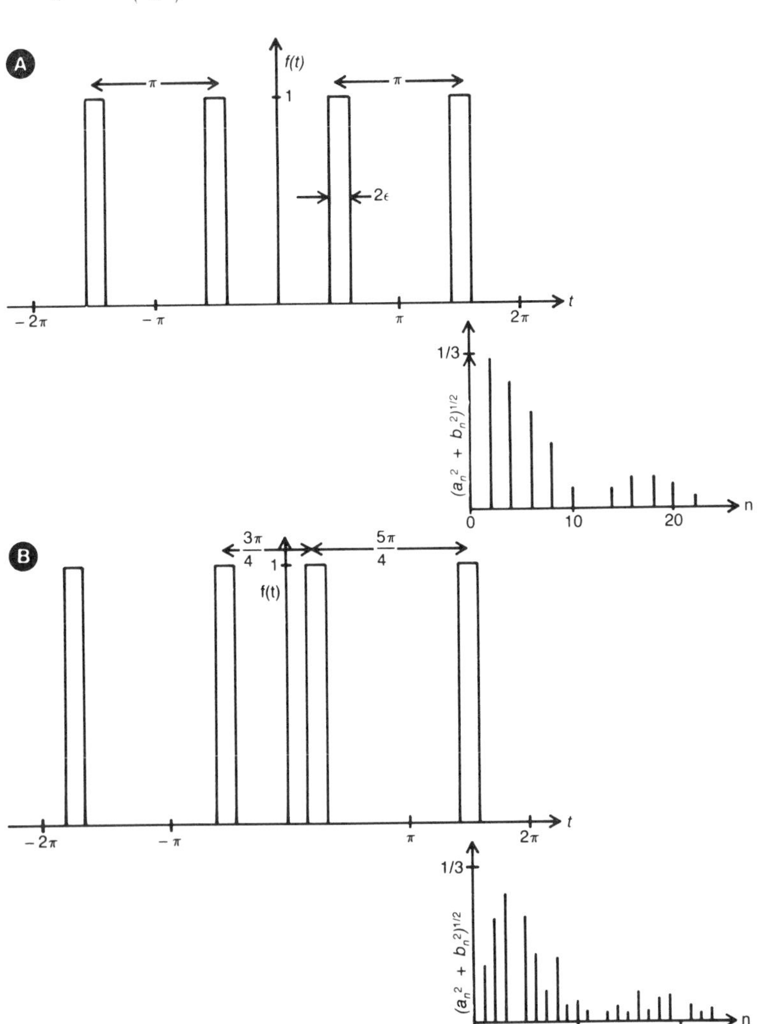

Fig. 2.4.2: Frequency spectrum for various tire tread configurations when $\epsilon = \pi/12$.

in the limit as $\epsilon \rightarrow 0$. This limit corresponds to a tire design with very narrow spikes. Because $f(t)$ is an even function, $b_n = 0$.

The amplitude of the Fourier coefficients is plotted as a function n next to the forcing function for $\epsilon = \pi/12$. Note that the periodic placement of the tire treads produces one loud tone plus strong harmonic overtones.

Let us now vary the interval between the treads so that the distances between any tread and its nearest neighbor is not equal. A simple example is plotted in part B of Fig. 2.4.2. Again we perform a Fourier analysis and obtain that

$$a_0 = \frac{1}{\pi}\left[\int_{-\pi/2-\epsilon}^{-\pi/2+\epsilon} 1\,dt + \int_{\pi/4-\epsilon}^{\pi/4+\epsilon} 1\,dt\right] = \frac{4\epsilon}{\pi} \qquad \textbf{(2.4.15)}$$

$$a_n = \frac{1}{\pi}\int_{-\pi/2-\epsilon}^{-\pi/2+\epsilon} \cos(nt)\,dt + \frac{1}{\pi}\int_{\pi/4-\epsilon}^{\pi/4+\epsilon} \cos(nt)\,dt$$

$$= \frac{1}{n\pi}\sin(nt)\Big|_{\pi/2-\epsilon}^{\pi/2+\epsilon} + \frac{1}{n\pi}\sin(nt)\Big|_{\pi/4-\epsilon}^{\pi/4+\epsilon}$$

$$= \frac{-1}{n\pi}\left[\sin(\frac{n\pi}{2} - n\epsilon) - \sin(\frac{n\pi}{2} + n\epsilon)\right]$$

$$+\frac{1}{n\pi}\left[\sin(\frac{n\pi}{4} + n\epsilon) - \sin(\frac{n\pi}{4} - n\epsilon)\right]$$

$$= \frac{2}{n\pi}\cos(\frac{n\pi}{2})\sin(n\epsilon) + \frac{2}{n\pi}\cos(\frac{n\pi}{4})\sin(n\epsilon)$$

$$= \frac{2}{n\pi}\left[\cos(\frac{n\pi}{2}) + \cos(\frac{n\pi}{4})\right]\sin(n\epsilon)$$

$$= \frac{2\epsilon}{\pi}\left[\cos(\frac{n\pi}{2}) + \cos(\frac{n\pi}{4})\right] \text{ in the limit of } \epsilon \rightarrow 0. \qquad \textbf{(2.4.16)}$$

$$b_n = \frac{1}{\pi}\int_{-\pi/2-\epsilon}^{-\pi/2+\epsilon} \sin(nt)\,dt + \frac{1}{\pi}\int_{\pi/4-\epsilon}^{\pi/4+\epsilon} \sin(nt)\,dt$$

$$b_n = \frac{-1}{n\pi}\left[\cos\left(\frac{n\pi}{2} - n\epsilon\right) - \cos\left(\frac{n\pi}{2} + n\epsilon\right) + \cos\left(\frac{n\pi}{4} + n\epsilon\right) - \cos\left(\frac{n\pi}{4} - n\epsilon\right)\right]$$

$$= \frac{-2}{n\pi}\left[\sin\left(\frac{n\pi}{2}\right) - \sin\left(\frac{n\pi}{4}\right)\right]\sin(n\epsilon)$$

$$= \frac{2}{n\pi}\sin(n\epsilon)\left[\sin\left(\frac{n\pi}{4}\right) - \sin\left(\frac{n\pi}{2}\right)\right]$$

$$= \frac{2\epsilon}{\pi}\left[\sin\left(\frac{n\pi}{4}\right) - \sin\left(\frac{n\pi}{2}\right)\right] \text{ in the limit of } \epsilon \to 0. \quad \textbf{(2.4.17)}$$

The amplitude of each harmonic in part B is plotted as a function of n in Fig. 2.4.2. The important point to notice is that some of the harmonics have been reduced or eliminated compared to the case of equally spaced treads. On the negative side we have excited some of the harmonics that were previously absent. However, the net effect is advantageous because the treads produce less noise at more frequencies rather than a lot at a few select frequencies.

If we were to extend this technique so that the treads occurred at completely random positions, then the treads would produce very little noise at many frequencies and the total noise would be comparable to that generated by other sources within the car. To find the distribution of treads with the whitest noise is a process of trial and error. Assuming a distribution, we can perform a Fourier analysis to obtain the frequency spectrum. If there are annoying peaks in the spectrum, we can then adjust the elements in the distribution that may contribute to the peak and analyze the revised distribution. You are finished when no peaks appear.

Summary of the Fourier Series

1. $$a_0 = \frac{1}{L}\int_{\tau}^{\tau+2L} f(x)\, dx$$

$$a_n = \frac{1}{L}\int_{\tau}^{\tau+2L} f(x)\cos\left(\frac{n\pi x}{L}\right) dx \qquad n \geq 1$$

$$b_n = \frac{1}{L}\int_{\tau}^{\tau+2L} f(x)\sin\left(\frac{n\pi x}{L}\right) dx \qquad n \geq 1$$

where

$$f(x) = a_0/2 + \sum_{n=1}^{\infty} a_n \cos\left(\frac{n\pi x}{L}\right) + b_n \sin\left(\frac{n\pi x}{L.}\right)$$

τ is the starting location for an interval of length $2L$.

2. $$\cos(n\pi) = (-1)^n \text{ and } \sin(n\pi) = 0$$

3. For an even function (function is symmetric about the y axis),

$$b_n = 0 \text{ for all } n.$$

4. For an odd function (function is asymmetric about the y axis),

$$a_0 = a_n = 0 \text{ for all } n.$$

5. For a function with discontinuities, the Fourier coefficients behave like $1/n$ for large n.

6. At discontinuities, the value given by the Fourier series equals the average of the function just to the left and right of the jump.

Exercises

Find the Fourier series for the following functions:

1.
$$f(x) = \begin{cases} -\pi & -\pi < x < 0 \\ x & 0 \leq x \leq \pi \end{cases}$$

2.
$$f(x) = \begin{cases} 0 & -\pi \leq x \leq 0 \\ x & 0 \leq x \leq \pi/2 \\ \pi - x & \pi/2 \leq x \leq \pi \end{cases}$$

3.
$$f(x) = \begin{cases} 1/2 + x & -1 \leq x \leq 0 \\ 1/2 - x & 0 \leq x \leq 1 \end{cases}$$

4.
$$f(x) = e^{ax} \quad -L < x < L$$

5.
$$f(x) = \begin{cases} 0 & -\pi \leq x \leq 0 \\ \sin(x) & 0 \leq x \leq \pi \end{cases}$$

6.
$$f(x) = x + x^2 \quad -L < x < L$$

7.
$$f(x) = \begin{cases} x & -1/2 < x < 1/2 \\ 1 - x & 1/2 < x < 3/2 \end{cases}$$

8.

$$f(x) = \begin{cases} 0 & -\pi < x < \pi/2 \\ \sin(2x) & -\pi/2 < x < \pi/2 \\ 0 & \pi/2 < x < \pi \end{cases}$$

9.

$$f(x) = \begin{cases} 0 & -a < x < 0 \\ 2x & 0 < x < a \end{cases}$$

To what values does the series converge at $x = -a$, $x = -a/2$, $x = 0$, $x = a$ and $x = 2a$?

10.

$$f(x) = (\pi - x)/2 \quad 0 < x < 2$$

11.

$$f(x) = x \cos\left(\frac{\pi x}{L}\right) \quad -L < x < L$$

12.

$$f(x) = \begin{cases} 0 & -\pi < x < 0 \\ x^2 & 0 < x < \pi \end{cases}$$

and use this result to establish

$$\frac{\pi^2}{6} = 1 + \frac{1}{2^2} + \frac{1}{3^2} + \frac{1}{4^2} + \dots$$

$$\frac{\pi^2}{12} = 1 - \frac{1}{2^2} + \frac{1}{3^2} - \frac{1}{4^2} + \dots$$

$$\frac{\pi^2}{8} = 1 + \frac{1}{3^2} + \frac{1}{5^2} + \frac{1}{7^2} + \dots$$

2.5 THE USE OF FOURIER SERIES IN THE SOLUTION OF ORDINARY DIFFERENTIAL EQUATIONS: THE THEORY OF SUSPENSION BRIDGES

An important application of Fourier series is its use in the solution of ordinary differential equations. We illustrate this by finding the particular solution to the ordinary differential equation

$$y'' + 9y = f(t) \tag{2.5.1}$$

where

$$f(t) = |t| \quad -\pi < t < \pi \tag{2.5.2}$$

and

$$f(t + 2\pi) = f(t). \tag{2.5.3}$$

This equation represents an oscillator being forced by a driver whose displacement is given by the saw-tooth function.

The approach that we shall take is to replace the function $f(t)$ by its Fourier series representation. This makes sense because the forcing function is periodic and it is far easier to work with its Fourier representation than to consider each interval from $n\pi < t < (n+1)\pi$ separately. Because the function is an even function, all the sine terms vanish and the Fourier series becomes (see Eq. (2.4.12)).

$$|t| = \pi/2 - \frac{4}{\pi} \sum_{n=1}^{\infty} \frac{1}{(2n-1)^2} \cos [(2n-1)t]. \qquad \textbf{(2.5.4)}$$

Our next task is to find a function $y_p(t)$ such that

$$y_p'' + 9y_p = f(t). \qquad \textbf{(2.5.5)}$$

In ordinary differential equations, if we had been asked to find the particular solution to the differential equation

$$y'' + 9y = -\frac{4}{\pi} \frac{\cos [(2n-1)t]}{(2n-1)^2}, \qquad \textbf{(2.5.6)}$$

we would have guessed the solution

$$y_p(t) = A_n \cos[(2n-1)t] + B_n \sin [(2n-1)t] \qquad \textbf{(2.5.7)}$$

by the method of undetermined coefficients. Hence, we anticipate that the particular solution for Eq. (2.5.5) is a sum of these particular solutions:

$$y_p(t) = \frac{A_0}{2} + \sum_{n=1}^{\infty} A_n \cos [(2n-1)t] + B_n \sin [(2n-1)t] \qquad \textbf{(2.5.8)}$$

because

$$y''_p(t) = \sum_{n=1}^{\infty} -(2n-1)^2 \{A_n \cos [(2n-1)t] + B_n \sin [(2n-1)t]\} \qquad \textbf{(2.5.9)}$$

we have

$$\sum_{n=1}^{\infty} -(2n-1)^2 \left[A_n \cos\,[(2n-1)t] + B_n \sin\,[(2n-1)t]\right] + \frac{9}{2} A_0$$

$$+9 \sum_{n=1}^{\infty} A_n \cos\ (2n-1)t] + B_n \sin\,[(2n-1)t] = \frac{\pi}{2} - \frac{4}{\pi} \sum_{n=1}^{\infty} \frac{\cos[(2n-1)t]}{(2n-1)^2}$$

(2.5.10)

or

$$\frac{9}{2} A_0 - \frac{\pi}{2} + \sum_{n=1}^{\infty} \left\{[9 - (2n-1)^2]\, A_n + \frac{4}{\pi} \frac{1}{(2n-1)^2}\right\} \cos\,[(2n-1)t]$$

$$+ \sum_{n=1}^{\infty} [9 - (2n-1)^2]\, B_n \sin\,[(2n-1)t] = 0. \qquad \textbf{(2.5.11)}$$

Because the A_n's and B_n's must be true for *all* time, each harmonic must vanish separately:

$$A_0 = \frac{\pi}{9} \qquad \textbf{(2.5.12)}$$

$$A_n = -\frac{4}{\pi} \frac{1}{(2n-1)^2\{9-(2n-1)^2\}} \qquad \textbf{(2.5.13)}$$

and

$$B_n = 0. \qquad \textbf{(2.5.14)}$$

Everything is all right except for $n = 2$. For $n = 2$, A_2 becomes undefined. The reason for this difficulty lies with the resonance of the driver's harmonic $\cos(3t)$ with the natural frequency of the system. We can avoid this problem by modifying our solution as follows:

$$y_p(t) = \frac{\pi}{18} - \frac{4}{\pi} \sum_{\substack{n=1 \\ n \neq 2}}^{\infty} \frac{\cos[(2n-1)t]}{(2n-1)^2\,[9-(2n-1)^2]} + Y(t) \qquad \textbf{(2.5.15)}$$

When this solution is substituted into the differential equation and the resulting equation simplified, we find that

$$Y'' + 9Y = -\frac{4}{9\pi} \cos(3t). \quad \textbf{(2.5.16)}$$

The solution to this equation is

$$Y(t) = -\frac{2}{27\pi} t \sin(3t). \quad \textbf{(2.5.17)}$$

This term, called the *secular term*, is the most important term in the solution. While the other terms merely represent simple oscillatory motion, the term $t \sin(3t)$ grows linearly with time and eventually becomes the dominant term in the expansion. Of course, the total solution is given by the sum of the homogenous solution $A \cos(3t) + B \sin(3t)$ plus the particular solution $y_p(t)$. In Fig. 2.5.1, the solution $y_p(t)$ has been graphed.

Even in cases when the forcing is not periodic, the use of Fourier series is important in the solution of ordinary differential equations. For example, S. Timoshenko (Timoshenko, S. P., 1943: Theory of suspension bridges. Part I and Part II. *J. of the Franklin Institute, 235,* 213-238 and 327-349.) used them in the so-called deflection theory of suspension bridges. The bridge structure is considered a combination of a string (the suspension cable) and a beam (a bridge truss). See Fig. 2.5.2. We wish to find what sag occurs in the bridge truss as an applied load $p(x)$ is placed on the bridge.

When an applied load is present, the bending moment M is given by

$$M = p(x) + hy(x) + (H+h)w(x) \quad \textbf{(2.5.18)}$$

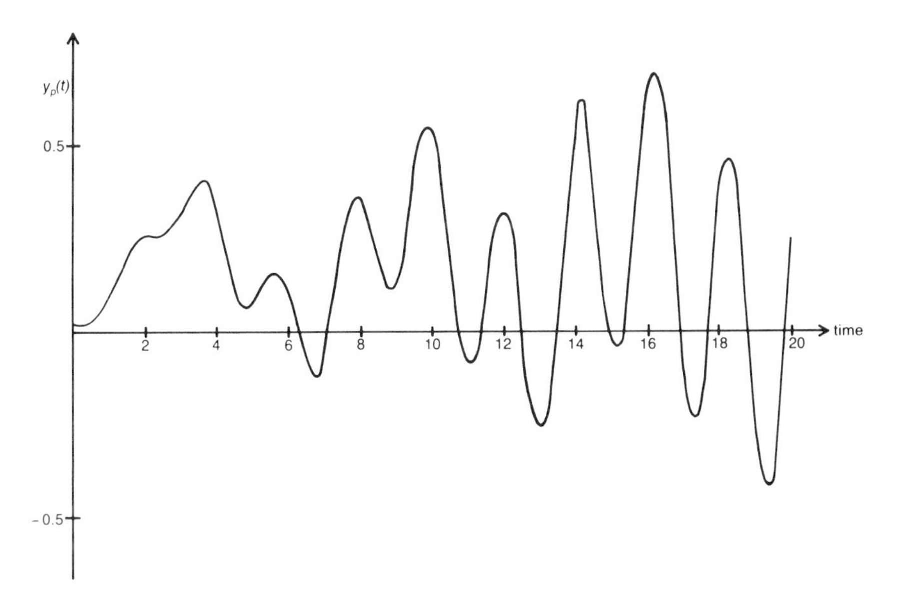

Fig. 2.5.1: Graph of the solution (2.5.15).

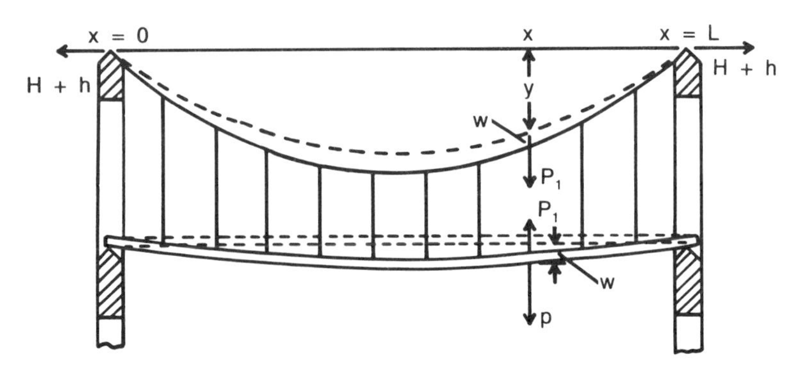

Fig. 2.5.2: Deflection of the truss and cable of a suspension bridge. Taken with permission from von Karman, Th. and M. A. Biot, 1940: *Mathematical Methods in Engineering*. New York: McGraw Hill Co., Inc., p. 277.

where $p(x)$ is the bending moment, due to the applied load, on the girder when treated as isolated and simply supported at its ends, $hy(x)$ the bending moment due to the presence of the suspension cable, and $(H+h)\,w(x)$ the bending moment due to the deflection w of the girder. The quantity $y(x)$ gives the initial cable shape at the point x. (See Pugsley, A., 1957: *The Theory of Suspension Bridges*. London: Edward Arnold Ltd., Chapter VII.). From the theory of elasticity, the deflection of a girder from its natural shape is

$$EI\,\frac{d^2w}{dx^2} = M = p(x) + hy(x) + (H+h)w(x) \tag{2.5.19}$$

or

$$EI\,\frac{d^2w}{dx^2} - (H+h)w = p(x) + hy(x) \tag{2.5.20}$$

where E is Young's modulus and I the moment of inertia.

Let us now consider the simple case of a single span bridge under a concentrated load P at $x = x_1$. The initial shape of the cable can be shown to be

$$y(x) = \frac{4d}{L^2}\,x\,(L-x) \tag{2.5.21}$$

where d is the maximum sag of the suspension cable from the horizon while the bending moment due to the applied load P is

$$p(x) = \begin{cases} \dfrac{-P\,x\,(L-x_1)}{L} & x < x_1 \\[2ex] \dfrac{-Px_1\,(L-x)}{L} & x > x_1. \end{cases} \tag{2.5.22}$$

Consequently,

$$p(x) + hy(x) = h \frac{4d}{L^2} x(L-x) - \frac{P}{L} x (L-x_1) \quad x < x_1 \quad \textbf{(2.5.23)}$$

and

$$p(x) + hy(x) = h \frac{4d}{L^2} x(L-x) - \frac{P}{L} x_1 (L-x). \quad x > x_1 \quad \textbf{(2.5.24)}$$

The compelling reason for using trigonometric series in this problem lies in the ability to write the right-hand side of Eq. (2.5.20) as a *single* Fourier series:

$$\sum_{n=1}^{\infty} b_n \sin(\frac{n\pi x}{L}). \quad \textbf{(2.5.25)}$$

We have only sine terms because $p(x)$ and $y(x)$ must vanish at $x = 0$ and $x = L$. The b_n's are given by

$$b_n = \frac{2}{L} \left\{ h \frac{4d}{L^2} \int_0^L x (L-x) \sin\left(\frac{n\pi x}{L}\right) dx \right.$$

$$- \frac{P(L-x_1)}{L} \int_0^{x_1} x \sin\left(\frac{n\pi x}{L}\right) dx$$

$$\left. - \frac{Px_1}{L} \int_{x_1}^L (L-x) \sin\left(\frac{n\pi x}{\mathrm{L}}\right) dx. \right\} \quad \textbf{(2.5.26)}$$

Upon integrating, this equation becomes

$$b_n = \frac{16hd}{n^3\pi^3} (1 - (-1)^n) - \frac{2PL}{n^2\pi^2} \sin\left(\frac{n\pi x_1}{L}\right). \quad \textbf{(2.5.27)}$$

We now rewrite Eq. (2.5.20) in terms of b_n as the single equation

$$EI \frac{d^2w}{dx^2} - (H+h) w = \sum_{n=1}^{\infty} b_n \sin\left(\frac{n\pi x}{L}\right). \quad \textbf{(2.5.28)}$$

Let us assume as the solution to Eq. (2.5.28) a Fourier series for $w(x)$:

$$w(x) = \sum_{n=1}^{\infty} a_n \sin\left(\frac{n\pi x}{L}\right). \quad \textbf{(2.5.29)}$$

If Eq. (2.5.29) is substituted into Eq. (2.5.28) and the coefficients of the resulting two series are equated,

$$a_n = -\frac{b_n L^2}{EIn^2\pi^2 + (H+\mathrm{h})\mathrm{L}^2} \,. \tag{2.5.30}$$

The final solution can be written as

$$w(x) = -\sum_{n=1}^{\infty} \frac{16hd[1-(-1)^n] - 2PLn\pi \sin\left(\frac{n\pi x_1}{L}\right)}{n^3\pi^3[EIn^2\pi^2 + (H+\mathrm{h})\mathrm{L}^2]} L^2 \sin\left(\frac{n\pi x}{L}\right). \tag{2.5.31}$$

In Fig. 2.5.3, $w(x)$ has been graphed from Eq. (2.5.31) for the typical values of $EI = 0.01\ HL^2$, $x_1 = 0.25\ L$, $h = 1.341\ P$ and $d = 0.1\ L$.

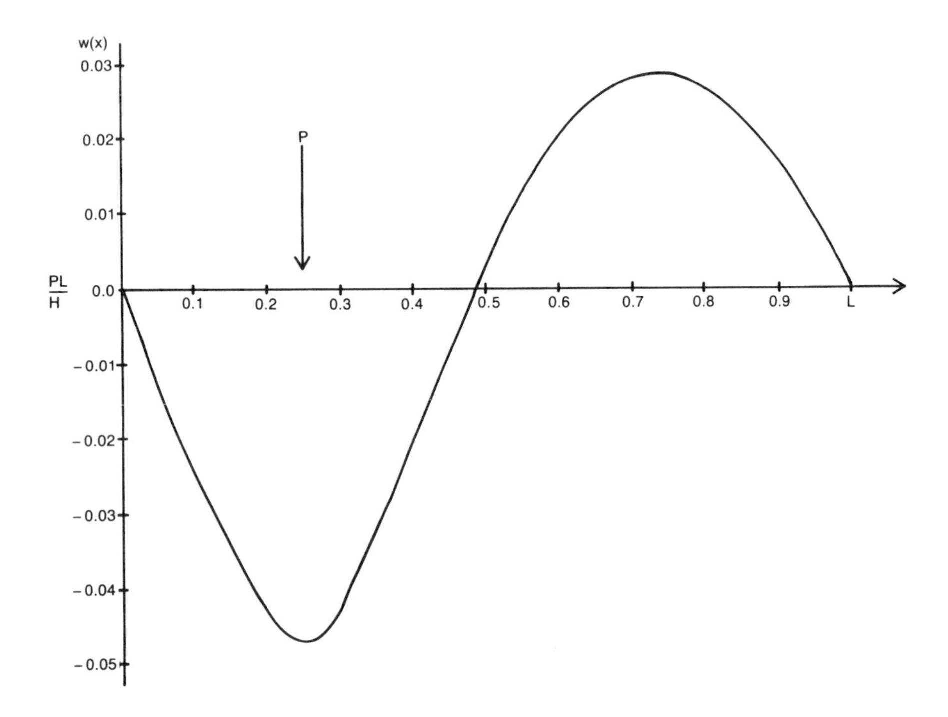

Fig. 2.5.3: The deflection of a girder in a suspension bridge when a constant load P is applied to it at $x_1 = 0.25$ L.

Exercises

1. If $F(t)$ is the periodic function whose definition over one period is

$$F(t) = \begin{cases} 0 & \pi < t < 2\pi, \\ 1 & 0 < t < \pi \end{cases}$$

find the particular solution of each of the following equations:

(a) $y'' - y = F(t)$
(b) $y'' + y = F(t)$
(c) $y'' - 3y' + 2y = F(t)$.

2. If $F(t)$ is the periodic function whose definition over one period is $f(t) = |t|$, $-\pi < t < \pi$, find the particular solution of each of the following equations:

(a) $y'' - y = F(t)$
(b) $y'' + 4y = F(t)$.

3. The linearized equation that describes the radiative cooling of an object into nocturnal surroundings is given by

$$\frac{dT}{dt} + aT = f(t)$$

$$a = \frac{U_c + U_R'}{M_c} \quad \text{and} \quad f(t) = \frac{(U_c T_a + U_R' T_S)}{M_c},$$

where $U_R' = U_R (T + T_S)(T^2 + T_S^2)$, U_R is the radiative heat loss coefficient, T_a the ambient dry bulb temperature, T_S the effective sky temperature, and M_c the heat capacity per unit area of the radiator. See Sodha, M.S., 1982: Transient radiative cooling. *Solar Energy, 28*, 541 for the deriviation.

If we assume that both T_a and T_s are periodic in time so that

$$f(t) = A_0 + \sum_{n=1}^{\infty} A_n \cos(n\omega t) + B_n \sin(n\omega t),$$

show that the temperature of radiator is

$$T(t) = T_0 e^{-at} + \frac{A_0}{a}(1 - e^{-at})$$

$$+ \sum_{n=1}^{\infty} A_n \left[a \cos(n\omega t) + n\omega \sin(n\omega t) - a\, e^{-at}\right]$$

$$+ \sum_{n=1}^{\infty} B_n \left[a \sin(n\omega t) - n\omega \cos(n\omega t) - n\omega\, e^{-at}\right]$$

where $T(0) = T_0$

4. In the early history of artificial satellites, it was important to determine the temperature distribution on the spacecraft's surface. An interesting special case is the temperature fluctuation in the skin due to the spinning of the vehicle. If we assume that the craft is thin-walled so that there is no radial dependence, Hrycak (Hrycak, P., 1963: Temperature distribution in a spinning spherical space vehicle. *AIAA Journal, 1*, 96-99.) showed that the non-dimensional temperature field at the equator of the rotating satellite could be approximately by

$$\frac{d^2T}{d\eta^2} + b\frac{dT}{d\eta} - c\left(T - \frac{3}{4}\right) = -\frac{\pi c}{4}\,\frac{f(\eta) + \beta/4}{1 + \beta\pi/4} \qquad \textbf{(1)}$$

where

$$b = 4\pi^2 r^2 f/a, \quad c = \frac{16\pi}{\gamma}\,\frac{S}{T_\infty}\,(1 + \beta\pi/4),$$

$$f(\eta) = \begin{cases} \cos(2\pi\eta) & 0 < \eta < 1/4 \\ 0 & 1/4 < \eta < 3/4, \\ \cos(2\pi\eta) & 3/4 < \eta < 1 \end{cases}$$

$$T_\infty = \left[\frac{S}{\pi\sigma\epsilon}\right]^{1/4}\left[\frac{1 + \varrho\,\beta/4}{1 + \beta}\right]^{1/4}$$

a the thermal diffusivity of the shell, f the rate of spin, r the radius of the spacecraft, S the direct solar heating, β the ratio of the emissivity of the interior surface to the emissivity of the exterior surface, ϵ the overall emissivity of the exterior surface, γ the satellite's skin conductance, and σ the Stefan-Boltzmann constant. The independent variable η is the longitude along the equator with the effect of rotation subtracted out ($2\pi\eta = \phi - 2\pi ft$). The reference temperature T_∞ equals the temperature that the spacecraft would have it spun with infinite angular speed so that the solar heating would essentially be uniform around the craft.

The solution of (1) is facilitated by introducing the new variables

$$y = T - \frac{3}{4} - \frac{\pi\beta}{16}\left(1 + \frac{\pi\beta}{16}\right)^{-1}$$

$$\varrho_0^2 = c$$

$$v_0 = \frac{2\pi^2 r^2 f}{a\varrho_0}$$

$$A_0 = -\frac{\pi}{4}\,\frac{\varrho_0^{\,2}}{1+\pi\beta/4}$$

so that

$$\frac{d^2y}{d\eta^2} + 2\varrho_0 v_0 \frac{dy}{d\eta} - \varrho_0^{\,2} y = A_0\, f(t). \qquad \textbf{(2)}$$

(a) Show that $f(\eta)$ can be written as

$$f(\eta) = \frac{1}{\pi} + 1/2 \cos(2\pi\eta) + \frac{2}{\pi} \sum_{n=1}^{\infty} \frac{(-1)^{n+1}}{4n^2-1} \cos(4n\pi\eta).$$

(b) Because the homogenous solution to (2) must die away with time, the only important part of the solution to (2) is the particular

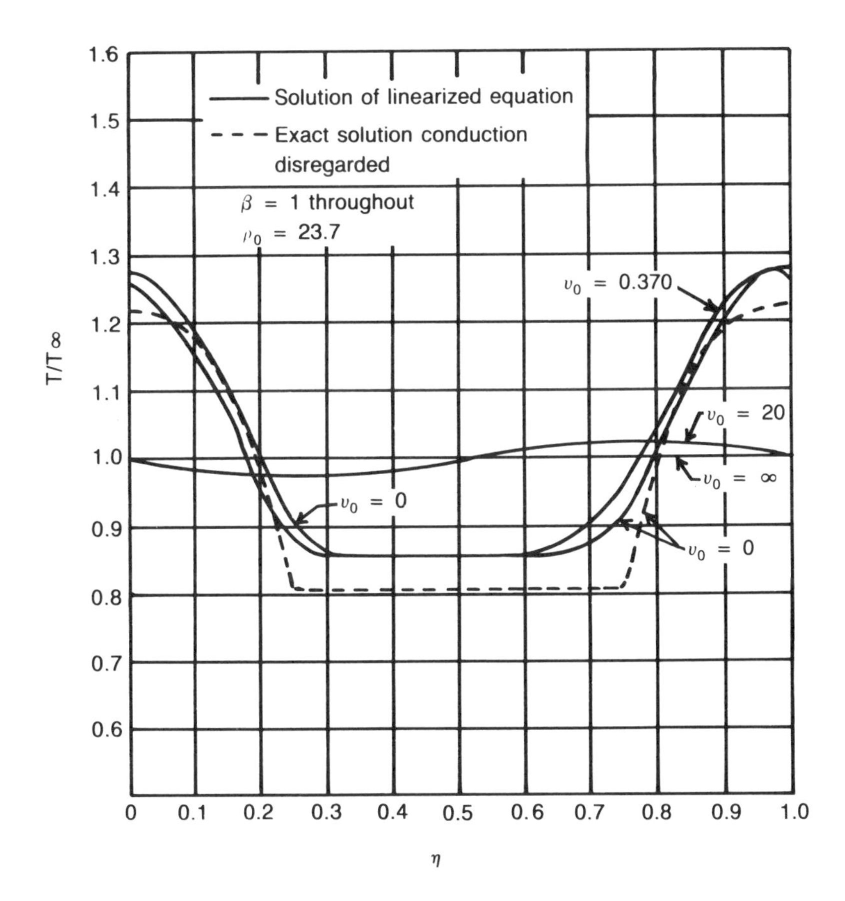

Fig. 2.5.4. Temperature distribution along the equator of a spinning spherical satellite. Taken from Hrycak, P., 1963: Temperature distribution in a spinning spherical space vehicle. *AIAA Journal, 1*, p. 97. (Copyright American Institute of Aeronautics and Astronautics, reprinted with permission.)

solution. Show that it is given by

$$y = \frac{1}{1 + \pi\beta/4}\left[\frac{1}{4} + \frac{\pi}{8A_1}\cos(2\pi\eta + \Phi_1)\right.$$

$$\left. + 1/2\sum_{n=1}^{\infty}\frac{(-1)^{n+1}}{4n^2 - 1}\frac{1}{A_{2n}}\cos(4n\pi\eta + \Phi_{2n})\right]$$

where

$$A_n = \left[(1 + 4n^2\frac{\pi^2}{\varrho_0})^2 + 16\frac{v^2_0}{\varrho^2_0}n^2\pi^2\right]^{1/2}$$

and

$$\Phi_n = \tan^{-1}\left(\frac{4v_0 n\pi\varrho_0}{\varrho_0^2 + 4n^2\pi^2}\right).$$

In Fig. 2.5.4, the results are presented for an aluminum skin, fully covered with glass-protected solar cells on a satellite of radius 2 ft. For a rotation of 0.3 rpm, the temperature fluctuations are typically $\pm$ 11.5° F. Consequently, even for a low spin rate the temperature fluctuations remain low for a satellite with a thin skin.

2.6 HALF-RANGE EXPANSIONS: FOURIER COSINE AND SINE SERIES

In many applications we must find a Fourier series representation for a function $f(x)$ that is defined only in the half Fourier interval $(0,L)$. Because we are completely free to define the function over the interval $(-L,0)$, the function is generally constructed so that the resulting series consists only of sines *or* cosines. In this section we shall show how we can obtain these so-called *half-range expansions*.

An important concept in our analysis revolves around the properties of odd and even functions. The function $g(x)$ is even if $g(-x) = g(x)$; $h(x)$ is odd if $h(-x) = -h(x)$. Most funtions are neither even nor odd; but any function may be written as the sum of an even and an odd function:

$$f(x) = 1/2[f(x) + f(-x)] + 1/2\,[f(x) - f(-x)]\,. \qquad \textbf{(2.6.1)}$$

The first term in Eq. (2.6.1) is an even function and the second is odd.

The important features about even and odd functions are

(1) The integral of an odd function over a symmetric interval is zero

$$\int_{-a}^{a} \text{odd } dx = 0 \tag{2.6.2}$$

(2) The integral of an even function over a symmetric interval is twice the integral over the right half of the interval

$$\int_{-a}^{a} \text{even } dx = 2\int_{0}^{a} \text{even } dx \tag{2.6.3}$$

(3) The following combinatorial properties hold:

$$\text{even} \times \text{odd} = \text{odd} \tag{2.6.4}$$
$$\text{even} \times \text{even} = \text{even} \tag{2.6.5}$$
$$\text{odd} \times \text{odd} = \text{even} \tag{2.6.6}$$
$$\text{even} + \text{even} = \text{even} \tag{2.6.7}$$
$$\text{odd} + \text{odd} = \text{odd} \tag{2.6.8}$$

If we now follow that

$$f_e(x) = 1/2[f(x) + f(-x)]$$

and

$$f_o(x) = 1/2[f(x) - f(-x)],$$

we obtain that

$$a_0 = \frac{1}{L}\int_{-L}^{L} f(x)\, dx = \frac{1}{L}\int_{-L}^{L} f_e(x)\, dx + \frac{1}{L}\int_{-L}^{L} f_o(x)\, dx$$

$$= \frac{2}{L}\int_{0}^{L} f_e(x)\, dx \tag{2.6.9}$$

$$a_n = \frac{1}{L}\int_{-L}^{L} f(x) \cos\left(\frac{n\pi x}{L}\right) dx$$

$$= \frac{1}{L}\int_{-L}^{L} f_e(x) \cos\left(\frac{n\pi x}{L}\right) dx + \frac{1}{L}\int_{-L}^{L} f_o(x) \cos\left(\frac{n\pi x}{L}\right) dx$$

$$= \frac{2}{L}\int_{0}^{L} f_e(x) \cos\left(\frac{n\pi x}{L}\right) dx \tag{2.6.10}$$

$$b_n = \frac{1}{L}\int_{-L}^{L} f(x) \sin\left(\frac{n\pi x}{L}\right) dx$$

$$b_n = \frac{1}{L}\int_{-L}^{L} f_e(x) \sin\left(\frac{n\pi x}{L}\right) dx + \frac{1}{L}\int_{-L}^{L} f_o(x) \sin\left(\frac{n\pi x}{L}\right) dx$$

$$= \frac{2}{L}\int_{0}^{L} f_o(x) \sin\left(\frac{n\pi x}{L}\right) dx \qquad \textbf{(2.6.11)}$$

after substituting into Eq. (2.3.2), (2.3.4), and (2.3.5). From Eq. (2.6.11) we immediately see that if $f(x)$ is an even function $(f_o(x) = 0)$, then $b_n = 0$ for *all n*. These series are referred to as *Fourier cosine series*. Similarly if $f(x)$ is an odd function $(f_e(x) = 0)$, then $a_n = 0$ and we have a *Fourier sine series*. Conversely, if $b_n = 0$ for *all n*, then $f(x)$ is an even function. If $a_n = 0$ for *all n*, then $f(x)$ is odd.

We now use these results to find a Fourier half-range expansion. We do this by extending the function defined over the interval $(0,L)$ as either an even or odd function into the interval $(-L,0)$. If we extend $f(x)$ as an even function, we will get a half-range cosine series; if we extend $f(x)$ as an odd function, we obtain a half-range sine series. This is possible because no matter what the answer is in the interval $(-L,0)$, we do not care. We still will get the correct result in the interval $(0,L)$, as well as in the intervals $(2L,3L)$, $(4L,5L)$, etc.

It is important to remember that half-range expansions are a special case of the general Fourier series. For any $f(x)$, we may construct either a Fourier sine or cosine series over the interval $(-L,L)$. Both of these series will give us the correct answer over the interval of $(0,L)$. Which one we choose to use depends upon whether we wish to deal with a cosine or sine series.

As an example, let us find the half-range sine expansion of

$$f(x) = 1 \qquad 0<x<\pi \qquad \textbf{(2.6.12)}$$

Because we want a half-range sine expansion, $a_n = 0$ and

$$b_n = \frac{2}{\pi}\int_{0}^{\pi} (1) \sin(nx)\, dx \qquad \textbf{(2.6.13)}$$

$$= \frac{-2}{n\pi} \cos(nx) \Big|_{0}^{\pi} = \frac{-2}{n\pi} [\cos(n\pi) - 1]$$

$$= \frac{-2}{n\pi} [(-1)^n - 1]. \qquad \textbf{(2.6.14)}$$

The Fourier series expansion of $f(x)$ is therefore given by

$$f(x) = \frac{2}{\pi} \sum_{n=1}^{\infty} \frac{[1-(-1)^n]}{n} \sin(nx) \tag{2.6.15}$$

$$= \frac{4}{\pi} \sum_{m=1}^{\infty} \frac{\sin[(2m-1)x]}{(2m-1)} . \tag{2.6.16}$$

In practice it is impossible to sum Eq. (2.6.16) exactly and only the first N terms are actually summed. In Fig. 2.6.1, $f(x)$ has been graphed when the Fourier series Eq. (2.6.16) has been truncated to N terms. As seen from the figure, the truncated series tries to achieve the infinite slope at $x = 0$, but in the attempt, it *overshoots*

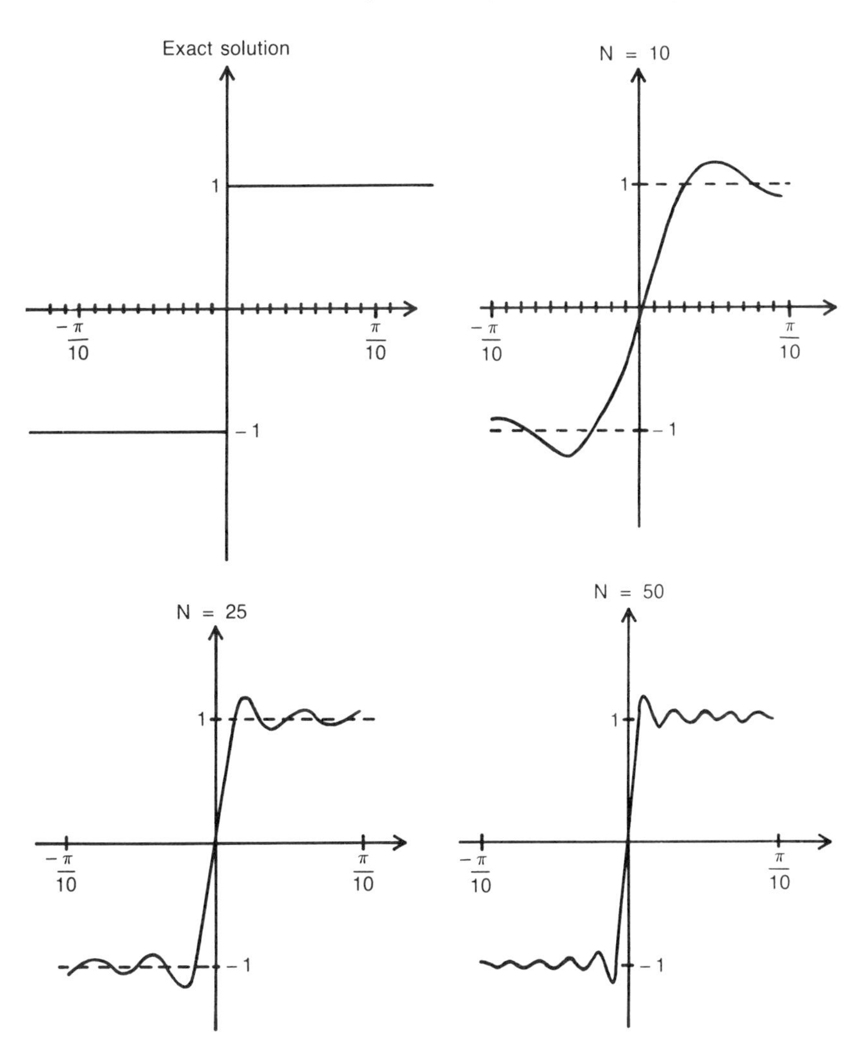

Fig. 2.6.1: Partial sum of N terms in the Fourier representation of a square wave.

the discontinuity by a certain amount (in this particular case, by 17.9%). This peculiarity of the sum of a finite number of terms of a Fourier series always occurs when the function has a simple discontinuity and is known as *Gibbs phenomena*. Increasing the number of terms does not remove this pecularity; it merely shifts it nearer to the discontinuity. The existence of this phenomena underscores the limitations in the process of representing functions by series of eigenfunctions.

2.7 AN EXAMPLE OF A FOURIER COSINE EXPANSION: ACOUSTIC VIBRATIONS AND INTERNAL COMBUSTION ENGINE PERFORMANCE

An important aspect of designing any gasoline engine involves the motion of the fuel air and exhaust gas mixture through the engine. Ordinarily this motion is considered to be a steady flow; but in the case of a four-stroke, single-cylinder gasoline engine, the steady flow of the gasoline-air mixture is interrupted for nearly three-quarters of the engine cycle, by the closing of the intake valve. This periodic interruption sets up standing waves in the intake pipe, waves which can build up an appreciable pressure amplitude just outside the input value.

When one of the harmonics of the engine frequency happens to equal one of the resonance frequency of the intake pipe, then the pressure fluctuations at the valve will be large. If the portion of the cycle when the intake valve is just closing coincides with

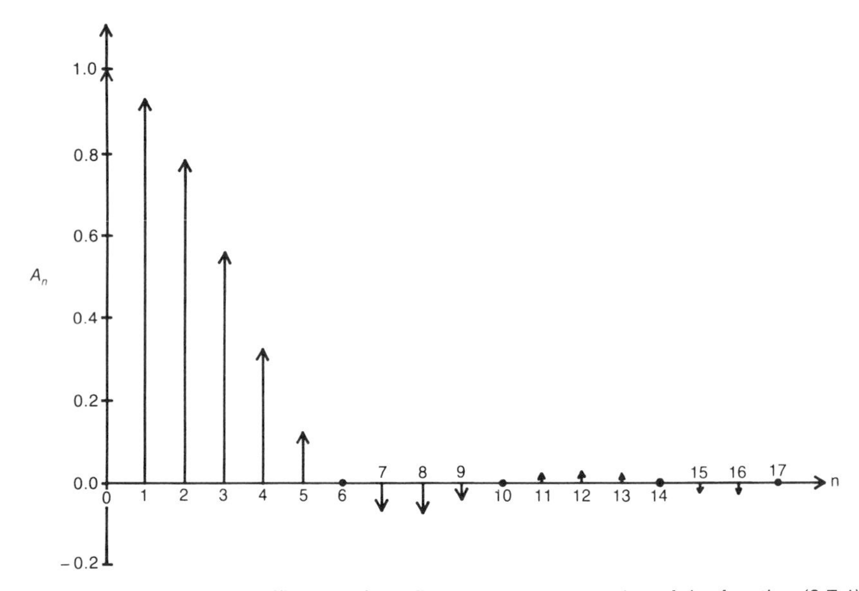

Fig. 2.7.1: The spectral coefficients of the Fourier cosine expansion of the function (2.7.1).

the time when the pressure is less than average, then the waves will reduce the power output; but if the valve is just closing when the pressure is greater than atmospheric, then the waves will have a supercharging effect and will produce an increase of power. This effect is called *inertia supercharging.*

While studying this problem, Morse *et al* (Philip M. Morse, R. H. Boden and Harry Schecter, 1938: Acoustic vibration and internal combustion engine performance. I. Standing waves in the intake pipe system. *J. Appl. Phys., 9,* 16-23) found it necessary to express the velocity of the air-gas mixture in the valve given by

$$v(t) = \begin{cases} 0 & -\pi < \omega t < -\pi/4 \\ \pi \cos(2\omega t) & -\pi/4 < \omega t < \pi/4 \\ 0 & \pi/4 < \omega t < \pi \end{cases} \quad \textbf{(2.7.1)}$$

in terms of a Fourier expansion.

Clearly $v(t)$ is an even function and its Fourier representation will be a cosine series. In this problem $\tau = \frac{-\pi}{\omega}$ and $L = \frac{\pi}{\omega}$. Therefore

$$a_0 = \frac{\omega}{\pi} \int_{-\pi/4\omega}^{\pi/4\omega} \pi \cos(2\omega t)\, dt = \sin(2\omega t/2) \Big|_{-\pi/4\omega}^{\pi/4\omega} = 1 \quad \textbf{(2.7.2)}$$

$$a_n = \frac{\omega}{\pi} \int_{-\pi/4\omega}^{\pi/4\omega} \pi \cos(2\omega t) \cos\left(\frac{n\pi t}{\pi/\omega} \right) dt \quad \textbf{(2.7.3)}$$

$$= \frac{\omega}{2} \int_{-\pi/4\omega}^{\pi/4\omega} \cos[(n+2)\omega t] + \cos[(n-2)\omega t]\, dt \quad \textbf{(2.7.4)}$$

$$= \begin{cases} \left. \dfrac{\sin[(n+2)\omega t]}{2(n+2)} + \dfrac{\sin[(n-2)\omega t]}{2(n-2)} \right|_{-\pi/4\omega}^{\pi/4\omega} & \text{if } n \neq 2 \quad \textbf{(2.7.5)} \\ \left. \dfrac{\omega t}{2} + \dfrac{\sin(4\omega t)}{4} \right|_{-\pi/4\omega}^{\pi/4\omega} & \text{if } n = 2 \quad \textbf{(2.7.6)} \end{cases}$$

$$= \begin{cases} \dfrac{-4}{n^2-4} \cos\left(\dfrac{n\pi}{4} \right) & \text{if } n \neq 2 \quad \textbf{(2.7.7)} \\ \pi/4 & \text{if } n = 2 \quad \textbf{(2.7.8)} \end{cases}$$

Because these Fourier coefficients become small rapidly (see Fig. 2.7.1), Morse *et al* were able to show that there are only about three resonances where we can use the acoustic properties of the intake pipe to enhance engine performance. These peaks occur

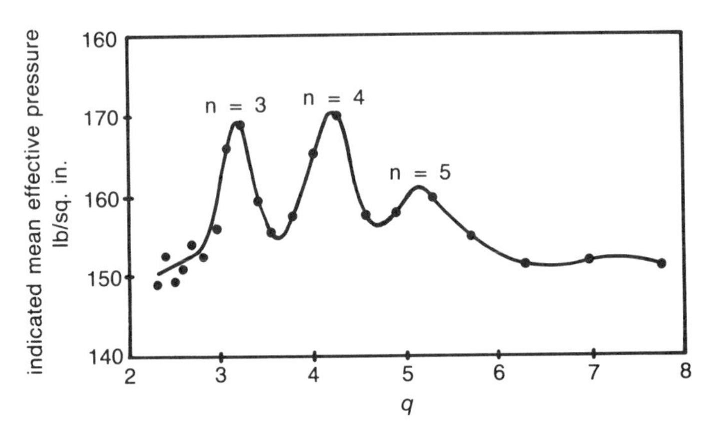

Fig. 2.7.2: Experimental verification of the resonance of the n = 3, 4 and 5 harmonics of the Fourier representation of the flow of the air-gas mixture with the intake pipe system. Taken with permission from Morse, P. M., R. H. Boden and H. Schecter, 1938: Acoustic vibrations and internal combustion engine performance. *J. of Appl. Phys.*, *9*, p. 17.

when $30\, c/NL = 3, 4$, or 5, where c is the velocity of sound in the air-gas mixture, L the effective length of the intake pipe and N the engine speed in rpm. See Fig. 2.7.2. These results were verified by experiments made later (Boden, R. H. and H. Schecter, 1944: Dynamics of the inlet system of a four-stroke engine. NACA Tech. Note 935.)

Exercises

Find the Fourier cosine and sine series of the following functions:

1.
$$f(x) = \begin{cases} x & 0 \le x \le 1/2 \\ 1-x & 1/2 \le x \le 1 \end{cases}$$

2.
$$f(x) = \pi^2 - x^2 \qquad 0 \le x \le \pi$$

3.
$$f(x) = \begin{cases} x & 0 \le x \le 1 \\ 1 & 1 \le x < 2 \end{cases}$$

4.
$$f(x) = \begin{cases} 0 & 0 < x < a/3 \\ x - a/3 & a/3 < x < 2a/3 \\ a/3 & 2a/3 < x < a \end{cases}$$

5.
$$f(x) = \begin{cases} 1/2 & 0 < x < a/2 \\ 1 & a/2 < x < a \end{cases}$$

6.
$$f(x) = \begin{cases} 2x/a & 0 < x < a/2 \\ (3a-2x)/2a & a/2 < x < a \end{cases}$$

7.
$$f(x) = \begin{cases} x & 0 < x < a/2 \\ a/2 & a/2 < x < a \end{cases}$$

8.
$$f(x) = (a-x)/a \qquad 0 < x < a$$

9.

$$f(x) = \begin{cases} 0 & 0<x<a/4 \\ 1 & a/4<x<3a/4 \\ 0 & 3a/4<x<a \end{cases}$$

10.

$$f(x) = x(a-x) \qquad 0<x<a$$

11.

$$f(x) = e^{kx} \quad 0<x<a$$

12.

$$f(x) = \begin{cases} 0 & 0<x<a/2 \\ 1 & a/2<x<a \end{cases}$$

13. Show that the Fourier sine series for

$$f(t) = 1 - (1+a)\frac{t}{\pi} + (a-1)\frac{t^2}{\pi^2} + (a+1)\frac{t^3}{\pi^3} - a\frac{t^4}{\pi^4}, \ (0<t<\pi)$$

is

$$f(t) = \frac{1}{\pi}\sum_{1}^{\infty}\frac{1}{n}\left[1 - \frac{3(a-1)}{2\pi^2 n^2}\right]\sin(2nt)$$

$$+ \frac{1}{\pi}\sum_{1}^{\infty}\frac{2}{2n-1}\left[1 + \frac{2(a-1)}{\pi^2(2n-1)^2} - \frac{48a}{\pi^4(2n-1)^4}\right]\sin[(2n-1)t]$$

2.8 FOURIER SERIES WITH PHASE ANGLES

Sometimes it is desirable to rewrite a general Fourier series a purely cosine or purely sine series with a phase angle. Suppose, for example, that we have a function $f(x)$ of period $2L$ given in the interval $(-L,L)$, whose Fourier series expansion is given by

$$f(x) = \frac{a_0}{2} + \sum_{n=1}^{\infty} a_n \cos(\frac{n\pi x}{L}) + b_n \sin(\frac{n\pi x}{L}). \qquad \textbf{(2.8.1)}$$

We wish to replace Eq. (2.8.1) by the series

$$f(x) = \frac{a_0}{2} + \sum_{n=1}^{\infty} c_n \sin(\frac{n\pi x}{L} + \phi_n). \qquad \textbf{(2.8.2)}$$

To do this we note that

$$c_n \sin(\frac{n\pi x}{L} + \phi_n) = a_n \cos(\frac{n\pi x}{L}) + b_n \sin(\frac{n\pi x}{L})$$

$$= c_n \sin(\frac{n\pi x}{L}) \cos(\phi_n) + c_n \sin(\phi_n) \cos(\frac{n\pi x}{L}) \qquad \textbf{(2.8.3)}$$

We equate coefficients of $\sin(n\pi x/L)$ and $\cos(n\pi x/L)$ on both sides and obtain

$$a_n = c_n \sin(\phi_n) \text{ and } b_n = c_n \cos(\phi_n). \qquad \textbf{(2.8.4)}$$

Hence, on squaring and adding

$$c_n = (a_n^2 + b_n^2)^{1/2} \qquad \textbf{(2.8.5)}$$

while taking the ratio gives

$$\phi_n = \tan^{-1}(a_n/b_n). \qquad \textbf{(2.8.6)}$$

Similarly we could rewrite Eq. (2.8.1) as

$$f(x) = \frac{a_0}{2} + \sum_{n=1}^{\infty} c_n \cos\left(\frac{n\pi x}{L} + \phi_n\right) \qquad \textbf{(2.8.7)}$$

where

$$c_n = (a_n^2 + b_n^2)^{1/2} \text{ and } \phi_n = \tan^{-1}(-b_n/a_n) \qquad \textbf{(2.8.8)}$$

$$a_n = c_n \cos(\phi_n) \qquad b_n = -c_n \sin(\phi_n) \qquad \textbf{(2.8.9)}$$

In both cases, care must be taken in computing ϕ_n because there are two possible determination of ϕ_n which satisfy Eq. (2.8.6) and (2.8.8). The value which must be selected is the one that satisfies both equations in (2.8.4) or (2.8.9).

2.9 IMPROVING THE CONVERGENCE OF FOURIER SERIES

In applications the most convenient Fourier series are those with rapidly *decreasing* coefficients. In this case the first few terms of the series give its sum quite accurately because the sum of *all* of the remaining terms is small. Slow convergence usually occurs when a discontinuity is present. The following method overcomes the slow convergence by removing the discontinuity. To understand this procedure, we consider first its application to a simpler problem—that of computing the sum of a slowly convergent series of constant terms.

Suppose that we want the sum of the series

$$\sum_{n=1}^{\infty} \frac{1}{1+n^2} = \frac{1}{2} + \frac{1}{5} + \frac{1}{10} + \frac{1}{17} + \ldots \qquad \textbf{(2.9.1)}$$

The direct determination of the sum with an accuracy of 10^{-3} requires 1000 terms. However, we note that we can rewrite the summation as

$$\sum_{n=1}^{\infty} \frac{1}{1+n^2} = \sum_{n=1}^{\infty} \frac{1}{n^2} - \frac{1}{n^4} + \frac{1}{n^4(1+n^2)} \quad \textbf{(2.9.2)}$$

$$= \sum_{n=1}^{\infty} \frac{1}{n^2} - \sum_{n=1}^{\infty} \frac{1}{n^4} + \sum_{n=1}^{\infty} \frac{1}{n^4+n^6} \quad \textbf{(2.9.3)}$$

The first two series, which converge slowly, are exactly summable, and have the values of $\pi^2/6$ and $\pi^4/90$. The last series converges so rapidly that in computing its sum we limit ourselves to the first four terms.

$$\sum_{n=1}^{\infty} \frac{1}{n^4+n^6} = \frac{1}{2} + \frac{1}{80} + \frac{1}{810} + \frac{1}{4352} \simeq 0.514. \quad \textbf{(2.9.4)}$$

Finally,

$$\sum_{n=1}^{\infty} \frac{1}{1+n^2} = \frac{\pi^2}{6} - \frac{\pi^4}{90} + 0.514 = 1.076. \quad \textbf{(2.9.5)}$$

Let us now apply this concept to improve the convergence of the Fourier series

$$f(x) = \sum_{n=2}^{\infty} (-1)^n \frac{n^3}{n^4-1} \sin(nx) \quad \textbf{(2.9.6)}$$

In this case we note that the quantity $n^3/(n^4-1)$ behaves as $1/n$ as n approaches infinity. Therefore we rewrite the Fourier coefficient as

$$\frac{n^3}{n^4-1} = \frac{1}{n} + \frac{1}{n^5-n} . \quad \textbf{(2.9.7)}$$

Consequently we have

$$f(x) = \sum_{n=2}^{\infty} (-1)^n \frac{\sin(nx)}{n} + \sum_{n=2}^{\infty} (-1)^n \frac{\sin(nx)}{n^5-n} . \quad \textbf{(2.9.8)}$$

However, the first summation may be replaced by

$$\sum_{n=1}^{\infty} \frac{(-1)^{n+1}}{n} \sin(nx) = \frac{x}{2} \quad (-\pi < x < \pi) \quad \textbf{(2.9.9)}$$

Hence,

$$f(x) = -\frac{x}{2} + \sin(x) + \sum_{n=2}^{\infty} (-1)^n \frac{\sin(nx)}{n^5 - n} \quad (-\pi < x < \pi) \quad \mathbf{(2.9.10)}$$

For a second example, let us try and improve the convergence of

$$f(x) = \sum_{n=1}^{\infty} \frac{n^4 - n^2 + 1}{n^2(n^4+1)} \cos(nx) \quad \mathbf{(2.9.11)}$$

In this case, the Fourier coefficients behave as $1/n^2$ as n approaches infinity. Therefore we have

$$\frac{n^4 - n^2 + 1}{n^2(n^4+1)} = \frac{1}{n^2} - \frac{1}{1+n^4} \quad \mathbf{(2.9.12)}$$

and

$$f(x) = \sum_{n=1}^{\infty} \frac{\cos(nx)}{n^2} - \sum_{n=1}^{\infty} \frac{\cos(nx)}{n^4+1} \quad \mathbf{(2.9.13)}$$

We can replace the first summation by

$$\sum_{n=1}^{\infty} \frac{\cos(nx)}{n^2} = \frac{3x^2 - 6\pi x + 2\pi^2}{12} \quad (0 \leq x \leq 2\pi) \quad \mathbf{(2.9.14)}$$

and the improved Fourier series becomes

$$f(x) = \frac{3x^2 - 6\pi x + 2\pi^2}{12} - \sum_{n=1}^{\infty} \frac{\cos(nx)}{n^4+1} . \quad (0 \leq x \leq 2\pi) \quad \mathbf{(2.9.15)}$$

The difficulty in this method is guessing the function of x that yields the Fourier series expansion that you want to eliminate.

2.10 FINITE FOURIER SERIES

In most applications the function for which we must construct a Fourier series is given to us as data points or in the form of a graph. In both cases we cannot resolve harmonics with a frequency higher than $\pi/\Delta s$ where Δs is the interval between observations. This is in sharp contrast with the situation where we have an analytic formula and there are an infinite number of "data points." Consequently we can no longer express the function with an infinite number of sines or cosines but must be satisfied with only a finite number of these periodic functions. We could find the coefficients a_n and b_n by integrating Eqs. (2.3.6) through (2.3.8) using Simpson's rule. However, as we shall shortly show, sines and cosines are orthogonal over both the continuous interval and sets of discrete, equally spaced points covering the period. This is extremely important because in computing we are usually given samples of the function at a set of equally spaced points. Consequently we can find the Fourier coefficients *exactly* (except for roundoff errors) in the discrete sampled problem, instead of approximately by numerical methods.

For our derivation we will assume that there are only an even number $2N$ of points. Let the $2N$ sample points be at

$$x_p = 0,\ P/2N,\ 2P/2N,\ \ldots,(2N-1)P/2N \qquad \textbf{(2.10.1)}$$

or more shortly,

$$x_p = \frac{Pp}{2N}, \quad p = 0, 1, 2, \ldots, 2N-1 \qquad \textbf{(2.10.2)}$$

where P is the period of the function. In this discrete representation of a continuous function, the value given at x_p actually represents the average value over the interval

$$(x_p - P/4N,\ x_p + P/4N).$$

We want to show that the $2N$ Fourier functions

$$1,\ \cos(2\pi x/P),\ \cos(4\pi x/P),\ \ldots,\ \cos(2\pi Nx/P),$$
$$\sin(2\pi x/P),\ \ldots,\ \sin(2\pi(N-1)x/P)$$

form an orthogonal set of functions. This is equivalent to showing that for $0 \le k \le N$ and $0 \le m \le N$

$$\sum_{p=0}^{2N-1} \cos\left(\frac{2\pi k}{P}\frac{Pp}{2N}\right)\cos\left(\frac{2\pi m}{P}\frac{Pp}{2N}\right) = \begin{cases} 0 & k \ne m \\ N & k = m \ne 0,N \\ 2N & k = m = 0,N \end{cases} \qquad \textbf{(2.10.3)}$$

$$\sum_{p=0}^{2N-1} \cos\left(\frac{2\pi k}{P}\frac{Pp}{2N}\right)\sin\left(\frac{2\pi m}{P}\frac{Pp}{2N}\right) = 0 \quad \text{for all } k \text{ and } m \qquad \textbf{(2.10.4)}$$

$$\sum_{p=0}^{2N-1} \sin\left(\frac{2\pi k}{P}\frac{Pp}{2N}\right)\sin\left(\frac{2\pi m}{P}\frac{Pp}{2N}\right) = \begin{cases} 0 & k \ne m \\ N & k = m \ne 0,N \\ 0 & k = m = 0,N \end{cases} \qquad \textbf{(2.10.5)}$$

These equations are the discrete analog to the orthogonality condition Eqs. (2.2.11), (2.3.12), and (2.3.14).

The derivation is best done by first considering the series

$$\sum_{p=0}^{2N-1} \exp\left(\frac{2\pi i}{P} k x_p\right) = \sum_{p=0}^{2N-1} \exp\left(\frac{i\pi k p}{N}\right) \text{ for integer } k \quad \textbf{(2.10.6)}$$

which is a geometric progression with ratio $\exp(i\pi k/N)$. The sum of the geometric progression is

$$1 - r^{2N}/(1-r) = 0 \qquad \text{if } r \neq 1 \qquad \textbf{(2.10.7)}$$

and

$$2N \qquad \text{if } r = 1. \qquad \textbf{(2.10.8)}$$

The upper sum is true because $r^{2N} = \exp(2\pi i k) = 1$. The lower value is true because we have a sum of $2N$ terms, each of which equals one. The condition $r = 1$ means $k = 0, \pm 2N, \pm 4N, \ldots$

We next show that

$$\sum_{p=0}^{2N-1} \exp\left(\frac{2\pi i}{P} k x_p\right) \exp\left(-\frac{2\pi i}{P} m x_p\right) = \sum_{p=0}^{2N-1} \exp\left(\frac{2\pi i}{P} (k-m) x_p\right)$$

$$= \begin{cases} 0 & |k-m| \neq 0, 2N, 4N, \ldots \\ 2N & |k-m| = 0, 2N, 4N, \ldots \end{cases}$$

(2.10.9)

where we have used the results from the previous paragraph after replacing the integer $k - m$ with another integer variable k'. Consequently the complex function $\exp\left(\frac{2\pi i}{P} k x_p\right)$ is orthogonal to the complex conjugate of the complex function $\exp\left(\frac{2\pi i}{P} m x_p\right)$.

We now return to the "real world" by using Euler's identity

$$\exp(ix) = \cos(x) + i \sin(x). \qquad \textbf{(2.10.10)}$$

The condition for the single exponential summed over the points x_p becomes two equations (the real and imaginary parts separately)

$$\sum_{p=0.}^{2N-1} \cos\left(\frac{2\pi k}{P} x_p\right) = \begin{cases} 0 & k \neq 0, \pm 2N, \pm 4N, \ldots \\ 2N & k = 0, \pm 2N, \pm 4N, \ldots \end{cases} \qquad \textbf{(2.10.11)}$$

$$\sum_{p=0}^{2N-1} \sin\left(\frac{2\pi k}{P} x_p\right) = 0 \qquad \text{for all } k \qquad \textbf{(2.10.12)}$$

With Eqs. (2.10.11) and (2.10.12) in hand, we can prove the orthogonality of the Fourier functions over the set of x_p equally spaced points. For example, the first of the three orthogonality equations, Eq. (2.10.3), can be written, by the use of the trigonometric identity

$$\cos(A)\cos(B) = 1/2[\cos(A+B) + \cos(A-B)], \qquad \textbf{(2.10.13)}$$

$$1/2 \sum_{p=0}^{2N-1} \cos\left[\pi(k+m)\frac{p}{N}\right] + \cos\left[\pi(k-m)\frac{p}{N}\right]$$

$$= \begin{matrix} 0 & |k-m| \text{ and } |k+m| \neq 0, 2N, 4N, \ldots \\ N & |k-m| \text{ or } |k+m| = 0, 2N, 4N, \ldots \\ 2N & |k-m| \text{ and } |k+m| = 0, 2N, 4N, \ldots \end{matrix} \qquad \textbf{(2.10.14)}$$

For the restricted set of functions we are now using, namely,

$$1, \cos(\frac{2\pi x}{P}), \cos(\frac{4\pi x}{P}), \ldots \cos(\frac{2\pi N}{P}x), \qquad \textbf{(2.10.15)}$$

$k+m$ will range from 0 to $2N$. The case $k+m = 0$ occurs when $k = m = 0$; $k+m = 2N$, when $k = m = N$. If $k \neq m$, then $k+m \neq 0$ or $2N$, $k-m \neq 0$ or $2N$, and Eq. (2.10.3) must equal zero. If $k = m \neq 0,N$, then $k+m \neq 0$ or $2N$ but $k-m = 0$ and Eq. (2.10.3) equals N. Finally if $k = m = 0,N$, then $k+m = 0$ or $2N$, $k-m = 0$, and Eq. (2.10.3) must equal $2N$. Thus, we have established the orthogonality condition (2.10.3) for the Fourier functions $\cos(\frac{2\pi k}{P}x_p)$. The remaining orthogonality conditions (2.10.4) and (2.10.5), are proven in a similar manner.

The method used to find the A_k's and B_k's is very similar to that used in the continuous case. We first assume that $f(x)$ can be written

$$f(x) = \frac{A_0}{2} + \sum_{k=1}^{N-1} A_k \cos(\frac{2\pi k}{P}x) + B_k \sin(\frac{2\pi k}{P}x) + \frac{A_N}{2}\cos(\frac{2\pi N}{P}x) \qquad \textbf{(2.10.16)}$$

To find A_k, for example, we multiply our Fourier representation by $\cos(\frac{2\pi m}{P}x)$ (m may take on the values of 0 to N) and sum from $p = 0$ to $2N-1$. Consequently at $x = x_p$

$$\sum_{p=0}^{2N-1} f(x_p)\cos(\frac{2\pi m}{P}x_p) = \frac{A_0}{2}\sum_{p=0}^{2N-1}\cos(\frac{2\pi m}{P}x_p) + ① + ② + ③$$

$$① = \sum_{k=1}^{N-1} A_k \sum_{p=0}^{2N-1} \cos(\frac{2\pi k}{P}x_p)\cos(\frac{2\pi m}{P}x_p)$$

$$\textcircled{2} = \sum_{k=1}^{N-1} B_k \sum_{p=0}^{2N-1} \sin(\frac{2\pi k}{P} x_p)\cos(\frac{2\pi m}{P} x_p)$$

$$\textcircled{3} = \frac{A_N}{2} \sum_{p=0}^{2N-1} \cos(\underset{P}{\overset{2\pi N}{\sum}} x_p) \cos(\frac{2\pi m}{P} x_p) \qquad \textbf{(2.10.17)}$$

If we assume that $m \neq 0$ or N, then the first summation on the right-hand side vanishes by Eq. (2.10.17), the third by Eq. (2.10.4), and the fourth by Eq. (2.10.3). In the second summation, the summation does *not* vanish if $k = m$ and is equal to N by Eq. (2.10.3). Similar considerations lead to the following formulas for the calculation of A_k and B_k:

$$A_k = \frac{1}{N} \sum_{p-0}^{2N-1} f(x_p) \cos(\frac{2\pi k}{P} x_p)\ k = 0,1,2,3, \ldots, N \qquad \textbf{(2.10.18)}$$

$$B_k = \frac{1}{N} \sum_{p-0}^{2N-1} f(x_p) \sin(\frac{2\pi k}{P} x_p)\ \ k = 1,2,3, \ldots, N-1 \qquad \textbf{(2.10.19)}$$

The function $f(x)$ is often given at $2N+1$ points, but still for $2N$ intervals, and it is assumed that $f(0) = f(L)$. If this is not so, then it it customary to use $[f(0)+f(L)]/2$ as the value of $f(0)$. In this case the formulas for the coefficients A_k and B_k are

$$A_k = \frac{1}{N} [(f(0) + f(L))/2 + \sum_{p=1}^{2N-1} f(x_p) \cos(\frac{2\pi k}{P} x_p)] \qquad \textbf{(2.10.20)}$$

$$B_k = \frac{1}{N} \sum_{p=1}^{2N-1} f(x_p) \sin(\frac{2\pi k}{P} x_p) \qquad \textbf{(2.10.21)}$$

These are effectively the trapezoid rule applied for the numerical integration of the corresponding integrals Eq. (2.3.2), (2.3.4), and (2.3.5).

It is important to note that $2N$ data points yields $2N$ Fourier coefficients A_k and B_k. Consequently the amount of information, whether in the form of data points or Fourier coefficients, will always be limited by our sampling frequency. It might be argued that from the Fourier series representation of $f(x)$ we could find the value of $f(x)$ for any given x, which is more than we can do with the data alone. This is not true. Although we can calculate a value for $f(x)$ at any x using the finite Fourier series, we simply do not know whether those values are correct or not. They are simply those given by a finite Fourier series which is forced to fit the given data points.

2.11 AN EXAMPLE OF A FINITE FOURIER SERIES; THE HARMONIC ANALYSIS OF THE DEPTH OF WATER AT BUFFALO, NEW YORK

Each entry in Table 2.11.1 gives the observed depth of water at Buffalo, NY (minus the low-water datum of 568.6 ft) on the fifteenth of the corresponding month during 1977 (National Ocean Survey, 1977: *Great Lakes Water Levels, 1977,* Rockville, MD, 107 pp.). This year was noted for a very severe winter. Assuming that the water level is a periodic function of period one year, and that the observations are taken at equal intervals, we want to construct a finite Fourier series from this data. This corresponds to computing the Fourier coefficients A_0, A_1, A_2, . . . B_1, B_2, . . . which gives the mean level and fluctuations of the depth of water for the periods of 12 mo, 6 mo, 4 mo, and so forth.

In this problem the period is 12 mo, N = 1/2(period) = 6 and

$$x_n = \frac{(period)\ n}{2N} = \frac{(12\ mo)\ n}{12} = n\ (mo) \qquad \textbf{(2.11.1)}$$

That is, there should be a data point for each month. From Eq. (2.10.12) and (2.10.13)

$$A_k = \frac{1}{6} \sum_{n=0}^{11} f(x_n) \cos(\frac{\pi}{6} k\ n), k = 0,1,2,3,4,5,6 \qquad \textbf{(2.11.2)}$$

$$B_k = \frac{1}{6} \sum_{n=0}^{11} f(x_n) \sin(\frac{\pi}{6} k\ n), k = 0,1,2,3,4,5 \qquad \textbf{(2.11.3)}$$

The substitution of the data into Eqs. (2.11.2) and (2.11.3) yields

A_0 = twice the mean level = 5.12 ft
A_1 = harmonic component with a period of 12 mo = −0.566 ft
B_1 = harmonic component with a period of 12 mo = −0.128 ft
A_2 = harmonic component with a period of 6 mo = −0.177 ft
B_2 = harmonic component with a period of 6 mo = −0.372 ft
A_3 = harmonic component with a period of 4 mo = −0.110 ft
B_3 = harmonic component with a period of 4 mo = −0.123 ft
A_4 = harmonic component with a period of 3 mo = 0.025 ft

Table 2.11.1: The Depth of Water in the Harbor at Buffalo, NY (Minus the Low-Water Datum of 568.6 Ft) on the Fifteenth Day of Each Month During 1977.

mo	n	depth	mo	n	depth	mo	n	depth
Jan	1	1.61	May	5	3.16	Sep	9	2.42
Feb	2	1.57	Jun	6	2.95	Oct	10	2.95
Mar	3	2.01	Jul	7	3.10	Nov	11	2.74
Apr	4	2.68	Aug	8	2.90	Dec	12	2.63

B_4 = harmonic component with a period of 3 mo = 0.052 ft
A_5 = harmonic component with a period of 2.4 mo = −0.079 ft
B_5 = harmonic component with a period of 2.4 mo = −0.131 ft
A_6 = harmonic component with a period of 2 mo = -0.107 ft

Figure 2.11.1 is a plot of our results using Eq. (2.10.10). Note that when all the harmonic terms are included, the finite Fourier series

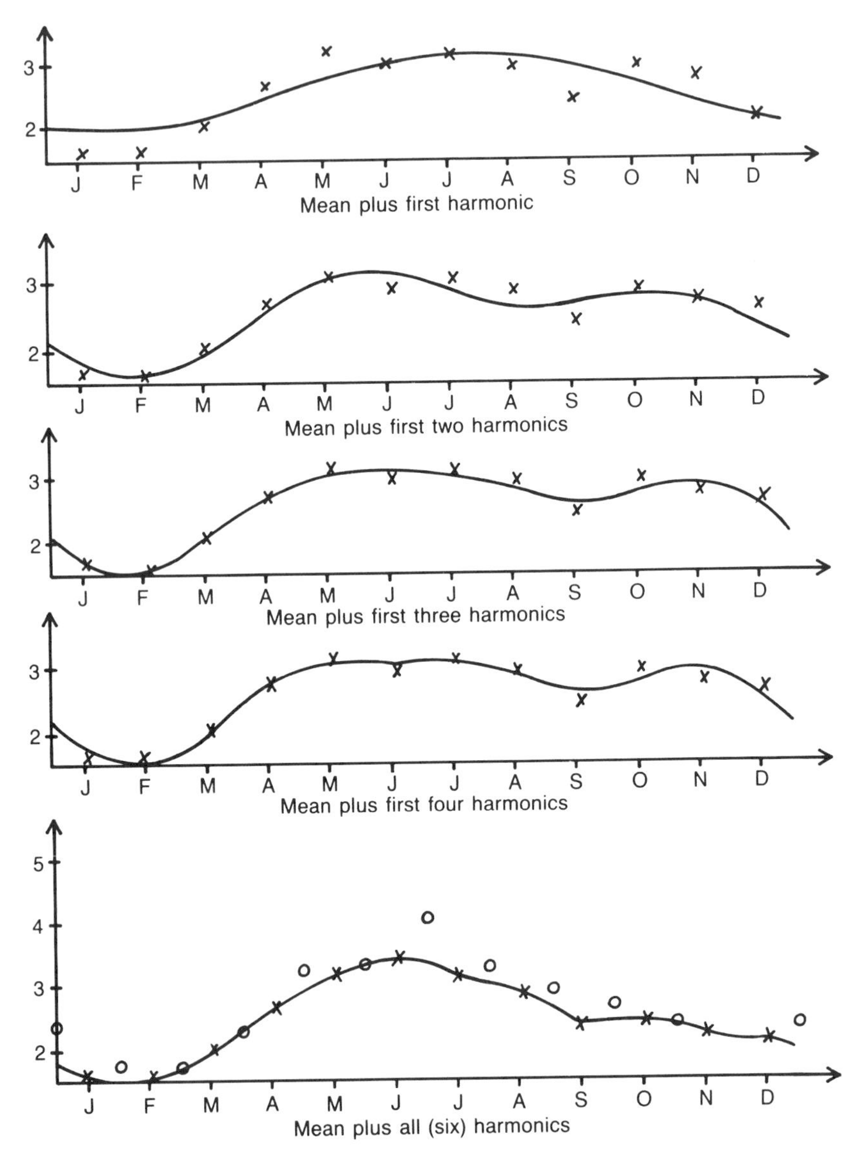

Fig. 2.11.1: Partial sum of the finite Fourier series for the depth of water in the harbor of Buffalo, NY during 1977. Crosses x indicate observations on the fifteenth of the month; circles o observations on the first.

fits the data points exactly. The values given by the series at points between the data points may be right or they may not. To illustrate this, the values for the first of each month have also been plotted. Sometimes the values given by the Fourier series and the data are quite different.

Let us now examine our results in terms of various physical processes. In the long run the depth of water in the harbor at Buffalo, NY depends upon the three-way balance between precipitation, evaporation and inflow-outflow of any rivers. Because the inflow and outflow of the rivers depends strongly upon precipitation, and evaporation is of secondary importance, the water level is strongly correlated to the precipitation rate. It is well known that more precipitation falls during the warmer months rather than the colder months. This effect is reflected in the large amplitude of the Fourier coefficients corresponding to the annual cycle $(k = 1)$.

Another important term in the harmonic analysis corresponds to the semiannual cycle $(k = 2)$. During the winter months around Lake Ontario, precipitation falls as snow. Therefore the inflow from rivers is greatly reduced. When spring comes, all the snow and ice melt and a jump in the water level occurs. Because periodic variations associated with seasonal variations is given by the second harmonic, this harmonic is absolutely necessary if we want to get the right answer.

Another curious result arises from our analyis. Although we have data for each of the twelve months, we can only resolve phenomena with a period of 2 months or greater. Figure 2.11.2 will help explain this phenomena. In case (a), we have quite a few data points over one cycle. Consequently, our picture, constructed from data, is fairly good. In case (b), we have only taken samples at the ridges and troughs of the wave. Although our picture of the real phenomena is poor, at least we know that there is a wave. From this picture we see that even if we are lucky enough to take our observations at the ridges and troughs of a wave, we need at least two data points per cycle (one for the ridge, the other for the trough) to resolve the highest frequency wave. In case (c), we have made a big mistake. We have taken a wave of frequency N Hz and misrepresented it as a wave of frequency $N/2$ Hz. This misrepresentation of a high frequency wave by a lower frequency is called *aliasing*. It arises because we are sampling a continuous signal at equal intervals. By comparing case (b) and case (c), we see that there is a cutoff between aliased and nonaliased frequencies. This frequency is called the *Nyquist* or *folding* frequency. It corresponds to the highest frequency resolved by our finite Fourier analysis.

Because most periodic functions require an infinite number of harmonics for their representation, aliasing of signals is a common problem. Consequently the question is not "can I avoid aliasing?"

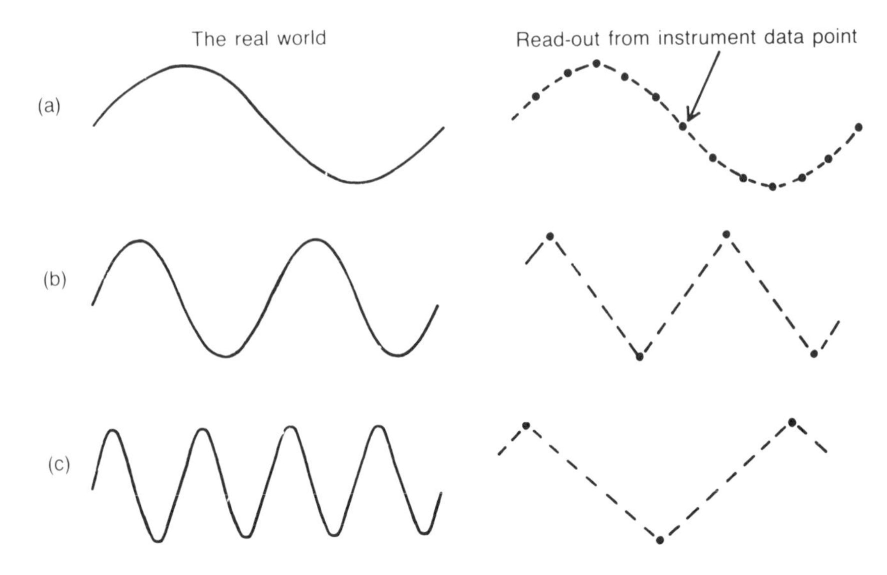

Fig. 2.11.2: The effect of sampling in the representation of periodic functions.

but "can I live with it?" Quite often, we can construct our experiment to say yes. An example where aliasing is unavoidable occurs in a Western at the movies when we see the rapidly rotating spokes of the stagecoach's wheel. A movie is a sampling of continuous motion where the data is presented as a succession of pictures. Consequently, the high rate of revolution of the stagecoach's wheel is aliased in such a manner so that it appears to be stationary or rotating very slowly.

Finally, to illustrate the finite Fourier series in a more complex problem, the Fourier spectrum has been found for the earth's orography along three latitude belts. Along each latitude the height above sea level was given at 144 points (see Fig. 2.11.3). The corresponding Fourier coefficients were found by means of the finite Fourier transform. A plot of $(A_n^2 + B_n^2)^{1/2}$ for each zonal wavenumber $n = 0{,}72$ (wavelength $= 2\pi/n$) is given in Fig. 2.11.4. From this figure it appears as if the Fourier coefficients tend towards zero more rapidly for the two latitude belts from the Northern Hemisphere rather than the one from the Southern Hemisphere. Actually, most of the differences arise from the different scaling of the ordinate.

Exercises

Find the finite Fourier series for

1. $x(0) = 0$, $x(1) = 1$, $x(2) = 2$, $x(3) = 3$, and $N = 2$.
2. $x(1/2) = 1$, $x(3/2) = 1$, $x(5/2) = -1$, $x(7/2) = -1$, and $N = 2$.

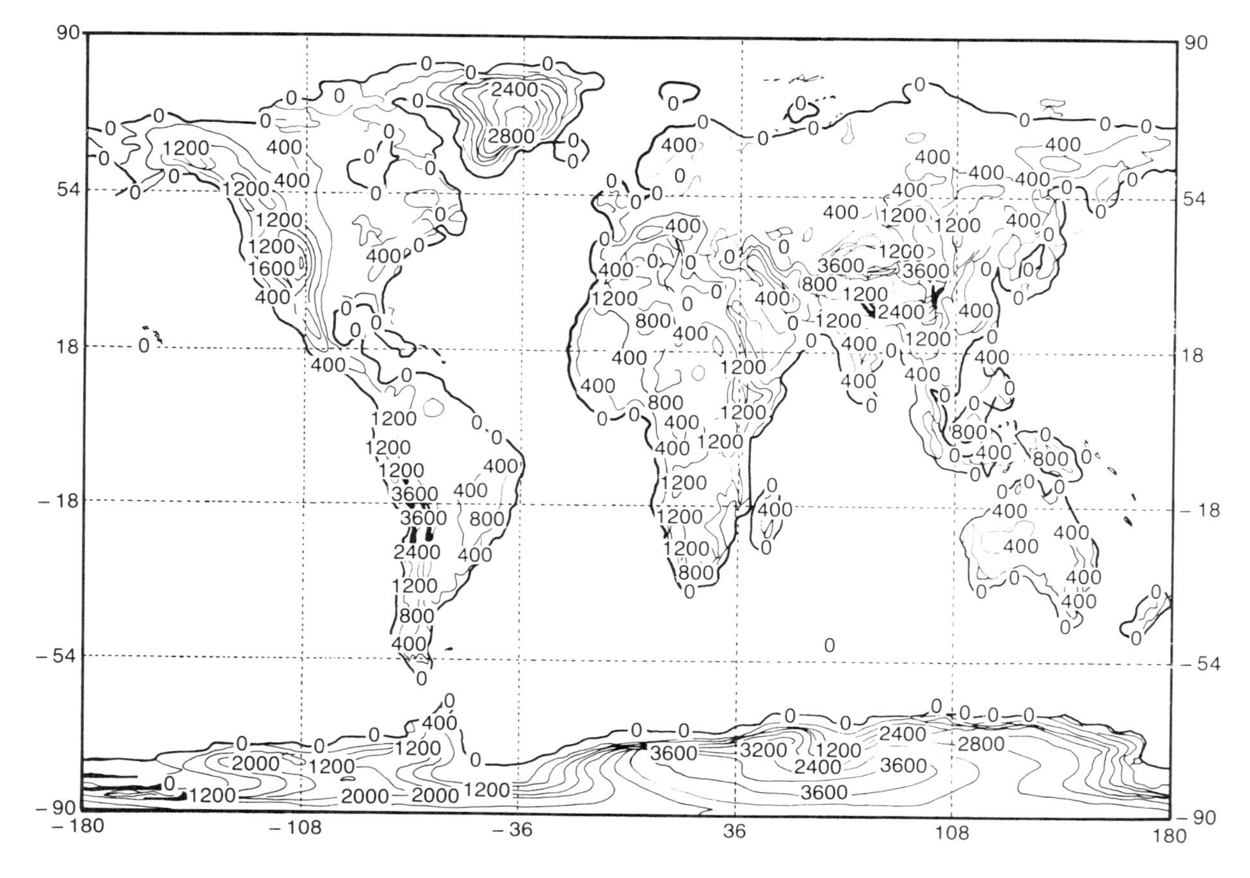

Fig. 2.11.3: The global topography field in meters given from data on 2° latitude by 2.5° longitude.

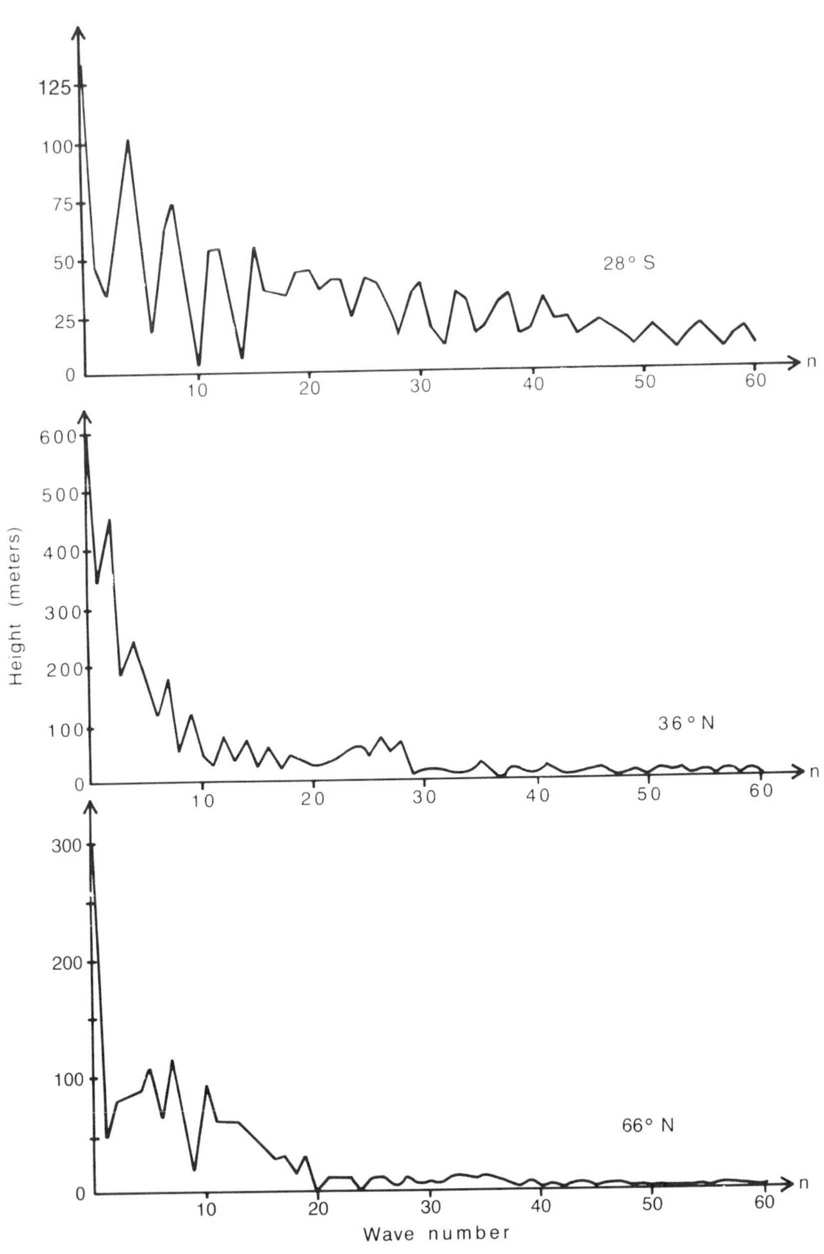

Fig. 2.11.4: The amplitude of the Fourier coefficients $(A_n^2 + B_n^2)^{1/2}$ for the earth's topography along three latitude belts for wavenumbers n = 0 to 72.

2.12 COMPLEX FOURIER SERIES

So far in our discussion, we have expressed Fourier series in terms of sines and cosines. We are now ready to reexpress Fourier series as a series of *complex* exponentials. We do this for two reasons. First, in many engineering and scientific applications of

Fourier series, the expansion of a function in terms of complex exponentials results in coefficients of considerable simplicity and clarity. Secondly, these complex Fourier series point the way to development of the Fourier integral.

For simplicity, let us introduce the variable

$$\omega_n = \frac{n\pi}{L} \quad \text{where } n = 0, \pm 1, \pm 2, \ldots \tag{2.12.1}$$

A famous formula of Euler states that

$$\cos(x) = \frac{(e^{ix} + e^{-ix})}{2} \quad \text{and} \quad \sin(x) = \frac{(e^{ix} - e^{-ix})}{2i} \tag{2.12.2}$$

where $i^2 = -1$. Consequently the Fourier series expansion becomes

$$f(x) = \frac{a_0}{2} + \sum_{n=1}^{\infty} a_n \frac{\exp(i\omega_n x) + \exp(-i\omega_n x)}{2}$$

$$+ \sum_{n=1}^{\infty} b_n \frac{\exp(i\omega_n x) - \exp(-i\omega_n x)}{2i} \tag{2.12.3}$$

$$= \frac{a_0}{2} + 1/2 \sum_{n=1}^{\infty} (a_n - ib_n) \exp(i\omega_n x)$$

$$+ 1/2 \sum_{n=1}^{\infty} (a_n + ib_n) \exp(-i\omega_n x) \tag{2.12.4}$$

If we define c_n as

$$c_n = a_n - ib_n, \tag{2.12.5}$$

then

$$c_n = a_n - ib_n = \frac{1}{L} \int_{\tau}^{\tau + 2L} f(x) \left[\cos(\omega_n x) - i \sin(\omega_n x)\right] dx \tag{2.12.6}$$

$$= \frac{1}{L} \int_{\tau}^{\tau + 2L} f(x) \exp(-i\omega_n x)\, dx \tag{2.12.7}$$

Similarly, $a_n + ib_n = c_n^*$, the complex conjugate of c_n, equals

$$c_n^* = a_n + ib_n = \frac{1}{L} \int_{\tau}^{\tau + 2L} f(x) \exp(i\omega_n x)\, dx \tag{2.12.8}$$

To simplify Eq. (2.12.4), we note that

$$\omega_{-n} = \frac{(-n)\pi}{L} = -\frac{n\pi}{L} = -\omega_n \tag{2.12.9}$$

which yields that

$$c_{-n} = \frac{1}{L}\int_{\tau}^{\tau+2L} f(x)\exp(-i\omega_{-n}x)\,dx = \frac{1}{L}\int_{\tau}^{\tau+2L} f(x)\exp(i\omega_n x)\,dx = c_n^* \tag{2.12.10}$$

so that Eq. (2.12.4) can be written

$$f(x) = \frac{a_0}{2} + 1/2\sum_{n=1}^{\infty} c_n \exp(i\omega_n x) + c_n^* \exp(-i\omega_n x) \tag{2.12.11}$$

$$= 1/2\left\{a_0 + \sum_{n=1}^{\infty} c_n \exp(i\omega_n x) + \sum_{n=1}^{\infty} c_{-n} \exp(-i\omega_n x)\right\} \tag{2.12.12}$$

Letting $n = -m$ in the second summation on the right-hand side, then

$$\sum_{n=1}^{\infty} c_{-n} e^{-i\omega_n x} = \sum_{m=-1}^{-\infty} c_m e^{-i\omega_{-m} x} = \sum_{m=-\infty}^{-1} c_m e^{i\omega_m x} = \sum_{n=-\infty}^{-1} c_n e^{i\omega_n x}. \tag{2.12.13}$$

Therefore,

$$f(x) = 1/2[a_0 + \sum_{n=1}^{\infty} c_n \exp(i\omega_n x) + \sum_{n=-\infty}^{-1} c_n \exp(\, i\omega_n x)]. \tag{2.12.14}$$

On the other hand,

$$a_0 = \frac{1}{L}\int_{\tau}^{\tau+2L} f(x)\,dx = c_0 = c_0 \exp(i\omega_0 x) \tag{2.12.15}$$

because $\omega_0 = \frac{0\pi}{L} = 0$, so that finally

$$\boxed{f(x) = 1/2\sum_{n=-\infty}^{\infty} c_n \exp(i\omega_n x)} \tag{2.12.16}$$

where

$$\boxed{c_n = \frac{1}{L}\int_{\tau}^{\tau+2L} f(x)\exp(-i\omega_n x)\,dx,\ n = 0, \pm 1, \pm 2, \ldots} \tag{2.12.17}$$

The set of coefficients c_n is often called the Fourier spectrum or simply the *spectrum* of $f(x)$.

Because $c_n = a_n - ib_n$, the c_n's for an even function will be purely real. On the other hand, the c_n's for an odd function are purely imaginary. It is important to note that we lose the advantage of even and odd functions in the sense that we cannot just integrate over the interval 0 to L and then double the result.

Summary of Complex Fourier Series

1. $$c_n = \frac{1}{L} \int_{\tau}^{\tau+2L} f(x) \exp(-\frac{in\pi x}{L})\, dx$$

where

$$f(x) = 1/2 \sum_{n=-\infty}^{\infty} c_n \exp(\frac{in\pi x}{L})$$

2. $$e^{ix} = \cos(x) + i \sin(x)$$

3. $$e^{in\pi} = (-1)^n \quad \text{and} \quad e^{2n\pi i} = 1.$$

2.13 AN EXAMPLE OF A COMPLEX FOURIER SERIES

Let us find the Fourier expansion of

$$f(x) = \begin{cases} 1 & 0<x<\pi \\ -1 & -\pi<x<0 \end{cases} \tag{2.13.1}$$

given in the full Fourier interval $(-\pi, \pi)$ using the complex Fourier series representation. With $L = \pi$ and $\tau = -\pi$,

$$\omega_n = \frac{n\pi}{L} = n. \tag{2.13.2}$$

Therefore

$$c_n = \frac{1}{\pi} \int_{-\pi}^{\pi} f(x)\, e^{-inx} dx = \frac{1}{\pi} \int_{-\pi}^{0} (-1) e^{-inx} dx + \frac{1}{\pi} \int_{0}^{\pi} (1)\, e^{-inx} dx \tag{2.13.3}$$

$$= \frac{-1}{\pi} (\frac{1}{-in})\, e^{-inx} \Big|_{-\pi}^{0} + \frac{1}{\pi} (\frac{1}{-in})\, e^{-inx} \Big|_{0}^{\pi} \quad \text{if } n \neq 0 \tag{2.13.4}$$

$$= \frac{-i}{n\pi} (1 - e^{n\pi i}) + \frac{i}{n\pi} (e^{-in\pi} - 1) \tag{2.13.5}$$

Because $\exp(n\pi i) = \cos(n\pi) + i\sin(n\pi) = (-1)^n$ and

$$\exp(-n\pi i) = \cos(-n\pi) + i\sin(-n\pi) = (-1)^n, \qquad \mathbf{(2.13.6)}$$

then

$$c_n = -\frac{2i}{n\pi}(1 - (-1)^n) = \begin{cases} 0 & n \text{ even} \\ -\dfrac{4i}{n\pi} & n \text{ odd} \end{cases} \qquad \mathbf{(2.13.7)}$$

with

$$f(x) = 1/2 \sum_{n=-\infty}^{\infty} c_n \exp(inx). \qquad \mathbf{(2.13.8)}$$

In this particular problem we must treat the case $n = 0$ specially because the integration that we preformed to obtain c_n is undefined for $n = 0$.

$$c_0 = \frac{1}{\pi}\int_{-\pi}^{\pi} f(x)\,dx = \frac{1}{\pi}\int_{-\pi}^{0} (-1)\,dx + \frac{1}{\pi}\int_{0}^{\pi} (1)\,dx \quad \mathbf{(2.13.9)}$$

$$= \frac{1}{\pi}\left(-x\Big|_{-\pi}^{0} + x\Big|_{0}^{\pi}\right) = \frac{1}{\pi}(0 - \pi + \pi - 0) = 0. \quad \mathbf{(2.13.10)}$$

Because $c_0 = 0$, the expansion for $f(x)$ can be written

$$f(x) = 1/2 \sum_{m=-\infty}^{\infty} \frac{-4i}{(2m-1)\pi} \exp\left[i(2m-1)x\right]$$

$$= \frac{-2i}{\pi} \sum_{m=\infty}^{\infty} \frac{1}{2m-1} \exp\left[i(2m-1)x\right] \qquad \mathbf{(2.13.11)}$$

because any odd integer n can be written as $2m - 1$ for any positive or negative value of m.

Exercises

1. $f(x) = e^x \quad 0<x<2$

2. $f(x) = \begin{cases} M & 0<x<\pi \\ -M & -\pi<x<0 \end{cases}$

3. $f(x) = \begin{cases} -x & -\pi \le x < 0 \\ x^2 & 0 \le x < \pi \end{cases}$

4. $f(x) = x \quad 0<x<2$

5. $f(x) = |x| \quad -\pi<x<\pi$

6. $f(x) = x^2 \quad -\pi<x<\pi$

7. $f(x) = \begin{cases} 0 & -\pi/2<x<0 \\ 1 & 0<x<\pi/2 \end{cases}$

8. $f(x) = x \quad -1<x<1$

9. $f(x) = \begin{cases} 1 & 0<x<1 \\ -1 & 1<x<2 \end{cases}$

2.14 THE FOURIER INTEGRAL

In the earlier sections of this chapter, we developed the representation of a periodic function in terms of sines and cosines with the same period. Periodic functions are fairly rare in practice and we now turn to the study of functions that are not necessarily periodic. This study will show that the method of Fourier series must be extended and the concept of the *Fourier integral* must be introduced.

To derive the Fourier integral, let us introduce a function $f_T(t)$ such that $f_T(t) = f(t)$ for $-T/2<t<T/2$ and is zero otherwise. From the complex representation of Fourier series we have

$$F(i\omega_n) = \int_{-T/2}^{T/2} f_T(t)\, e^{-i\omega_n t}\, dt \qquad (2.14.1)$$

$$f_T(t) = \frac{1}{T} \sum_{n=-\infty}^{\infty} F(i\omega_n)\, e^{i\omega_n t} \qquad (2.14.2)$$

where $\omega_n = 2\pi n/T$ and $\Delta\omega_n = \omega_{n+1} - \omega_n = 2\pi/T$. Note that the normalized factor $1/T$ has been placed in the equation for $f_T(t)$ rather than in $F(i\omega_n)$.

We may rewrite the equation for $f_T(t)$ as

$$f_T(t) = \frac{1}{2\pi} \sum_{n=-\infty}^{\infty} F(i\omega_n)\, e^{i\omega_n t}\Delta\omega_n. \qquad (2.14.3)$$

It is reasonable to suppose that as $T \to \infty$ the summation will become an integral provided the function $f(t)$ is reasonably well behaved so that we have

$$F(i\omega) = \int_{-\infty}^{\infty} f(t)\, e^{-i\omega t}\, dt \qquad (2.14.4)$$

and

$$f(t) = \frac{1}{2\pi} \int_{-\infty}^{\infty} F(i\omega)\, e^{i\omega t}\, d\omega. \qquad (2.14.5)$$

The function $F(i\omega)$ is called the spectrum or *Fourier transform* of $f(t)$, and $f(t)$ is called the *inverse Fourier transform* of $F(i\omega)$. $F(i\omega)$ *and* $f(t)$ are also called *Fourier transform pairs*. The quantity $F(i\omega)\, d\omega$ corresponds, loosely, to the c_n in the complex Fourier series.

R. W. Hamming has given the beautiful analogy between the

Fourier integral and a prism. We can regard a function $f(t)$ as if it were a light beam being broken up by a prism into a spectrum of colors—the Fourier transform breaks up the function into a spectrum of frequencies ω.

As an example, let us find the Fourier transform of the unit rectangular function

$$f(x) = \begin{cases} 1 & |x| < 1/2 \\ 0 & |x| > 1/2 \end{cases} \qquad \textbf{(2.14.6)}$$

By definition

$$F(i\omega) = \int_{-\infty}^{\infty} f(x)\, e^{-i\omega x}\, dx = \int_{-1/2}^{1/2} (1)\, e^{-i\omega x}\, dx \qquad \textbf{(2.14.7)}$$

$$= -\frac{1}{i\omega}\, e^{-i\omega x} \Bigg|_{1/2}^{1/2} = \frac{2}{\omega}\, \frac{e^{i\omega/2} - e^{-i\omega/2}}{2i}$$

$$= \frac{2}{\omega} \sin\left(\frac{\omega}{2}\right) \qquad \textbf{(2.14.8)}$$

In Fig. 2.14.1, this Fourier transform is graphed. Because it is in

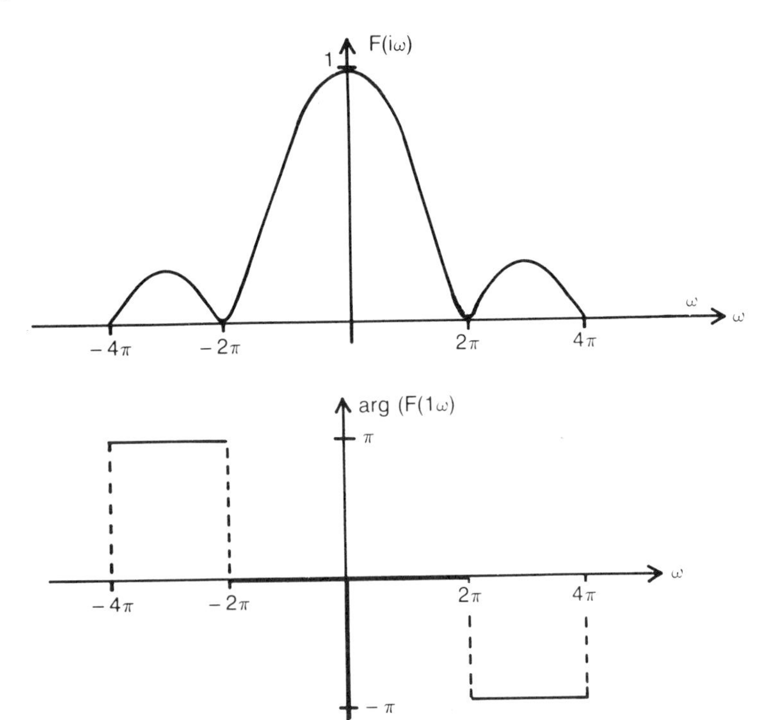

Fig. 2.14.1: The Fourier transform (both amplitude and phase) of the unit rectangular function.

general a complex number, we need both a graph of its amplitude and phase of $F(i\omega)$.

A fundamental relation involving Fourier transforms is known as the *convolution theorem*. Suppose that we have two functions $f(t)$ and $g(t)$. The convolution $h(t)$ of $f(t)$ with $g(t)$ is defined as

$$h(t) = \int_{-\infty}^{\infty} f(s) g(t-s)\, ds \quad \textbf{(2.14.9)}$$

$$= \int_{-\infty}^{\infty} f(t-s)\, g(s)\, ds \quad \textbf{(2.14.10)}$$

By definition the transform of $h(t)$ is

$$H(i\omega) = \int_{-\infty}^{\infty} f(s)\, e^{-i\omega s} \int_{-\infty}^{\infty} g(t-s)\, e^{-i(t-s)\omega} dt\, ds \quad \textbf{(2.14.11)}$$

$$= \int_{-\infty}^{\infty} f(s)\, e^{-i\omega s}\, G(i\omega)\, ds \quad \textbf{(2.14.12)}$$

$$= F(i\omega)\, G(i\omega). \quad \textbf{(2.14.13)}$$

Thus we have the result: *The transform of the convolution of two functions is the product of their transforms.*

Another important identity follows by writing Eq. (2.14.4) and (2.14.5) as one equation:

$$f(x) = \frac{1}{2\pi} \int_{-\infty}^{\infty} e^{i\omega x} \int_{-\infty}^{\infty} e^{-i\omega t} f(t)\, dt\, d\omega \quad \textbf{(2.14.14)}$$

$$= \frac{1}{2\pi} \int_{-\infty}^{\infty} \int_{-\infty}^{\infty} f(t)\, e^{-i\omega(t-x)}\, dt\, d\omega \quad \textbf{(2.14.15)}$$

$$= \frac{1}{2\pi} \int_{-\infty}^{0} \int_{-\infty}^{\infty} f(t)\, e^{-i\omega(t-x)}\, dt\, d\omega$$

$$+ \frac{1}{2\pi} \int_{0}^{\infty} \int_{-\infty}^{\infty} f(t)\, e^{-i\omega(t-x)}\, dt\, d\omega \quad \textbf{(2.14.16)}$$

In the first integral, let us replace ω by $-\omega$.

$$f(x) = \frac{1}{2\pi} \Bigg[\int_{0}^{\infty} \int_{-\infty}^{\infty} f(t)\, e^{i\omega(t-x)}\, dt\, d\omega$$

$$+ \int_{0}^{\infty} \int_{-\infty}^{\infty} f(t)\, e^{-i\omega(t-x)}\, dt\, d\omega \Bigg] \quad \textbf{(2.14.17)}$$

Using the Euler's identity that $\cos(x) = (e^{ix} + e^{-ix})/2$, we obtain

$$f(x) = \frac{1}{\pi} \int_{0}^{\infty} \int_{-\infty}^{\infty} f(t) \cos[(\omega(t-x)] \, dt \, d\omega. \qquad \textbf{(2.14.18)}$$

Equation (2.14.18) is known as the *complete Fourier integral* representation of $f(x)$ and represents $f(x)$ for all values of x if $f(x)$ is piecewise differentiable in every finite interval and if the integral $\int_{-\infty}^{\infty} |f(x)| dx$ exists. Incidently all periodic functions, except zero, violate this later condition. Fourier integrals converge to $[f(x+0) + f(x-0)]/2$.

An important result that follows from Eq. (2.14.18) is

$$f(x) = \frac{1}{\pi} \int_{0}^{\infty} A(\omega) \cos(\omega x) + B(\omega) \sin(\omega x) \, d\omega \qquad \textbf{(2.14.19)}$$

if we define

$$A(\omega) = \int_{-\infty}^{\infty} f(t) \cos(\omega t) \, dt$$

and

$$B(\omega) = \int_{-\infty}^{\infty} f(t) \sin(\omega t) \, dt. \qquad \textbf{(2.14.20)}$$

The functions $A(\omega)$ and $B(\omega)$ are called the cosine and sine transform or Fourier transforms or spectrum of the function $f(t)$. Conversely, $f(t)$ is often called the inverse Fourier transform of $A(\omega)$ and $B(\omega)$. We note in passing that if $f(t)$ is an even function, $B(\omega) = 0$. If $f(t)$ is an odd function, $A(\omega) = 0$.

$A(\omega)$ and $B(\omega)$ are particularly useful because

$$F(i\omega) = A(\omega) - i\,B(\omega). \qquad \textbf{(2.14.21)}$$

Summary of the Fourier Integral

1. In the complex domain:

$$\text{Fourier transform of } f(t) = \int_{-\infty}^{\infty} f(t)\, e^{-i\omega t} \, dt$$

$$\text{Inverse of the Fourier transform} = \frac{1}{2\pi} \int_{-\infty}^{\infty} F(i\omega)\, e^{i\omega t} \, d\omega$$

2. In the real domain:

$$f(x) = \frac{1}{\pi} \int_{0}^{\infty} A(\omega) \cos(\omega x) + B(\omega) \sin(\omega x) \, d\omega$$

where

$$A(\omega) = \int_{-\infty}^{\infty} f(t) \cos(\omega t) \, dt$$

$$B(\omega) = \int_{-\infty}^{\infty} f(t) \sin(\omega t) \, dt$$

3. Convolution:

If
$$h(t) = \int_{-\infty}^{\infty} f(s) \, g(t-s) \, ds$$

then
$$H(i\omega) = F(i\omega) \, G(i\omega).$$

4.
$$F(i\omega) = A(\omega) - i \, B(\omega).$$

2.15 AN EXAMPLE OF A FOURIER INTEGRAL: THE SAMPLING THEOREM

One of the most important results from information theory, the sampling theorem, is based upon Fourier series and the Fourier integral. (See Shannon, C. E., 1949: Communications in the Presence of Noise. *Proc. I.R.E., 37,* 10-21.) Let $f(t)$, the signal, be integrable. Then there is a Fourier integral representation for it:

$$f(t) = \frac{1}{2\pi} \int_{-\infty}^{\infty} C(i\omega) \, e^{i\omega t} \, d\omega \qquad \textbf{(2.15.1)}$$

$$C(i\omega) = \int_{-\infty}^{\infty} f(t) \, e^{-i\omega t} \, dt. \qquad \textbf{(2.15.2)}$$

A signal is called *band-limited* if its Fourier transform is zero except in a finite interval; that is, if $C(i\omega) = 0$ for $|\omega| > \Omega$ where Ω is called the cutoff frequency. If f is band limited, we can write it in the form

$$f(t) = \frac{1}{2\pi} \int_{-\Omega}^{\Omega} C(i\omega) \, e^{i\omega t} \, d\omega \qquad \textbf{(2.15.3)}$$

because $C(i\omega)$ is zero outside the interval $-\Omega < \omega < \Omega$.

Next we express $C(i\omega)$ in a Fourier series

$$C(i\omega) = 1/2 \sum_{n=-\infty}^{\infty} c_n \exp\left(\frac{in\pi\omega}{\Omega}\right) \quad -\Omega < \omega < \Omega \qquad \textbf{(2.15.4)}$$

where

$$c_n = \frac{1}{\Omega} \int_{-\Omega}^{\Omega} C(i\omega) \exp\left(-\frac{in\pi\omega}{\Omega}\right) d\omega \qquad \textbf{(2.15.5)}$$

The point of the sampling theorem is to observe that the integral for the c_n's actually is one value of $f(t)$ at a particular time. From Eq. (2.15.3) we see that

$$c_n = \frac{2\pi}{\Omega} f(-\frac{n\pi}{\Omega}). \qquad \textbf{(2.15.6)}$$

Consequently using this result in Eq. (2.15.4), we have

$$C(i\omega) = \frac{\pi}{\Omega} \sum_{n=-\infty}^{\infty} f(-\frac{n\pi}{\Omega}) \exp(\frac{in\pi\omega}{\Omega})$$

$$= \frac{\pi}{\Omega} \sum_{n=-\infty}^{\infty} f(\frac{n\pi}{\Omega}) \exp(-\frac{in\pi\omega}{\Omega}) \quad -\Omega<\omega<\Omega \qquad \textbf{(2.15.7)}$$

By utilizing Eq. (2.14.5) again, we can construct $f(t)$:

$$f(t) = \frac{1}{2\pi} \int_{-\infty}^{\infty} C(i\omega)\, e^{i\omega t} d\omega \qquad \textbf{(2.15.8)}$$

$$= \frac{1}{2\pi} \sum_{n=-\infty}^{\infty} f(\frac{n\pi}{\Omega}) \int_{-\Omega}^{\Omega} \exp(-\frac{in\pi\omega}{\Omega})\, e^{i\omega t}\, d\omega. \qquad \textbf{(2.15.9)}$$

Carrying out the integration and using the identity

$$\sin(x) = \frac{1}{2i} (e^{ix} - e^{-ix}),$$

we find

$$f(t) = \sum_{n=-\infty}^{\infty} f(\frac{n\pi}{\Omega}) \frac{\sin(\Omega t - n\pi)}{\Omega t - n\pi}. \qquad \textbf{(2.15.10)}$$

This is the main result of the sampling theorem. The band limited function may be reconstructed from the samples of $f(t)$ taken at $t = 0, \pm\pi/\Omega, \ldots$

One point must be mentioned. Strictly speaking a band-limited signal is physically impossible to obtain. For all physical signals begin at some time and are zero before that time, or they are nonzero over a finite interval. The Fourier integral for such physically attainable signals contain components at *all* frequencies, although with greatly diminished amplitudes over certain frequency ranges.

For example, the Fourier transform of a retangular pulse of width $2a$:

$$f(t) = \begin{cases} 1 & |x|<a \\ 0 & \text{otherwise} \end{cases} \qquad \textbf{(2.15.11)}$$

is

$$F(i\omega) = \frac{2a\sin(\omega a)}{\omega a}. \qquad \textbf{(2.15.12)}$$

Consequently, the Fourier integral requires all of the frequencies from negative to positive infinity to exactly give us back $f(t)$.

Because it is impossible to have systems with infinite bandwidths, the actual transform is approximated by chopping off the higher frequencies at some point in the hope that the discarded portion of the transform is negligible. To illustrate the effect of doing this, let us approximate $F(i\omega)$ by

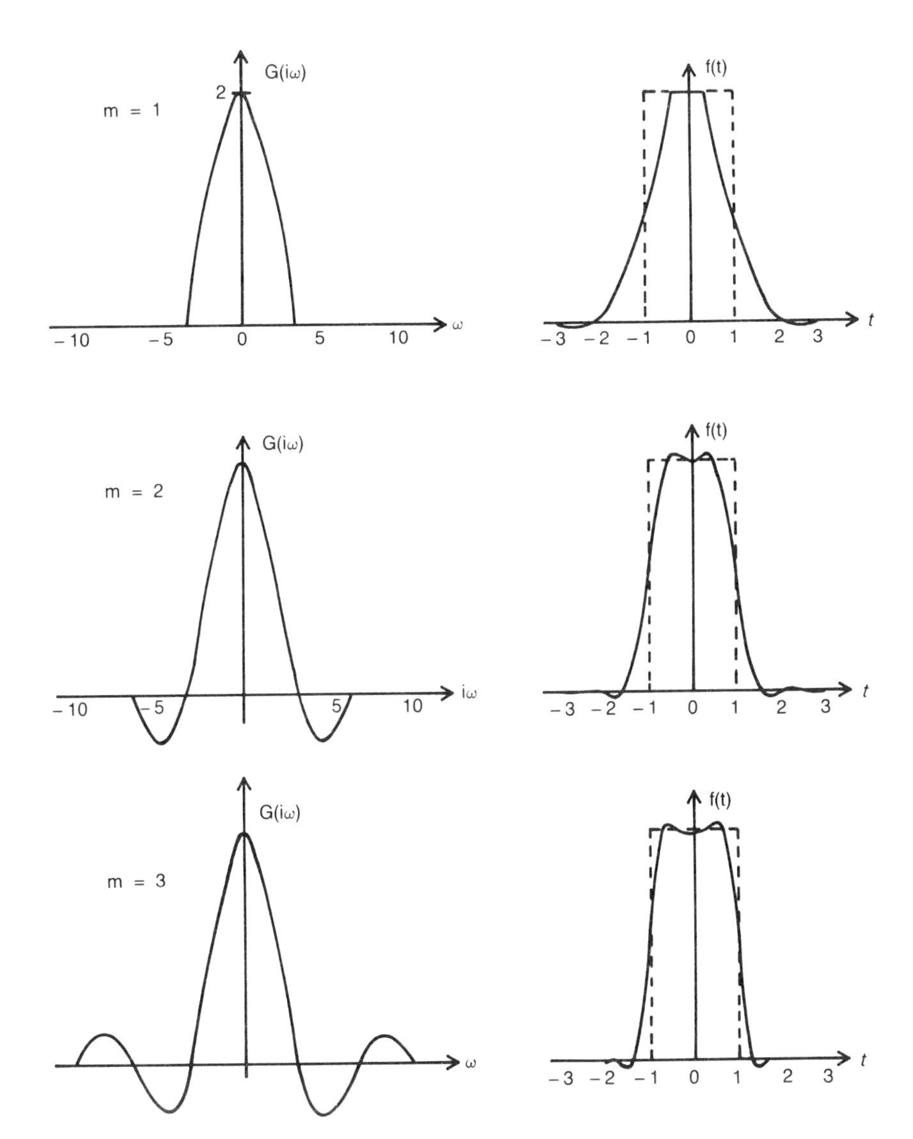

Fig. 2.15.1: The reconstruction (on the right) of a unit rectangular pulse Eq. (2.15.11) using the sampling theorem (2.15.10) where the band-limited Fourier transform is given on the left.

$$G(i\omega) = \begin{cases} 2a \dfrac{\sin(\omega a)}{\omega a} & |\omega a| \leq m\pi \\ 0 & \text{otherwise.} \end{cases} \tag{2.15.13}$$

for $m = 1$, 2, and 3. This approximate transform is shown in the left-hand column of Fig. 2.15.1. Then $f(t)$ is reconstructed using the sampling theorem for these three approximate transforms in the right-hand column of Fig. 2.15.1. For $m = 1$, there are only three nonzero coefficients, $n = -1,0,1$; for $m = 2$, $n = -2, -1,0,1,2$; and $m = 3$, $n = -3,-2,-1,0,2,3$. As expected the wider the bandwidth is, the better the reconstruction of $f(t)$.

2.16 THE SPECTRUM OF A DAMPED VIBRATION

Finding the spectrum of a damped vibration provides another example of a Fourier integral. We assume that it begins at a definite instant $t = 0$ and continues indefinitely. Consequently

$$f(t) = \begin{cases} 0 & t<0 \\ e^{-at} \sin(bt) & t>0 \end{cases} \tag{2.16.1}$$

In our problem, Eq. (2.14.18) simplifies to

$$f(x) = \frac{1}{\pi} \int_0^\infty d\omega \int_{-\infty}^\infty \sin(bt) \cos[\omega(t-x)]\, dt. \tag{2.16.2}$$

Carrying out the integration in time, we obtain

$$\begin{aligned} \phi(\omega) &= \frac{1}{\pi} \int_0^\infty e^{-at} \sin(bt) \cos(\omega t)\, dt \\ &= \frac{1}{2\pi} \left[\frac{b+\omega}{(b+\omega)^2 + a^2} + \frac{b-\omega}{(b-\omega)^2 + a^2} \right] \end{aligned} \tag{2.16.3}$$

$$\begin{aligned} \psi(\omega) &= \frac{1}{\pi} \int_0^\infty e^{-at} \sin(bt) \sin(\omega t)\, dt \\ &= \frac{1}{2\pi} \left[\frac{a}{(b-\omega)^2 + a^2} - \frac{a}{(b+\omega)^2 + a^2} \right] \end{aligned} \tag{2.16.4}$$

with

$$f(x) = \int_0^\infty \phi(\omega) \cos(\omega x) + \psi(\omega) \sin(\omega x)\, d\omega. \tag{2.16.5}$$

Thus, the Fourier representation of a damped vibration includes an infinite range of frequencies.

In Fig. 2.16.1, the integrand has been graphed for several values of a/b. We see that the smaller the damping a, the more concentrated the spectrum about the value $\omega = b$, whereas the greater the damping the broader the spectral line. Consequently, the less exact the tuning is, the greater the response will be at all frequencies. This result is important in radio communications where in order to reduce interference with foreign receivers, the transmitted signal must only be slightly damped.

Exercises

1. Find the Fourier transform

$$x(t) = \begin{cases} h \quad (1 - \dfrac{|t|}{a}) & |t| < a \\ 0 & |t| > a \end{cases}$$

and show that

$$x(t) = \frac{1}{2\pi} \int_{-\infty}^{\infty} ah \frac{\sin^2(\omega a/2)}{a^2 \omega^2} e^{i\omega t} d\omega$$

2. Let $F(\omega)$ satisfy the equation

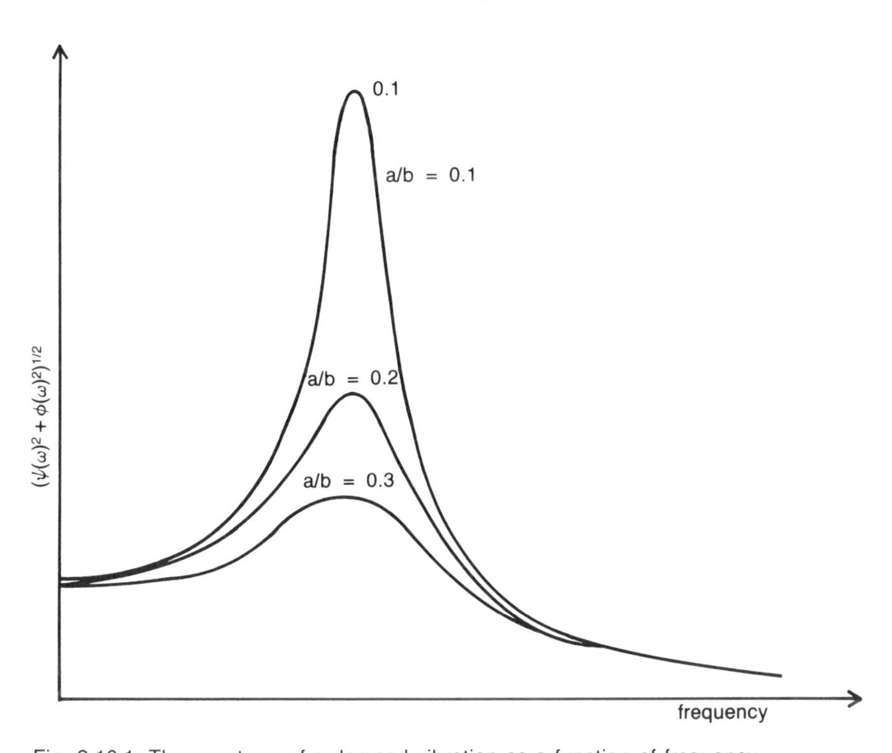

Fig. 2.16.1: The spectrum of a damped vibration as a function of frequency.

$$\frac{d^2F}{d\omega^2} + \phi(\omega)\, F(\omega) = 0$$

and let $F(\omega)$ be the Fourier transform of $f(t)$ and $\phi(\omega)$ be the transform of $x(t)$. Show that $f(t)$ satisfies the equation

$$t^2 f(t) - \int_{-\infty}^{\infty} x(u)\, f(t-u)\, du = 0$$

Find the Fourier transform of the following functions:

3. $f(x) = \begin{cases} \exp(-(1+i)x) & x>0 \\ -\exp(+(1-i)x) & x<0 \end{cases}$

4. $f(x) = e^{-x}\cos(x) \quad 0 \leq x < \infty$

$0 < x <$

5. $f(x) = \begin{cases} \sin(x) & 0 \leq x < 1 \\ 0 & \text{otherwise} \end{cases}$

6. $f(x) = \begin{cases} x & 0 \leq x < 1 \\ 0 & \text{otherwise} \end{cases}$

7. $f(x) = xe^{-x} \quad 0 \leq x \leq \infty$

Find the Fourier integral representation of

8. $f(x) = \begin{cases} (a-x)/a & 0<x<a \\ 0 & \text{otherwise} \end{cases}$

9. $f(x) = \begin{cases} 1 & 0<x<1 \\ 2-x & 1<x<2 \\ 0 & \text{otherwise} \end{cases}$

10. $f(x) = \begin{cases} 1-x & |x|<1 \\ 0 & \text{otherwise} \end{cases}$

2.17 THE SOLUTION OF LAPLACE'S EQUATION ON THE UPPER-HALF PLANE

The equation for the steady-state temperature distribution in two dimensions is $T_{xx} + T_{yy} = 0$. The same equation, called Laplace's equation describes the equilibrium (time-independent) displacement of a two-dimensional membrane. Many other physical phenomena—gravitational and electrostatic potentials, certain fluid flows—are described by this equation, thus making it one of the most important of mathematics, physics, and engineering.

We wish to solve it using Fourier integrals and convolution where the domain is the upper-half plane. We shall assume that the solution remains bounded over the entire domain and the value of the solution will be specified, $T(x,0) = f(x)$, along the x-axis. Laplace's equation can then be integrated to yield

$$\int_{-\infty}^{\infty} T_{xx}\, e^{-i\omega x}\, dx + \int_{-\infty}^{\infty} T_{yy}\, e^{-i\omega x}\, dx = 0 \qquad \textbf{(2.17.1)}$$

The first integral in Eq. (2.17.1) may be simplified by successive integration by parts and gives

$$\int_{-\infty}^{\infty} T_{xx}\, e^{-i\omega x}\, dx = T_x\, e^{-i\omega x}\Big|_{-\infty}^{\infty} + i\omega \int_{-\infty}^{\infty} T_x\, e^{-i\omega x}\, dx \qquad \textbf{(2.17.2)}$$

$$= i\omega T\, e^{-i\omega x}\Big|_{-\infty}^{\infty} - \omega^2 \int_{-\infty}^{\infty} T(x,y)\, e^{-i\omega x}\, dx \qquad \textbf{(2.17.3)}$$

$$= -\omega^2\, \hat{T}(w,y) \qquad \textbf{(2.17.4)}$$

where $\hat{T}(\omega,y) = \int_{-\infty}^{\infty} T(x,y)\, e^{-i\omega x}\, dx.$

The second integral becomes

$$\int_{-\infty}^{\infty} T_{yy}\, e^{-i\omega x}\, dx = \frac{d^2}{dy^2} \int_{-\infty}^{\infty} T(x,y)\, e^{-i\omega x}\, dx \qquad \textbf{(2.17.5)}$$

$$= \frac{d^2 \hat{T}(\omega,y)}{dy^2} \qquad \textbf{(2.17.6)}$$

along with the boundary condition:

$$F(\omega) = T(\omega,0) = \int_{-\infty}^{\infty} f(x)\, e^{-i\omega x}\, dx. \qquad \textbf{(2.17.7)}$$

Consequently we have reduced Laplace's equation, a partial differential equation, to an ordinary differential equation in y where ω is merely a parameter

$$\frac{d^2 \hat{T}}{dy^2} - \omega^2\, \hat{T} = 0 \qquad \textbf{(2.17.8)}$$

with the boundary condition $\hat{T}(\omega, 0) = F(\omega)$. The solution to Eq. (2.17.8) is

$$\hat{T}(\omega,y) = A(\omega)\, e^{|\omega| y} + B(\omega)\, e^{-|\omega| y} \quad y \geq 0. \qquad \textbf{(2.17.9)}$$

We must discard the $\exp(|\omega| y)$ term because it becomes infinite as we go to infinity along the y-axis. The boundary condition results in $B(\omega) = F(\omega)$.

Consequently

$$\hat{T}(\omega,y) = F(\omega)\, e^{-|\omega| y}. \qquad (2.17.10)$$

The inverse of the Fourier transform $\exp(-|\omega| y)$ equals

$$\frac{1}{2\pi}\int_{-\infty}^{\infty} \exp(-|\omega| y)\, e^{i\omega x}\, d\omega = \frac{1}{2\pi}\int_{-\infty}^{0} \exp(\omega y)\, e^{i\omega x}\, d\omega + \frac{1}{2\pi}\int_{0}^{\infty} \exp(-\omega y)\, e^{-i\omega x}\, d\omega \qquad (2.17.11)$$

$$= \frac{1}{2\pi}\int_{0}^{\infty} \exp(-\omega y)\, e^{-iwx}\, d\omega + \frac{1}{2\pi}\int_{0}^{\infty} \exp(-\omega y)\, e^{i\omega x}\, d\omega \qquad (2.17.12)$$

$$= \frac{1}{\pi}\int_{0}^{\infty} \exp(-\omega y)\cos(\omega x)\, d\omega \qquad (2.17.13)$$

$$= \frac{1}{\pi}\frac{e^{-\omega y}}{x^2+y^2}\Big[-y\cos(\omega x) + x\sin(\omega x)\Big]_0^{\infty} \qquad (2.17.14)$$

$$= \frac{1}{\pi}\frac{y}{x^2+y^2}. \qquad (2.17.15)$$

Furthermore, because Eq. (2.17.10) represents the convolution of two Fourier transforms, the inverse of Eq. (2.17.10) may be written

$$T(x,y) = \frac{1}{\pi}\int_{-\infty}^{\infty} \frac{y\, f(t)}{(x-t)^2+y^2}\, dt. \qquad (2.17.16)$$

Equation (2.17.16) is known as the *Poisson integral formula* for the half plane $y>0$ or the *Schwarz integral formula*.

As an example, let $T(x,0) = 1$ if $|x|<1$ and zero otherwise.

$$T(x,y) = \frac{1}{\pi}\int_{-1}^{1} \frac{y}{(x-t)^2+y^2}\, dt = \frac{1}{\pi}\left[\tan^{-1}\left(\frac{1-x}{y}\right) + \tan^{-1}\frac{x+1}{y}\right] \qquad (2.17.17)$$

Exercises

Find the solution to Laplace's equation on the upper-half plane for the following boundary conditions:

1. $T(x,0) = \begin{cases} 1 & 0<x<1 \\ 0 & \text{otherwise} \end{cases}$

2. $T(x,0) = \begin{cases} 1 & x>0 \\ -1 & x<0 \end{cases}$

3. $T(x,0) = \begin{cases} T_0 & x<0 \\ 0 & \text{otherwise} \end{cases}$

4. $T(x,0) = \begin{cases} 2T_0 & x<-1 \\ T_0 & -1<x<1 \\ 0 & x>1 \end{cases}$

5. $T(x,0) = \begin{cases} T_0 & -1<x<0 \\ (T_1 - T_0)x + T_0 & 0<x<1 \\ 0 & \text{otherwise} \end{cases}$

6. $T(x,0) = \begin{cases} V_0 & x<a_1 \\ V_1 & a_1<x<a_2 \\ V_2 & a_2<x<a_3 \\ \cdot & \\ \cdot & \\ \cdot & \\ V_n & a_n<x \end{cases}$

Answers

P. 77

1. $$f(x) = -\frac{\pi}{4} - \frac{2}{\pi}\sum_{n=1}^{\infty}\frac{\cos\,[(2n-1)\,x]}{(2n-1)^2} + \frac{2}{\pi}\sum_{n=1}^{\infty}\frac{1 - 2(-1)^n}{n}\sin(nx)$$

2. $$f(x) = \frac{\pi}{8} + \frac{4}{\pi}\sum_{n=1}^{\infty}\frac{\cos(n\pi/2)\sin^2(n\pi/4)}{n}\cos(nx) + \frac{2}{\pi}\sum_{n=1}^{\infty}\frac{\sin(n\pi/2)}{n^2}\sin(nx)$$

3. $$f(x) = -\frac{4}{\pi^2}\sum_{n=1}^{\infty}\frac{\cos[(2n-1)\pi x]}{(2n-1)^2}$$

4. $$f(x) = \frac{\sinh(L)}{L} + 2L\ \sinh(L) \sum_{n=1}^{\infty} \frac{(-1)^n}{n^2\pi^2 + L^2} \cos\left(\frac{n\pi x}{L}\right)$$
$$- 2\ \sinh(L) \sum_{n=1}^{\infty} \frac{n\pi\ (-1)^n}{n^2\pi^2 + L^2} \sin\left(\frac{n\pi x}{L}\right)$$

5. $$f(x) = \frac{1}{\pi} + 1/2\ \sin(x) + \frac{2}{\pi} \sum_{n=2}^{\infty} \frac{\cos\ [(2n-1)x]}{2n-1}$$

6. $$f(x) = \frac{L^2}{3} + \frac{4L^2}{\pi^2} \sum_{n=1}^{\infty} \frac{(-1)^n}{n^2} \cos\left(\frac{n\pi x}{L}\right)$$
$$- \frac{2L}{\pi} \sum_{n=1}^{\infty} \frac{(-1)^n}{n} \sin\left(\frac{n\pi x}{L}\right)$$

7. $$f(x) = \frac{4}{\pi^2} \sum_{n=1}^{\infty} \frac{(-1)^{n+1}}{(2n-1)^2} \sin\ [(2n-1)\pi x]$$

8. $$f(x) = \frac{4}{3\pi} \sin(x) + 1/2\ \sin(2x)$$
$$+ \frac{1}{\pi} \sum_{n=2}^{\infty} \left\{ \frac{\sin[(n-2)\pi/2]}{n-2} - \frac{\sin[(n+2)\pi/2]}{n+2} \right\} \sin(nx)$$

9. $$f(x) = a/2 - \frac{4a}{\pi^2} \sum_{n=1}^{\infty} \frac{\cos\left[\frac{(2n-1)\pi x}{a}\right]}{(2n-1)^2}$$
$$+ \frac{2a}{\pi} \sum_{n=1}^{\infty} \frac{(-1)^n}{n} \sin\left(\frac{n\pi x}{a}\right);\ a,0,0,a,0$$

10. $$f(x) = \frac{\pi - 1}{2} + \sum_{n=1}^{\infty} \frac{\sin(n\pi x)}{n\pi}$$

11. $$f(x) = -\frac{L}{2\pi} \sin\left(\frac{\pi x}{L}\right) + \frac{2L}{\pi} \sum_{n=2}^{\infty} \frac{(-1)^n\ n}{n^2 - 1} \sin\left(\frac{n\pi x}{L}\right)$$

12. $$f(x) = \frac{\pi^2}{6} + 2 \sum_{n=1}^{\infty} \frac{(-1)^n}{n^2} \cos(nx)$$
$$- \frac{1}{\pi} \sum_{n=1}^{\infty} \left[\frac{(n^2\pi^2 - 2)(-1)^n - 2}{n^3} \right] \sin(nx)$$

P. 84

1. (a) $$y(t) = -1/2 - \frac{2}{\pi} \sum_{n=1}^{\infty} \frac{\sin[(2n-1)t]}{(2n-1) + (2n-1)^3}$$

(b) $$y(t) = -1/2 - \frac{t\cos(t)}{\pi} - \frac{2}{\pi} \sum_{n=2}^{\infty} \frac{\sin[(2n-1)t]}{(2n-1)^2 - (2n-1)}$$

(c) $$y(t) = -1/2 + \sum_{n=1}^{\infty} A_n \cos[(2n-1)t] + \sum_{n=1}^{\infty} B_n \sin[(2n-1)t]$$

$$A_n = \frac{2}{(2n-1)\pi} \quad \frac{3(2n-1)}{[2 - (2n-1)^2]^2 + 9(2n-1)^2}$$

$$B_n = \frac{2}{(2n-1)\pi} \quad \frac{2 - (2n-1)^2}{[2 - (2n-1)^2]^2 + 9(2n-1)^2}$$

2. (a) $$y(t) = -\frac{\pi}{2} + \frac{4}{\pi} \sum_{n=1}^{\infty} \frac{\cos[(2n-1)t]}{(2n-1)^4 + (2n-1)^2}$$

(b) $$y(t) = \frac{\pi}{8} - \frac{4}{\pi} \sum_{n=1}^{\infty} \frac{\cos[(2n-1)t]}{4(2n-1)^2 - (2n-1)^4}$$

P. 93

1. $$f(t) = \frac{1}{4} - \frac{2}{\pi^2} \sum_{n=1}^{\infty} \frac{1}{(2n-1)^2} \cos[2(2n-1)\pi t]$$

$$f(t) = \frac{4}{\pi^2} \sum_{n=1}^{\infty} \frac{(-1)^{n+1}}{(2n-1)^2} \sin[(2n-1)\pi t]$$

2. $$f(t) = \frac{2\pi^2}{3} - 4 \sum_{n=1}^{\infty} \frac{(-1)^n}{n^2} \cos(nt)$$

$$f(t) = 2\pi \sum_{n=1}^{\infty} \frac{\sin(nt)}{n} + \frac{8}{\pi} \sum_{n=1}^{\infty} \frac{\sin[(2n-1)t]}{(2n-1)^3}$$

3. $$f(t) = \frac{3}{4} - \frac{8}{\pi^2} \sum_{n=1}^{\infty} \frac{\sin^2(n\pi/4)}{n^2} \cos(n\pi t/2)$$

$$f(t) = \frac{4}{\pi^2} \sum_{n=1}^{\infty} \frac{(-1)^{n+1}}{(2n-1)^2} \sin[(2n-1)\pi t/2] - \frac{2}{\pi} \sum_{n=1}^{\infty} \frac{(-1)^n}{n} \cos(n\pi t/2)$$

4. $$f(t) = \frac{a}{6} - \frac{4a}{\pi^2} \sum_{n=1}^{\infty} \frac{(-1)^n \sin\left[\frac{(2n-1)\pi}{6}\right]}{(2n-1)^2} \cos\left[\frac{(2n-1)\pi t}{a}\right]$$

$$f(t) = \frac{a}{\pi^2} \sum_{n=1}^{\infty} \frac{(-1)^n \sin(n\pi/3)}{n^2} \sin\left(\frac{2n\pi t}{a}\right)$$

$$- \frac{2a}{3\pi} \sum_{n=1}^{\infty} \frac{(-1)^n}{n} \sin\left(\frac{n\pi t}{a}\right)$$

5. $$f(t) = \frac{3}{4} - \frac{1}{\pi} \sum_{n=1}^{\infty} \frac{(-1)^n}{(2n-1)} \cos\left[\frac{(2n-1)\pi t}{a}\right]$$

$$f(t) = \sum_{n=1}^{\infty} \frac{1}{n\pi}\left[1 + \cos(n\pi/2) - 2\cos(n\pi)\right] \sin\left(\frac{n\pi t}{a}\right)$$

6. $$f(t) = \frac{5}{8} + \frac{2}{\pi^2} \sum_{n=1}^{\infty} \left[\frac{3\cos(n\pi/2) - 2 - (-1)^n}{n^2}\right] \cos\left(\frac{n\pi t}{a}\right)$$

$$f(t) = \sum_{n=1}^{\infty} \left[\frac{6}{n^2\pi^2} \sin(n\pi/2) - \frac{(-1)^n}{n\pi}\right] \sin\left(\frac{n\pi t}{a}\right)$$

7. $$f(t) = \frac{3a}{8} + \frac{2a}{\pi^2} \sum_{n=1}^{\infty} \left[\frac{\cos(n\pi/2) - 1}{n^2}\right] \cos\left(\frac{n\pi t}{a}\right)$$

$$f(t) = \frac{a}{\pi} \sum_{n=1}^{\infty} \left[\frac{2\sin(n\pi/2)}{n^2\pi} - \frac{(-1)^n}{n}\right] \sin\left(\frac{n\pi t}{a}\right)$$

8. $$f(t) = 1/2 + \frac{4}{\pi^2} \sum_{n=1}^{\infty} \frac{\cos[(2n-1)\pi t/a]}{(2n-1)^2}$$

$$f(t) = \frac{2}{\pi} \sum_{n=1}^{\infty} \frac{\sin(n\pi t/a)}{n}$$

9. $$f(t) = 1/2 + \frac{2}{\pi} \sum_{n=1}^{\infty} \frac{(-1)^n \sin(n\pi/2)}{n} \cos\left(\frac{2n\pi t}{a}\right)$$

$$f(t) = \frac{4}{\pi} \sum_{n=1}^{\infty} \frac{(-1)^n \sin\left[\frac{(2n-1)\pi}{4}\right]}{(2n-1)} \sin\left[\frac{(2n-1)\pi t}{a}\right]$$

10. $f(t) = \frac{a^2}{6} - \frac{a^2}{\pi^2} \sum_{n=1}^{\infty} \frac{1}{n^2} \cos\left(\frac{2n\pi t}{a}\right)$

$f(t) = -\frac{a^2}{\pi^3} \sum_{n=1}^{\infty} \frac{1}{n^3} \sin\left(\frac{2n\pi t}{a}\right)$

11. $f(t) = \frac{1}{ak}(e^{ak} - 1) + \sum_{n=1}^{\infty} \frac{2ka\,[e^{ka}(-1)^n - 1]}{k^2a^2 + n^2\pi^2} \cos\left(\frac{n\pi t}{a}\right)$

$f(t) = \sum_{n=1}^{\infty} \frac{-2n\pi\,[e^{ka}(-1)^n - 1]}{k^2a^2 + n^2\pi^2} \sin\left(\frac{n\pi t}{a}\right)$

12. $f(t) = 1/2 + \frac{2}{\pi} \sum_{n=1}^{\infty} \frac{(-1)^n}{2n-1} \cos\left[\frac{(2n-1)\pi t}{a}\right]$

$f(t) = \frac{2}{\pi} \sum_{n=1}^{\infty} \frac{\cos(n\pi/2) - (-1)^n}{n} \sin\left(\frac{n\pi t}{a}\right)$

P. 106

1. $f(t) = \frac{3}{2} - \cos(\pi x/2) - \cos(\pi x)/2 - \sin(\pi x/2)$

2. $f(t) = 2^{1/2} \sin(\pi x/2)$

P. 112

1. $\mathrm{f(t)} = \frac{e^2 - 1}{2} \sum_{n=-\infty}^{\infty} \frac{1 + in}{1 + n^2} e^{in\pi t}$

2. $f(t) = \frac{2M}{i\pi} \sum_{n=-\infty}^{\infty} \frac{1}{2n-1} e^{i(2n-1)t}$

3. $f(t) = \frac{\pi}{4} + \frac{\pi^2}{6}$

$$+ \sum_{\substack{n=-\infty \\ n\neq 0}}^{\infty} \left[\frac{(-1)^n(1+\pi)}{2n} + \frac{(-1)^n - 1 + 2(-1)^n\pi}{2\pi n^2} + \frac{i(1-(-1)^n)}{\pi n^3}\right] e^{int}$$

4. $f(t) = 1 + \frac{i}{\pi} \sum_{\substack{n=-\infty \\ n\neq 0}}^{\infty} \frac{1}{n} e^{in\pi x}$

5. $f(t) = \dfrac{\pi}{2} - \dfrac{2}{\pi} \sum_{n=-\infty}^{\infty} \dfrac{e^{i(2n-1)t}}{(2n-1)^2}$

6. $f(t) = \dfrac{\pi^2}{3} + \sum_{\substack{n=-\infty \\ n \neq 0}}^{\infty} \dfrac{4(-1)^n}{n} e^{int}$

7. $f(t) = 1/2 - \dfrac{2i}{\pi} \sum_{n=-\infty}^{\infty} \dfrac{1}{(2n-1)} e^{2i(2n-1)t}$

8. $f(t) = \dfrac{i}{\pi} \sum_{\substack{n=-\infty \\ n \neq 0}}^{\infty} \dfrac{(-1)^n}{n} e^{in\pi t}$

9. $f(t) = -\dfrac{4i}{\pi} \sum_{n=-\infty}^{\infty} \dfrac{1}{(2n-1)} e^{i(2n-1)\pi t}$

P. 121

1. $F(i\omega) = \dfrac{2[1 - \cos(\omega a)]}{\omega^2 a}$

3. $F(i\omega) = \dfrac{-2i(1+\omega)}{1 + (1+\omega)^2}$

4. $F(i\omega) = \dfrac{1}{1 + (1+\omega)^2}$

5. $F(i\omega) = -1/2 \left[\dfrac{\cos(1-\omega) - 1}{1-\omega} + \dfrac{\cos(1+\omega) - 1}{1+\omega} \right]$

$- \dfrac{i}{2} \left[\dfrac{\sin(1-\omega)}{1-\omega} - \dfrac{\sin(1+\omega)}{1+\omega} \right]$

6. $F(i\omega) = \dfrac{\cos(\omega) + \omega\sin(\omega) - 1}{\omega^2} + i \dfrac{\omega\cos(\omega) - \sin(\omega)}{\omega^2}$

7. $F(i\omega) = \dfrac{1 - \omega^2 - 2i\omega}{(1-\omega)^2 + 4\omega^2}$

8. $F(i\omega) = \dfrac{1 - \cos(\omega a)}{a\omega^2} + i \left[\dfrac{\sin(a\omega) - a\omega}{a\omega^2} \right]$

9. $$F(i\omega) = \frac{\cos(\omega) - \cos(2\omega)}{\omega^2} + i\,\frac{\sin(2\omega) - \sin(\omega) - \omega}{\omega^2}$$

10. $$F(i\omega) = 2\,\frac{\sin(\omega)}{\omega} + 2i\,\frac{\sin(\omega) - \omega\cos(\omega)}{\omega^2}$$

P. 125

1. $$T(x,y) = \frac{1}{\pi}\left[\tan^{-1}\left(\frac{1-x}{y}\right) + \tan^{-1}\left(\frac{x}{y}\right)\right]$$

2. $$T(x,y) = \frac{2}{\pi}\tan^{-1}\left(\frac{x}{y}\right)$$

3. $$T(x,y) = \frac{T_0}{\pi}\left[\frac{\pi}{2} - \tan^{-1}\left(\frac{x}{y}\right)\right]$$

4. $$T(x,y) = \frac{T_0}{\pi}\left[\pi - \tan^{-1}\left(\frac{1+x}{y}\right) + \tan^{-1}\left(\frac{1-x}{y}\right)\right]$$

5. $$T(x,y) = \frac{T_0}{\pi}\left[\tan^{-1}\left(\frac{1-x}{y}\right) + \tan^{-1}\left(\frac{1+x}{y}\right)\right]$$

$$+ \frac{T_1 - T_0}{\pi}\left\{\frac{y}{2}\,\ell n\left[\frac{(x-1)^2 + y^2}{x^2 + y^2}\right]\right.$$

$$\left. - xy\left[\tan^{-1}\left(\frac{x-1}{y}\right) - \tan^{-1}\left(\frac{x}{y}\right)\right]\right\}$$

6. $$T(x,y) = \frac{V_0}{\pi}\left[\tan^{-1}\left(\frac{a_1 - x}{y}\right) + \frac{\pi}{2}\right] + \frac{V_1}{\pi}\left[\tan^{-1}\left(\frac{a_2 - x}{y}\right)\right.$$

$$\left. + \tan^{-1}\left(\frac{x - a_1}{y}\right)\right] + \frac{V_2}{\pi}\left[\tan^{-1}\left(\frac{a_3 - x}{y}\right) + \tan^{-1}\left(\frac{x - a_2}{y}\right)\right]$$

$$+ \ . \ . \ . + \frac{V_n}{\pi}\left[\frac{\pi}{2} - \tan^{-1}\left(\frac{x - a_n}{y}\right)\right]$$

Chapter 3

Laplace Transforms

IN THE PREVIOUS CHAPTER, WE INTRODUCED THE CONCEPT of the Fourier integral. In the case when the function is nonzero only when $t>0$, we can define a similar transform, the *Laplace transform*, which shares many of the general properties with the Fourier integral. It is particularly useful in solving initial-value problems with ordinary and partial differential equations where the solutions and initial conditions have discontinuities. In later chapters we will show exactly how Laplace transforms are used in solving partial differential equations. The present chapter is devoted to developing the general properties and techniques of transform methods.

3.1 DEFINITION AND ELEMENTARY PROPERTIES

Let $f(t)$ be sectionally continuous[1] in every interval $0<t<T$. The Laplace transform of $f(t)$, written $\mathscr{L}[f(t)]$ or $F(s)$, is defined by the integral

$$\mathscr{L}[f(t)] = F(s) = \int_0^\infty f(t)\, e^{-st}\, dt \tag{3.1.1}$$

[1]A function $f(t)$ is sectionally continuous in a finite range if it is possible to divide that range into a finite number of intervals such that $f(t)$ is continuous inside each interval and approaches finite values as either end of any interval is approached from the interior.

Thus, the Laplace transform converts a function of t into a function of the transform variable s. This parameter s may be thought of as real in the elementary applications of Laplace transform; but, for the inversion theorem s must be regarded as a complex variable.

Not every function of t has a Laplace transform, because the defining integral may fail to converge. The simple functions t^{-1}, $\tan(t)$ and $\exp(t^2)$ do not possess Laplace transforms. A large class of functions that possess a Laplace transform are of "exponential order." By exponential order we mean that some constants, M and k, exist for which

$$|f(t)| \leq M e^{kt} \tag{3.1.2}$$

when $t>0$. Then, the Laplace transform of $f(t)$ certainly exists if s, or just the real part of s, is greater than k.

The direct calculation of Laplace transforms may be used to obtain the following fundamental transforms

$$\mathcal{L}(1) = \int_0^\infty e^{-st}\,dt = -\frac{e^{-st}}{s}\Big|_0^\infty = \frac{1}{s} \quad (s>0) \tag{3.1.3}$$

$$\mathcal{L}(e^{at}) = \int_0^\infty e^{at} e^{-st}\,dt = \int_0^\infty e^{-(s-a)t}dt \tag{3.1.4}$$

$$= \frac{-1}{s-a} e^{-(s-a)t}\Big|_0^\infty = \frac{1}{s-a} \quad (s>a) \tag{3.1.5}$$

$$\mathcal{L}[\sin(at)] = \int_0^\infty e^{-st} \sin(at)\,dt \Big|_0^\infty \tag{3.1.6}$$

$$= \frac{-e^{-st}}{s^2+a^2}(s \sin(at) + a\cos(at)) \tag{3.1.7}$$

$$= \frac{a}{s^2+a^2} \quad (s>0) \tag{3.1.8}$$

$$\mathcal{L}[\cos(at)] = \int_0^\infty e^{-st} \cos(at)\,dt \tag{3.1.9}$$

$$= \frac{e^{-st}}{a^2+s^2}[-s\cos(at) + a \sin(at)]\Big|_0^\infty \tag{3.1.10}$$

$$= \frac{s}{a^2+s^2} \quad (s>0) \tag{3.1.11}$$

$$\mathcal{L}(t^\alpha) = \int_0^\infty t^\alpha e^{-st}\,dt = \frac{\Gamma(\alpha+1)}{s^{\alpha+1}} \quad (s>0) \tag{3.1.12}$$

where $\Gamma(n+1)$ is the gamma function

$$\Gamma\ (n) = \int_{0}^{\infty} x^{n-1}\, e^{-x}\, dx \qquad \textbf{(3.1.13)}$$

The inequalities with regard to the parameter s are dictated by the requirement that the integrals converge.

In general the gamma function cannot be evaluated in closed form and must be computed numerically. Figure 3.1.1 presents a graph of the gamma function as a function of x. In the special case when n is a *positive integer*, $\Gamma(n) = (n-1)!$ Consequently in this special case

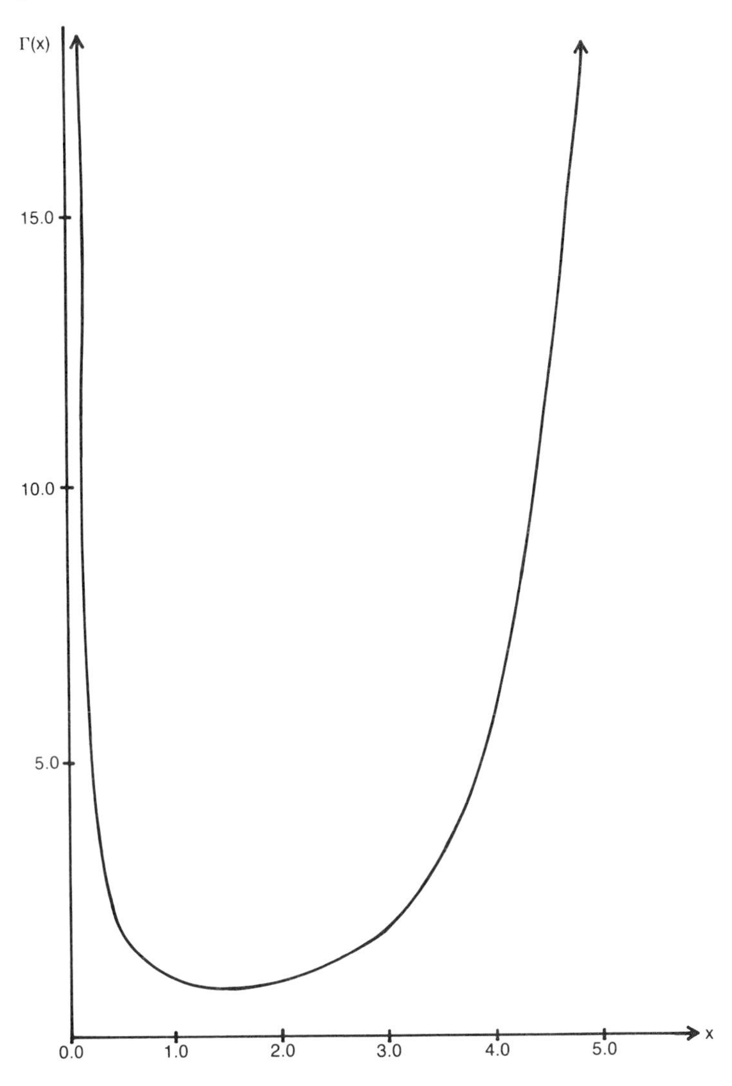

Fig. 3.1.1: Graph of the gamma function as a function of x.

$$\mathcal{L}(t^n) = \frac{n!}{s^{n+1}} . \qquad (3.1.14)$$

The Laplace transform inherits two important properties from the integral used in its definition. The first one is the transform of a sum is equal to the sum of the transforms:

$$\mathcal{L}[c_1 f(t) + c_2 g(t)] = c_1 \mathcal{L}[f(t)] + c_2 \mathcal{L}[g(t)] . \qquad (3.1.15)$$

This linearity property holds with complex numbers and functions as well.

The real virtue of the Laplace transform is revealed by its effect on derivatives. Suppose $f(t)$ is continuous and has a sectionally continuous derivative $f'(t)$. Then

$$\mathcal{L}[f'(t)] = \int_0^\infty f'(t)\, e^{-st}\, dt \qquad (3.1.16)$$

$$= e^{-st} f(t) \Big|_0^\infty + s \int_0^\infty f(t)\, e^{-st}\, dt \qquad (3.1.17)$$

by integration by parts. If $f(t)$ is of exponential order, $e^{-st} f(t)$ must go to zero as t tends to infinity (for large enough s), so that

$$\mathcal{L}[f'(t)] = s\,\mathcal{L}[f(t)] - f(0) \qquad (3.1.18)$$

Similarly, if $f(t)$ and $f'(t)$ are continuous, $f''(t)$ is sectionally continuous, and if all three functions are of exponential order, then

$$\mathcal{L}[f''(t)] = s\,\mathcal{L}[f'(t)] - f'(0) \qquad (3.1.19)$$

$$= s^2 \mathcal{L}[f(t)] - f'(0) - s\,f(0) \qquad (3.1.20)$$

In general we find

$$\boxed{\begin{aligned}\mathcal{L}[f^{(n)}(t)] &= s^n \mathcal{L}[f(t)] - s^{n-1} f(0) - \ldots \\ &\quad - s\,f^{(n-2)}(0) - f^{(n-1)}(0)\end{aligned}} \qquad (3.1.21)$$

on the assumption that $f(t)$ and its first $n-1$ derivatives are continuous, $f^{(n)}(t)$ is sectionally continuous, and all are of exponential order.

The converse of Eq. (3.1.18) is also of some importance. If

$$v(t) = \int_0^t u(\tau)\, d\tau \qquad (3.1.22)$$

then

$$\mathcal{L}[v(t)] = \int_0^\infty e^{-st} \left(\int_0^t u(\tau)\, d\tau \right) dt \qquad (3.1.23)$$

$$\mathscr{L}[v(t)] = \frac{e^{-st}}{-s} \int_0^t u(\tau)\, d\tau \Bigg|_0^\infty + \frac{1}{s} \int_0^\infty e^{-st} f(t)\, dt \qquad \textbf{(3.1.24)}$$

$$\boxed{\mathscr{L}\left[\int_0^t u(\pi)\, d\pi\right] = \frac{\overline{u}(s)}{s}} \qquad \textbf{(3.1.25)}$$

where the bar denotes a Laplace transform and it is assumed that $v(0) = 0$.

One of the great advantages of Laplace transforms lies in the ease with which it deals with discontinuous time functions. We now introduce two discontinuous functions that will be shown to be of considerable practical and theoretical significance.

The *Heaviside step function* is defined as

$$H(t) = \begin{cases} 1 & t>0 \\ 0 & t<0 \end{cases} \qquad \textbf{(3.1.26)}$$

From this definition, we have

$$\mathscr{L}[H(t)] = \int_0^\infty H(t)\, e^{-st}\, dt = \int_0^\infty e^{-st}\, dt = \frac{1}{s} \quad (s>0) \qquad \textbf{(3.1.27)}$$

It should be noticed immediately that this transform is the one that we derived for the transform of unity. This should not surprise us. As pointed out earlier, it is always understood that $f(t)$ is zero for all $t<0$. Consequently, to avoid confusion we shall generally associate the transform $1/s$ with the Heaviside step function.

Heaviside's step function is essentially a bookkeeping device that gives us the ability to "switch on" or "switch off" a given function. For example, if we wanted a function $f(t)$ to become nonzero at time $t = a$, this process is represented by the product

$$H(t-a)\, f(t).$$

On the other hand, if we only wanted the function to be "turned on" when $a<t<b$, the desired expression is then

$$f(t)\, [H(t-a) - H(t-b)].$$

For $t<a$, both step functions in the brackets have the value of zero. Between $t = a$ and $t = b$, the first has the value of unity and the second zero, so that we have $f(t)$; while for $t<b$, both step functions are equal to unity, so that their difference is zero.

The second discontinuous function is the *unit impulse* or *Dirac delta function*. From a mathematical point of view, it is a highly irregular function, but its use under the proper conditions has been justified. It is certainly an extremely useful function and has fundamental significance where the Laplace transform is concerned.

The impulse function is defined by

$$\delta(t-a) = \begin{matrix} \infty & t \neq a \\ 0 & t = a \end{matrix} \quad \textbf{(3.1.28)}$$

and

$$\int_{a-\epsilon}^{a+\epsilon} \delta\,(t-a)\,dt = 1 \quad \textbf{(3.1.29)}$$

where ϵ is any positive number, however small.

The delta function is best regarded as the limit of the rectangular pulse:

$$\delta(t-a) = \lim_{\epsilon \to 0} \begin{cases} 1/\epsilon & a - \epsilon/2 < t < a + \epsilon/2 \\ 0 & \text{otherwise} \end{cases} \quad \textbf{(3.1.30)}$$

The pulse is of width ϵ, height $1/\epsilon$ and centered at $t = a$, so that its area is unity. Now as this pulse shrinks in width, we require that its height increases so that it remains centered at $t = a$ and its area equals unity. If we continue this process, always keeping the area unity and the pulse symmetric about $t = a$, eventually an extremely narrow, very large amplitude pulse at $t = a$ will be obtained. If we proceed to the limit, where the width approaches zero and the height approaches infinity (but still with unit area), we obtain the delta function $\delta(t-a)$.

To find its Laplace transform, we calculate

$$\mathscr{L}[\delta(t-a)] = \int_0^\infty \delta(t-a)\, e^{-st}\, dt$$

$$= \lim_{\epsilon \to 0} \frac{1}{\epsilon} \int_{a-\epsilon/2}^{a+\epsilon/2} e^{-st}\, dt \quad \textbf{(3.1.31)}$$

$$= \lim_{\epsilon \to 0} -\frac{1}{s\epsilon} \left(e^{-as-\epsilon s/2} - e^{-as+\epsilon s/2}\right) \quad \textbf{(3.1.32)}$$

$$= e^{-as} \lim_{\epsilon \to 0} \left[-\frac{1}{s\epsilon} \left(1 - \epsilon s/2 + \frac{1}{8}\epsilon^2 s^2 - 1 - \epsilon s/2 - \frac{1}{8}\epsilon^2 s^2 - \ldots \right) \right] \quad \textbf{(3.1.33)}$$

$$= e^{-as} \quad \textbf{(3.1.34)}$$

In the special case when $a = 0$, $\mathscr{L}[\delta(t)] = 1$. This property of $\delta(t)$ will make the delta function very important in subsequent sections.

If we integrate the impulse function, we find that

$$\int_0^t \delta(t'-a)\,dt' = \begin{cases} 0 & t<a \\ 1 & t>a \end{cases} \tag{3.1.35}$$

according to whether the impulse does or does not come within the range of integration. This integral gives a result that is precisely the definition of the Heaviside step function:

$$\int_0^t \delta(t'-a)\,dt' = H(t-a) \tag{3.1.36}$$

so that the delta function can be recognized as the derivative of the step function:

$$\frac{d}{dt}(H(t-a)) = \delta(t-a) \tag{3.1.37}$$

Actually, the derivative can hardly be said to exist at a point of obvious discontinuity. What we have done is to extend the meaning of the derivative.

Exercises

Obtain the Laplace transform of the following functions from the definition by integration:

1. $e^{-t}\cos(5t)$
2. $t\,e^{-t}$
3. $t\sin(at)$
4. $\sinh(at)$
5. $\cosh(at)$
6. $t\,(t+1)\,(t+2)$
7. $t^2\cos(t)$
8. $\sin^3(t)$
9. $\begin{cases} t & 0<t<a \\ a\,e^{-(t-a)} & t>a \end{cases}$

10. Given
$$v(t) = \begin{cases} \sin(t) & 0<t<\pi \\ 0 & t>\pi \end{cases}$$

show that

$$v(t) = \sin(t) + \sin(t-\pi)\,H(t-\pi)$$

and find its transform.

11. Write

$$v(t) = \begin{cases} t & 0<t<a \\ 0 & t>a \end{cases}$$

in a form involving Heaviside step functions and find its transform.

3.2 SOME USEFUL THEOREMS

Although at first sight there would appear to be a bewildering number of transforms to either memorize or tabulate, there are several useful theorems which can reduce the amount of work. The first of these is called the *first shifting theorem*.

Consider the transform of the function $e^{-bt} f(t)$. Then by definition,

$$\mathcal{L}[e^{-bt} f(t)] = \int_0^\infty e^{-bt} e^{-st} f(t)\, dt = \int_0^\infty e^{-(s+b)t} f(t)\, dt$$

$$\boxed{\mathcal{L}(e^{-bt} f(t)) = F(s+b)} \qquad (3.2.1)$$

That is, *if $F(s)$ is the transform of $f(t)$ and b is a constant, then $F(s+b)$ is the transform of $e^{-bt} f(t)$.*

For example, this formula can be used to compute the transform of $e^{-at} \sin(bt)$:

$$\mathcal{L}[e^{-at} \sin(bt)] = \frac{b}{(s+a)^2 + b^2} \qquad (3.2.2)$$

where we have simply replaced s by $s+a$ in the transform for $\sin(bt)$.

The *second shifting theorem* states that *if $F(s)$ is the transform of $f(t)$, then $e^{-bs} F(s)$ is the transform of $f(t-b)\, H(t-b)$, where b is real and positive.* To show this, consider the Laplace transform of $f(t-b)\, H(t-b)$:

$$\mathcal{L}[f(t-b)\, H(t-b)] = \int_0^\infty f(t-b)\, H(t-b)\, e^{-st}\, dt \qquad (3.2.3)$$

$$= \int_b^\infty e^{-st} f(t-b)\, dt = \int_0^\infty e^{-bs} e^{-sx} f(x) dx \qquad (3.2.4)$$

$$= e^{-bs} \int_0^\infty e^{-sx} f(x)\, dx$$

$$\boxed{\mathcal{L}[f(t-b)\, H(t-b)] = e^{-bs} F(s)} \qquad (3.2.5)$$

where we have used the definitions that $H(t-b) = 0$ if $t<b$ and $x = t-b$. This theorem is of fundamental importance because it allows us to write down the transforms for "delayed" time functions. That is, functions which "turn on" b units after the initial time.

For example, let us find the inverse of the transform $\dfrac{1 - e^{-s}}{s}$.

Because $\dfrac{1 - e^{-s}}{s} = \dfrac{1}{s} - \dfrac{e^{-s}}{s}$, we have

$$\mathscr{L}^{-1}\left(\frac{1}{s} - \frac{e^{-s}}{s}\right) = \mathscr{L}^{-1}\left(\frac{1}{s}\right)\mathscr{L}^{-1}\left(\frac{e^{-s}}{s}\right) = H(t) - H(t-1) \quad \textbf{(3.2.6)}$$

In addition to these shifting theorems, there are two theorems involving the derivative and integral of the transform $F(s)$. For example, if we write

$$F(s) = \mathscr{L}[f(t)] = \int_0^\infty f(t)\, e^{-st}\, dt \qquad \textbf{(3.2.7)}$$

and differentiate with respect to s, then

$$F'(s) = \int_0^\infty -t\, f(t)\, e^{-st}\, dt = -\mathscr{L}[t\, f(t)]. \qquad \textbf{(3.2.8)}$$

Also

$$\int_s^\infty F(z)\, dz = \int_s^\infty \left[\int_0^\infty f(t)\, e^{-zt}\, dt\right] dz \qquad \textbf{(3.2.9)}$$

Upon interchanging the order of integration,

$$\int_s^\infty F(z)\, dz = \int_0^\infty f(t) \left(\int_s^\infty e^{-zt}\, dz\right) dt$$

$$= \int_0^\infty f(t) \left.\frac{e^{-zt}}{-t}\right|_s^\infty dt = \int_0^\infty \frac{f(t)}{t}\, e^{-st} dt$$

$$\boxed{\int_s^\infty F(z)dz = \mathscr{L}\left(\frac{f(t)}{t}\right)} \qquad \textbf{(3.2.10)}$$

To illustrate these theorems, let us find the transform of $t \sin(at)$ and $\frac{1 - \cos(at)}{t}$. First off,

$$\mathscr{L}\{t \sin(at)\} = -\frac{d}{ds}\mathscr{L}\{\sin(at)\} = -\frac{d}{ds}\left(\frac{a}{s^2 + a^2}\right).$$

$$= \frac{2as}{(s^2 + a^2)^2} \qquad \textbf{(3.2.11)}$$

To solve the second problem, we have from Eq. (3.2.10)

$$\mathscr{L}\left(\frac{1 - \cos(at)}{t}\right) = \int_s^\infty \mathscr{L}[1 - \cos(at)]\, dz \qquad \textbf{(3.2.12)}$$

$$= \int_s^\infty \frac{1}{z} - \frac{z}{z^2 + a^2}\, dz \qquad \textbf{(3.2.13)}$$

$$= \ln(z) - 1/2 \ln(z^2 + a^2) \Big|_s^\infty \qquad \textbf{(3.2.14)}$$

$$\mathcal{L}\left(\frac{1-\cos(at)}{t}\right) = \ln\left(\frac{z}{\sqrt{z^2+a^2}}\right)\Bigg|_s^\infty = \ln(1) - \ln\left(\frac{s}{\sqrt{s^2+a^2}}\right) \quad (3.2.15)$$

$$= -\ln\left(\frac{s}{\sqrt{s^2+a^2}}\right) \quad (3.2.16)$$

Exercises

Find the Laplace transform of the following functions:

1. $t^2\, H(t-1)$
2. $\cos(t)\, H(t-2)$
3. $\cos\,(t-1)\, H(t-1)$
4. $(t^2-1)\, H(t-1)$
5. $e^{2t}\, H(t-2)$
6. $t\int_0^t e^{-3\tau}\sin(2\tau)\, d\tau$
7. $t\, e^{-3t}\sin(2t)$
8. $\int_0^t \frac{\exp(\tau)-\cos(2\tau)}{\tau}\, d\tau$
9. $\frac{\exp(2t)-1}{t}$
10. $\frac{\exp(-3t)\sin(2t)}{t}$

Find the inverse of the following transform:

11. $\frac{1}{(s+2)^2}$
12. $\frac{1}{s\,(s+2)^2}$
13. $\frac{e^{-2s}}{s^2+4}$
14. $\frac{e^{-3s}}{s^2-9}$
15. $\frac{e^{-s}}{(s+1)^3}$
16. $\frac{e^{-s}+e^{-2s}}{s^2-3s+2}$

3.3 THE LAPLACE TRANSFORM OF A PERIODIC FUNCTION

Periodic functions frequently occur in engineering problems and we shall now show how to calculate their transform. This is in sharp contrast with the Fourier transform where the transform of a periodic function does not exist because the integral does not converge as $t \to \infty$.

When we speak of a periodic function in connection with the Laplace transform, we mean a function $f(t)$ with the properties that

$$\begin{aligned} &(1)\ f(t) \text{ is defined for } t>0 \\ &(2)\ f(t+T) = f(t) \text{ for } t>0 \\ \text{and}\quad &(3)\ f(t) = 0 \text{ for } t<0 \end{aligned} \quad (3.3.1)$$

where T is the period of the function $f(t)$. If we define $x(t)$ as

$$x(t) = f(t) \qquad 0<t<T$$
$$x(t) = 0 \qquad \text{otherwise,} \tag{3.3.2}$$

then we shall show that the Laplace transform $F(s)$ of $f(t)$ may be written in terms of the Laplace transform $X(s)$ of $x(t)$.

By definition

$$F(s) = \int_0^\infty f(t)\, e^{-st}\, dt \tag{3.3.3}$$

$$= \int_0^T f(t)\, e^{-st}\, dt + \int_T^{2T} f(t)\, e^{-st}\, dt + \ldots$$

$$+ \int_{kT}^{(k+1)T} f(t)\, e^{-st}\, dt + \ldots \tag{3.3.4}$$

Now let $z = t - kT$ where $k = 0, 1, 2, 3, \ldots$ in the kth integral and $F(s)$ becomes

$$F(s) = \int_0^T f(z)\, e^{-sz}\, dz + \int_0^T f(z+T)\, e^{-s(z+T)}\, dz + \ldots$$

$$+ \int_0^T f(z+kT)\, e^{-s(z+kT)}\, dz + \ldots \tag{3.3.5}$$

However,

$$x(z) = f(z) = f(z+T) = \ldots = f(z+kT) = \ldots \tag{3.3.6}$$

because the range of integration in each integral is from 0 to T. Thus, $F(s)$ becomes

$$F(s) = \int_0^T x(z)\, e^{-sz}\, dz + e^{-sT} \int_0^T x(z)\, e^{-sz}\, dz$$

$$+ \ldots + e^{-ksT} \int_0^T x(z)\, e^{-sz}\, dz + \ldots \tag{3.3.7}$$

or

$$F(s) = X(s)\,(1 + e^{-sT} + e^{-2sT} + e^{-3sT} + \ldots + e^{-ksT} + \ldots) \tag{3.3.8}$$

The coefficients of $X(s)$ is a geometric series of ratios e^{-sT} and $|e^{-sT}| < 1$. Hence

$$\boxed{F(s) = \frac{X(s)}{1 - e^{-sT}}\,.} \tag{3.3.9}$$

For example, let us find the Laplace transform of the square wave with period T:

$$f(t) = \begin{cases} h & 0<t< T/2 \\ -h & T/2 <t<T \end{cases} \quad \textbf{(3.3.10)}$$

By definition, $x(t)$ is given by

$$x(t) = \begin{cases} h & 0<t< T/2 \\ -h & T/2<t<T \\ 0 & \text{otherwise} \end{cases} \quad \textbf{(3.3.11)}$$

$$X(s) = \int_0^{\infty} x(t)\, e^{-st}\, dt = \int_0^{T/2} h\, e^{-st}\, dt + \int_{T/2}^{T} (-h)\, e^{-st}\, dt \quad \textbf{(3.3.12)}$$

$$= \frac{h}{s} (1 - 2\, e^{-sT/2} + e^{-sT}) = \frac{h}{s} (1 - e^{-sT/2})^2 \quad \textbf{(3.3.13)}$$

and

$$F(s) = \frac{\frac{h}{s} (1 - e^{-sT/2})^2}{(1 - e^{-sT})} = \frac{\frac{h}{s} (1 - e^{-sT/2})}{1 + e^{-sT/2}}. \quad \textbf{(3.3.14)}$$

If we multiply numerator and denominator by exp $(sT/4)$ and recall that

$$\tanh(u) = \frac{e^u - e^{-u}}{e^u + e^{-u}} \quad \textbf{(3.3.15)}$$

we have

$$F(s) = \frac{h}{s} \tanh(sT/4). \quad \textbf{(3.3.16)}$$

As a second example, let us find the Laplace transform of the periodic function

$$f(t) = \begin{cases} \sin\left(\frac{2\pi t}{T}\right) & 0<t<T/2 \\ 0 & T/2<t<T \end{cases} \quad \textbf{(3.3.17)}$$

Here

$$x(t) = \begin{cases} \sin\left(\frac{2\pi t}{T}\right) & 0<t< T/2 \\ 0 & \text{otherwise} \end{cases} \quad \textbf{(3.3.18)}$$

$$X(s) = \int_0^{T/2} \sin\left(\frac{2\pi t}{T}\right) e^{-st}\, dt$$

$$= \frac{2\pi T}{s^2T^2 + 4\pi^2}\left(1 + e^{-sT/2}\right) \qquad \textbf{(3.3.19)}$$

Hence

$$F(s) = \frac{X(s)}{1 - e^{-sT}} = \frac{2\pi T}{s^2T^2 + 4\pi^2}\,\frac{1 + e^{-sT/2}}{1 - e^{-sT}} \qquad \textbf{(3.3.20)}$$

$$= \frac{2\pi T}{s^2T^2 + 4\pi^2}\,\frac{1}{1 - e^{-sT/2}} \qquad \textbf{(3.3.21)}$$

Exercises

Find the Laplace transform of the periodic functions whose definition in one period are

1. $$f(t) = \sin(t) \qquad 0 \leq t \leq \pi$$

2. $$f(t) = \begin{cases} \sin(t) & 0 \leq t \leq \pi \\ 0 & \pi \leq t \leq 2\pi \end{cases}$$

3. $$f(t) = \begin{cases} t & 0 < t < a \\ 0 & a < t < 2a \end{cases}$$

4. $$f(t) = \begin{cases} 1 & 0 < t < a \\ 0 & a < t < 2a \\ -1 & 2a < t < 3a \\ 0 & 3a < t < 4a \end{cases}$$

Summary of the Laplace Transforms

1. Definition

$$\mathcal{L}\{f(t)\} = \int_0^{\infty} f(t)\, e^{-st}\, dt$$

2. Some fundamental transforms

$$\mathcal{L}(e^{at}) = \frac{1}{s - a}$$

$$\mathcal{L}[\sin(at)] = \frac{a}{s^2 + a^2}$$

$$\mathcal{L}[\cos(at)] = \frac{s}{s^2 + a^2}$$

$$\mathscr{L}(t^n) = \frac{n!}{s^{n+1}} \quad \text{where } n \text{ is a non-negative integer}$$

$$\mathscr{L}[H(t)] = \frac{1}{s}$$

$$\mathscr{L}[\delta(t)] = 1$$

3. $\mathscr{L}[f^{(n)}(t)] = s^n \mathscr{L}[f(t)] - s^{n-1} f(0) - \ldots - s f^{(n-2)}(0) - f^{(n-1)}(0).$

4. $$\mathscr{L}\left[\int_0^t u(\tau)\, d\tau\right] = \frac{\mathscr{L}[u(t)]}{s}$$

5. First shifting theorem

$$\mathscr{L}[e^{-bt} f(t)] = F(s+b)$$

6. Second shifting theorem

$$\mathscr{L}[f(t-b)\, H(t-b)] = e^{-bs} F(s) \qquad b>0$$

7. $$\mathscr{L}\left[\frac{f(t)}{t}\right] = \int_s^\infty F(z)\, dz$$

8. $$\mathscr{L}[t\, f(t)] = -\frac{dF(s)}{ds}$$

9. Laplace transform of a periodic function

$$F(s) = \frac{X(s)}{1 - e^{-sT}}$$

where $X(s) = \int_0^T e^{-st} x(t)\, dt$ and $x(t)$ describes the periodic function over one period.

3.4 THE SOLUTION OF ORDINARY DIFFERENTIAL EQUATIONS WITH PERIODIC FORCING

In Section 2.5 we showed how to solve an ordinary differential equation with periodic forcing by Fourier series. We expanded the forcing function $f(t)$ in a Fourier series and determined the particular solution for each particular Fourier component of $f(t)$ and then summed over all the harmonics. In this section we shall show how Laplace transforms may be used for the same purpose.

Consider the simple harmonic oscillator

$$y'' + \omega^2 y = f(t) \qquad \textbf{(3.4.1)}$$

subject to the initial conditions $y(0) = y'(0) = 0$ and the forcing

$$f(t) = \begin{cases} \dfrac{2t}{T} & 0 \le t \le T/2 \\ 2(1 - \dfrac{t}{T}) & T/2 \le t \le T \end{cases} \quad (3.4.2)$$

$$f(t) = f(t+T).$$

The forcing corresponds to a triangle wave.

The Laplace transform of both sides of Eq. (3.4.1) is

$$s^2\, \overline{y}(s) + \omega^2\, \overline{y}(s) = F(s) \quad (3.4.3)$$

where

$$F(s) = \frac{\int_0^{T/2} \frac{2t}{T} e^{-st}\, dt + \int_{T/2}^{T} 2(1 - \frac{t}{T}) e^{-st}\, dt}{1 - e^{-sT}} \quad (3.4.4)$$

$$= \frac{\frac{2}{T}\left[\frac{e^{-st}}{s^2}(-st-1)\right]_0^{T/2} - \frac{2}{s} e^{-st}\Big|_{T/2}^{T}}{1 - e^{-sT}} - \frac{\frac{2}{T}\left[\frac{e^{-st}}{s^2}(-st-1)\right]_{T/2}^{T}}{1 - e^{-sT}}$$

$$(3.4.5)$$

$$= \frac{\frac{2}{T}\left[\frac{e^{-sT/2}}{s^2}(-\frac{sT}{2} - 1) + \frac{1}{s^2}\right] - \frac{2}{s}(e^{-sT} - e^{-sT/2})}{1 - e^{-st}}$$

$$+ \frac{2(1 - sT)e^{-sT} + 2e^{-sT/2}}{s^2T(1 - e^{-sT})} \quad (3.4.6)$$

$$= \frac{2}{s^2T}\,\frac{1 - 2e^{-sT/2} + e^{-sT}}{1 - e^{-sT}} = \frac{2}{s^2T}\,\frac{1 - e^{-sT/2}}{1 + e^{-sT/2}} \quad (3.4.7)$$

or

$$F(s) = \frac{2}{s^2T}\left[1 + 2\sum_{n=1}^{\infty} (-1)^n \exp -(\frac{nsT}{2})\right]$$

$$= \frac{2}{s^2T} \tanh(sT/4) \quad (3.4.8)$$

Solving for $\overline{y}(s)$, we have

$$\overline{y}(s) = \frac{F(s)}{s^2 + \omega^2} = \frac{2}{s^2(s^2 + \omega^2)T}\left[1 + 2\sum_{n=1}^{\infty} (-1)^n \exp(-\frac{nsT}{2})\right]$$

$$(3.4.9)$$

$$\bar{y}(s) = \frac{2}{T\omega^2}\left(\frac{1}{s^2} - \frac{1}{s^2+\omega^2}\right)\left[1 + 2\sum_{n=1}^{\infty}(-1)^n \exp\left(-\frac{nsT}{2}\right)\right]$$

(3.4.10)

Employing the shifting theorem, we can invert Eq. (3.4.10) to obtain

$$y(t) = \frac{2}{T\omega^2}\left[x(t) + 2\sum_{n=1}^{\infty}(-1)^n x(t - nT/2)\, H(t - nT/2)\right]$$

(3.4.11)

where

$$x(t) = t - \frac{1}{\omega}\sin(\omega t). \qquad \textbf{(3.4.12)}$$

This solution has been graphed in Fig. 3.4.1.

Equation (3.4.12) is especially advantageous for relatively small values of t, but it is not satisfactory for $t >> T$ because of the large number of terms that must be retained in the series. To find an inverse for large time, we make use of the formula

$$\tanh(x) = 2x \sum_{n=1,3,5,\ldots}^{\infty} [(n\pi/2)^2 + x^2]^{-1} \qquad \textbf{(3.4.13)}$$

(Gradshteyn, I. S. and I. M. Ryzhik, 1965: *Table of Integrals, Series, and Products*. New York: Academic Press. Formula 1.421.2.) Eq. (3.4.8) can then be written as

$$F(s) = \frac{16}{T^2}\sum_{n=1,3,5,\ldots}^{\infty} \frac{1}{s}\left(\frac{1}{s^2+\omega^2}\right)\left(\frac{1}{s^2+\omega_s^2}\right) \qquad \textbf{(3.4.14)}$$

where $\omega_s = 2\pi n/T$. Therefore,

$$y(s) = \frac{16}{T^2}\sum_{n=1,3,5,\ldots}^{\infty} \frac{1}{s}\,\frac{1}{\omega_s^2-\omega^2}\left(\frac{1}{s^2+\omega^2} - \frac{1}{s^2+\omega_s^2}\right)$$

(3.4.15)

which yields the inverse

$$y(t) = \frac{16}{T^2}\sum_{n=1,3,5,\ldots}^{\infty} \frac{1}{\omega_s^2-\omega^2}\left(\frac{1-\cos(\omega t)}{\omega^2} - \frac{1-\cos(\omega_s t)}{\omega_s^2}\right)$$

(3.4.16)

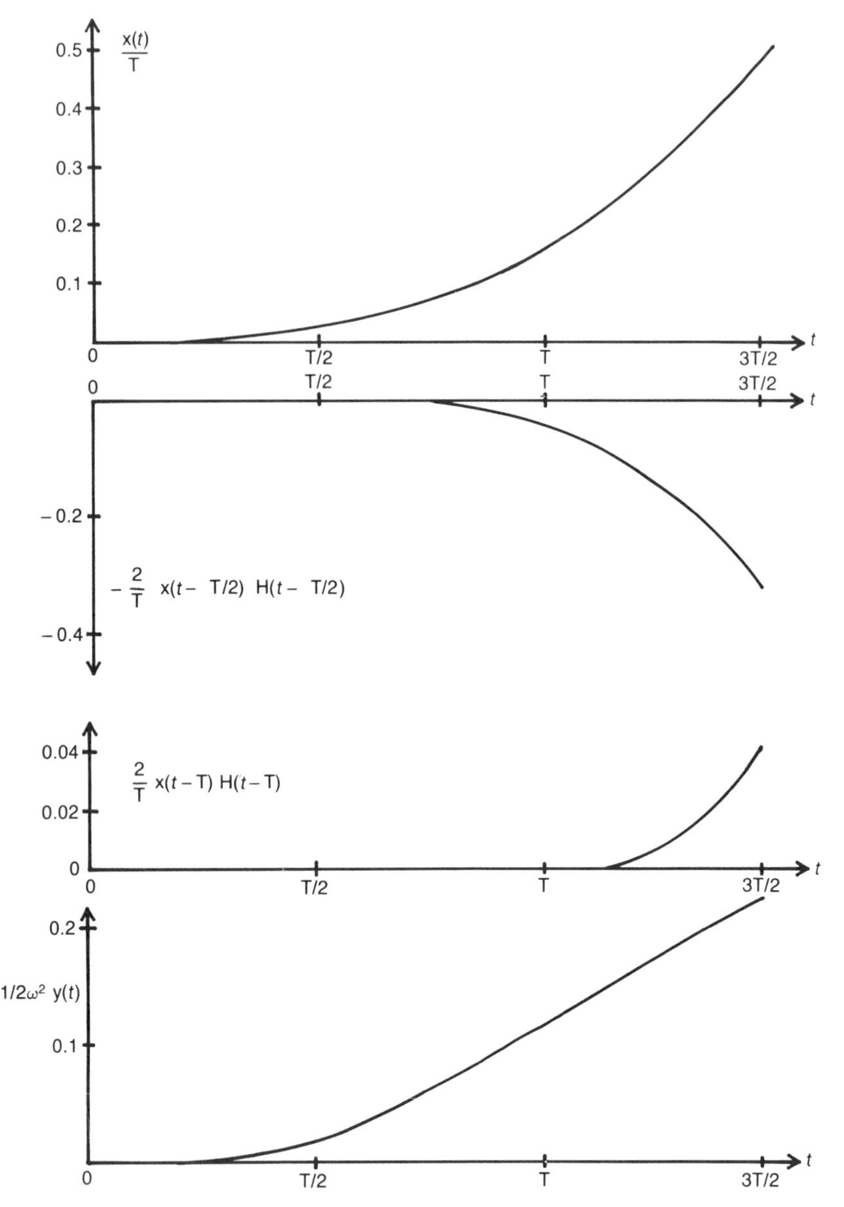

Fig. 3.4.1: A graph of the solution y(t) for times up to 3T/2 given by Eq. (3.4.11) for the parameter $\omega T = 1$.

Because

$$\frac{16}{T^2} \sum_{n=1,3,5,\ldots}^{\infty} \frac{1}{\omega_s^2 - \omega^2} = \sum_{n=1,3,5,\ldots}^{\infty} [(n\pi/2)^2 - (\omega T/4)^2]^{-1} \tag{3.4.17}$$

$$\frac{16}{T^2} \sum_{n=1,3,5,\ldots}^{\infty} \frac{1}{w_s^2 - w^2} = \frac{2}{\omega T} \tan(\omega T/4) \qquad \textbf{(3.4.18)}$$

(Gradshteyn, I. S. and I. M. Ryzhik, 1965: *Tables of Integrals, Series, and Products*. New York: Academic Press. Formula 1.421.1.), we can simplify Eq. (3.4.16) to

$$y(t) = \frac{2}{\omega T} \tan(\omega T/4) \frac{1 - \cos(\omega t)}{\omega^2}$$

$$- \frac{16}{T^2} \sum_{n=1,3,5,\ldots}^{\infty} \frac{1}{\omega_s^2} \frac{1 - \cos(\omega_s t)}{\omega_s^2 - \omega^2} \qquad \textbf{(3.4.19)}$$

However, because

$$\sum_{n=1,3,5,\ldots}^{\infty} \frac{1}{\omega_s^2} = \frac{T^2}{32} \qquad \textbf{(3.4.20)}$$

we have

$$\frac{16}{T^2} \sum_{n=1,3,5,\ldots}^{\infty} \frac{1}{\omega_s^2} \frac{1}{\omega_s^2 - \omega^2} = \frac{16}{T^2\omega^2} \sum_{n=1,3,5,\ldots}^{\infty} \frac{1}{\omega_s^2 - \omega^2} - \frac{1}{\omega_s^2}$$

$$\textbf{(3.4.21)}$$

$$= \frac{2}{\omega^3 T} \tan(\omega T/4) - \frac{1}{2\omega^2} \qquad \textbf{(3.4.22)}$$

and

$$y(t) = \frac{1}{2\omega^2} - \frac{2}{\omega^3 T} \tan(\omega T/4) \cos(\omega t)$$

$$+ \left(\frac{2T}{\pi}\right)^2 \sum_{n=1,3,5,\ldots}^{\infty} \frac{1}{n^2} \frac{1}{(2\pi n)^2 - (\omega T)^2} \cos\left(\frac{2\omega n t}{T}\right) \qquad \textbf{(3.4.23)}$$

This is the result that we would have obtained if we had used a Fourier series. The term cos (ωt) in Eq. (3.4.16), (3.4.19) and (3.4.23) represents the free oscillation that is necessary so that the solution satisfies the initial conditions. The remaining terms represent the forced oscillations produced by the periodic forcing. Resonance occurs if $\omega T/2\pi$ is an odd integer m, in which case both $\tan(\omega T/4)$ and the mth term in each of the series are infinite. Using l′ Hospital's rule, the mth term then becomes

$$y_m(t) = \frac{16}{\omega^4 T^2} [1 - \cos(\omega t) - (\omega t/2) \sin(\omega t)] \qquad \textbf{(3.4.24)}$$

where $\omega T = 2\pi m$.

Exercises

Find the solution to the following differential equations with the periodic function *f(t)*:

1. $$y' + 4y + 3\int_0^t y\,d\tau = \begin{cases} 1 & 0<t<2 \\ -1 & 2<t<4 \end{cases} \qquad y(0)=1$$

2. $$y'' + 4y' + 4y = \begin{cases} 1 & 0<t<1 \\ 0 & 1<t<2 \end{cases} \qquad y(0) = y'(0) = 0$$

3. $$y'' + y = \begin{cases} 1 & 0<t<\pi \\ 0 & \pi<t<2\pi \end{cases} \qquad y(0) = y'(0) = 0$$

3.5 INVERSION BY PARTIAL FRACTIONS: HEAVISIDE'S EXPANSION THEOREM

In the previous sections, we have devoted our efforts to calculating the Laplace transform of a given function. Obviously we must have a method for going the other way. Given a transform, we must find the corresponding function. This is often a very formidable task. In the next few sections we shall present some general techniques for the inversion of a Laplace transform.

Quite often, the transform can be written as the ratio of two polynomials: $F(s) = q(s)/p(s)$. We shall assume that the order of $q(s)$ is *less* than $p(s)$. In principle we know that $p(s)$ has n zeros where n is the order of the $p(s)$ polynomial. Some of the zeros may be complex, some of them may be real, and some of them may be duplicates of other zeros. It is customary (for reasons that will appear in Section 3.13) to call a nonrepeating root (either purely real or complex) of $p(s)$ a *simple pole* of the transform $F(s)$. On the other hand, if the root repeats m times, then the root is called a *multiple pole of order m* of the transform $F(s)$. In the case when $p(s)$ has n simple poles, a simple method of inverting the transform can be found.

We want to rewrite $F(s)$ in the fraction form:

$$\begin{aligned} F(s) &= \frac{A_1}{s - r_1} + \frac{A_2}{s - r_2} + \cdots + \frac{A_n}{s - r_n} \\ &= \frac{q(s)}{p(s)} \end{aligned} \qquad \textbf{(3.5.1)}$$

where r_1, r_2, . . ., and r_n are the n simple roots of $p(s)$. $F(s)$ then has n simple poles. We now multiply both sides of (3.5.1) by $s - r_1$, so that

$$\frac{(s - r_1)q(s)}{p(s)} = A_1 + \frac{A_2 (s - r_1)}{s - r_2} + \ldots + \frac{A_n (s - r_1)}{s - r_n} \quad \textbf{(3.5.2)}$$

If we set $s = r_1$, the right-hand side of Eq. (3.5.2) becomes simply A_1. The left-hand side becomes 0/0 and two cases must be considered. If we can write $p(s) = (s - r_1)g(s)$, then

$$A_1 = q(s_1)/g(s_1) \quad \textbf{(3.5.3)}$$

In general this cannot be done and we use l′Hospital's rule to give us

$$A_1 = \lim_{s \to r_1} \frac{(s - r_1)q(s)}{p(s)} = \lim_{s \to r_1} \frac{(s - r_1)q'(s) + q(s)}{p'(s)}$$

$$= \frac{q(r_1)}{p'(r_1)} \quad \textbf{(3.5.4)}$$

In a similar manner, all the A_k's $(k = 1, 2, 3, \ldots, n)$ can be computed. Therefore

$$\mathscr{L}^{-1}[F(s)] = \mathscr{L}^{-1}\left[\frac{q(s)}{p(s)}\right]$$

$$= \mathscr{L}^{-1}\left(\frac{A_1}{s - r_1} + \frac{A_2}{s - r_2} + \ldots + \frac{A_n}{s - r_n}\right) \quad \textbf{(3.5.5)}$$

$$= A_1 e^{r_1 t} + A_2 e^{r_2 t} + \ldots + A_n e^{r_n t} \quad \textbf{(3.5.6)}$$

This is known as *Heaviside's expansion theorem* for simple poles.

For example, let us invert the transform $\frac{s}{(s+2)(s^2+1)}$. It has three simple poles at $s = -2$ and $\pm i$. From our earlier discussion, $q(s) = s$, $p(s) = (s+2)(s^2+1)$ and $p'(s) = 3s^2 + 4s + 1$. Therefore

$$\mathscr{L}^{-1}\left\{\frac{s}{(s+2)(s^2+1)}\right\} = \frac{-2}{12-8+1} e^{-2t} + \frac{i}{-3+4i+1} e^{it}$$

$$+ \frac{-i}{-3-4i+1} e^{-it} \quad \textbf{(3.5.7)}$$

$$= -\frac{2}{5} e^{-2t} + \frac{i}{2+4i} e^{-it} - \frac{i}{-2-4i} e^{it} \quad \textbf{(3.5.8)}$$

$$\mathcal{L}^{-1}\left(\frac{s}{(s+2)\,(s^2+1)}\right) = -\frac{2}{5}\,e^{-2t} + i\,\frac{-2-4i}{4+16}\,e^{it} - \frac{-2+4i}{4+16}\,e^{-it} \tag{3.5.9}$$

$$= -\frac{2}{5}\,e^{-2t} + \frac{1}{5}\,\sin(t) + \frac{2}{5}\,\cos(t) \tag{3.5.10}$$

On the other hand, suppose we want to find the inverse of

$$F(s) = \frac{1}{(s-1)(s-2)(s-3)} \tag{3.5.11}$$

There are three simple poles at $s = 1, 2$, and 3. In this case, it is easier to calculate A_1, A_2, and A_3 as follows:

$$A_1 = \lim_{s\to 1} \frac{\cancel{s-1}}{\cancel{(s-1)}\,(s-2)\,(s-3)} = 1/2 \tag{3.5.12}$$

$$A_2 = \lim_{s\to 2} \frac{\cancel{s-2}}{(s-1)\,\cancel{(s-2)}\,(s-3)} = -1 \tag{3.5.13}$$

$$A_3 = \lim_{s\to 3} \frac{\cancel{s-3}}{(s-1)\,(s-2)\,\cancel{(s-3)}} = 1/2 \tag{3.5.14}$$

Therefore

$$\mathcal{L}^{-1}\left[\frac{1}{(s-1)(s-2)(s-3)}\right] = e^{t}/2 - e^{2t} + e^{3t}/2 \tag{3.5.15}$$

3.6 THE USE OF LAPLACE TRANSFORMS IN THE DESIGN OF THE FILM PROJECTOR

For good motion-picture audio fidelity, the speed at which the film passes by the electric eye must remain as nearly constant as possible; otherwise, a frequency modulation of the reproduced sound results. Figure 3.6.1 shows a diagram of the scanner used in most motion-picture projectors. In this section we shall show that this particular design filters out variations in the film speed caused by irregularities either in the driving gear trains or in the engagement of the sprocket teeth with the holes in the film.

The film head is a hollow drum of small moment of inertia J_1. Within it is a concentric inner flywheel of moment of inertia J_2, which is large compared with J_1. (See Fig. 3.6.1(B).) The remainder of the space within the drum is filled with oil. The inner flywheel rotates on precision ball bearings on the drum shaft. The only coupling between drum and flywheel is through fluid friction

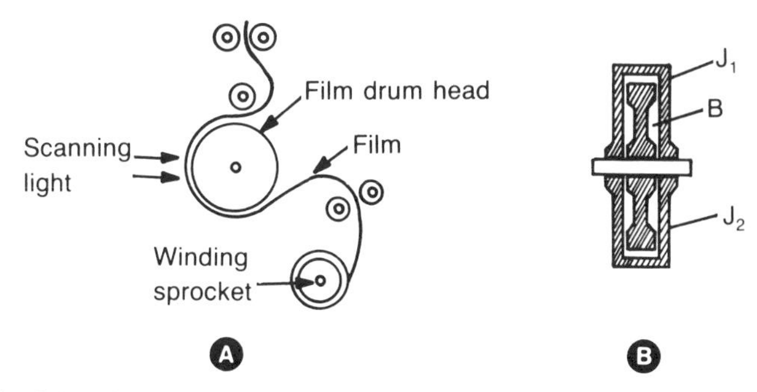

Fig. 3.6.1: A rotational-mechanical filter used in a motion-picture projector. Taken with permission from Gardner, M. F. and J. L. Barnes, 1942. *Transients in Linear Systems. Vol. I.* New York: John Wiley and Sons, Inc., p. 187.

and the very small friction in the ball bearings. The necessary spring restoring force for the system is provided by the flection of the film loops between drum head and idler pulleys when the film is running rapidly through the system.

From Fig. 3.6.1, we see that the dynamical equations governing the outer case and inner flywheel are

rate of change of the outer casing of the film head	=	frictional torque given to casing from inner flywheel
	+	restoring torque due to the flection of the film
rate of change of inner flywheel	= −	frictional torque given to outer casing from inner flywheel.

Assuming that the frictional torque between the two flywheels is proportional to the difference in their angular velocities, the frictional torque given to the casing from the inner flywheel is $B(\omega_2 - \omega_1)$ where B is the frictional resistance, ω_1 and ω_2 the deviation of the drum and inner flywheel from their normal angular velocity. If r is the ratio of the diameter of the winding sprocket to the diameter of the drum, the restoring torque due to the flections of the film and its corresponding angular twist is given by

$$K\int_0^t (r\omega_0 - \omega_1)\, dt'$$

where K is the rotational stiffness and ω_0 the deviation of the winding sprocket from its normal angular velocity. The quantity $r\omega_0$ gives the angular velocity at which the film is running through the projector because the winding sprocket is the mechanism that pulls the film. Consequently the equations governing this mechanical system are

$$J_1 \frac{d\omega_1}{dt} = K \int_0^t (r\omega_0 - \omega_1)\, d\tau + B(\omega_2 - \omega_1) \qquad \textbf{(3.6.1)}$$

$$J_2 \frac{d\omega_2}{dt} = B(\omega_1 - \omega_2). \qquad \textbf{(3.6.2)}$$

For details, see Cook, E.D., 1935: The technical aspects of the high-fidelity reproducer. *Soc. Motion Pict. Eng., Journal, 25,* 289-312.

With the winding sprocket, the drum and the flywheel running at their normal uniform angular velocities, let us assume that the winding sprocket introduces a disturbance equivalent to a unit increase in its angular velocity for 0.15 sec, followed by a resumption of its normal velocity. It is assumed that the film in contact with the drum cannot slip. The initial conditions are given by $\omega_1(0) = \omega_2(0) = 0$.

Taking the Laplace transform of Eq. (3.6.1) and (3.6.2), we have

$$(J_1 s + B + \frac{K}{s})\ \bar{\omega}_1(s) - B\,\bar{\omega}_2(s) = T_0(s) \qquad \textbf{(3.6.3)}$$

$$-B\,\bar{\omega}_1(s) + (J_2 s + B)\bar{\omega}_2(s) = 0 \qquad \textbf{(3.6.4)}$$

where

$$T_0(s) = \frac{rK}{s}\,\bar{\omega}_0(s) = rK\,\mathscr{L}\left[\int_0^t \bar{\omega}_0(\tau)\, d\tau\right]. \qquad \textbf{(3.6.5)}$$

The solution of Eq. (3.6.3) and (3.6.4) for $\bar{\omega}_1(s)$ is

$$\bar{\omega}_1(s) = \frac{rK}{J_1}\ \frac{(s + a_0)\bar{\omega}_0(s)}{s^3 + b_2 s^2 + b_1 s + b_0} \qquad \textbf{(3.6.6)}$$

where typical values are

$$\frac{rK}{J_1} = 90.8 \qquad b_1 = \frac{K}{J_1} = 157.$$

$$a_0 = \frac{B}{J_2} = 1.47 \qquad b_2 = \frac{B(J_1 + J_2)}{J_1 J_2} = 8.20$$

$$b_0 = \frac{BK}{J_1 J_2} = 231.$$

The transform $\bar{\omega}_1(s)$ has three simple poles located at $s_1 = -1.58$, $s_2 = -3.32 + 11.6\,i$ and $s_3 = -3.32 - 11.6\,i$.

Because the sprocket angular-velocity deviation $\omega_0(t)$ is a pulse of unit amplitude and 0.15 second duration, it can be expressed as the difference of two unit step functions:

$$\omega_0(t) = H(t) - H(t-0.15). \qquad \textbf{(3.6.7)}$$

Its Laplace transform is

$$\bar{\omega}_0(s) = \frac{1}{s} - \frac{1}{s} e^{-0.15s} \qquad \textbf{(3.6.8)}$$

so that Eq. (3.6.6) becomes

$$\bar{\omega}_1(s) = \frac{rK}{J_1} \frac{s + a_0}{s(s-s_1)(s-s_2)(s-s_3)} (1 - e^{-0.15s}). \qquad \textbf{(3.6.9)}$$

The inversion of Eq. (3.6.9) follows directly from the second shifting theorem and the Heaviside expansion theorem:

$$\omega_1(t) = K_0 + K_1e^{s1t} + K_2e^{s2t} + K_3e^{s3t}$$

$$- (K_0 + K_1e^{s1T} + K_2e^{s2T} + K_3e^{s3T})H(T) \qquad \textbf{(3.6.10)}$$

where

$$T = t - 0.15$$

$$K_0 = \frac{rK}{J_1} \frac{\cancel{s}(s + a_0)}{\cancel{s}(s - s_1)(s - s_2)(s - s_3)}\bigg|_{s=0} = 0.578$$

$$K_1 = \frac{rK}{J_1} \frac{\cancel{(s - s_1)}(s + a_0)}{s\cancel{(s - s_1)}(s - s_2)(s - s_3)}\bigg|_{s=s_1} = 0.046$$

$$K_2 = \frac{rK}{J_1} \frac{\cancel{(s - s_2)}(s + a_0)}{s(s - s_1)\cancel{(s - s_2)}(s - s_3)}\bigg|_{s=s_2} = 0.326\, e^{165°i}$$

$$K_3 = \frac{rK}{J_1} \frac{\cancel{(s - s_3)}(s + a_0)}{s(s - s_1)(s - s_2)\cancel{(s - s_3)}}\bigg|_{s=s_3} = 0.326\, e^{-165°i}$$

Because $\cos(x) = 1/2(e^{ix} + e^{-ix})$, Eq. (3.6.10) can be rewritten as

$$\omega_1(t) = 0.578 + 0.046\, e^{-1.58t} + 0.652\, e^{-3.32t} \cos(11.6\, t + 165°)$$

$$- [0.578 + 0.046\, e^{-1.58T} + 0.652\, e^{-3.32T} \cos(11.6\, T + 165°)\, H(T)$$

$$\textbf{(3.6.11)}$$

The results from Eq. (3.6.11) are graphed in Fig.

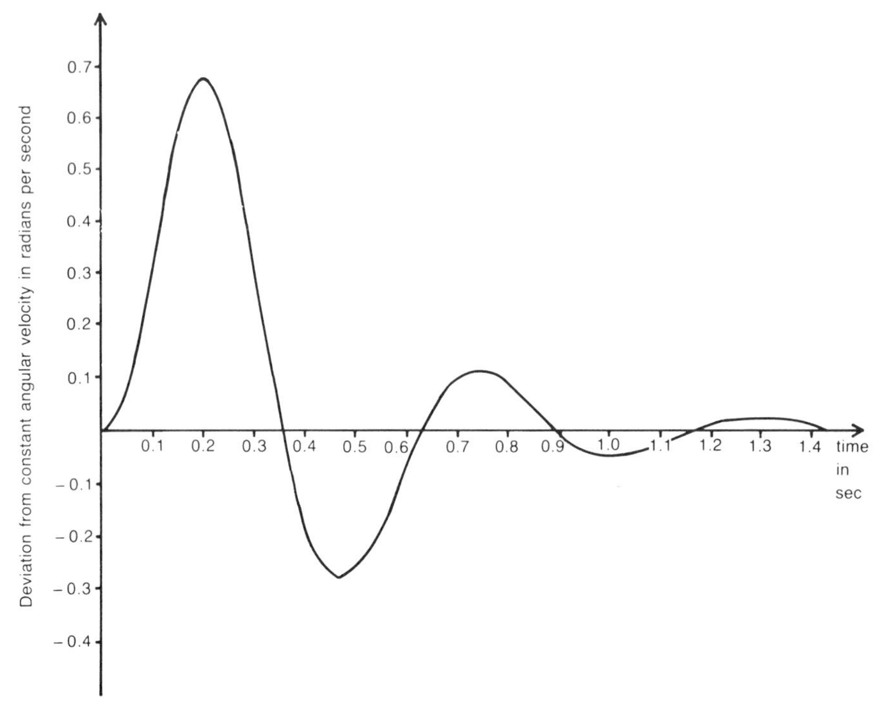

Fig. 3.6.2: Response of the film head to rectangular pulse in the angular velocity of the winding sprocket.

3.6.2. Note that the fluctuation in $\omega_1(t)$ is damped out by the particular design of this film projector. Because this mechanical device damps out unwanted fluctuations (or noise) in the motion-picture projector, this particular device is given the special name of *mechanical filter.*

Exercises

Find the inverse of the following transforms by partial fractions.

1. $$\frac{s^2 + 2}{(s^2 + 5s + 4)(s^2 + 7s + 10)}$$

2. $$\frac{s^2 - s + 3}{s^3 + 6s^2 + 11s + 6}$$

3. $$\frac{1}{s^2 + 3s + 2}$$

4. $$\frac{s + 3}{(s+4)(s-2)}$$

5. $$\frac{4s - 3}{2s^2 - 3s + 1}$$

3.7 THE COMPLEMENTARY ERROR FUNCTION AND ASSOCIATED FUNCTIONS

An important transform in solving the heat equation is $s^{-1} \exp(-k\, s^{1/2})$. We wish to show that

$$\mathscr{L}^{-1}\left[s^{-1} \exp(-ks^{1/2})\right] = \operatorname{erfc}\left(\frac{k}{2t^{1/2}}\right),\ k \geq 0, \qquad \textbf{(3.7.1)}$$

where erfc(x) is the complementary error function defined as

$$\operatorname{erfc}(x) = \frac{2}{\sqrt{\pi}} \int_x^{\infty} e^{-y^2}\, dy. \qquad \textbf{(3.7.2)}$$

From its definition it is already shown that erfc(0) = 1. The complementary error function is graphed in Fig. 3.7.1.

For the case $k = 0$, we have $\mathscr{L}^{-1}(1/s) = 1 = \operatorname{erfc}(0)$. For $k \neq 0$, we begin our demonstration by introducing the function

$$y_1(t) = \frac{1}{\sqrt{\pi t}} \exp\left(-\frac{k^2}{4t}\right). \qquad \textbf{(3.7.3)}$$

Taking the Laplace transform of both sides, we obtain

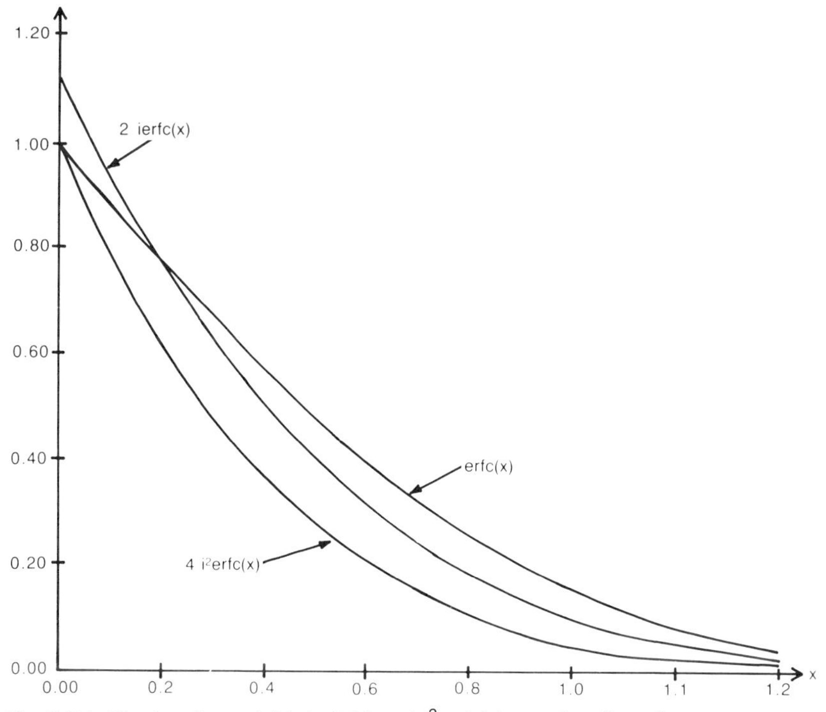

Fig. 3.7.1: The function erfc(x), ierfc(x) and i^2 erfc(x) as a function of x.

$$\sqrt{\pi}\,\overline{y}_1(s) = \int_0^\infty \frac{1}{\sqrt{t}} \exp\left(-st - \frac{k^2}{4t}\right)\, dt \qquad \textbf{(3.7.4)}$$

$$= e^{-ks^{1/2}} \int_0^\infty \exp\left(-\left(\sqrt{st} - \frac{k}{2\sqrt{t}}\right)^2\right). \frac{dt}{\sqrt{t}} \qquad \textbf{(3.7.5)}$$

which becomes

$$\sqrt{\pi}\,\overline{y}_1(s) = e^{-ks^{1/2}} \int_0^\infty \exp\left[-\left(\frac{\sqrt{s}}{2}\,\tau - \frac{k}{\tau}\right)^2\right] d\tau \qquad \textbf{(3.7.6)}$$

if we substitute τ for $2\sqrt{t}$ or

$$\sqrt{\pi}\, e^{ks^{1/2}}\, \overline{y}_1(s) = \frac{2k}{\sqrt{s}} \int_0^\infty \exp\left[-\left(\frac{\sqrt{s}}{2}\,\lambda - \frac{k}{\lambda}\right)^2\right] \frac{d\lambda}{\lambda^2}$$

(3.7.7)

if $\sqrt{s}\,\tau/2 = -k/\lambda$.

By adding Eq. (3.7.6) and Eq. (3.7.7) and substituting x for $s^{1/2}\lambda/2 - k/\lambda$, we find that

$$2\sqrt{\pi}\, e^{ks^{1/2}}\, \overline{y}_1(s) = \frac{2}{\sqrt{s}} \int_0^\infty \exp\left[-\left(\frac{\sqrt{s}}{2}\,\lambda - \frac{k}{\lambda}\right)^2\right] \left(\frac{\sqrt{s}}{2} + \frac{k}{\lambda^2}\right) d\lambda$$

(3.7.8)

$$= \frac{2}{\sqrt{s}} \int_{-\infty}^{\infty} \exp(-x^2)\, dx = 2\sqrt{\pi/s}. \qquad \textbf{(3.7.9)}$$

Therefore, we obtain our first important result that

$$\boxed{\mathcal{L}\left[\frac{1}{\sqrt{\pi t}} \exp\left(-\frac{k^2}{4t}\right)\right] = \frac{e^{-ks^{1/2}}}{\sqrt{s}},\ s>0,\ k \geq 0.} \qquad \textbf{(3.7.10)}$$

We next use the fact that $\mathcal{L}[f(t)/t] = \int_s^\infty F(z)\, dz$ to establish another important transform, namely that

$$\mathcal{L}\left[\frac{1}{\sqrt{\pi t^3}} \exp\left(-\frac{k^2}{4t}\right)\right] = \int_s^\infty \frac{e^{-kz^{1/2}}}{\sqrt{z}}\, dz \qquad \textbf{(3.7.11)}$$

$$= \frac{2}{k}\, e^{-ks^{1/2}} \qquad \textbf{(3.7.12)}$$

Finally, we use Eq. (3.7.12) and the result that

$$F(s)/s = \mathcal{L}\left[\int_0^t f(\tau)\, d\tau\right]$$

to complete our demonstration by noting that

$$\mathcal{L}^{-1}\left(s^{-1}e^{-ks^{1/2}}\right) = \frac{k}{2\sqrt{\pi}}\int_0^t e^{-k^2/(4\tau)}\,\tau^{-3/2}\,d\tau \qquad (3.7.13)$$

$$= \frac{2}{\sqrt{\pi}}\int_{k/(2\sqrt{t})}^{\infty} e^{-\lambda^2}\,d\lambda \qquad (3.7.14)$$

$$\boxed{\mathcal{L}^{-1}\left(s^{-1}e^{-ks^{-1/2}}\right) = \text{erfc}\left[\frac{k}{(2\sqrt{t})}\right] \quad k \geq 0,\ s \geq 0.} \qquad (3.7.15)$$

In addition to the three transforms given by Eqs. (3.7.9), (3.7.12) and (3.7.15), two valuable transforms follow from the formula $F(s)/s = \mathcal{L}[\int_0^t f(\tau)\,d\tau]$. The first of these is

$$\mathcal{L}^{-1}\left(s^{-3/2}e^{-ks^{1/2}}\right) = \int_0^t \frac{1}{\sqrt{\pi\tau}}\exp\left(-\frac{k^2}{4\tau}\right)d\tau \qquad (3.7.16)$$

$$= 2\left[\frac{t}{\pi}\right]^{1/2}\exp(-k^2/4t) - k\,\text{erfc}\left[\frac{k}{(2\sqrt{t})}\right] \qquad (3.7.17)$$

$$= 2t^{1/2}\left[\frac{1}{\sqrt{\pi}}\exp\left(-\frac{k^2}{4t}\right) - \frac{k}{2\sqrt{t}}\,\text{erfc}\left(\frac{k}{2\sqrt{t}}\right)\right] \qquad (3.7.18)$$

$$\boxed{\mathcal{L}^{-1}\left(s^{-3/2}e^{-ks^{1/2}}\right) = 2t^{1/2}\,i^1\text{erfc}\left(\frac{k}{2\sqrt{t}}\right).} \qquad (3.7.19)$$

The general class of functions $i^n\text{erfc}(x)$ appears frequently in the heat conduction literature (see Carslaw, H.S. and J.C. Jaeger, 1959: *Conduction of Heat in Solids*. London: Oxford University Press. Appendix II). It is defined by

$$i^n\text{erfc}(x) = \int_x^{\infty} i^{n-1}\text{erfc}(z)\,dz, \qquad n = 1,2,3,\ldots \qquad (3.7.20)$$

with

$$i^0\text{erfc}(x) = \text{erfc}(x). \qquad (3.7.21)$$

or it may be computed from the general recurrence formula

$$2n\,i^n\text{erfc}(x) = i^{n-2}\text{erfc}(x) - 2x\,i^{n-1}\text{erfc}(x) \qquad (3.7.22)$$

where

$$i^1\text{erfc}(x) = \frac{1}{\sqrt{\pi}}e^{-x^2} - x\,\text{erfc}(x) \qquad (3.7.23)$$

and

$$i^0\text{erfc}(x) = \text{erfc}(x). \qquad \textbf{(3.7.24)}$$

These formulas associated with the complementary error function are useful in finding the inverse to the following transforms:

$$\boxed{\mathcal{L}^{-1}\left(\frac{e^{-ks^{1/2}}}{s^{1+n/2}}\right) = (4t)^{n/2}\, i^n\text{erfc}\,\left(\frac{kx}{2\sqrt{t}}\right)} \qquad \textbf{(3.7.25)}$$

where $n = 0, 1, 2, 3, \ldots$ The function $i^1\text{erfc}(x)$ and $i^2\text{erfc}(x)$ have been graphed along with the complementary error function $\text{erfc}(x)$ in Fig. 3.7.1.

3.8 CONVOLUTION

In this section, we turn to the most fundamental concept in Laplace transforms: convolution. We shall restrict ourselves to its use in finding the inverse of a transform when that transform is composed of the product of two simpler transforms. In subsequent sections we will show how it is used in solving integral equations and ordinary differential equations.

We begin our analysis by formally introducing the mathematical operation of the *convolution product*:

$$f(t)*g(t) = \int_0^t f(t-x)\, g(x)\, dx$$

$$= \int_0^t g(t-x)\, f(x)\, dx. \qquad \textbf{(3.8.1)}$$

In most cases the operations required by Eq. (3.8.1) are straightforward. For example, let us find the convolution between $\cos(t)$ and $\sin(t)$.

$$\cos(t)*\sin(t) = \int_0^t \sin(t-x)\cos(x)\, dx \qquad \textbf{(3.8.2)}$$

$$= \int_0^t 1/2\, \sin(t) + 1/2\, \sin(t-2x)\, dx \qquad \textbf{(3.8.3)}$$

$$= 1/2 \int_0^t \sin(t)\, dx + 1/2 \int_0^t \sin(t-2x)\, dx \qquad \textbf{(3.8.4)}$$

$$= 1/2\, \sin(t)\, x \Big|_0^t + 1/4\, \cos(t-2x) \Big|_0^t \qquad \textbf{(3.8.5)}$$

$$= 1/2\, t \sin(t) \qquad \textbf{(3.8.6)}$$

Similarly, the convolution between t^2 and $\sin(t)$ is given by

$$t^2 * \sin(t) = \int_0^t (t-x)^2 \sin(x)\, dx \tag{3.8.7}$$

$$= -(t-x)^2 \cos(x) \Big|_0^t - 2\int_0^t (t-x)\cos(x)\, dx \tag{3.8.8}$$

$$= t^2 - 2(t-x)\sin(x) \Big|_0^t + \int_0^t \sin(x)\, dx \tag{3.8.9}$$

$$= t^2 - 2\cos(x) \Big|_0^t = t^2 + 2\cos(t) - 2. \tag{3.8.10}$$

On the other hand, convolution involving discontinuous functions require special treatment. For example, let us find the convolution between e^t and the pulse $H(t-1) - H(t-2)$:

$$\left[e^t * H(t-1) - H(t-2)\right] = \int_0^t e^{t-x}\left[H(x-1) - H(x-2)\right] dx \tag{3.8.11}$$

$$= e^t \int_0^t e^{-x}\left[H(x-1) - H(x-2)\right] dx \tag{3.8.12}$$

In order to evaluate the integral (3.8.12), we must examine various cases. If $t<1$, then both of the unit step functions are zero and we obtain zero. However, when $1<t<2$, the first Heaviside is equal to one while the second one is zero. Therefore,

$$e^t * \left[H(t-1) - H(t-2)\right] = e^t \int_1^t e^{-x}\, dx \tag{3.8.13}$$

$$= e^t\,(-e^{-x}) \Big|_1^t$$

$$= e^t\,(-e^{-t} + e^{-1}) \tag{3.8.14}$$

$$= e^{t-1} - 1 \qquad 1<t<2 \tag{3.8.15}$$

because the portion of the integral from zero to one is equal to zero. Finally, when $t>2$, the integrand is only nonzero for that portion of the integration when x runs from one to two. Therefore,

$$e^t * \left[H(t-1) - H(t-2))\right] = e^t \int_0^t e^{-x}\left[(H(x-1) - H(x-2))\right] dx \tag{3.8.16}$$

$$= e^t \int_1^2 e^{-x} dx = e^t (-e^{-x}) \Big|_1^2 \qquad (3.8.17)$$

$$= e^t (e^{-1} - e^{-2}) = e^{t-1} - e^{t-2}, \; t > 2. \qquad (3.8.18)$$

Therefore, the convolution of e^t with the pulse $H(t-1) - H(t-2)$ may be summarized by

$$e^t * \Big[H(t-1) - H(t-2)\Big] = \begin{cases} 0 & 0 \leq t \leq 1 \\ e^{t-1} - 1 & 1 \leq t \leq 2 \\ e^{t-1} - e^{t-2} & t \geq 2 \end{cases} \qquad (3.8.19)$$

The reason why we have introduced the convolution integral follows from this *fundamental* theorem (often called *Borel's theorem*) that if

$$w(t) = u(t) * v(t) \qquad (3.8.20)$$

then

$$\bar{w}(s) = \bar{u}(s)\, \bar{v}(s). \qquad (3.8.21)$$

In other words a complicated transform can be found through the convolution of two simpler functions.

The proof is as follows:

$$\overline{w}(s) = \int_0^\infty \left(\int_0^t u(x)\, v(t-x)\, dx \right) e^{-st}\, dt \qquad (3.8.22)$$

$$= \int_0^\infty \left(\int_x^\infty u(x)\, v(t-x)\, e^{-st}\, dt \right) dx \qquad (3.8.23)$$

$$= \int_0^\infty u(x) \left(\int_0^\infty v(r)\, e^{-s(r+x)}\, dr \right) dx \qquad (3.8.24)$$

$$= \int_0^\infty u(x)\, e^{-sx}\, dx \int_0^\infty v(r)\, e^{-sr}\, dr = \bar{u}(s)\, \bar{v}(s) \qquad (3.8.25)$$

where $t = r + x$.

Convolution is used for three purposes: (1) the inversion of complicated transforms, (2) the solution of integral equations, and (3) the solution of nonhomogeneous differential equations.

To illustrate the use of the convolution theorem in the inversion of complicated transform, let us find the inverse of the transform

$$\frac{s}{(1+s^2)^2} = \frac{s}{1+s^2}\, \frac{1}{1+s^2} = \mathscr{L}\Big[\cos(t)\Big] \mathscr{L}\Big[\sin(t)\Big] \qquad (3.8.26)$$

Therefore,

$$\frac{s}{(1+s^2)^2} = \mathscr{L}\left[\cos(t) * \sin(t)\right] = \mathscr{L}\left[1/2\ t \sin(t)\right]. \qquad \textbf{(3.8.27)}$$

As a second example, consider

$$\frac{1}{(s^2+a^2)^2} = \frac{1}{a^2}\left(\frac{a}{s^2+a^2} \times \frac{a}{s^2+a^2}\right)$$

$$= \frac{1}{a^2}\mathscr{L}\left[\sin(at)\right]\mathscr{L}\left[\sin(at)\right]. \qquad \textbf{(3.8.28)}$$

$$\mathscr{L}^{-1}\left(\frac{1}{(s^2+a^2)^2}\right) = \frac{1}{a^2}\int_0^t \sin\left[a(t-x)\right]\sin(ax)\,dx \qquad \textbf{(3.8.29)}$$

$$= \frac{1}{2a^2}\int_0^t \cos\left[a(t-2x)\right]dx$$

$$- \frac{1}{2a^2}\int_0^t \cos(at)\,dx \qquad \textbf{(3.8.30)}$$

$$= -\frac{1}{4a^3}\sin\left[a(t-2x)\right]\Big|_0^t$$

$$- \frac{\cos(at)}{2a^2}\,x\Big|_0^t \qquad \textbf{(3.8.31)}$$

$$= \frac{1}{2a^3}\sin(at) - at\cos(at). \qquad \textbf{(3.8.32)}$$

Exercises

Determine the following convolutions.

1. $1 * \sin(at)$
2. $e^{at} * e^{bt}$
3. $t * e^{at}$
4. $\sin(at) * \sin(bt)$

Use Borel's theorem to find the inverse of the following transforms.

5. $\dfrac{1}{(s^2+4)^2}$

6. $\dfrac{s}{(s^2+9)^3}$

7. $\dfrac{1}{s^2+5s+6}$

8. Compute $t^2 * \sin(at)$ and find its transform.

9. Evaluate the convolution $H(t-a)*H(t-b)$, where a and b are positive, and check the results using Borel's theorem.

10. If

$$\text{erf}(x) = \frac{2}{\sqrt{\pi}} \int_0^x e^{-t^2}\, dt,$$

show that

$$\mathcal{L}\{\text{erf}(t^{1/2})\} = \frac{1}{s\sqrt{1+s}}$$

3.9 INTEGRAL EQUATIONS

An equation which contains a dependent variable under an integral sign is called an *integral equation*. The convolution theorem provides an excellent tool for solving a very special class of integral equations.

To illustrate this, let us find $f(t)$ from the integral equation

$$f(t) = 4t - 3\int_0^t f(x)\sin(t-x)\, dx. \tag{3.9.1}$$

The integral in Eq. (3.9.1) is in precisely the right form to permit the use of the convolution theorem. Then, because

$$\mathcal{L}\{\sin(t)\} = \frac{1}{s^2+1}, \tag{3.9.2}$$

the convolution theorem yields

$$\mathcal{L}\left[\int_0^t f(x)\sin(t-x)\, dx\right] = \frac{F(s)}{s^2+1}. \tag{3.9.3}$$

Therefore, the Laplace operator converts Eq. (3.9.1) into

$$F(s) = \frac{4}{s^2} - \frac{3F(s)}{s^2+1}. \tag{3.9.4}$$

$$\left(1 + \frac{3}{s^2+1}\right) F(s) = \frac{4}{s^2} \tag{3.9.5}$$

or

$$F(s) = \frac{4(s^2+1)}{s^2(s^2+4)}. \tag{3.9.6}$$

By partial fractions, or by inspection,

$$F(s) = \frac{1}{s^2} + \frac{3}{s^2+4} \tag{3.9.7}$$

Therefore

$$f(t) = t + \frac{3}{2} \sin(2t). \tag{3.9.8}$$

It is important to realize that the original equation (3.9.1) could equally well have been encountered in the equivalent form

$$f(t) = 4t - 3\int_0^t f(t-x) \sin(x)\, dx. \tag{3.9.9}$$

The essential ingredient is that the integral involved be in exactly the convolution integral form.

As a second illustration, we solve the equation

$$g(x) = x^2/2 - \int_0^x (x-y)\, g(y)\, dy \tag{3.9.10}$$

Again the integral is one of the convolution type with x playing the role of the independent variable. Let the Laplace transform of $g(x)$ be $G(s)$. Because $\mathscr{L}(1/2x^2) = 1/s^3$ and $\mathscr{L}(x) = 1/s^2$, we may apply the operator $\mathscr{L}$ throughout Eq. (3.9.10) and obtain

$$G(s) = \frac{1}{s^3} - \frac{G(s)}{s^2}, \tag{3.9.11}$$

from which

$$\left(1 + \frac{1}{s^2}\right) G(s) = \frac{1}{s^3}. \tag{3.9.12}$$

or

$$G(s) = \frac{1}{s(s^2+1)} = \frac{s^2+1-s^2}{s(s^2+1)} = \frac{1}{s} - \frac{s}{s^2+1}. \tag{3.9.13}$$

Then

$$g(x) = 1 - \cos(x) \tag{3.9.14}$$

Exercises

Solve the given equations:

1. $f(t) = 1 + 2\int_0^t f(t-x)\, e^{-2x}\, dx$

2. $$f(t) = 1 + \int_0^t f(x) \sin(t-x)\, dx$$

3. $$f(t) = t + \int_0^t f(t-x)\, e^{-x}\, dx$$

4. $$f(t) = 4t^2 - \int_0^t f(t-x)\, e^{-x}\, dx$$

5. $$f(t) = t^3 - \int_0^t f(x) \sin(t-x)\, dx$$

6. $$f(t) = 8t^2 - 3 \int_0^t f(x) \sin(t-x)\, dx$$

7. $$f(t) = t^2 - 2 \int_0^t f(t-x) \sinh(2x)\, dx$$

8. $$f(t) = 1 + 2 \int_0^t f(t-x) \cos(x)\, dx$$

9. $$f(t) = 9e^{2t} - 2 \int_0^t f(t-x) \cos(x)\, dx$$

10. $$f(t) = t^2 + \int_0^t f(x) \sin(t-x)\, dx$$

11. $$f(t) = e^{-t} - 2 \int_0^t f(x) \cos(t-x)\, dx$$

12. $$f(t) = 6t + 4 \int_0^t (x-t)^2 f(x)\, dx$$

13. Solve the following equation for $f(t)$ with the condition that $f(0) = 4$:

$$f'(t) = t + \int_0^t f(t-x) \cos(x)\, dx.$$

14. Solve the following equation for $f(t)$ with the condition that $f(0) = 0$:

$$f'(t) = \sin(t) + \int_0^t f(t-x) \cos(x)\, dx.$$

3.10 THE APPLICATION OF LAPLACE TRANSFORMS TO LINEAR DIFFERENTIAL EQUATIONS WITH CONSTANT COEFFICIENTS: TRANSFER FUNCTIONS, GREEN'S FUNCTIONS AND INDICIAL ADMITTANCE

Laplace transforms are particularly well suited to the solution of ordinary differential equations when they are linear with constant coefficients, the forcing term is of exponential order, and the solution is made definite by initial conditions. An important class of these differential equations arises in electrical circuits because various things are switching on and off. The effect of applying the Laplace transform is then to replace the differential equation by a linear algebraic equation of the transformed function, as in the following example.

The charge $Q(t)$ carried on a capacitor within a simple circuit consisting of a resistor R, capacitor C and electromotive force $E(t)$ is governed by the differential equation:

$$RC \frac{dQ}{dt} + Q = C\,E(t) \tag{3.10.1}$$

We wish to find the charge on the capacitor for any time $t > 0$ given $E(t)$.

The Laplace transform of both sides of Eq. (3.10.1) is

$$RCs\,\bar{Q}(s) - RC\,Q(0) + \bar{Q}(s) = C\,\bar{E}(s) \tag{3.10.2}$$

or

$$\bar{Q}(s) = \frac{C}{1 + RCs}\,\bar{E}(s) = \bar{T}(s)\,\bar{E}(s) \tag{3.10.3}$$

because the circuit is initially dead and $Q(0) = 0$. Consequently the transform of the charge $\bar{Q}(s)$ is given by the product of the transform $\bar{T}(s)$ and the transform of the electromotive force $\bar{E}(s)$. $\bar{T}(s)$ depends only upon s and the parameters of the system and is completely independent of the forcing. In this special case when all the initial conditions are equal to zero, $\bar{T}(s)$ is given the special name of *transfer* (or *system*) *function*. It is a fundamental quantity in the study of linear systems.

If we denote the inverse of $\bar{T}(s)$ by the function $G(t)$, then $G(t)$ is only a function of time and the properties of the system. In our particular example, it is equal to

$$G(t) = \frac{1}{R} \exp\left(-\frac{t}{RC}\right) H(t) \tag{3.10.4}$$

We note that if $\bar{E}(s)$ equals unity (i.e., $E(t)$ is the delta function), then $Q(t) = G(t)$ and $G(t)$ is frequently called the *impulse response* because $G(t)$ is the response of the system to the impulse function. $G(t)$ can be determined experimentally by measuring the response of a system to an impulse.

In general, $\bar{E}(s) \neq 1$ and $\bar{Q}(s)$ is given by the product of two Laplace transforms. Consequently $Q(t)$ can be written

$$Q(t) = G(t) * E(t) = \int_0^t G(t-x)\, E(x)\, dx \qquad \textbf{(3.10.5)}$$

by the convolution theorem. Equation (3.10.5) shows that we can convert a differential equation, with suitable initial conditions, into an integral equation. In the study of ordinary differential equations, solutions to differential equations when forced by a delta function are known as a *Green's function*. As seen from Eq. (3.10.5), Green's function provides us with the possibility, once and for all, of finding solutions to an inhomogeneous equation for an arbitrary forcing function in terms of the convolution integral.

Despite the fundamental importance of the impulsive response or Green's function for a given linear system, it is often quite difficult to determine, especially experimentally. Consequently it is often more convenient in practice to deal with the response to the unit step $H(t)$, and this is called the *indicial admittance* $A(t)$. Because $H(t)] = \mathscr{L} 1/s$, the transfer function can be determined from the indicial admittance because $\mathscr{L}[A(t)] = T(s)\mathscr{L}[H(t)]$ or $s\,\bar{A}(s) = \bar{T}(s)$. Furthermore, because

$$\mathscr{L}[G(t)] = \bar{T}(s) = \frac{\mathscr{L}[A(t)]}{\mathscr{L}[H(t)]} = s\mathscr{L}[A(t)] \qquad \textbf{(3.10.6)}$$

then

$$G(t) = \frac{dA(t)}{dt} \qquad \textbf{(3.10.7)}$$

from Eq. (3.1.18).

In the example that we have been working through, the indicial admittance is given by

$$\mathscr{L}[A(t)] = \frac{\bar{T}(s)}{s} = \frac{1}{s}\,\frac{C}{1 + RCs} \qquad \textbf{(3.10.8)}$$

$$= \left(\frac{1}{s} - \frac{RC}{1 + RCs}\right) C \qquad \textbf{(3.10.9)}$$

and

$$A(t) = C\,H(t)\left[1 - \exp\left(-\frac{t}{RC}\right)\right] \tag{3.10.10}$$

Finally, let us calculate the charge $Q(t)$ given the electromotive force

$$E(t) = \frac{At}{CT}\,[H(t) - H(t-T)] \tag{3.10.11}$$

Because $E(t)$ is a discontinuous, we must consider two cases: (1) $t < T$ and (2) $t > T$. In both cases,

$$Q(t) = \int_0^t G(t-x)\,E(x)\,dx \tag{3.10.12}$$

$$= \int_0^t \frac{1}{RC} \exp\left(-\frac{t-x}{RC}\right) \frac{Ax}{T}\,[H(x) - H(x-T)]\,dx \tag{3.10.13}$$

For the first case,

$$Q(t) = \frac{A}{RCT} \exp\left(-\frac{t}{RC}\right) \int_0^t x \exp\left(\frac{x}{RC}\right) dx \tag{3.10.14}$$

$$= A\left\{\frac{t}{T} - \frac{RC}{T}\left[1 - \exp\left(-\frac{t}{RC}\right)\right]\right\} \quad 0 < t < T \tag{3.10.15}$$

In the second case,

$$Q(t) = \frac{A}{RCT} \exp\left(-\frac{t}{RC}\right) \int_0^T x \exp\left(\frac{x}{RC}\right) dx \tag{3.10.16}$$

$$= A\left(1 - \frac{RC}{T} + \frac{RC}{T} \exp\left(-\frac{T}{RC}\right)\right) \exp\left(-\frac{t-T}{RC}\right) \quad t > T \tag{3.10.17}$$

In Fig. 3.10.1, the solution Eq. (3.10.16) and Eq. (3.10.17) have been graphed for the case $T = RC$.

3.11 SUPERPOSITION: DUHAMEL'S INTEGRAL

We shall now show that given the indicial admittance $A(t)$ of a system, it is possible to determine the response of the system of any arbitrary excitation.

Consider first a system that is dormant until a certain time

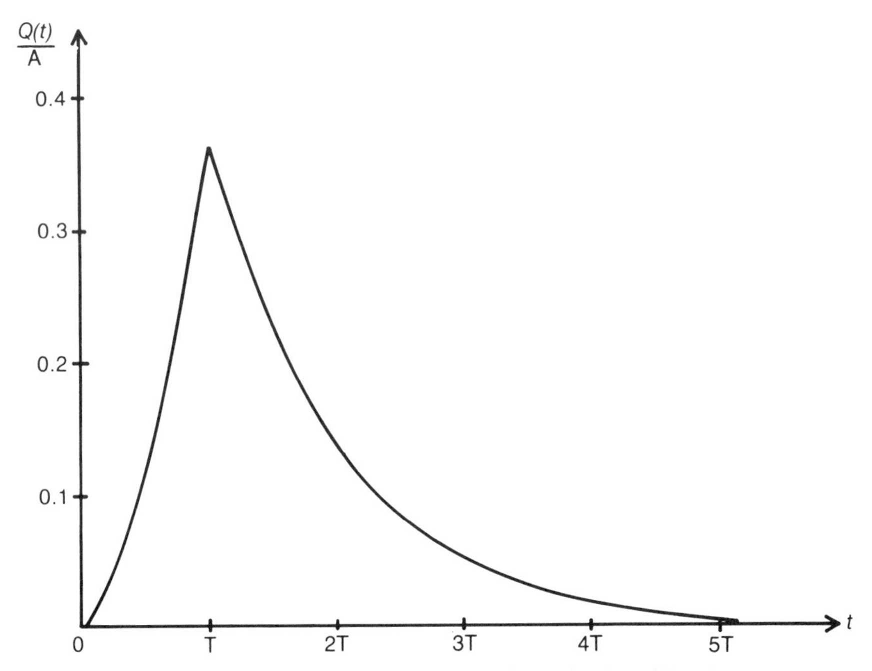

Fig. 3.10.1: The charge on a capacitor in a simple RC circuit when driven by the electromotive force of A(t) [H(t) − H(t − T)]/CT.

$t = \tau_1$. At that instant the system is subjected to a forcing $H(t-\tau_1)$. Then it is easily seen that the response will be zero if $t<\tau_1$ and will be given by the indicial admittance $A(t-\tau_1)$ when $t>\tau_1$ because the indicial admittance is the response of a system to the step function. Here $t-\tau_1$ is the time measured from the instant of change.

Next, suppose instead that the system is forced abruptly at the value $f(0)$ when $t = 0$ and held at the value until $t = \tau_1$, then is again abruptly raised by an amount $f(\tau_1) - f(0)$ to the value $f(\tau_1)$ at the time τ_1 and held at that value until $t = \tau_2$, then is abruptly raised by an amount $f(\tau_2) - f(\tau_1)$ at the time τ_2 and so on (see Fig. 3.11.1). From the linearity of the problem it is seen that at the instant following $t = \tau_n$, the response is given by the sum

$$\begin{aligned} y(t) &= f(0)A(t) + [f(\tau_1) - f(0)]\, A(t-\tau_1) \\ &+ [f(\tau_2) - f(\tau_1)]\, A(t-\tau_2) + \ldots \\ &+ [f(\tau_n) - f(\tau_{n-1})]\, A(t-\tau_n) \end{aligned} \tag{3.11.1}$$

If we write

$$f(\tau_k) - f(\tau_{k-1}) = \Delta f_k \tag{3.11.2}$$

$$\tau_k - \tau_{k-1} = \Delta\tau_k, \tag{3.11.3}$$

Eq. (3.11.1) can be written in the form

$$y(t) = f(0)A(t) + \sum_{k=1}^{n} A(t-\tau_k) \frac{\Delta f}{\Delta \tau_k} \Delta \tau_k \qquad \textbf{(3.11.4)}$$

Finally, proceeding to the limit as the number n of jumps becomes infinite, in such a way that all jumps and intervals between successive jumps tend to zero, this sum has the limit

$$y(t) = f(0)\, A(t) + \int_0^t f'(\tau)\, A(t - \tau)\, d\tau \qquad \textbf{(3.11.5)}$$

which is referred to as the *superposition integral*, or *Duhamel's integral*. It is referred to as superposition because the total response of the system is given by the weighted sum [the weights being $A(t)$] of the forcing from the initial moment up to the time t.

Equation (3.11.5) can be expressed in several different forms. Integrating by parts yields

$$y(t) = f(t)\, A(0) + \int_0^t f(\tau)\, A'(t-\tau)\, d\tau \qquad \textbf{(3.11.6)}$$

$$= \frac{d}{dt} \int_0^t f(\tau)\, A(t-\tau)\, d\tau \qquad \textbf{(3.11.7)}$$

For example, in the previous section we found that the indicial admittance for an RC circuit is

$$A(t) = C\left[1 - \exp - \left(\frac{t}{RC}\right)\right] \qquad \textbf{(3.11.8)}$$

When an electromotive force $E(t) = \exp(-at)$, $t \geq 0$, is applied

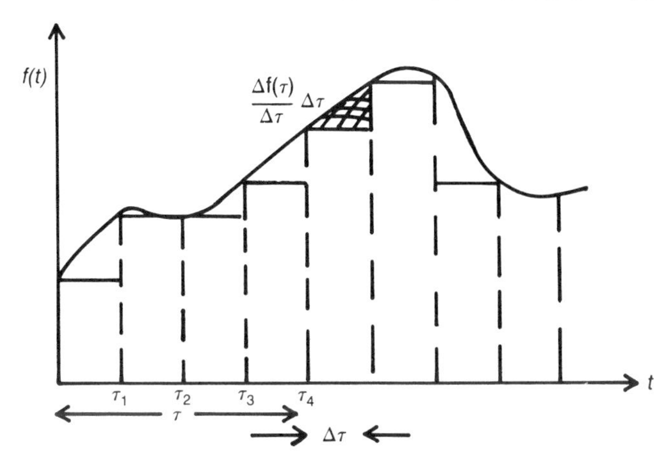

Fig. 3.11.1: Diagram used in the derivation of Duhamel's integral.

to a RC circuit which has no initial charge, the charge response can be found by the superposition integral:

$$Q(t) = E(0)A(t) + \int_0^t E'(x)\, A(t-x)\, dx \qquad (3.11.9)$$

$$= C\left[1 - \exp\left(-\frac{t}{RC}\right)\right]$$

$$- aC\int_0^t e^{-ax}\left(1 - \exp\left(-\frac{(t-x)}{RC}\right)\right) dx \qquad (3.11.10)$$

$$= C\left[1 - \exp\left(-\frac{t}{RC}\right)\right] - aC\int_0^t e^{-ax}\, dx$$

$$+ aC\, e^{-t/RC}\int_0^t e^{-ax + x/RC}\, dx \qquad (3.11.11)$$

$$= \frac{1}{R}\left(\frac{1}{1/RC - a}\right)\left[e^{-at} - \exp\left(-\frac{t}{RC}\right)\right] \quad t \geq 0 \qquad (3.11.12)$$

Exercises

For the following nonhomogeneous differential equations, find the transfer function, Green's function, and indicial admittance. Assume that all the necessary initial conditions are zero.

1. $y' + ky = f(t)$
2. $y'' - 2y' - 3y = f(t)$
3. $y'' + 4y' + 3y = f(t)$
4. $y'' - 2y' + 5y = f(t)$
5. $y'' - 3y' + 2y = f(t)$
6. $y'' + 4y' + 4y = f(t)$
7. $y'' - 9y = f(t)$
8. $y'' + y = f(t)$
9. $y''' + 2y'' - y' - 2y = f(t)$
10. $y''' - 6y'' + 11y' - 6y = f(t)$
11. $y''' - 2y'' - y' + 2y = f(t)$
12. $y''' + 3y'' + 3y' + y = f(t)$

3.12 A DYNAMICAL APPLICATION: THE RESPONSE OF A SIMPLE OSCILLATOR TO A BOMB BLAST

In Section 3.10 we discussed the use of Laplace transforms in electrical circuits where devices are turning on and off. Laplace transforms are also useful in dynamical problems when the system is forced by an impulsive force. To illustrate this let us determine the displacement of a mass m attached to a spring that is excited by a bomb blast, the forcing given by

$$F(t) = mA\left(1 - \frac{t}{T}\right) e^{-t/T} \qquad \textbf{(3.12.1)}$$

(See Glasstone, S., 1963: *The Effects of Nuclear Weapons*. Washington, D.C.: Department of Defense and Atomic Energy, pp. 124-128 for the derivation of Eq. (3.12.1).) This driving force is illustrated in Fig. 3.12.1.

The dynamical equation governing this system is

$$y'' + \omega^2 y = A\left(1 - \frac{t}{T}\right) e^{-t/T} \qquad \textbf{(3.12.2)}$$

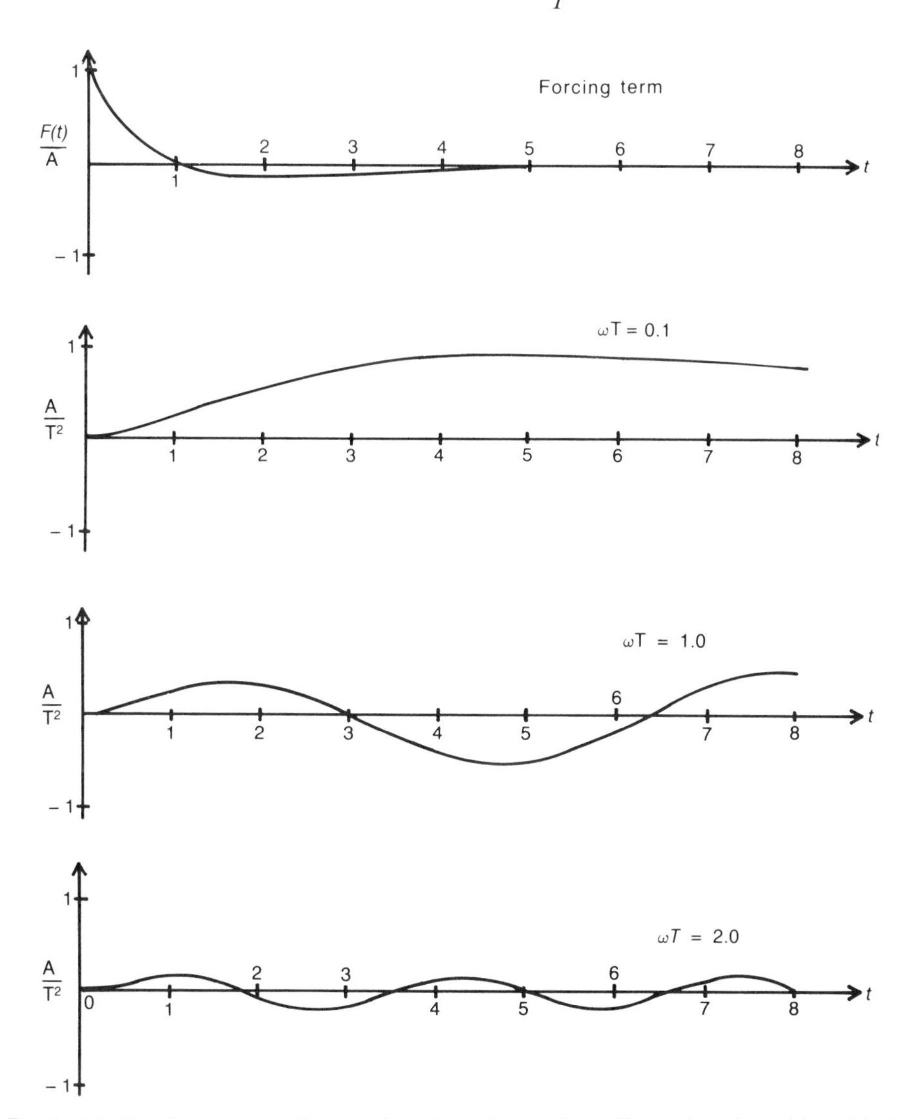

Fig. 3.12.1: The displacement of a mass in a simple harmonic oscillator when forced by a blast wave (given in the top graph).

where $\omega^2 = k/m$ and k is the spring constant. We assume that initially $y(0) = y'(0) = 0$. The corresponding transformed equation is

$$s^2 \bar{y} + \omega^2 \bar{y} = A \frac{1}{s + \frac{1}{T}} - \frac{A}{T} \frac{1}{(s + \frac{1}{T})^2} \qquad \textbf{(3.12.3)}$$

or

$$\bar{y} = \frac{A}{(s^2 + \omega^2)(s + \frac{1}{T})} - \frac{A}{T} \frac{1}{(s^2 + \omega^2)(s + \frac{1}{T})^2} \qquad \textbf{(3.12.4)}$$

Partial fractions yield

$$\bar{y} = \frac{A}{\omega^2 + \frac{1}{T^2}} \left[\frac{1}{s + \frac{1}{T}} - \frac{s - \frac{1}{T}}{s^2 + \omega^2} \right]$$

$$- \frac{A}{T} \frac{1}{(\omega^2 + \frac{1}{T^2})^2} \left[\frac{\frac{1}{T^2} - \omega^2}{s^2 + \omega^2} - \frac{\frac{2s}{T}}{s^2 + \omega^2} + \frac{\omega^2 + \frac{1}{T^2}}{(s + \frac{1}{T})^2} + \frac{\frac{2}{T}}{s + \frac{1}{T}} \right]$$

$$\textbf{(3.12.5)}$$

Inverting Eq. (3.12.5), term by term, gives

$$y(t) = \frac{A}{\omega^2 + \frac{1}{T^2}} \left[(e^{-t/T} - \cos(\omega t) + \frac{1}{\omega T} \sin(\omega t) \right] - \frac{A}{T} \frac{1}{(\frac{1}{T^2} + \omega^2)^2} \times$$

$$\left\{ (\frac{1}{T^2} - \omega^2) \frac{\sin(\omega t)}{\omega} + \frac{2}{T} [e^{-t/T} - \cos(\omega t) + (\omega^2 + \frac{1}{T^2}] te^{-t/T} \right\}$$

$$\textbf{(3.12.6)}$$

The solution to this problem consists of two parts. The exponential functions are those terms resulting from the direct forcing. The sinusoidal terms are those natural oscillations that are necessary so that the total solution will satisfy the initial conditions. In Fig. 3.12.1, the solution is graphed when $\omega T = 0.1$, 1 and 2. Note how the displacement decreases in magnitude as the natural frequency of the oscillator becomes larger.

Exercises

Show that the solution of

$$y'' + \omega^2 y = \omega^2 \sin(\pi t/t_i)\,[H(t) - H(t-t_i)]$$

is

$$u(t) = \frac{1}{(T/2t_i) - (2t_i/T)}\left[\sin(\omega t) - \frac{2t_i}{T}\sin(\pi t/t_i)\right] \qquad 0<t<t_i$$

$$= \frac{1}{(T/2t_i) - (2t_i/T)}\left\{[1+\cos(\omega t_i)]\sin(\omega t) - \sin(\omega t_i)\cos(\omega t)\right\} \qquad t>t_i$$

if $u(0) = u'(0) = 0$ and $T = 2\pi/\omega$.

3.13 THE INVERSION INTEGRAL

In Sections 3.5 and 3.8, we have shown how partial fractions and convolution can be used to find the inverse of the Laplace transform $F(s)$. In many instants these methods fail simply because of the complexity of the transform to be inverted. In this section we shall show how inversions may be found through the powerful method of contour integration.

Consider the piecewise differentiable function $f(x)$ which vanishes for $x<0$. We can express the function $e^{-cx} f(x)$ by the complex Fourier representation of

$$f(x)\, e^{-cx} = \frac{1}{2\pi} \lim_{R\to\infty} \int_{-R}^{R} e^{i\omega x} \left[\int_0^\infty e^{-i\omega t} e^{-ct} f(t)\, dt\right] d\omega \qquad \textbf{(3.13.1)}$$

according to Eq. (2.14.14) for any value of the real constant c where the integral

$$I = \int_0^\infty e^{-ct}\, |f(t)|\, dt \qquad \textbf{(3.13.2)}$$

exists. By multiplying both sides of Eq. (3.13.1) by e^{cx} and bringing e^{cx} inside the first integral, we obtain

$$f(x) = \frac{1}{2\pi} \lim_{R\to\infty} \int_{-R}^{R} e^{(i\omega + c)x} \left[\int_0^{\infty} e^{-(i\omega + c)t} f(t)\, dt \right] d\omega \tag{3.13.3}$$

With the substitution $i\omega + c = s$, where s is a new, complex variable of integration, we have

$$f(x) = \frac{1}{2\pi i} \lim_{R\to\infty} \int_{c-iR}^{c+iR} e^{sx} \left[\int_0^{\infty} e^{-st} f(t)\, dt \right] ds \tag{3.13.4}$$

The quantity inside the integral is equal to the Laplace transform $F(s)$. Therefore $f(t)$ can be expressed in terms of its transform by the *complex* integral

$$f(t) = \frac{1}{2\pi i} \lim_{R\to\infty} \int_{c-iR}^{c+iR} e^{tz} F(z)\, dz \tag{3.13.5}$$

This integral is taken along the infinite line $x = c$ parallel to the imaginary axis and c units to the right of it. The value of c is chosen sufficiently large so that the integral (3.13.2) exists. Under these conditions Eq. (3.13.5) is known as *Bromwich's integral.*

The power behind Eq. (3.13.5) is the ability to write down, as a contour integral, a formula for $f(t)$ as soon as $F(s)$ is given. Therefore, Eq. (3.13.5) yields the answer to a problem as a time function straight away, once the transform has been found. The answer in this form requires, of course, a sufficient knowledge of complex variables so that the contour integral can be evaluated. Because of the power of the residue theorem (see Appendix B.8) in complex variables, the contour integral present in Eq. (3.13.5) is usually transformed into the closed contour shown in Fig. 3.13.1. This transformation requires that the contribution around the infinite semicircle C_1 of Fig. 3.13.1 to be zero. Because most transforms arising in engineering and scientific problems tend to zero as $|z| \to \infty$, the contribution from this path at infinite is indeed zero and Eq. (3.13.5) may be evaluated around the closed contour shown in Fig. 3.13.1 provided that all the singular points are to the *left*

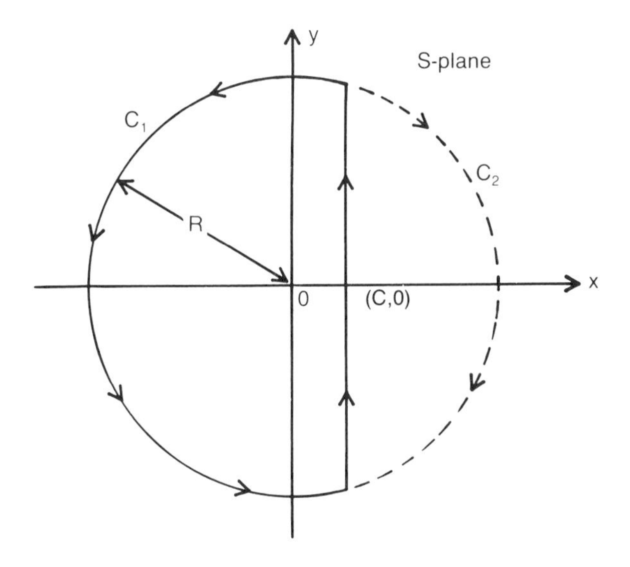

Fig. 3.13.1: Contour used in the inversion integral.

of the line and there are no branch points. If branch points are present, the inversion integral still holds but with a modified contour. The following examples will illustrate the proper use of Eq. (3.13.5).

Let us find the inverse of the transform

$$F(s) = \frac{1}{s} \tanh(as/2) = \frac{1}{s} \frac{\sinh(as/2)}{\cosh(as/2)} \tag{3.13.6}$$

The transform appears to have poles at $s = 0$ and $\cosh(as/2) = 0$. Considering the case of $s = 0$ first, the application of *l'Hospital's* rule to this transform shows that $s = 0$ is not a pole. On the other hand, $\cosh(as/2) = 0$ does give an infinite number of simple poles because

$$\cosh(as/2) = \cos(ias/2) = 0 \tag{3.13.7}$$

at

$$ias/2 = \pm(2n+1)\frac{\pi}{2} \tag{3.13.8}$$

or

$$s = \pm \frac{(2n+1)\pi}{a} i, \; n = 0, 1, 2, 3, \ldots \tag{3.13.9}$$

The residues are most readily computed from the formula for simple poles:

$$\text{Residue at } (s = k) = \frac{\text{Numerator}}{\text{Derivative of Denominator}}\bigg|_{s=k} \qquad \textbf{(3.13.10)}$$

Thus

$$Res\left(s = \pm (2n+1)\frac{\pi i}{a}\right) = \frac{e^{st}\sin(as/2)}{\frac{d}{ds}(s\cosh(as/2))}\bigg|_{s = \pm(2n+1)\frac{\pi i}{a}} \qquad \textbf{(3.13.11)}$$

$$= \frac{e^{st}\sinh(as/2)}{as/2\ \sinh(as/2) + \cosh(as/2)}\bigg|_{s = \pm(2n+1)\frac{\pi i}{a}} \qquad \textbf{(3.13.12)}$$

$$= \frac{\exp(\pm i\ (2n+1)\frac{\pi t}{a})}{\pm(2n+1)\ \frac{\pi i}{2}} \qquad \textbf{(3.13.13)}$$

because $\cosh(as/2) = 0$ and $\sinh\left((2n+1)\frac{\pi i}{2}\right) = (-1)^n i$.

By the residue theorem, the value of the integral (3.13.5) is given by the sum of the residues over all the poles. Therefore

$$\mathcal{L}^{-1}\left\{\frac{1}{s}\left[\tanh(as/2)\right]\right\} = \sum_{n=0}^{\infty} \frac{\exp\left[(2n+1)\ \frac{i\pi t}{a}\right]}{i(2n+1)\pi/2} + \sum_{n=0}^{\infty} \frac{\exp\left(-(2n+1)\frac{i\pi t}{a}\right)}{i[-(2n+1)\pi/2]} \qquad \textbf{(3.13.14)}$$

$$= \frac{2}{\pi}\sum_{n=0}^{\infty} \frac{e^{i(2n+1)\pi t/a} - e^{-i(2n+1)\pi t/a}}{(2n+1)i} \qquad \textbf{(3.13.15)}$$

$$= \frac{4}{\pi}\sum_{n=0}^{\infty} \frac{\sin[(2n+1)\pi t/a]}{(2n+1)} \qquad \textbf{(3.13.16)}$$

In Section 3.7, we found the inverse of $s^{1/2}$. We now find the inverse by the inversion integral. This problem is of particular interest because there is a branch point (see Appendix B.11 for discussion of branch points) at the origin so that the residue method may not be applied. To circumvent this difficulty we take the branch cut along the negative x axis and distort our contour as shown in

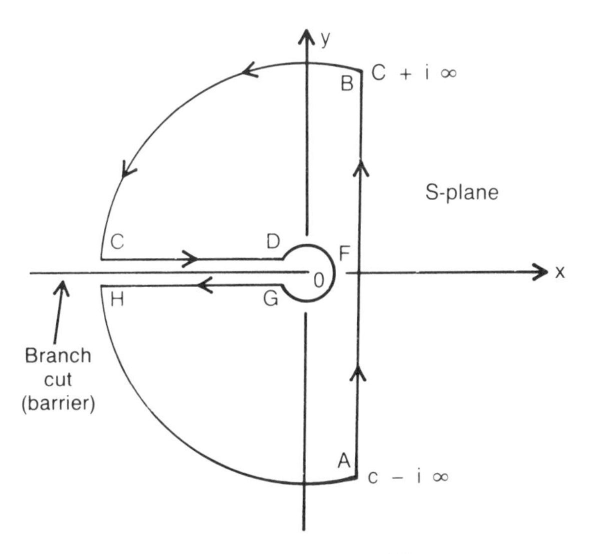

Fig. 3.13.2: Contour used in the inversion of $s^{1/2}$.

Fig. 3.13.2. In particular *DFG* is a small circle surrounding the origin whose radius ϵ will be allowed to approach zero.

Because the entire contour does not include any singularities, Cauchy's theorem states that

$$\int_{ABCDFGH} = \int_{AB} + \int_{BC} + \int_{CD} + \int_{DFG} + \int_{GH} + \int_{HA} = 0 \quad \mathbf{(3.13.17)}$$

Next we note that the contribution from the paths *BC* and *HA* become negligibly small as the outer semicircle expands to infinity. Consequently, we are left with

$$\int_{AB} \frac{e^{st}}{s^{1/2}} \, ds = \int_{HG} \frac{e^{st}}{s^{1/2}} \, ds + \int_{GFD} \frac{e^{st}}{s^{1/2}} \, ds + \int_{DC} \frac{e^{st}}{s^{1/2}} \, ds \quad \mathbf{(3.13.18)}$$

Along *HG*, $s = x\, e^{-i\pi} = -x$, $ds = -\, dx$, $s^{1/2} = x^{1/2}\, e^{-\pi i/2} = -i\, x^{1/2}$ and

$$\int_{HG} \frac{e^{st}}{s^{1/2}} \, ds = -i \int_{\infty}^{0} x^{-1/2}\, e^{-xt}\, dx$$

$$= i \int_{0}^{\infty} x^{-1/2}\, e^{-xt}\, dt \quad \mathbf{(3.13.19)}$$

Along *DC*, $s = x\, e^{i\pi} = -x$, $ds = -dx$, $s^{1/2} = x^{1/2}\, e^{\pi i/2} = i\, x^{1/2}$, giving

$$\int_{DC} \frac{e^{st}}{s^{1/2}} \, ds = i \int_0^\infty x^{-1/2} e^{-xt} \, dx \tag{3.13.20}$$

For the integral around the small circle GFD, $s = r\, e^{i\theta}$, $ds = i\, r\, e^{i\theta}\, d\theta$, $s^{1/2} = r^{1/2}\, e^{\theta i/2}$,

$$\int_{GFD} \frac{e^{st}}{s^{1/2}} \, ds = i \lim_{r \to 0} \int_{-\pi}^{\pi} \exp[r(\cos(\theta) + i \sin(\theta)]\, t)\, r^{1/2}\, e^{i\theta/2}\, d\theta$$

$$= 0 \tag{3.13.21}$$

Consequently, we are left with

$$\int_{AB} \frac{e^{st}}{s^{1/2}} \, ds = \frac{4i}{t^{1/2}} \int_0^\infty x^{-1/2} e^{-xt} \, dx \tag{3.13.22}$$

$$= \frac{4i}{t^{1/2}} \int_0^\infty e^{-\lambda^2} \, d\lambda \tag{3.13.23}$$

$$= \frac{4i}{t^{1/2}} \left[\frac{\pi^{1/2}}{2} \operatorname{erf}(\infty) \right] \tag{3.13.24}$$

$$= \frac{2i\pi^{1/2}}{t^{1/2}} \tag{3.13.25}$$

Finally,

$$\mathscr{L}^{-1}(s^{-1/2}) = \frac{1}{2\pi i} \int_{c-i\infty}^{c+i\infty} \frac{e^{st}}{s^{1/2}} \, ds \tag{3.13.26}$$

$$= \frac{1}{2\pi i} \int_{AB} \frac{e^{st}}{s^{1/2}} \, ds = \frac{1}{\sqrt{\pi t}} \tag{3.13.27}$$

Exercises

In problems 1 through 8, obtain the inverse transform of the following by the method of residues.

1. $\dfrac{s}{s^2 + 9}$

2. $\dfrac{1}{s^2 + 5s + 6}$

3. $\dfrac{1}{s(s+1)(s+2)}$

4. $\dfrac{1}{s^2(s-a)}$

5. $\dfrac{s+1}{s^2(s-1)}$

6. $\dfrac{1}{s^3(s+1)^2}$

7. $\dfrac{s+1}{s(s^2+1)^2}$

8. $\dfrac{s+2}{s(s-a)(s^2+4)}$

9. Obtain the inverse transform of $\tanh(s)/s^2$ as a Fourier series by the residue method.

10. Find the inverse of $[1 - \cosh(\gamma\xi)/\cosh(\gamma)]/\gamma^2$ where $\gamma = \sqrt{Rs + M^2/\alpha}$.

Answers

P. 138

1. $\dfrac{s+1}{(s+1)^2+25}$

2. $\dfrac{1}{(s+1)^2}$

3. $\dfrac{2as}{(s^2+a^2)^2}$

4. $\dfrac{a}{s^2-a^2}$

5. $\dfrac{s}{a^2-s^2}$

6. $\dfrac{6}{s^4} + \dfrac{6}{s^3} + \dfrac{2}{s^2}$

7. $\frac{2s(s^2 - 3)}{(s^2 + 1)^3}$

8. $\frac{6}{(s^2 + 1)(s^2 + 9)}$

9. $\frac{1}{s^2} + \left(\frac{a}{s + 1} - \frac{1}{s^2} - \frac{a}{s}\right) e^{-as}$

10. $\frac{1 + e^{-s\pi}}{1 + s^2}$

11. $\frac{1}{s^2} - \left(\frac{1}{s^2} + \frac{a}{s}\right) e^{-as}$

P. 141

1. $\frac{e^{-s}(2 + 2s + s^2)}{s^3}$

2. $\frac{e^{-2s}}{s^2+1} [s \cos(2) - \sin(2)]$

3. $\frac{se^{-s}}{s^2+1}$

4. $\frac{2e^{-s}}{s^3} + \frac{2e^{-s}}{s^2}$

5. $e^4 \frac{e^{-2s}}{s-2}$

6. $\frac{6s^2 + 24s + 26}{(s^3 + 6s^2 + 13s)^2}$

7. $\frac{4s + 12}{(s^2 + 6s + 13)^2}$

8. $-\frac{1}{s} \ln \left(\frac{s-1}{\sqrt{s^2+4}}\right)$

9. $-\ln \left(\frac{s-1}{s}\right)$

10. $\dfrac{\pi}{2} - \tan^{-1}\left(\dfrac{s+3}{2}\right)$

11. te^{2t}

12. $e^{2t}\,(2t-1)/4 + 1/4$

13. $1/2 \sin[2(t-2)]\, H(t-2)$

14. $\sinh[3(t-3)]\, H(t-3)/3$

15. $(t-1)^2\, e^{1-t}\, H(t-1)/2$

16. $(e^{2(t-1)} - e^{t-1})H(t-1) + (e^{2(t-2)} - e^{(t-2)})\; H(t-2)$

P. 144

1. $\dfrac{\coth(\pi s/2)}{1 + s^2}$

2. $\dfrac{1}{(1 + s^2)\,(1 - e^{-\pi s})}$

3. $\dfrac{1 - (1 + sa)\, e^{-sa}}{s^2\,(1 - e^{-2as})}$

4. $\dfrac{1}{s(1 + e^{-as})}$

P. 150

1. $$y(t) = 1/2(e^{-t} - e^{-3t}) - H(t-2)\,(e^{-t+2} - e^{-3t+6}) + H(t-4)\;(e^{-t+4} - e^{-3t+12}) - \ldots$$

$$= \sum_{n=1}^{\infty} \frac{3}{a_n}\,\frac{3\sin(a_n t) - a_n\cos(a_n t) + a_n e^{-3t}}{a_n^{\,2} + 9} - \sum_{n=1}^{\infty} \frac{1}{a_n}\,\frac{\sin(a_n t) - a_n\cos(a_n t) + a_n e^{-t}}{a_n^{\,2} + 1}$$

where $a_n = (2n-1)\pi/2$.

2. $$y(t) = 1/4\,[1-(1+2t)e^{-2t}] - 1/4\,\{1-[1+2(t-1)]e^{-2(t-1)}\}\; H(t-1) + \ldots$$

$$= \sum_{n=1}^{\infty} \frac{te^{-2t}}{a^2_{\;n}+4} - \frac{4e^{-2t}\cos(2a_n t)}{(a_n^{\,2} + 4)^2} + ①$$

$$① = \sum_{n=1}^{\infty} \frac{(4 - a_n^2)\sin(a_n t) + 4a_n \cos(a_n t)}{a_n(a_n^2 + 4)^2}$$

where $a_n = (2n - 1)\pi$.

3. $y(t) = 1 - \cos(t) - [1 - \cos(t - \pi)]\,H(t - \pi) + [1 - \cos(t - 2\pi)]H(t - 2\pi) - \ldots$

$$= \frac{2}{\pi}[\sin(t) - t\cos(t)]$$

$$+ \frac{2}{\pi} \sum_{n=2}^{\infty} \frac{\sin(nt)\cos[(n-1)t]}{n} - \frac{\cos(nt)\sin[(n-1)t}{n - 1}$$

P. 156

1. $e^{-t}/4 - e^{-2t} + 3e^{-4t} - \frac{9}{4}e^{-5t}$

2. $\frac{5}{2}e^{-t} - 9e^{-2t} + \frac{15}{2}e^{-3t}$

3. $e^{-t} - e^{-2t}$

4. $\frac{1}{6}e^{-4t} + \frac{5}{6}e^{2t}$

5. $3e^{t} - 2e^{t/2}$

P. 163

1. $[1 - \cos(at)]/a$
2. $(e^{at} - 1 - at)/a^2$
3. $(e^{at} - e^{bt})/(a - b)$
4. $[b\sin(at) - a\sin(bt)]/(b^2 - a^2)$
5. $[\sin(2t) - 2t\cos(2t)]/16$
6. $\sin(3t)/216 - t^2\cos(3t)/72$
7. $e^{-2t} - e^{-3t}$
8. $t^2/a - 4\sin^2(at/2)/a^3$
9. $e^{(a+b)s}/s^2$; $t\,H(t - a - b)$

P. 165

1. $f(t) = 1 + 2t$
2. $f(t) = 1 + t^2/2$
3. $f(t) = t + t^2/2$
4. $f(t) = -1 + 2t + 2t^2 + e^{-2t}$
5. $f(t) = t^3 + t^5/20$
6. $f(t) = 2t^2 + 3 - 3\cos(2t)$

7. $f(t) = t^2 - t^4/3$
8. $f(t) = 1 + 2t\, e^t$
9. $f(t) = 5e^{2t} + 4e^{-t} - 6t\, e^{-t}$
10. $f(t) = t^2 + t^4/12$
11. $f(t) = e^{-t}(1 - t)^2$
12. $f(t) = e^{2t} - e^{-t}\,[\cos(3^{1/2}t) - 3^{1/2}\sin(3^{1/2}t)]$
13. $f(t) = 4 + \dfrac{5}{2}\, t^2 + \dfrac{t^4}{24}$
14. $f(t) = t^2/2$

P. 172

1. $\dfrac{1}{s+k}$, e^{-kt}, $\dfrac{1}{k}\,(1 - e^{-kt})$

2. $\dfrac{1}{s^2 - 2s + 3}$, $(e^{3t} - e^{-t})/4$, $-\dfrac{1}{3} + \dfrac{e^{3t}}{12} + e^{-t}/4$

3. $\dfrac{1}{s^2 + 4s + 3}$, $(e^{-t} - e^{-3t})/2$, $\dfrac{1}{3} + \dfrac{e^{-3t}}{6} - e^{-t}/2$

4. $\dfrac{1}{s^2 - 2s + 5}$, $e^t \sin(2t)/2$, $\dfrac{1}{10}\,(e^t[\sin(2t) - 2\cos(2t)] + 2)$

5. $\dfrac{1}{s^2 - 3s + 2}$, $e^{2t} - e^t$, $1/2 + e^{2t}/2 - e^t$

6. $\dfrac{1}{s^2 + 4s + 4}$, te^{-2t}, $[e^{-2t}(-2t - 1) + 1]/4$

7. $\dfrac{1}{s^2 - 9}$, $\dfrac{1}{6}\,(e^{3t} - e^{-3t})$, $-\dfrac{1}{9} + \dfrac{e^{3t}}{18} + \dfrac{e^{-3t}}{18}$

8. $\dfrac{1}{s^2 + 1}$, $\sin(t)$, $1 - \cos(t)$

9. $\dfrac{1}{s^3 + 2s^2 - s - 2}$, $\dfrac{1}{3}\, e^{-2t} - e^{-t}/2 + \dfrac{1}{6}\, e^t$

10. $\dfrac{1}{s^3 - 6s^2 + 11s - 6}$, $-e^{2t} + e^{3t}/2 + e^t/2$,

$-\dfrac{1}{6} - e^{2t}/2 + \dfrac{1}{6}\, e^{3t} + e^t/2$

11. $\dfrac{1}{s^3 - 2s^2 - s + 2}$, $\dfrac{1}{3} e^{2t} - e^t/2 + \dfrac{1}{6} e^{-t}$, $1/2 + \dfrac{1}{6} e^{2t} - e^t/2 - \dfrac{1}{6} e^{-t}$

12. $\dfrac{1}{s^3 + 3s^2 + 3s + 1}$, $t^2 e^{-t}/2$, $1 - (t^2 + 2t + 2)e^{-t}/2$

P. 180

1. $\cos(3t)$
2. $e^{-2t} - e^{-3t}$
3. $1/2 - e^{-t} + e^{-2t}/2$
4. $\dfrac{1}{a^2}(e^{at} - at - 1)$
5. $2e^t - t - 2$
6. $t^2/2 - 2t + 3 - (t+3)e^{-t}$
7. $1 - (t-1)\sin(t)/2 - (t+2)\cos(t)/2$
8. $\dfrac{a+2}{a(a^2+4)} e^{at} + \dfrac{1}{2a} + \dfrac{(a+2)\cos(2t) + (a^2+2a+6)\sin(2t)}{2a(a^2+4)}$
9. $1 - \dfrac{8}{\pi^2} \sum_{n=0}^{\infty} \dfrac{\cos[(2n+1)\pi t/2]}{(2n+1)^2}$
10. $\dfrac{2}{R} \sum_{n=0}^{\infty} \dfrac{(-1)^n}{\lambda_n} \exp\left[-(\lambda_n^2 + \dfrac{M^2}{\alpha}) \dfrac{t}{R}\right] \cos(\lambda_n \xi)$ where $\lambda_n = \dfrac{2n+1}{2}\pi$

Chapter 4

First-Order Partial Differential Equations

4.1 INTRODUCTION

This chapter treats various aspects of the solution of first-order partial differential equations (PDEs). Of particular importance is the concept of characteristics because of its appearance in the solution of the wave equation. Consequently we shall dwell on this topic in considerable detail.

4.2 THE SOLUTION OF THE QUASILINEAR EQUATION; THE CAUCHY PROBLEM

The most general first-order, quasilinear partial differential equation is

$$P(x,y,z)\, z_x + Q(x,y,z)\, z_y = R(x,y,z) \tag{4.2.1}$$

This is the simplest case. There are only two independent variables x and y and the dependent variable z. (All that we may say, however, can be extended to n- dimensions.) If P and Q are independent of z and R is a linear function of z, $R_1(x,y)\, Z + R_2(x,y)$, then Eq. (4.2.1) is linear. It should be note that particular difficulties occur at points where $P = Q = 0$.

Suppose that

$$u(x,y,z) = c \tag{4.2.2}$$

Geometrically Eq. (4.2.2) represents a surface in the (x,y,z) coordinate system. Eq. (4.2.2) is an *integral* of Eq. (4.2.1) in the sense that it gives z as a function of x and y that satisfies Eq. (4.2.1). Finding this integral surface, which passes through some specified curve (the initial conditions) is called the *Cauchy problem*.

From the chain rule we obtain

$$u_x + u_z z_x = 0 \text{ and } u_y + u_z z_y = 0. \qquad \textbf{(4.2.3)}$$

In the derivatives u_x, u_y and u_z the variables x, y and z are considered as independent. Therefore

$$z_x = -u_x/u_z \text{ and } z_y = -u_y/u_z \qquad \textbf{(4.2.4)}$$

if $u_z \neq 0$. If these expressions are introduced into Eq. (4.2.1), an equivalent equation is obtained in the form

$$P\,u_x + Q\,u_y + R\,u_z = 0. \qquad \textbf{(4.2.5)}$$

The nature of this equation can be made slightly more explicit by introducing the vector $P\boldsymbol{i} + Q\boldsymbol{j} + R\boldsymbol{k}$ where $\boldsymbol{i}, \boldsymbol{j}$, and $\boldsymbol{k}$ are the usual Cartesian unit vectors. Consequently Eq (4.2.5) becomes

$$(P\,\mathbf{i} + Q\,\mathbf{j} + R\,\mathbf{k})\cdot\nabla u = 0 \qquad \textbf{(4.2.6)}$$

Thus we see that at any point on the integral surface,the vector (P,Q,R) is tangent to the curve in the integral surface $u = c$ which passes through that given point.

Let us now introduce the position vector $\boldsymbol{r}$. From vector calculus we know that the unit tangent vector to the surface $u(x,y,z) = c$ is given by

$$\frac{d}{ds}(\mathbf{r}) = \frac{dx}{ds}\,\boldsymbol{i} + \frac{dy}{ds}\,\boldsymbol{j} + \frac{dz}{ds}\,\mathbf{k} \qquad \textbf{(4.2.7)}$$

where s represents the arc length along a given curve in the integral surface. The requirement that this vector have the same direction as the vector $P\boldsymbol{i} + Q\boldsymbol{j} + R\boldsymbol{k}$ is

$$P = \mu\frac{dx}{ds},\quad Q = \mu\frac{dy}{ds} \text{ and } R = \mu\frac{dz}{ds} \qquad \textbf{(4.2.8)}$$

where μ may be a function of x, y, and z. This can be written

$$\frac{dx}{P} = \frac{dy}{Q} = \frac{dz}{R} \qquad \textbf{(4.2.9)}$$

Equation (4.2.9) yields two ordinary differential equations. Let $u_1(x,y,z) = c_1$ and $u_2(x,y,z) = c_2$ be the solutions of these two independent equations. Then because Eq. (4.2.9) and Eq. (4.2.8) are equivalent, we have

$$du_1 = \frac{\partial u_1}{\partial x}\, dx + \frac{\partial u_1}{\partial y}\, dy + \frac{\partial u_1}{\partial z}\, dz$$

$$= (P\frac{\partial u_1}{\partial x} + Q\frac{\partial u_1}{\partial y} + R\frac{\partial u_1}{\partial z})\,\frac{ds}{\mu} = 0. \qquad \textbf{(4.2.10)}$$

Hence $u = u_1$ and (similarly) $u = u_2$ are solutions of Eq. (4.2.5). These are also solutions to Eq. (4.2.1) because $u_1 = c_1$ and $u_2 = c_2$ determine z as functions of x and y. We next replace the independent variables x,y,z in Eq. (4.2.5) by the new variables $r = u_1(x,y,z)$, $s = u_{21}(x,y,z)$ and $t = \phi(x,y,z)$ where ϕ is *any* function that is independent of u_1 and u_2 and which does *not* satisfy Eq. (4.2.5). There then follows

$$u_x = u_r\, r_x + u_s\, s_x + u_t\, t_x = u_r \frac{\partial u}{\partial x}{}_1 + u_s \frac{\partial u}{\partial x}2 + u_t \frac{\partial \phi}{\partial x} \qquad \textbf{(4.2.11)}$$

together with similar expressions for u_y and u_z. With these substitutions, Eq. (4.2.5) takes the form

$$(P\frac{\partial u_1}{\partial x} + Q\frac{\partial u_1}{\partial y} + R\frac{\partial u_1}{\partial z})u_r + (P\frac{\partial u_2}{\partial x} + Q\frac{\partial u_2}{\partial y} + R\frac{\partial u_2}{\partial z})u_s$$

$$+ (P\frac{\partial \phi}{\partial x} + Q\frac{\partial \phi}{\partial y} + R\frac{\partial \phi}{\partial z})u_t = 0 \qquad \textbf{(4.2.12)}$$

where u is now considered as a function of the independent variables $r = u_1$, $s = u_2$ and $t = \phi$. But because u_1 and u_2 satisfy Eq. (4.2.5) identically, the coefficients of u_r and u_s vanish. Because ϕ does *not* satisfy Eq. (4.2.5), $u_t = 0$. Hence the most general solution of Eq. (4.2.5) is $u = F(r,s) = F(u_1,u_2)$ where F is an arbitrary differentiable function. Thus finally, if $u_z \neq 0$, the most general solution of Eq. (4.2.1) is $u = c$ or $F(u_1,u_2) = c$. Because c can be incorporated into F, c can be replaced by zero and $F(u_1,u_2) = 0$ can be rewritten as $u_2 = f(u_1)$.

To illustrate what we have been discussing, let us solve the problem

$$x\, u_x + (x^2 + y)\, u_y + (\frac{y}{x} - x)\, u = 1 \qquad \textbf{(4.2.13)}$$

with the initial condition that $u(1,y) = 0$. (This boundary condition is often called Cauchy data.) In this case $P = x$, $Q = x^2 + y$, and an $R = 1 - (\frac{y}{x} - x)u$. Consequently

$$\frac{dx}{P(x,y,u)} = \frac{dx}{x} = \frac{dy}{x^2+y} = \frac{dy}{Q(x,y,u)} \qquad (4.2.14)$$

Integrating Eq. (4.2.14), which is equivalent to solving the ordinary differential equation

$$x\frac{dy}{dx} - y = x^2 \frac{d}{dx}\left(\frac{y}{x}\right) = x^2, \qquad (4.2.15)$$

we obtain

$$y = x^2 + u_1 x \qquad (4.2.16)$$

or

$$\frac{y-x^2}{x} = u_1 = c_1. \qquad (4.2.17)$$

To find $u(x,y)$ we solve

$$\frac{du}{1 - \left(\frac{y}{x} - x\right)u} = \frac{du}{R(x,y,z)} = \frac{dx}{x} = \frac{dx}{P(x,y,z)} \qquad (4.2.18)$$

or

$$\frac{du}{dx} = \frac{1}{x}\left[1 - \left(\frac{y}{x} - x\right)u\right] = \frac{1}{x}(1 - c_1 u) \qquad (4.2.19)$$

or

$$\frac{du}{(1-c_1 u)} = \frac{dx}{x}. \qquad (4.2.20)$$

Integrating this equation gives us

$$\left[1 - \left(\frac{y - x^2}{x}\right)u\right] x^{(y-x^2)/x} = u_2 = c_2 \qquad (4.2.21)$$

$$= f(u_1) = f\left(\frac{y-x^2}{x}\right). \qquad (4.2.22)$$

To determine f, we use the Cauchy data $u = 0$ on $x = 1$. With these values we find that $f(y-1) = 1$. Consequently

$$u = \frac{x}{y - x^2}\left[1 - x^{-(y-x^2)/x}\right]. \qquad (4.2.23)$$

From the technique outlined above, it would appear that we can obtain an unique solution for a first-order partial differential equation for any specified boundary condition. This is not true. If we had chosen the initial condition that $u = 0$ along the line $y - x^2 = x$, we would have obtained the absurd result that $f(1) = x$. The explanation of this result will be given in Section 4.4.

4.3 AN EXAMPLE OF A FIRST-ORDER PARTIAL DIFFERENTIAL EQUATION: INCOMING CALLS AT A TELEPHONE EXCHANGE

Partial differential equations of the first order occur in the theory of stochastic processes—in Brownian motion, radioactive disintegrations, noise in communication systems, population growth and many problems dealing with telephone traffic. With certain assumptions regarding the duration of the calls, the time of day and whether we are considering a business day or a holiday, the problem of determining the probable number of incoming calls at a telephone exchange may be formulated as follows.

The probability that n telephone lines at an exchange will be busy is an example of a birth and death stochastic process. Each busy line is born, has a finite life-time that is independent of the other phone conversations, and eventually dies. In the simplest model the probability that n lines will be busy, $P_n(t)$, is directly proportional to the probability that a new call will be received and the probability that a conversation will be ended. If n lines are busy, the probability that one of them will become free within the time interval Δt is $na\ \Delta t + O(\Delta t^2)$ where "a" is the probability that a conversation will end. On the other hand, the probability of a new call arriving is $b\Delta t + O(\Delta t^2)$ where "b" is the probability that a new call arrives. Consequently the probability $P_n(t)$ at the time $t + \Delta t$ may be computed from the Taylor expansion

$$P_n(t+\Delta t) = P_n(t)(1 - b\Delta t - na\Delta t) + b\Delta t\, P_{n-1}(t) + na\Delta t\, P_{n+1}(t) + \text{higher order terms.} \quad \textbf{(4.3.1)}$$

Transposing the $P_n(t)$, dividing the equation by Δt and letting $\Delta t \to 0$, we get

$$P_n'(t) = -(b+na)\, P_n(t) + b\, P_{n-1}(t) + (n+1)\, a\, P_{n+1}(t). \quad \textbf{(4.3.2)}$$

Of course, we must have the equation

$$P_0'(t) = -b\, P_0(t) + a\, P_1(t) \quad \textbf{(4.3.3)}$$

for $n = 0$ because there is no $P_{-1}(t)$.

Equations (4.3.2) and (4.3.3) are finite difference equations and may be solved as such (see von Karman and M. A. Biot, 1940: *Mathematical Methods in Engineering*. New York: McGraw Hill Book Co.) However, if we invent the multivariable function

$$P(s,t) = \sum_{n=0}^{\infty} P_n(t)\, s^n, \qquad (4.3.4)$$

then Eqs. (4.3.2) and (4.3.3) may be transformed into the first-order partial differential equation

$$\frac{\partial P}{\partial t} = (1-s)\left(-bP + a\frac{\partial P}{\partial s}\right) \qquad (4.3.5)$$

because

$$\frac{\partial P}{\partial t} = \sum_{n=0}^{\infty} s^n P_n'(t) \qquad (4.3.6)$$

$$\frac{\partial P}{\partial s} = \sum_{n=0}^{\infty} (n+1)s\, P_{n+1}(t) \qquad (4.3.7)$$

$$s\frac{\partial P}{\partial s} = \sum_{n=0}^{\infty} n\, s^n P_n(t). \qquad (4.3.8)$$

The auxiliary equation is

$$\frac{ds}{a(s-1)} = \frac{dt}{1} = \frac{dP}{b(s-1)P} \qquad (4.3.9)$$

or

$$P\, e^{-bs/a} = c_1 \qquad (4.3.10)$$

and

$$(s-1)\, e^{-at} = c_2 \qquad (4.3.11)$$

The initial condition is given by $P(s,0) = s^k$ (corresponding to $P_k(0) = 1$) where k is the number of phone calls that are taking place at $t = 0$. Therefore

$$s^k = f(s-1)\, e^{bs/a} \qquad (4.3.12)$$

or

$$f(x) = e^{-b(1+x)/a}\,(1+x)^k. \qquad (4.3.13)$$

Consequently $P(s,t)$ becomes

$$P(s,t) = \exp\left[-\frac{b}{a}(1 - e^{-at})\right] \exp\left[\frac{bs}{a}(1 - e^{-at})\right]$$

$$\times\, [1 + (s-1)e^{-at}]^k. \qquad (4.3.14)$$

To obtain $P_n(t)$ again, Eq. (4.3.14) must be expanded out in terms of powers on s and all of the terms of common powers of s collected together:

$$P_n(t) = \sum_{i=0}^{n} \frac{\left[\frac{b}{a}(1 - e^{-at})\right]^i}{i!} \exp\left[-\frac{b}{a}(1 - e^{-at})\right]$$

$$\times \binom{k}{n-i} e^{-at(n-i)} (1 - e^{-at})^{k-n+i} \qquad \textbf{(4.3.15)}$$

where $\binom{n}{m}$ is the binomial coefficient for the integers n and m. In the limit of sufficiently long time, the limiting or equilibrium probability is given by the Poisson distribution with the parameter b/a—the traffic intensity—as

$$P_n(t = \infty) = \frac{1}{n!}\left(\frac{b}{a}\right)^n e^{-b/a}, \quad n = 0,1,2,3,\ldots \qquad \textbf{(4.3.16)}$$

In Fig. 4.3.1, the probabilities for no phone calls, one phone call, two phone calls are graphed for the simple case $k = 1$.

Exercises

Find the general solution for each of the following equations:

1. $u_x + u_y - u = 0$
2. $2u_x - 3u_y = x$
3. $u_x - 2u_y + u = \sin(x) + y$
4. $u_x - u_y - 2u = e^{2x}\cos(3y)$

Find the solution of the following partial differential equations satisfying the prescribe conditions:

5. $u_x + u_y = 1$, $u = e^x$ when $y = 0$.
6. $u_x + u_y - u = 0$, $u = 1 + \cos(x)$ when $y = 2x$.
7. $u_x + \cos(x)\, u_y = \sin(x)$, $u = y - 1 + \cos(x)$ when $x = 0$.
8. $2u_x - 5u_y + 4u = x^2$, $u = \sin(y) + e^y + 1/8$ when $x = 0$.
9. $x u_x - y u_y + u = x$, $u = x^2$ when $y = x$.
10. $y u_x - x u_y + x u = 0$, $u = y$ when $x^2 + 2y^2 = 4$.
11. $y u_x - x u_y + 2xy\, u = 0$, $u = e^x \sin(1 + x)$ when $y^2 = 2x + 1$.

12. A problem similar to that which we solved involving telephone exchanges occurs in predicting the demand on an electric circuit by N welders who work independently of each other. The probability that n welders are using the current is given by (see Feller, W., 1957: *An Introduction to Probability Theory and Its*

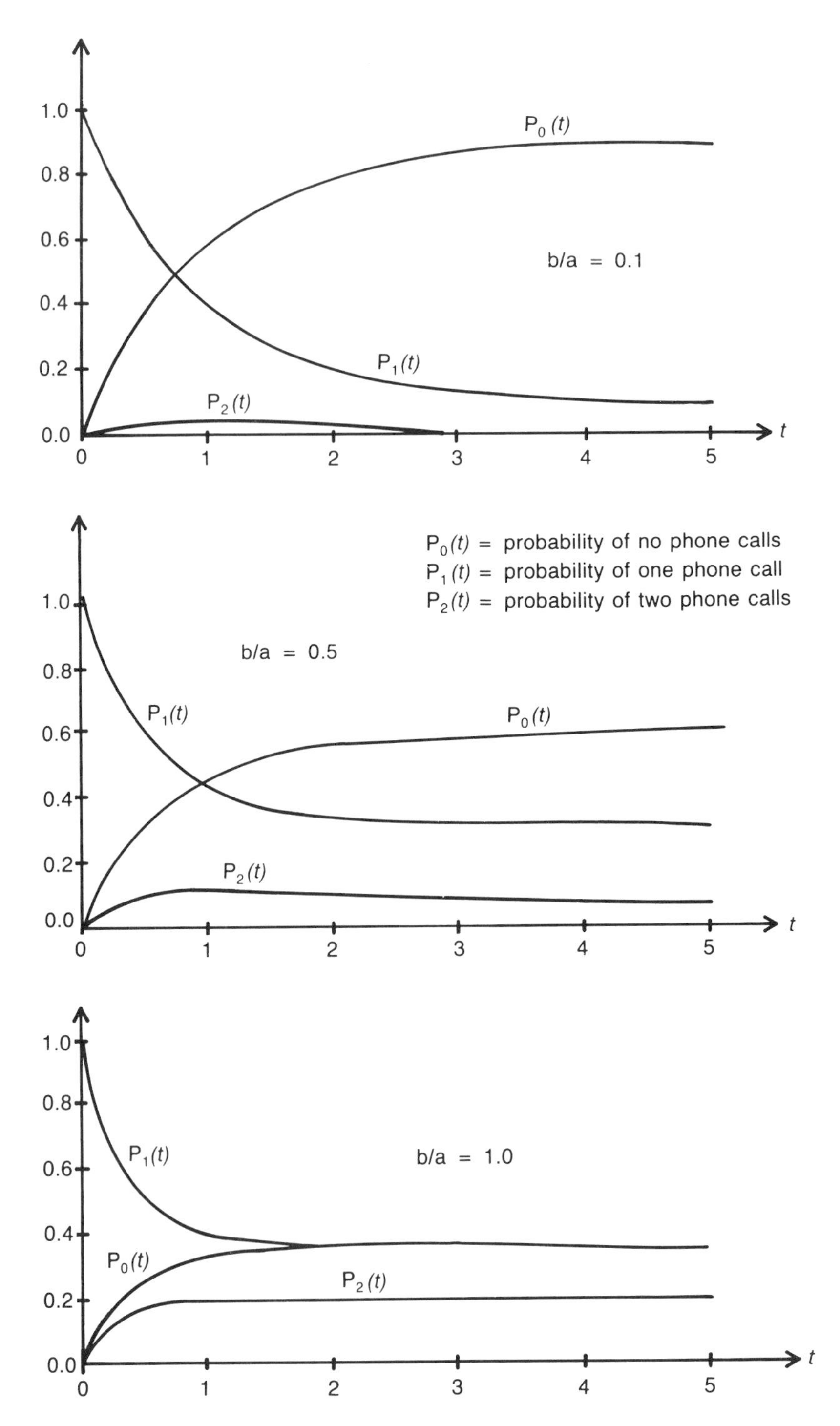

Fig. 4.3.1: The probability of phone calls at a telephone exchange as a function of time for various ratios of b/a.

Application. Vol. 1. 2nd Ed. New York: John Wiley and Sons, Inc. pp. 420-421.)

$$P_0'(t) = -N\lambda P_0(t) + \mu P_1(t)$$

$$P_n'(t) = -(n\mu + (N-n)\lambda)P_n(t) + (n+1)\mu P_{n+1}(t) + (N-n+1)\lambda P_{n-1}(t) \qquad 1 \leq n \leq N-1$$

$$P_N'(t) = -N\mu P_N(t) + \lambda P_{N-1}(t)$$

where μ denotes the probability for a new call for current within the time interval Δt and λ, the probability that one of the n welders ceases using the current during the interval Δt.

(a) Show that these ordinary differential equations can be converted into the single, first-order partial differential equation

$$P_t = (1 - s)((\mu + \lambda s)P_s - N\lambda P)$$

where

$$P(s,t) = \sum_{n=0}^{N} P_n(t) s^n.$$

(b) Show that the solution is

$$P(s,t) = \left(\frac{\mu + \lambda s}{\mu + \lambda}\right)^N (1 - \lambda y)^N \text{ where } y = \frac{s-1}{\mu + \lambda s} e^{-(\mu+\lambda)t}$$

and $P_0(0) = 1$

and that the limiting probabilities ($t \to \infty$) are given by

$$P_n(\infty) = \binom{N}{n}\left(\frac{\lambda}{\lambda+\mu}\right)^n \left(\frac{\mu}{\lambda+\mu}\right)^{N-n}$$

4.4 CHARACTERISTICS

In Section 4.2 we hinted that solutions to quasilinear, first-order partial differential equations may be interpreted geometrically. In this section we shall explore this facet of these partial differential equations.

We have already shown that the first-order partial differential equation

$$x u_x + (x^2 + y)u_y + \left(\frac{y}{x} - x\right)u = 1 \qquad \textbf{(4.4.1)}$$

could be written

$$\frac{dx}{x} = \frac{dy}{y + x^2} = \frac{du}{1 - \left(\frac{y}{x} - x\right)u} \qquad \textbf{(4.4.2)}$$

This system of equations was then broken into two ordinary differential equation Eq. (4.2.15) and (4.2.19) and then solved.

An alternative method of solving Eq. (4.4.2) is to write it in the parametric form

$$\frac{dx}{dt} = x \tag{4.4.3}$$

$$\frac{dy}{dt} = y + x^2 \tag{4.4.4}$$

$$\frac{du}{dt} = 1 - (\frac{y}{x} - x)u. \tag{4.4.5}$$

Equations (4.4.3) through (4.4.5) define a family of curves in (x,y,u) space that are known as the *characteristic traces* of Eq. (4.4.1) and this technique is called the *method of characteristics*. See Section 4.6.

The solutions of Eqs. (4.4.3) through (4.4.5) are

$$x(s,t) = x_0(s)\, e^t \tag{4.4.6}$$

$$y(s,t) = y_0(s)\, e^t + x_0^2(s)\, e^{2t} \tag{4.4.7}$$

$$u(s,t) = u_0(s) \exp(-y_0(s)t/x_0(s)) + x_0(s)/y_0(s) \tag{4.4.8}$$

where the initial data for x,y and u is given by the curve (arc)

$$x(s,0) = x_0(s),\ y(s,0) = y_0(s) \text{ and } u_0(s,0) = u_0(s).$$

In the present problem, $y_0(s) = s$, $x_0(s) = 1$ and $u_0(s) = -1/s$ so that

$$x(s,t) = e^t \tag{4.4.9}$$

$$y(s,t) = s\, e^t + e^{2t} \tag{4.4.10}$$

and

$$u(s,t) = \frac{1}{s}(1 - e^{-st}) \tag{4.4.11}$$

If we eliminate t from Eq. (4.4.11), we have

$$\frac{y - x^2}{x} = s \tag{4.4.12}$$

and recover our earlier result that

$$u(x,y) = \frac{x}{y - x^2}[1 - x^{(x^2-y)/x}] \tag{4.4.13}$$

from Eq. (4.4.11).

The use of parametric equations to solve first-order partial differential equations requires the inversion of $x = x(s,t)$ and $y = y(s,t)$ to give s and t as functions of x and y. In order that the inversion can be carried out at a given point, the Jacobian $x_s y_t - y_s x_t$ must be nonzero at time $t = 0$. That is,

$$x_s(s,0)\, y_t(s,0) - y_s(s,0)\, x_t(s,0) = x_s(s,0)\, b - y_s(s,0)\, a \neq 0 \quad \textbf{(4.4.14)}$$

which is the condition on the initial data that it cannot be specified along a characteristic trace or tangent to one. Consequently no solution will exist if we choose our initial data to lie along

$$y - x^2 = sx$$

(for any s) as we did in the closing paragraph of Section 4.2.

All of this discussion suggests that characteristics are of fundamental importance in our understanding of the mathematical theory of first-order partial differential equations. In addition to these theoretical considerations, characteristics may represent important physical concepts.

Consider the following first-order partial differential equation:

$$u_t + c\, u_x = 0 \quad \textbf{(4.4.15)}$$

where c is a constant. Equation (4.4.15) is called the *advection equation* because it describes the concentration of an inert substance (such as water vapor or ozone) as it is blown (advected) by the fluid with speed c. In this case, the characteristic is given by

$$x - ct = c_1 \quad \textbf{(4.4.16)}$$

and the general solution is

$$u(x,t) = f(x - ct) \quad \textbf{(4.4.17)}$$

where the arbitrary, differentiable function $f(x)$ is given by the initial conditions.

We are particularly interested in the characteristic $x - ct$ because u remains invariant as long as $x - ct$ remains constant. Consequently the characteristic is a propagation path: a path followed by some entity (namely, u in the present discussion) when that entity is propagated. For each new c_1 we have a new member in the family of characteristic $x - ct$, each member being analogous to a street in the north-south direction within a large metropolitan city along which vehicles "propagate."

Another analogy consists of a column of vehicles moving from left to right along a road. At successive times, this column is photographed and the photographs subsequently arranged so that the distance between successive snapshots is proportional to the time interval between them. See Fig. 4.4.1. We have, in effect, a

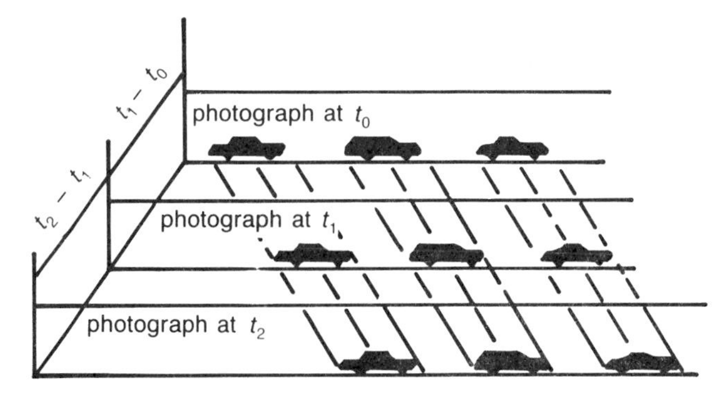

Fig. 4.4.1: Characteristics as propagation paths of "geometrical entities" - in this case motor vehicles - in time.

picture of the displacement of the individual vehicles as though they moved along lines in a space formed by the plane of the photographs and time. These lines connecting successive positions of the individual vehicles then represent propagation paths in space, and these propagation paths are in every sense—physical and mathematical—characteristics.

In the case of a single column of vehicles there is only one family of characteristics, corresponding to the ensemble of lines representing the motions of the individual vehicles. On a more normal road, however, where there is usually two columns moving in opposite directions, we have two families of characteristics. If the speeds of the column are v_1 and $-v_2$, so that v_1 and v_2 are themselves positive and left to right is taken as the positive direction, then the slopes of the characteristics are given by

$$\frac{dx}{dt}_{+} = v_1 \text{ and } \frac{dx}{dt}_{-} = -v_2 \qquad \textbf{(4.4.18)}$$

where the positive suffix refers to the forward facing characteristic (where dx/dt is positive), while the negative refers to the backward facing characteristic. The former characteristic is often called the c_+ characteristic, while the latter is then called the c_- characteristic. When all the vehicles in each column travel at the same constant speed, so that all the characteristics of each family have the same slope, the region covered by the resulting straight and parallel characteristics is called a region of *constant state.* In Section 4.7 we illustrate one way in which these characteristics may arise.

The example of a column of motor vehicles can be utilized to describe a particularly important arrangement of characteristics—the shock. Let us suppose that one of the vehicles in a column suddenly decelerates. The resulting situation is familiar enough: Because of the over-long reaction times of the drivers, each one applies his brakes a little later, so that within the one column the characteristics will converge and a discontinuity or "shock front" will form. At this front something violent takes place, with its own rather particular "physics." On the other hand, characteristics may diverge from each other and form a *fan wave*. In terms of our car analogy, this is similar to the spreading out of cars after they have passed an accident.

This discussion of shocks naturally leads to the question of whether a discontinuity in derivatives may propagate along a characteristic. To demonstrate that it can, consider the equation $u_x = 1$ with the initial condition $u(0,y) = |y|$. The solution to this problem is $u(x,y) = x + |y|$ and the equation is satisfied for all x and y. Because u_y is discontinuous across $y = 0$, a discontinuity in the derivative is propagated along the characteristic $y = 0$. This is not the case of a shock because there is no convergence of the characteristics.

4.5 THE LAPLACE TRANSFORM TECHNIQUE

In the case when the first-order partial differential equation is linear, Laplace transforms may prove to be a powerful method of solving them. Normally Laplace transforms are used to reduce ordinary differential equations to algebraic equations. In a similar way, a partial differential equation in two variables x and t may be reduced to an ordinary differential equation in x by means of a Laplace transform with respect to t. In addition to the condition that the equation is linear, we require that the coefficients of the unknown function and its derivatives must be independent of t.

To illustrate this technique, let us consider the first-order partial differential equation

$$u_t + u_x = xt. \quad \textbf{(4.5.1)}$$

We wish to solve Eq. (4.5.1) subject to the conditions that

$$u(x,0) = \sin(x) \text{ and } u(0,t) = 0. \quad \textbf{(4.5.2)}$$

Introducing the Laplace transform

$$\overline{u}(x,s) = \int_0^\infty u(x,t)\, e^{-st}\, dt \quad \textbf{(4.5.3)}$$

leads to the ordinary differential equation

$$\frac{d\overline{u}}{dx} + s\,\overline{u} = \frac{x}{s^2} + \sin(x) \quad \textbf{(4.5.4)}$$

assuming that we can differentiate with respect to x under the integral sign and

$$\overline{u}(0,s) = 0. \quad \textbf{(4.5.5)}$$

After multiplying Eq. (4.5.4) by the integrating factor exp(sx), we find

$$\bar{u}(x,s) = A(s)\, e^{-sx} + e^{-sx} \int_0^x e^{sy} \left[\frac{y}{s^2} + \sin(y) \right] dy \qquad \textbf{(4.5.6)}$$

The constant of integration is the function $A(s)$ and the boundary condition (4.5.5) forces

$$A(s) = 0. \qquad \textbf{(4.5.7)}$$

We may now integrate the second term in Eq. (4.5.6) to give

$$\bar{u}(x,s) = e^{-sx} \left[\frac{1}{s^2} \left(\frac{e^{sy}}{s^2} (sy - 1) \right) \Bigg|_0^x + \frac{e^{sy}\,(s \sin(y) - \cos(y))}{s^2+1} \Bigg|_0^x \right] \qquad \textbf{(4.5.8)}$$

$$= e^{-sx} \left[\frac{e^{sx}}{s^4} (sx + 1) + \frac{1}{s^4} + \frac{e^{sx}\,(s \sin(x) - \cos(x))}{s^2+1} + \frac{1}{s^2+1} \right] \qquad \textbf{(4.5.9)}$$

$$= \frac{sx-1}{s^4} + \frac{s \sin(x) - \cos(x)}{s^2+1} - \frac{e^{-sx}}{s^4} + \frac{e^{-sx}}{s^2+1}. \qquad \textbf{(4.5.10)}$$

Inverting the Laplace transform, we find the solution

$$u(x,t) = x\,\frac{t^2}{2} - \frac{t^3}{6} + \sin(x-t) - \frac{(x-t)^3}{6} H(t-x)$$

$$- \sin(x-t)\, H(t-x) \qquad \textbf{(4.5.11)}$$

where $H(t)$ is the Heaviside step function.

Exercises

Solve the following set of first-order partial differential equations by Laplace transforms.

1. $u_t + (x+1)\, u_x = e^{-t}$ $\qquad u(x,0) = 0 \qquad u(0,t) = t$
2. $x\, u_t + u_x = x$ $\qquad u(x,0) = 1 \qquad u(0,t) = e^t$

4.6 THE METHOD OF CHARACTERISTICS

Quasilinear first-order partial differential equations occur in such diverse fields as compressible gas dynamics, traffic flow theory, and wave motion in shallow water theory. In Section 4.2 we showed how these partial differential equations can be written as a set of ordinary differential equations. Sometimes these equations cannot be integrated in closed form. In this section we introduce a technique that may be used to construct graphical solutions in those instances. The essential point is that we shall integrate this set *along a given characteristic.* This technique may also be employed to construct numerical solutions.

Consider the first-order partial differential equation

$$\frac{\partial u}{\partial t} + c(x,t,u)\frac{\partial u}{\partial x} = 0 \tag{4.6.1}$$

If we consider an observer moving in some prescribed fashion $x(t)$, the value of u changes in time as the observer moves by

$$\frac{du}{dt} = \frac{\partial u}{\partial t} + \frac{dx}{dt}\frac{\partial u}{\partial x} \tag{4.6.2}$$

By comparing Eq. (4.6.1) to Eq. (4.6.2), u will remain constant from the observer's viewpoint

$$\frac{du}{dt} = 0 \tag{4.6.3}$$

if the observer moves at the "velocity"

$$\frac{dx}{dt} = c(x,t,u). \tag{4.6.4}$$

Equation (4.6.4) also gives the characteristic curve for this particular problem.

Consider the characteristic that is initially at the position $x = a$. Along the curve $dx/dt = c(x,t,u)$, $du/dt = 0$ or u is a constant. Initially u is specified by the initial conditions at $x = a$. Thus along this one particular characteristic

$$u = u(a,0) = u_a \tag{4.6.5}$$

which is a known constant. To find the characteristic for u_a, we must integrate

$$\frac{dx}{dt} = c(x,t,u_a). \tag{4.6.6}$$

If the function $c(x,t,u_a)$ is a simple function of x and t, we can find the characteristic in closed form. If $c(x,t,u_a)$ is a complicated function, then numerical integration must be used. Similarly, for a characteristic initially emanating from $x = b$, we can find a different characteristic, corresponding to the solution of

$$\frac{dx}{dt} = c(x,t,u_b). \tag{4.6.7}$$

In this manner the value of u at any future time and location can be predicted. To determine u at some later time $t = t_*$ at a particular place $x = x_*$, we simply follow the characteristic that passes through the point (x_*, t_*) back to the appropriate x-intercept and $u(x_*, t_*) = u(x_0, 0)$. This graphical method of constructing a solution for first-order partial differential equations is called the *method of characteristics*.

4.7 THE COLLAPSE OF A DAM

An interesting application of the method of characteristics is the problem of determining the flow which results from the sudden destruction of a dam. Because we are primarily interested in the mathematical technique rather than the realism of the model, we shall take the simplest possible case. We shall assume that an infinitesimally thin dam separates a reservoir of depth H (located to the right of the dam) from a dry river bed. The water is assumed to be at rest initially. At the time $t = 0$, the dam is suddenly destroyed, and our problem is to determine the subsequent motion of water for all x and t.

The equations governing this problem are the shallow-water equations. The shallow water equations govern phenomena where the horizontal length scale is large compared to the depth of the fluid. They may be written in the form

$$\left(\frac{\partial}{\partial t} + (u+c)\frac{\partial}{\partial x}\right)(u - 2c) = 0 \qquad \textbf{(4.7.1)}$$

$$\left(\frac{\partial}{\partial t} + (u-c)\frac{\partial}{\partial x}\right)(u + 2c) = 0 \qquad \textbf{(4.7.2)}$$

where x denotes the distance downstream from the dam, t time, u the Eulerian velocity in the x direction, $c = (gh)^{1/2}$, g the gravitational attraction, and h the depth of the fluid at any position or time. The reservoir and river bed have a level bottom.

Equations (4.7.1) and (4.7.2) can be replaced by the system:

$$\frac{dx}{dt} = (u \pm c) \text{ and } \frac{d}{dt}(u \pm 2c) = 0. \qquad \textbf{(4.7.3)}$$

If we define the c_+ characteristic by the ordinary differential equation $\frac{dx}{dt} = u + c$, then $u + 2c = \text{constant} = J_+$. Similarly the c_- characteristic is defined by the ordinary differential equation $\frac{dx}{dt} = u - c$ so that $u - 2c = \text{constant} = J_-$. The constants J_+ and J_- have been traditionally called the *Riemann invariants*. This name has its origin in compressible gas dynamics where similar dynamical equations occur.

Physically we know that the solution must consist of three distinct regions. The first region consists of the dry river bed where the water has *not* yet advanced. The third region is far upstream where the information about the dam's collapse has not reached and the second region lies between these two extremes.

Characteristics going toward the left (see Fig. 4.7.1) are given by

$$\frac{dx}{dt} = u - c = c - 2c_0 = u/2 - c_0 \tag{4.7.4}$$

because $u - 2c = -2c_0$ where $c_0 = (gH)^{1/2}$. At the leading edge, the depth of water is zero and $c = 0$. Therefore that characteristic

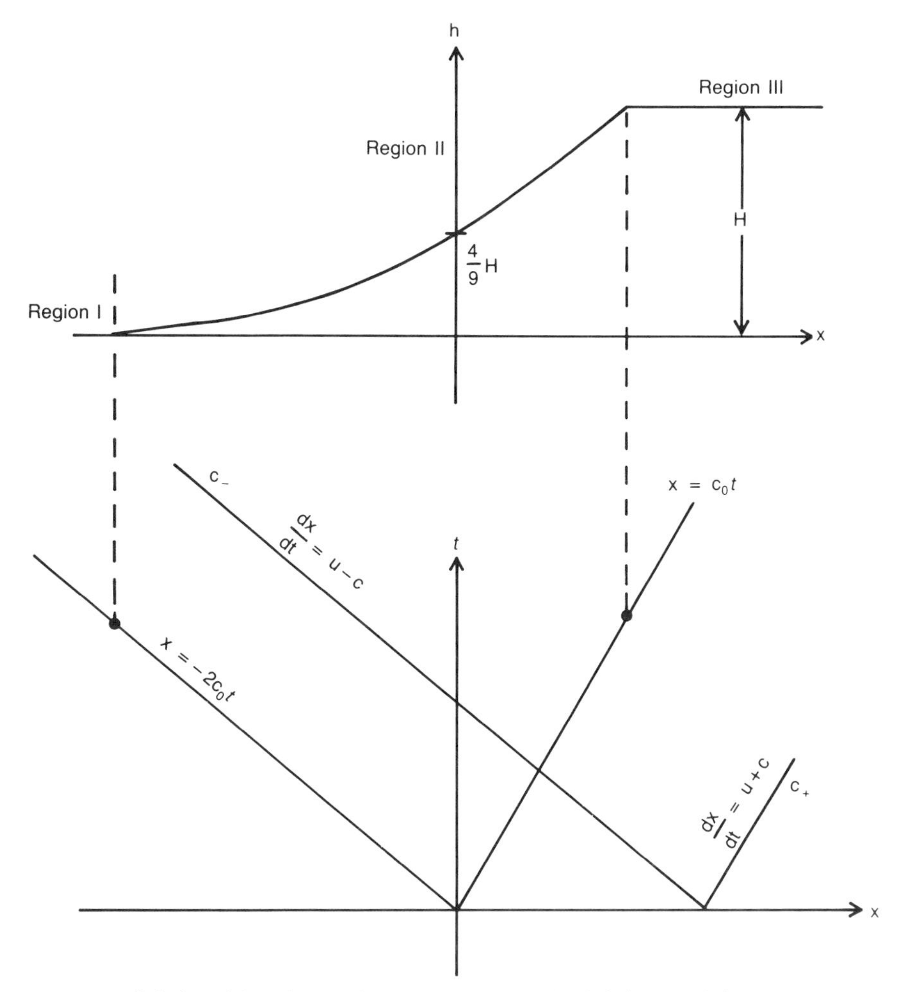

Fig. 4.7.1: Solution of the collapse of a dam using the method of characteristics.

is given by $\frac{dx}{dt} = -2c_0$ or $x = -2c_0t$. The arbitrary constant of integration must be zero because the characteristic emanates from the origin. On the other hand, information about the collapse of dam propagates upstream along the characteristic

$$\frac{dx}{dt} = u + c = 2c_0 - c = u/2 + c_0 \qquad \textbf{(4.7.5)}$$

because $u + 2c = 2c_0$. The dividing line between the portion of the reservoir that knows about the collapse of the dam from that portion that does not is given by a characteristic that emanates from the origin with $u = 0$ and $c = c_0$. Hence that characteristic is given by

$$\frac{dx}{dt} = u/2 + c_0 = c_0 \text{ or } x = c_0t. \qquad \textbf{(4.7.6)}$$

So far the easiest part of the problem has been calculated, namely the regions where the water level is either zero or H. We seemed to have utilized the method of characteristics to its total extend because the initial water level consisted of only the two values zero and H. We must now find a way to calculate u and c in the region

$$-2c_0t < x < c_0t. \qquad \textbf{(4.7.7)}$$

Clearly all these characteristics emanate from the origin.

Before we can proceed further, we must prove that if u and c on any characteristic are constant, then the characteristic is a straight line and is embedded in a family of straight-line characteristics along each of which u and c are constant. To show this, consider Fig. 4.7.2.
We assume that u and c are constant along the characteristic $\overline{DE}$. Because $\frac{dx}{dt} = u + c$ along $\overline{DE}$, the slope must be constant and the

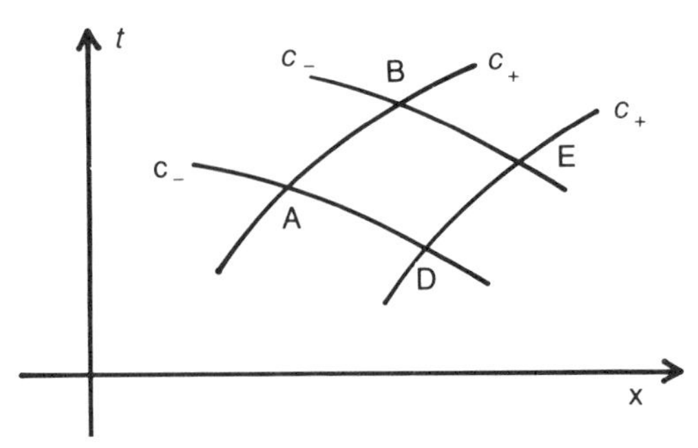

Fig. 4.7.2: Diagram used in the proof of the construction of the characteristics in the dam problem.

characteristic $\overline{DE}$ is a straight line. From the Riemann invariants we can immediately write

$$u_A + 2c_A = u_B + 2c_B \tag{4.7.8}$$

$$u_A - 2c_A = u_D - 2c_D \tag{4.7.9}$$

$$u_B - 2c_B = u_E - 2c_E = u_D - 2c_D \tag{4.7.10}$$

because $u_E = u_D$ and $c_E = c_D$. Eliminating u_D and c_D from these equations, we have $u_A = u_B$ and $c_A = c_B$. Finally u and c must be constant along the characteristic AB because the points A and B are completely arbitrary and must be independent of x and t.

We have already shown that the characteristics at the leading and trailing edges of the flood are straight lines. Consequently all the characteristics are straight lines with constant slope dx/dt. The equation for this straight line is

$$x = \frac{dx}{dt} t \text{ or } \frac{dx}{dt} = \frac{x}{t} = u + c = \frac{3}{2} u + c_0. \tag{4.7.11}$$

Solving for u,

$$u = \frac{2}{3} \left(\frac{x}{t} - c_0 \right) \tag{4.7.12}$$

and

$$c = u/2 + c_0 = \frac{1}{3} \left(\frac{x}{t} + 2c_0 \right). \tag{4.7.13}$$

Because $c^2 = gh$

$$h = \frac{1}{9g} \left(\frac{x}{t} - 2\, c_0 \right)^2. \tag{4.7.14}$$

From Eq. (4.7.14) we see that the curve of the water surface at any time t is a parabola from the leading edge of the flood to the point $x = c_0 t$, after which it is horizontal. As a numerical example, if the water behind the dam were 200 ft. high, our results predict the front of the flood would roar down the valley at a speed of about 110 mph.

4.8 TRAFFIC FLOW: UNIFORM TRAFFIC STOPPED BY A RED LIGHT

The method of characteristics is also an important tool in the solution of problems in deterministic traffic flow theory. The flow of traffic on a single lane highway is governed by the equation:

$$\frac{\partial \varrho}{\partial t} + u_{max}\left(1 - \frac{2\varrho}{\varrho_{max}}\right)\frac{\partial \varrho}{\partial x} = 0 \qquad \textbf{(4.8.1)}$$

where ϱ is the density of cars on the road (the number of vehicles per mile of highway), ϱ_{max} the maximum density of cars that the road can handle (i.e., bumper to bumper traffic), and u_{max} the speed limit. We will investigate what happens to a uniform stream of moving traffic (with density ϱ_0) as it is suddenly halted by a red light at $x = 0$. We will ignore the slowing down associated with a yellow light and restrict ourselves to what occurs in front of the light.

We envision a situation where traffic is moving along a single lane with a uniform density $\varrho_0 < \varrho_{max}$ when the traffic is suddenly stopped and becomes bumper to bumper at $x = 0$ for all time $t > 0$. To solve Eq. (4.8.1) with these initial and boundary conditions, we replace Eq. (4.8.1) with the ordinary differential equations

$$\frac{d\varrho}{dt} = 0 \qquad \textbf{(4.8.2)}$$

and

$$\frac{dx}{dt} = u_{max}\left(1 - \frac{2\varrho}{\varrho_{max}}\right). \qquad \textbf{(4.8.3)}$$

Because the density along a characteristic remains constant, we can integrate Eq. (4.8.3) to give the equation describing a characteristic emanating from the abscissa

$$x = u_{max}\left(1 - \frac{2\varrho_0}{\varrho_{max}}\right)t + x_0. \qquad \textbf{(4.8.4)}$$

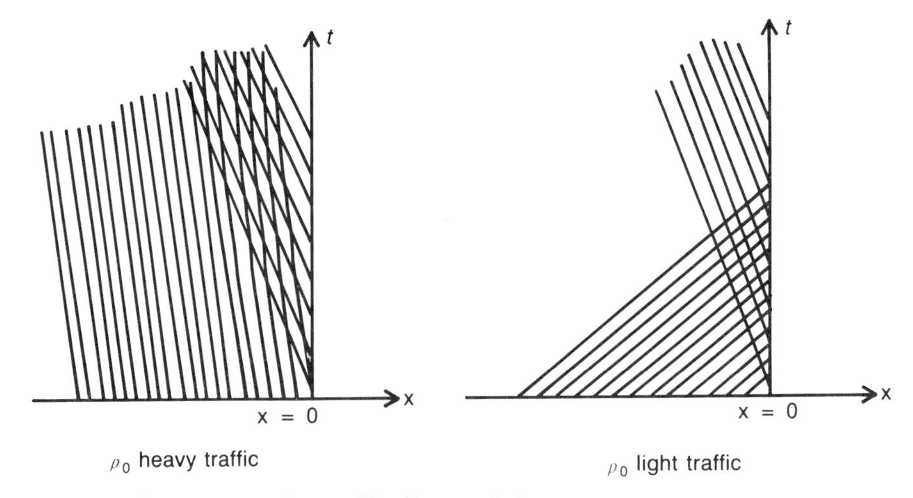

Fig. 4.8.1: Stopping of uniform traffic: Characteristics.

If $\varrho_0 > \varrho_{max}/2$ (i.e., heavy traffic), the slope of the characteristic is negative; if $\varrho_0 < \varrho_{max}/2$ (i.e., light traffic), the slope is positive. Along the ordinate $\varrho = \varrho_{max}$ (i.e., stopped traffic) and the characteristics are described by

$$x = -u_{max}t + x_0. \tag{4.8.5}$$

The solution for heavy and light traffic is given in Fig. 4.8.1. In either case there is a cross-hatched region indicating that the method of characteristics yields a multivalued solution to the partial differential equation. This is clearly unacceptable.
This convergence of the characteristics associated with

$$\varrho = \varrho_0 \text{ and } \varrho = \varrho_{max}$$

is the mathematical description of a car traveling in traffic which must suddenly stop because of the line of cars that are waiting at the stop light. This rapid and discontinuous change from characteristics representing traffic of density $\varrho = \varrho_0$ to bumper to bumper traffic occurs along a line called a *shock*. The name "shock" has its origin in gas dynamics where discontinuities in density and pressures are found to occur in explosions and supersonic flow.

In those cases where shocks form, we must find an equation that describes its location as a function of time. To the right of the shock, the density of cars equals ϱ_{max}; to the left of the shock, ϱ_0. Because we do not know what is happening within the shock, we must use conservation laws in order to derive the conditions across any shock. These conservation equations may include conservation of mass, momentum, and energy.

In traffic flow theory, the fundamental conservation law is conservation of cars. To the left of the shock the rate of cars entering the shock is

$$q(x_s^-) - \varrho(x_s^-)\frac{dx_s}{dt} \tag{4.8.6}$$

where $q = \varrho u$ and u is the speed of the vehicle. The second term is required because the shock boundary is moving. On the right side of the shock, the rate of cars appearing is

$$q(x_s^+) - \varrho(x_s^+)\frac{dx_s}{dt}. \tag{4.8.7}$$

Conservation of cars requires that these two quantities are the same:

$$\frac{dx_s}{dt} = \frac{q(x_s^-) - q(x_s^+)}{\varrho(x_s^-) - \varrho(x_s^+)}. \tag{4.8.8}$$

The denominator is nonzero because the densities on the two sides of the shock are different. In our problem $q(x_s^-) = \varrho_0 u(\varrho_0)$, $q(x_s^+) = \varrho_{max}u(\varrho_{max}) = 0$, $\varrho(x_s^-) = \varrho_0$, and $\varrho(x_s^+) = \varrho_{max}$ and

$$\frac{dx_s}{dt} = \frac{-\varrho_0 u(\varrho_0)}{\varrho_{max} - \varrho_0} \tag{4.8.9}$$

or

$$x_s = -\frac{\varrho_0 u(\varrho_0)}{\varrho_{max} - \varrho_0}\, t. \tag{4.8.10}$$

The resulting space-time diagram is sketched in Fig. 4.8.2. For any time $t>0$, the traffic density is discontinuous. This shock separates cars standing still from cars moving forward at velocity $u(\varrho_0)$. Cars must decelerate from $u(\varrho_0)$ to zero instantaneously and an accident is predicted. Obviously, this does not happen often and we must modify our theory to conform more closely to reality near the area of the predicted shock.

Exercises

1. Using method of characteristics, find the solution to $u_t + u_x = 0$

with $u(x,0) = \begin{matrix} 1 & 0<x<1 \\ 0 & \text{otherwise} . \end{matrix}$ Sketch the solution on the x,t plane.

2. Consider the nonlinear partial differential equation

$$\varrho_t - \varrho^2 \varrho_x = 0 \qquad -\infty < x < \infty .$$

Solve this partial differential equation by the method of characteristics subject to the initial conditions:

$$\varrho(x,0) = \begin{matrix} 1 & x<0 \\ 1-x & 0<x<1. \\ 0 & x>1 \end{matrix}$$

3. Solve the density equation from traffic flow theory

$$\varrho_t + u_{max}(1 - \frac{2\varrho}{\varrho_{max}})\, \varrho_x = 0$$

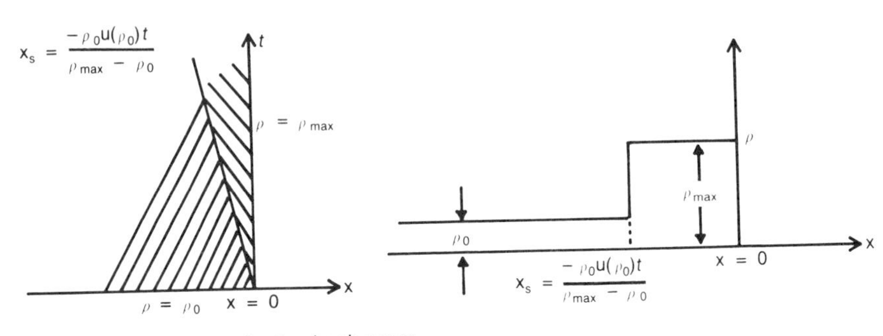

Fig. 4.8.2: Constant velocity shock wave.

by the method of characteristics if

$$\varrho(x,0) = \begin{cases} \varrho_0 & x<0 \\ \varrho_0 \quad (L-x)/L & 0<x<L. \\ 0 & x>L \end{cases}$$

Sketch $\varrho(x,t)$ for the special cases $\varrho_0 < \varrho_{max}/2$, $\varrho_0 = \varrho_{max}/2$, and $\varrho_0 > \varrho_{max}/2$.

4. Repeat problem 3 with the exception that

$$\varrho(x,0) = \begin{cases} \varrho_{max}/5 & x<0 \\ 3\varrho_{max}/5 & x>0 \end{cases}$$

(a) Sketch the initial density.
(b) Determine and sketch the density at later times.

Answers

P. 193

1. $u(x,y) = e^x f(x-y)$
2. $u(x,y) = f(3x + 2y) + x^2/4$
3. $u(x,y) = \dfrac{\sin(x) - \cos(x)}{2} + y + 2 + e^{-x}f(2x + y)$
4. $u(x,y) = - e^{2x} \sin(3y)/2 + e^{2x} f(x + y)$
5. $u(x,y) = y + e^{x-y}$
6. $u(x,y) = e^{2x-y} [1 + \cos(x-y)]$
7. $u(x,y) = -\cos(x) + \cos[\sin(x) - y] - \sin(x) + y$
8. $u(x,y) = e^{-2x} \sin\left(\dfrac{5x+2y}{2}\right) + e^{(x+y)/2} + x^2/2 - x/4 + 1/8$
9. $u(x,y) = x/2 + x^{1/2}y^{3/2} - y/2$
10. $u(x,y) = e^y(4 - x^2 - y^2)^{1/2} \exp(-(4-x^2-y^2)^{1/2})$
11. $u(x,y) = \exp[-y^2-(x^2+y^2)^{1/2}] \sin[(x^2 + y^2)^{1/2}]$

P. 200

1. $u(x,t) = 1 - e^{-t} + [t - 1 - \ell n(1+x)e^{-t} + (1+x)e^{-t}) H(t - \ell n(1+x)]$

2. $u(x,t) = 1 + t + (x^2/2 + \exp(t - x^2/2) - t - 1)H(t - x^2/2)$

Chapter 5

The Wave Equation

IN THIS CHAPTER WE SHALL STUDY PROBLEMS THAT ARE associated with the equation

$$\frac{\partial^2 u}{\partial t^2} - c^2 \frac{\partial^2 u}{\partial x^2} = F(x,t)$$

where x and t are the two independent variables and c is a constant. This equation, called the *wave equation*, serves as the prototype for a general class of partial differential equations known as hyperbolic. It arises in the study of many important physical problems involving wave propagation, such as the transverse vibrations of an elastic string, and the longitudinal vibrations or torsional oscillations of a rod. The wave equation is certainly one of the few equations of mathematical physics that truly deserves the name classical.

5.1 THE VIBRATING STRING

A simple example of the wave equation is provided by the motion of a string of length L and constant density ϱ, which is stretched between two supports. See Fig. 5.1.1. The equilibrium position of the string lies along the interval $[0,L]$ on the x-axis. The displacement $u(x,t)$ of the string is the unknown. The equation of motion for the string is found by considering a short piece whose ends are at x and $x + \Delta x$ and applying Newton's second law of mo-

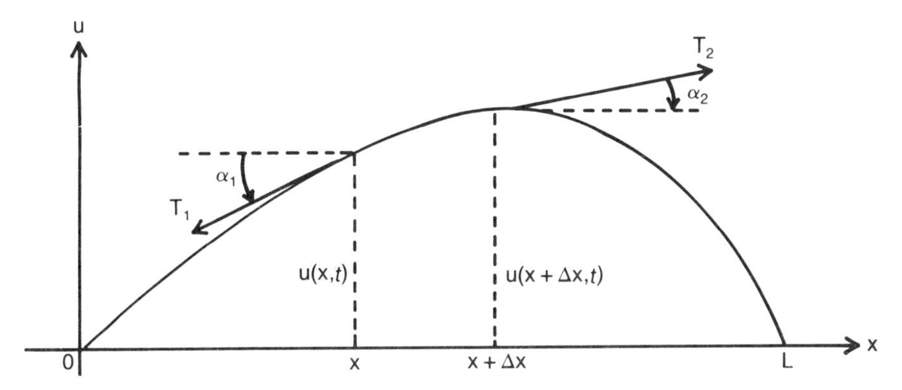

Fig. 5.1.1: The vibrating string.

tion to it. The portion of the string just to the right and left of our element exerts forces on it that cause accelerations.

If we assume that the string is perfectly flexible and offers no resistance to the bending, the forces in the element are shown in Fig. 5.1.1. Applying Newton's second law in the x-direction, we have for the sum of forces

$$-T(x)\cos(\alpha_1) + T(x+\Delta x)\cos(\alpha_2). \tag{5.1.1}$$

If we assume that a point on the string moves only in the vertical direction, the sum of forces in Eq. (5.1.1) equals zero and the horizontal component of the tension is constant. In this case we have

$$-T(x)\cos(\alpha_1) + T(x+\Delta x)\cos(\alpha_2) = 0 \tag{5.1.2}$$

and

$$T(x)\cos(\alpha_1) = T(x+\Delta x)\cos(\alpha_2) = T, \text{ a constant.} \tag{5.1.3}$$

If gravity is the only external force, Newton's law in the vertical direction gives

$$-T(x)\sin(\alpha_1) + T(x+\Delta x)\sin(\alpha_2) - mg = m\,\frac{\partial^2 u}{\partial t^2} \tag{5.1.4}$$

because u_{tt} is the acceleration if the displacement is only in the vertical direction.

Because

$$T(x) = \frac{T}{\cos(\alpha_1)} \text{ and } T(x+\Delta x) = \frac{T}{\cos(\alpha_2)} \tag{5.1.5}$$

then

$$-T\tan(\alpha_1) + T\tan(\alpha_2) - \varrho\Delta x\, g = \varrho\Delta x\,\frac{\partial^2 u}{\partial t^2} \tag{5.1.6}$$

because ϱ equals mass per unit length. The quantity $\tan(\alpha_1)$ is the slope of the string at x, and $\tan(\alpha_2)$ the slope of the string at $x + \Delta x$; that is

$$\tan(\alpha_1) = \frac{\partial u(x,t)}{\partial x} \quad \text{and} \quad \tan(\alpha_2) = \frac{\partial u(x+\Delta x,t)}{\partial x} \tag{5.1.7}$$

Substituting Eq. (5.1.7) into Eq. (5.1.6), we have

$$T \left[\frac{\partial u(x+\Delta x,t)}{\partial x} - \frac{\partial u(x,t)}{\partial x} \right] = \varrho \, \Delta x \, \left(\frac{\partial^2 u}{\partial t^2} + g \right) \tag{5.1.8}$$

After dividing through by Δx, we have the difference quotient on the left:

$$\frac{T}{\Delta x} \left[\frac{\partial u(x+\Delta x,t)}{\partial x} - \frac{\partial u(x,t)}{\partial x} \right] = \left(\frac{\partial^2 u}{\partial t^2} + g \right). \tag{5.1.9}$$

In the limit as $\Delta x \to 0$, the difference quotient becomes a partial derivative with respect to x, leaving Newton's second law in the form

$$T \, u_{xx} = \varrho u_{tt} + \varrho g \tag{5.1.10}$$

or

$$u_{xx} = \frac{1}{c^2} u_{tt} + \frac{g}{c^2} \tag{5.1.11}$$

where $c^2 = T/\varrho$. Because u_{tt} is generally found to be much larger than g, we can neglect the last term, giving the equation of the vibrating string:

$$u_{xx} = \frac{1}{c^2} u_{tt}. \tag{5.1.12}$$

As a second example of how the wave equation arises in the physical sciences, we derive the threadline equation, which describes how a thread of yard vibrates as it is drawn between two eyelets spaced a distance L apart. Again we assume that the tension in the string is constant, the vibrations are small, the string is assumed perfectly flexible, the effects of gravity and air drag are negligible and the mass of the string per unit length is constant. Unlike the case of a string vibrating between two fixed ends, the threadline is being drawn through the eyelets at a speed V so that a segment of string experiences motion in both the x and y direction as it vibrates about its equilibrium position. The eyelets may move in the vertical direction.

From Newton's second law

$$\frac{d}{dt}\left(m\frac{dy}{dt}\right) = \text{the sum of forces} \tag{5.1.13}$$

where m is the mass of the thread. But

$$dy = \frac{\partial y}{\partial t}\,dt + \frac{\partial y}{\partial x}\,dx \tag{5.1.14}$$

or

$$\frac{dy}{dt} = \frac{\partial y}{\partial t} + \frac{dx}{dt}\frac{\partial y}{\partial x}. \tag{5.1.15}$$

Because $\frac{dx}{dt} = V$, we have

$$\frac{dy}{dt} = \frac{\partial y}{\partial t} + V\frac{\partial y}{\partial x} \tag{5.1.16}$$

and

$$\frac{d}{dt}\left(m\frac{dy}{dt}\right) = \frac{\partial}{\partial t}\left(m\frac{dy}{dt}\right) + V\frac{\partial}{\partial x}\left(m\frac{dy}{dt}\right) \tag{5.1.17}$$

$$= \frac{\partial}{\partial t}\left[m\left(\frac{\partial y}{\partial t} + V\frac{\partial y}{\partial x}\right)\right] + V\frac{\partial}{\partial x}\left[m\left(\frac{\partial y}{\partial t} + V\frac{\partial y}{\partial x}\right)\right]. \tag{5.1.18}$$

Because both m and V are constant,

$$\frac{d}{dt}\left(m\frac{dy}{dt}\right) = m\frac{\partial^2 y}{\partial t^2} + 2mV\frac{\partial y^2}{\partial x\partial t} + mV^2\frac{\partial^2 y}{\partial t^2}. \tag{5.1.19}$$

The sum of the forces is once again equal to

$$T\frac{\partial^2 y}{\partial x^2}\,\Delta x \tag{5.1.20}$$

so that the threadline equation is

$$T \frac{\partial^2 y}{\partial x^2} \Delta x = m V^2 \frac{\partial^2 y}{\partial x^2} + 2mV \frac{\partial^2 y}{\partial x \partial t} + m \frac{\partial^2 y}{\partial t^2} \qquad (5.1.21)$$

or

$$\frac{\partial^2 y}{\partial t^2} + 2V \frac{\partial^2 y}{\partial x \partial t} + \left(V^2 - \frac{gT}{\varrho}\right) \frac{\partial^2 y}{\partial x^2} = 0 \qquad (5.1.22)$$

where ϱ is the density of the thread.

5.2 INITIAL CONDITIONS: THE CAUCHY CONDITION

Any physical problem must state not only the differential equation that is to be solved but also any imposed conditions that the solutions must satisfy. Out of the mass of possible solutions we must pick the unique answer for our particular problem. Because our equation will correspond to some sort of physical reality, the solution for a given equation cannot satisfy *any* sort of condition, but must satisfy conditions that correspond to that particular physical problem. In the case of partial differential equations this task of satisfying these conditions may be as difficult as solving the equation.

In the case of the wave equation, some physical entity is propagating from one location to another in space and time. This implies an evolution from some given initial state or condition. We either toss a rock into the water or sound a horn or throw a switch. This imposed initial condition, called the *Cauchy problem*, is correctly posed if we specify the value of the function *and* its time derivative along some time line, say $t = 0$. This particular boundary condition is then known as the *Cauchy boundary condition*.

Another important but less obvious condition involving the wave equation requires us to impose *no* boundary condition at infinite time. Physically this is saying that the initial conditions determine the entire evolution of the problem and we do not have to satisfy any condition at some later date.

In addition to this initial condition, we may specify boundary conditions in the x direction. For example, we may require that the end of the string be fixed. In this chapter, we will concentrate on the correct formulation of the initial condition, reserving our examination of boundary conditions for the next chapter.

Exercises

1. Verify that the function $u(x,t) = \sin(2x)\sin(2t)$ satisfies the

equation $u_{tt} = u_{xx}$ and the initial conditions $u(x,0) = 0$ and $u_t(x,0) = 2\sin(2x)$.

2. Verify that the function $u(x,t) = \sin(2x)\cos(2t)$ satisfies the equation $u_{tt} = u_{xx}$ and the initial conditions $u(x,0) = 2\sin(x)\cos(x)$ and $u_t(x,0) = 0$.

3. Show that $u(x,t) = t\sin(\pi x)$ is a solution to the initial-boundary value problem

$$\begin{aligned} u_{tt} - u_{xx} &= \pi^2 t\sin(\pi x) && 0<x<1,\ t>0 \\ u(x,0) &= 0,\ u_t(x,0) = \sin(\pi x) && 0<x<1 \\ u(0,t) &= 0,\ u(1,t) = 0 && t\geq 0 \end{aligned}$$

4. Show that $u(x,t) = xt^2 + \sin(x+t)$ is a solution of the initial value problem

$$\begin{aligned} u_{tt} - u_{xx} &= 2x && |x|<\infty \\ u(x,0) = \sin(x), \quad u_t(x,0) &= \cos(x) && |x|<\infty \end{aligned}$$

5. Show that $u(x,t) = (e^{x-t} + e^{x+t})/2$ is a solution to the initial-boundary value problem

$$\begin{aligned} u_{tt} - u_{xx} &= 0 && x>0 \\ u(x,0) = e^x, \quad u_t(x,0) &= 0 && x>0 \\ u(0,t) &= \cosh(t) && t\geq 0 \end{aligned}$$

6. Show that $u(x,t) = \cos(x)\sin(t)$ is a solution of the initial-boundary problem

$$\begin{aligned} u_{tt} - u_{xx} &= 0 && x>0 \\ u(x,0) &= 0,\ u_t(x,0) = \cos(x) && x>0 \\ u_x(0,t) &= 0. && t\geq 0 \end{aligned}$$

5.3 SEPARATION OF VARIABLES: THE SOLUTION OF THE VIBRATING STRING PROBLEM

In Section 5.1 we saw that the vibration of a string is described by the equation $u_{tt} = c^2u_{xx}$. In Section 5.2 we stated that this problem could be solved uniquely only if our initial condition was a Cauchy problem: $u(x,0) = f(x)$ and $u_t(x,0) = g(x)$. We will now solve the wave equation for the special case when the string is clamped in position at $x = 0$ and $x = L$. Mathematically this physical condition is written $u(0,t) = u(L,t) = 0$ for $t>0$.

As a first attempt in solving (5.1.12), we assume that the solution $u(x,t)$ can be written as the product of $X(x)$ and $T(t)$, i.e.,

$$u(x,t) = X(x)\, T(t) \tag{5.3.1}$$

(T no longer denotes tension.) Because

$$u_{tt} = X(x)\, T''(t) \tag{5.3.2}$$

and

$$u_{xx} = X''(x)\, T(t), \tag{5.3.3}$$

the wave equation (5.1.12) becomes

$$c^2 X''T = T''X \tag{5.3.4}$$

or

$$\frac{X''}{X} = \frac{T''}{c^2 T} \qquad 0<x<L,\ t>0 \tag{5.3.5}$$

after dividing through by $c^2X(x)T(t)$. Because the right-hand side of (5.3.5) depends only on x and the left-hand side only on t, the only way that they can equal each other is to be equal to a constant. We write this constant $-k^2$ and separate Eq. (5.3.5) into two ordinary differential equations by this *separation constant*:

$$T'' + k^2c^2T = 0 \qquad t>0 \tag{5.3.6}$$

and

$$X'' + k^2 X = 0. \qquad 0<x<L \tag{5.3.7}$$

We assume *a priori* that the separation constant is a negative constant. For completeness we should have checked the cases when the separation constant is positive or zero. When Eq. (5.3.5) is solved in these cases, trival solutions are found for the given boundary conditions.

We now rewrite the boundary conditions in terms of $X(x)$ and $T(t)$ by noting that the boundary conditions become

$$u(0,t) = X(0)\, T(t) = 0 \tag{5.3.8}$$

and

$$u(L,t) = X(L)\, T(t) = 0 \tag{5.3.9}$$

for $t>0$. If we were to choose $T(t) = 0$, then we would have a trivial solution for $u(x,t)$. Consequently we must have

$$X(0) = X(L) = 0. \tag{5.3.10}$$

The solution of

$$X'' + k^2X = 0 \tag{5.3.11}$$

with

$$X(0) = X(L) = 0 \tag{5.3.12}$$

is

$$X(x) = C \cos(kx) + D \sin(kx). \tag{5.3.13}$$

The condition $X(0) = 0$ gives $C = 0$ and the condition $X(L) = 0$ yields

$$D \sin(kL) = 0 \tag{5.3.14}$$

or

$$kL = n\pi \tag{5.3.15}$$

with $n = 1,2,3, \ldots$ because $D = 0$ would imply $X(x) = 0$. Viewing Eqs. (5.3.11)-(5.3.12) as a regular Sturm-Liouville problem, the x dependence is given by the eigenvalues

$$\lambda_n = \frac{n^2\pi^2}{L^2} \tag{5.3.16}$$

and eigenfunctions

$$X_n(x) = \sin\left(\frac{n\pi x}{L}\right) \tag{5.3.17}$$

with $n = 1,2,3, \ldots$ we have dropped the amplitude D from the $X_n(x)$ solution because in the product solution $X(x)T(t)$ the constant D and the coefficient from the $T(t)$ solution may be combined together to form a single constant. Consequently, no less of generality results by taking $D = 1$.

The solution of the $T(t)$ equation is

$$T_n(t) = A_n \cos(k_n ct) + B_n \sin(k_n ct) \tag{5.3.18}$$

where A_n and B_n are arbitrary. For each $n = 1,2,3, \ldots$, we now have the particular solution

$$u_n(x,t) = \sin(k_n x) \left[A_n \cos(k_n ct) + B_n \sin(k_n ct) \right]. \tag{5.3.19}$$

For any choice of A_n and B_n, Eq. (5.3.19) is a solution of the partial differential equation Eq. (5.1.12) and also satisfies the end boundary conditions. Therefore any linear combination of the $u_n(x,t)$ also satisfies the partial differential equation and the boundary conditions. In making our linear combination we need no new constants because A_n and B_n are still arbitrary. We have,then,

$$u(x,t) = \sum_{n=1}^{\infty} \sin\left(\frac{n\pi x}{L}\right)\left[A_n \cos\left(\frac{n\pi ct}{L}\right) + B_n \sin\left(\frac{n\pi ct}{L}\right)\right] \tag{5.3.20}$$

This method of building up the general solution from the particular solutions is due to Daniel Bernoulli (1775). His method appeared to d'Alembert and Euler to be impossible because such a series, having a period of $2L$, could not possibly represent such a function as a straight line when $t = 0$. A controversy arose between these mathematicians and it was not until the work of Fourier on heat conduction that Bernoulli was proven correct.

Our example of using particular solutions to build up the general solution illustrates the powerful *principle of linear superposition*, which is applicable to any *linear* system. This principle states that in a linear system any complicated phenomena may be described by the sum of simpler functions of phenomena. It is extremely important because it allows us to break down any complex problem into simpler problems which can hopefully be solved. The final answer is then constructed from these simpler solutions.

Our final task remains to determine A_n and B_n. At $t = 0$, we have

$$u(x,0) = \sum_{n=1}^{\infty} A_n \sin\left(\frac{n\pi x}{L}\right) = f(x) \tag{5.3.21}$$

and

$$u_t(x,0) = \sum_{n=1}^{\infty} B_n \frac{n\pi c}{L} \sin\left(\frac{n\pi x}{L}\right) = g(x) \tag{5.3.22}$$

Both of these series are Fourier half-range sine expansions over the interval $(0,L)$. Applying the results from Section 2.6, we have

$$A_n = \frac{2}{L}\int_0^L f(x) \sin\left(\frac{n\pi x}{L}\right) dx \tag{5.3.23}$$

$$\frac{n\pi c}{L} B_n = \frac{2}{L}\int_0^L g(x) \sin\left(\frac{n\pi x}{L}\right) dx \tag{5.3.24}$$

or

$$B_n = \frac{2}{n\pi c}\int_0^L g(x) \sin\left(\frac{n\pi x}{L}\right) dx \tag{5.3.25}$$

As an example, let us take as the initial conditions

$$f(x) = \begin{cases} 0 & 0 < x < \frac{L}{4} \\ 4h\left(\frac{x}{L} - \frac{1}{4}\right) & \frac{L}{4} < x < \frac{L}{2} \\ 4h\left(\frac{3}{4} - \frac{x}{L}\right) & \frac{L}{2} < x < \frac{3L}{4} \\ 0 & \frac{3L}{4} < x < L \end{cases} \qquad \textbf{(5.3.26)}$$

and

$$g(x) = 0 \qquad 0 < x < L \qquad \textbf{(5.3.27)}$$

A graph of $f(x)$ for the case $L = \pi$ is given in Fig. 5.3.1. for time $t = 0$. The peak value is h.

In this particular example, $B_n = 0$ for all n because $g(x) = 0$. On the other hand.

$$A_n = \frac{8h}{L}\int_{L/4}^{L/2} \left(\frac{x}{L} - \frac{1}{4}\right) \sin\left(\frac{n\pi x}{L}\right) dx$$

$$+ \frac{8h}{L}\int_{L/2}^{3L/4} \left(\frac{3}{4} - \frac{x}{L}\right) \sin\left(\frac{n\pi x}{L}\right) dx \qquad \textbf{(5.3.28)}$$

$$= \frac{8h}{n^2\pi^2}\left[(2 \sin\left(\frac{n\pi}{2}\right) - \sin\left(\frac{3n\pi}{4}\right) - \sin\left(\frac{n\pi}{4}\right)\right] \qquad \textbf{(5.3.29)}$$

$$= \frac{8h}{n^2\pi^2}\, [\,(2 \sin(n\pi/2) - 2\sin(n\pi/2)\cos(n\pi/4)] \qquad \textbf{(5.3.30)}$$

$$= \frac{16h}{n^2\pi^2} \sin(n\pi/2)\,(1 - \cos(n\pi/4))$$

$$= \frac{32h}{n^2\pi^2} \sin \frac{n\pi}{2}\,)\, \sin^2\left(\frac{n\pi}{8}\right) \qquad \textbf{(5.3.31)}$$

because $\sin(A) + \sin(B) = 2 \sin\left(\frac{A+B}{2}\right) \cos\left(\frac{A-B}{2}\right)$ and $1 - \cos(2A) = 2 \sin^2(A)$.

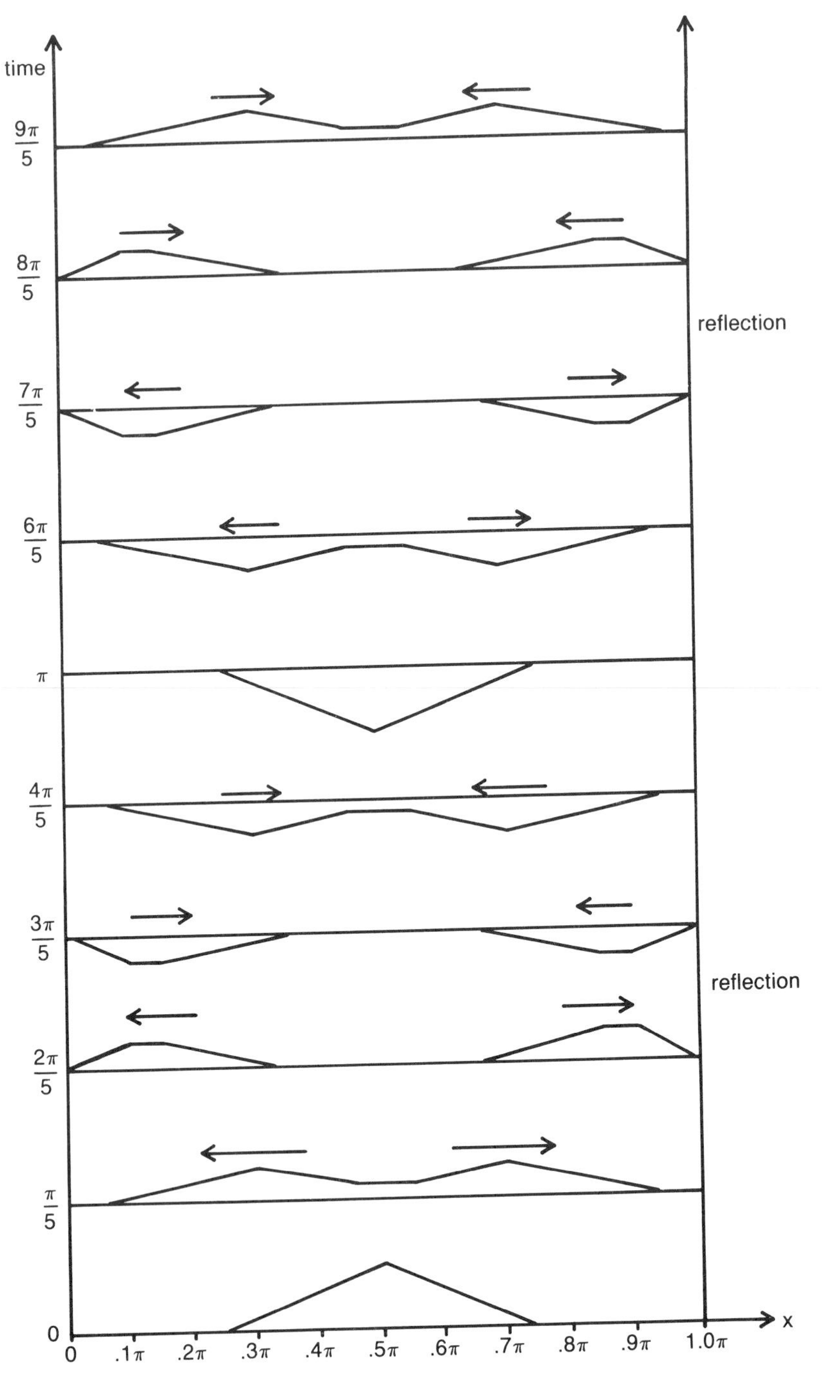

Fig. 5.3.1: The vibration of a string at various times and position when $L = \pi$ and $c = 1$, as given by Eq. (5.3.33).

Therefore

$$u(x,t) = \frac{32h}{\pi^2} \sum_{n=1}^{\infty} \sin\left(\frac{n\pi}{2}\right) \sin^2\left(\frac{n\pi}{8}\right) \frac{1}{n^2} \sin\left(\frac{n\pi x}{L}\right) \cos\left(\frac{n\pi ct}{L}\right) \quad \textbf{(5.3.32)}$$

Because $\sin(n\pi/2)$ vanishes for n even, A_n will be zero for n even. Consequently it is preferable to rewrite Eq. (5.3.32) so that we eliminate these vanishing terms. The most convenient way is to introduce the general expression for an odd integer $n = 2m - 1$ $(m = 1,2,3, \ldots)$ and note that $\sin[(2m-1)\pi/2] = (-1)^{m+1}$. Therefore Eq. (5.3.32) becomes

$$u(x,t) = \frac{32h}{\pi^2} \sum_{m=1}^{\infty} \frac{(-1)^{m+1}}{(2m-1)^2} \sin^2\left(\frac{(2m-1)\pi}{8}\right) \times \sin\left(\frac{(2m-1)\pi x}{L}\right) \cos\left(\frac{(2m-1)\pi ct}{L}\right) \quad \textbf{(5.3.33)}$$

Although we have completely solved the problem, several useful points can be made by rewriting Eq. (5.3.33) as

$$u(x,t) = 1/2 \sum_{n=1}^{\infty} A_n \left\{ \sin\left[\frac{n\pi}{L}(x-ct)\right] + \sin\left[\frac{n\pi}{L}(x+ct)\right]\right\} \quad \textbf{(5.3.34)}$$

by applying the trigonometric identity

$$\sin(A)\cos(B) = [\sin(A-B) + \sin(A+B)]/2. \quad \textbf{(5.3.35)}$$

From general physics we find expressions like $\sin[k_n(x-ct)]$ or $\sin(kx-\omega t)$ arising in studies of simple wave motions. $\mathrm{Sin}[k_n(x-ct)]$ is the mathematical description of a propagating wave in the sense that we must move to the right at the speed c if we wish to keep in the same position relative to the nearest crest and trough. The quantities k, ω and c are called the wave-number, frequency and phase speed or wave-velocity, respectively. The relationship $\omega = kc$ holds between the frequency and phase speed.

It may seem paradoxical that we are talking about traveling waves in a problem dealing with waves confined on a string of length L. Actually we are dealing with standing waves because at the same time that a wave is propagating to the right its mirror image is running to the left so there is no resultant progressive wave motion. This is shown in Fig. 5.3.1 for the case when $L = \pi$ and $c = 1$. Furthermore, unlike many wave problems, the wave-number k_n is not a free parameter but has been restricted to the values of $n\pi/L$. This restriction on wave-number is common in wave problems dealing with limited domains (for example, a building, ship, lake, or planet) and these oscillations are given the special name of *normal modes* or *natural vibrations*.

In our problem of the vibrating string we have found that all the components propagate with the same phase speed. That is, all the waves regardless of wave-number k_n will move the distance $c\Delta t$ or $-c\Delta t$ after the time interval Δt has elapsed. We now turn to the case of the damped wave equation where this will not be true.

5.4 THE KLEIN-GORDON EQUATION: DISPERSION

In the preceding section, we saw that the solution to the vibrating string problem consisted of two simple waves, each propagating with a phase speed c to the right and left. In problems where the equations of motion are a little more complicated than Eq. (5.1.12), all the harmonics no longer propagate with the same phase speed but at a speed that depends upon wave-number. In such systems the phase relation between harmonics is altered and these systems are referred to as *dispersive*.

A simple illustration is provided by modifying the vibrating string problem so that each element of the string is now subject to an additional applied force which is proportional to its displacement

$$u_{tt} = c^2 u_{xx} - hu \tag{5.4.1}$$

where $h \geq 0$ is constant. For example, if the string is embedded in a thin sheet of rubber, then in addition to the restoring force due to tension, there will be a restoring force due to the rubber on each portion of the string. From its use in the quantum mechanics of the "scalar" meson, Eq. (5.4.1) is often referred to as the Klein-Gordon equation.

Once again, we shall look for particular solutions of the form

$$u(x,t) = X(x)T(t) \tag{5.4.2}$$

This time, however, we find that

$$XT'' - c^2 X''T + h\,XT = 0 \tag{5.4.3}$$

or

$$\frac{T''}{c^2 T} + \frac{h}{c^2} = \frac{X''}{X} = -k^2 \tag{5.4.4}$$

which leads to the two ordinary differential equations

$$X'' + k^2 X = 0 \tag{5.4.5}$$

and

$$T'' + (k^2c^2 + h)T = 0 \tag{5.4.6}$$

If we again require that the string is attached at $x = 0$ and $x = L$, we find for the $X(x)$ solution

$$X_n(x) = \sin(k_n x) \tag{5.4.7}$$

with

$$k_n = \frac{n\pi}{L} \tag{5.4.8}$$

On the other hand, the $T(t)$ solution becomes

$$T_n(t) = A_n \cos(\sqrt{k_n^2c^2 + h}\ t) + B_n \sin(\sqrt{k_n^2c^2 + h}\ t) \tag{5.4.9}$$

so that

$$u_n(x,t) = \sin \frac{n\pi x}{L} \left[A_n \cos(\sqrt{k_n^2c^2 + h}\ t) + B_n \sin(\sqrt{k_n^2c^2 + h}\ t) \right]. \tag{5.4.10}$$

Finally, the general solution becomes

$$u(x,t) = \sum_{n=1}^{\infty} \sin(k_n x) \left[A_n \cos(\sqrt{k_n^2c^2 + h}\ t) + B_n \sin(\sqrt{k_n^2c^2 + h}\ t) \right] \tag{5.4.11}$$

from the principle of linear superposition. Let us consider the case when $B_n = 0$. Then Eq. (5.4.11) can be written

$$u(x,t) = \sum_{n=1}^{\infty} A_n/2 \left[\sin(k_n x + \sqrt{k_n^2c^2 + h}\ t) + \sin(k_n x - \sqrt{k_n^2c^2 + h}\ t) \right]. \tag{5.4.12}$$

Comparing our results with Eq. (5.3.28) we see that the distance that a particular mode k_n moves during the time interval Δt depends not only upon external parameters such as h, the tension and the density of the string but also upon its wave-number (or equivalently, wavelength). Furthermore, the allowed frequencies are all larger than those when $h = 0$. This result is not surprising, because the added stiffness of the medium should increase the natural frequencies.

The importance of dispersion lies in the fact that if the solution $u(x,t)$ is a superposition of progressive waves in the same direction, then the phase relationship between the different harmonics will change with time. Because most signals consist of an infinite series of these progressive waves, dispersion causes the signal to

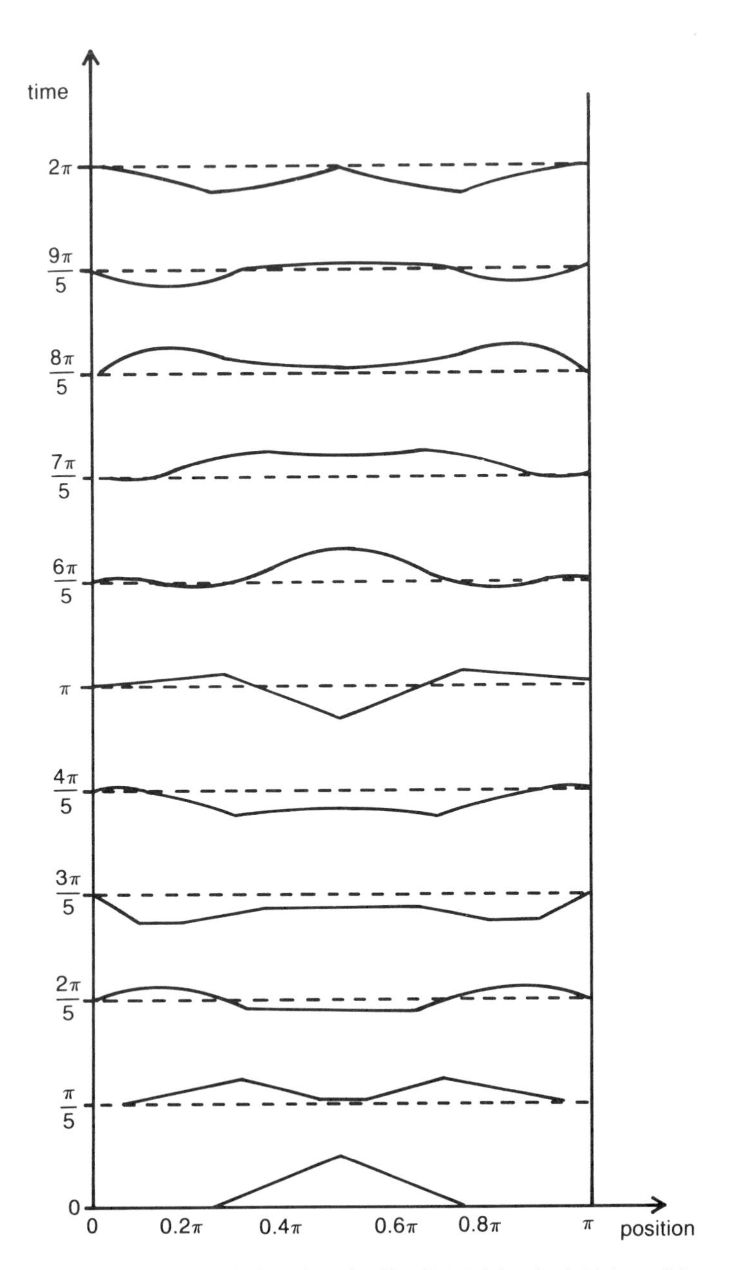

Fig. 5.4.1: The wave solution given by Eq. (5.4.11) for the initial conditions (5.3.27) – (5.3.28) for various positions and times when L = π, c = 1 and h = 1.

become garbled. This is clearly shown by comparing the solution Eq. (5.4.11) given in Fig. 5.4.1 for the initial conditions (5.3.26) and (5.3.27) with $L = \pi$, $c = 1$, and $h = 1$ to the results given in Fig. 5.3.1.

5.5 THE EFFECT OF FRICTION: THE SKIN EFFECT

In the previous section, we showed that a slight modification of the wave equation resulted in a wave solution where each Fourier harmonic propagates with its own particular phase speed. In this section we again introduce a modification to the wave equation that will result not only in dispersive waves but also in the exponential decay of the amplitude as the wave propagates.

So far we have neglected the reaction of the surrounding medium (air or water) on the motion of the string. For small-amplitude motions this reaction is opposed to the motion of each element of length and is proportional to its velocity. The equation of motion, when we take into account tension and friction of the medium but not stiffness or internal friction, is

$$u_{tt} + 2h\, u_t = c^2\, u_{xx}. \tag{5.5.1}$$

Because Eq. (5.5.1) first arose in the mathematical description of the telegraph, it is generally known as the *equation of telegraphy*. The effect of friction is, of course, to damp out the free vibration.

As before we assume a solution of the form

$$u(x,t,) = X(x)T(t) \tag{5.5.2}$$

and separate the variables to obtain the two ordinary differential equations:

$$X'' + k^2X = 0 \tag{5.5.3}$$

$$T'' + 2h\, T' + k^2c^2T = 0 \tag{5.5.4}$$

with

$$X(0) = X(L) = 0. \tag{5.5.5}$$

The shapes of the normal modes are not affected by the friction, being still

$$X(x) = \sin\left(\frac{n\pi x}{L}\right) \tag{5.5.6}$$

and

$$k_n = \frac{n\pi}{L} \tag{5.5.7}$$

The solution for the $T(t)$ equation is

$$T(t) = e^{-ht}\left[A_n \cos(\sqrt{k_n^2c^2 - h^2}\, t) + B_n \sin(\sqrt{k_n^2c^2 - h^2}\, t)\right] \quad \textbf{(5.5.8)}$$

with the condition that $k_n c > h$. If this condition is violated, the solutions are two exponentially damping functions in time. Because this condition is usually fulfilled in most physical problems, we will concentrate on this solution.

From the principle of linear superposition, the general solution is

$$u(x,t) = \sum_{n=1}^{\infty} \sin\frac{n\pi x}{L}\; e^{-ht}\left[A_n \cos(\sqrt{k_n^2c^2 - h^2}\, t) + B_n \sin(\sqrt{k_n^2c^2 - h^2}\, t)\right] \quad \textbf{(5.5.9)}$$

From Eq. (5.5.9) we see two important effects. First, all of the harmonics are slowed down by the presence of friction. Furthermore, all of the harmonics are damped. In Fig. 5.5.1 the solution (5.5.9) is illustrated using the initial conditions given by Eq. (5.3.26) and (5.3.27) with $c = 1$, $L = \pi$ and $h = 0.5$. This is a rather large coefficient of friction and Fig. 5.5.1 shows the rapid dispersion and damping that results.

This damping and dispersion of waves also occurs in solutions to the equation of telegraphy where the solutions are progressive waves. Because early telegraph lines were so short, time delay effects were negligible. However, when the first transoceanic cables were laid in the 1850s, the time delay became seconds and differences in the velocity of propagation of different frequencies, as predicted by Eq. (5.5.9), became noticeable to the operators. This resulted in distortion and restricted the signaling speed of the early cables to about 60 characters per second.

When long-distance telephony was instituted just before the turn of the century, this difference in velocity between frequencies made it necessary to limit the circuits to a few tens of miles. Prof. Michael Pupin, at Columbia University, obtained patents covering the addition of inductors ("loading coils") in series with a line at regular intervals to equalize the velocity for the various frequencies. Thus, adding resistance and inductance, which would seem to make things worse, actually made possible long-distance telephony. Pupin's work was done in 1899; by 1901, it was already in use, and long-distance telephony was on its way. Similar suggestions made earlier by Heaviside had not been put to practical use.

The equation of telegraphy is also met in the electromagnetic theory of waves propagating within a homogeneous, infinite conductor. If the electric field in the y-direction is denoted by E, then

$$\frac{K\mu}{c^2} E_{tt} + \frac{4\pi\sigma\mu}{c^2} E_t = E_{xx} \tag{5.5.10}$$

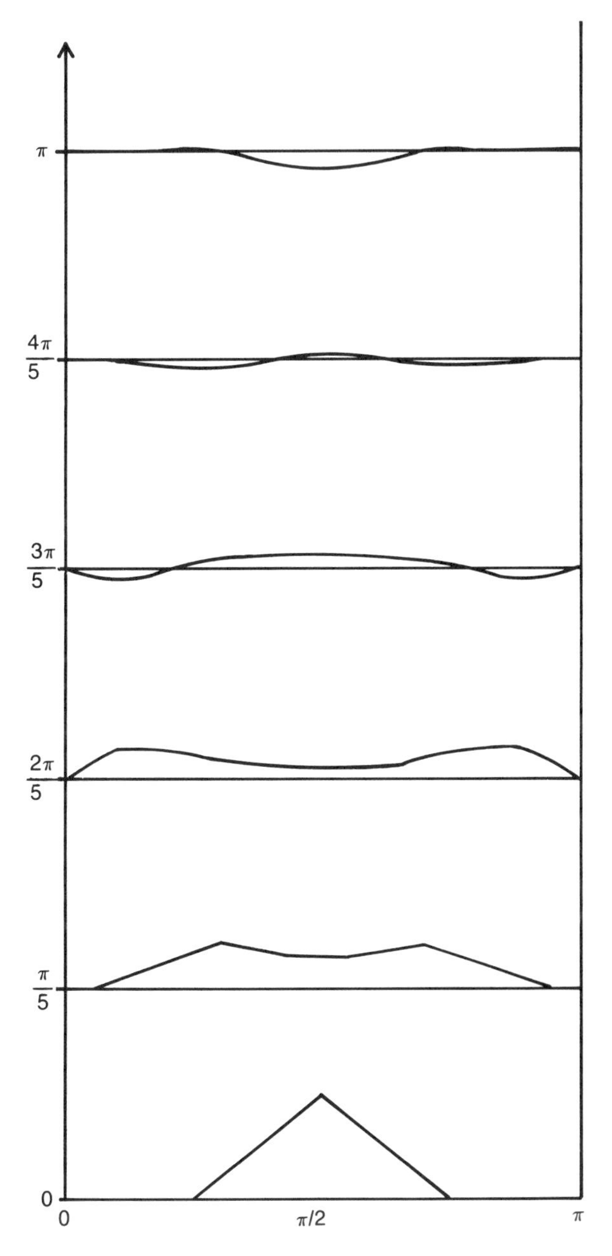

Fig. 5.5.1: The wave solution (5.5.9) for the damped wave equation (5.5.1) for various times and positions when c = 1, L = π and h = 0.5.

where c is the speed of light, K the dielectric constant, μ the permeability, and σ the conductivity. A detailed derivation of this equation may be found in Born, M. and E. Wolf, 1975: *Principles of Optics*. New York: Pergamon Press. Section 13.1.

We now seek a solution to Eq. (5.5.10) of the form

$$E(x,t) = \chi(x)\cos(\omega t) + \psi(x)\sin(\omega t). \qquad \textbf{(5.5.11)}$$

Substituting this solution into Eq. (5.5.10), we have

$$\left(\frac{K\mu}{c^2}\omega^2\chi - \frac{4\pi\sigma\mu}{c^2}\omega\psi + \frac{d^2\chi}{dx^2}\right)\cos(\omega t)$$

$$+\left(\frac{K\mu}{c^2}\omega^2\psi + \frac{4\pi\sigma\mu}{c^2}\omega\chi + \frac{d^2\psi}{dx^2}\right)\sin(\omega t) = 0 \qquad \textbf{(5.5.12)}$$

Because Eq. (5.5.12) must hold for all time, each of the bracketed terms must equal zero:

$$\frac{d^2\chi}{dx^2} - \frac{4\pi\sigma\mu}{c^2}\omega\psi + \frac{K\mu}{c^2}\omega^2\chi = 0 \qquad \textbf{(5.5.13)}$$

$$\frac{d^2\psi}{dx^2} + \frac{4\pi\sigma\mu}{c^2}\omega\chi + \frac{K\mu}{c^2}\omega^2\psi = 0 \qquad \textbf{(5.5.14)}$$

Eliminating ψ between these two equations, we obtain

$$\left(\frac{d^2}{dx^2} + \frac{K\mu\omega^2}{c^2}\right)^2\chi = -\frac{16\pi^2\sigma^2\mu^2}{c^4}\chi \qquad \textbf{(5.5.15)}$$

Unlike our analysis of wave motion on a vibrating string, we assume that the electromagnetic wave is propagating in the positive x-direction. Consequently, the general solution is

$$\chi(x) = e^{-\gamma x}[A\cos(kx) + B\sin(kx)]$$
$$+ e^{\gamma x}[C\cos(kx) + D\sin(kx)]. \qquad \textbf{(5.5.16)}$$

If we assume that $\gamma > 0$, then we must take $C = D = 0$ because the solution would blow up as x increases. Substituting this solution into Eq. (5.5.16) we obtain

$$k = \frac{\omega}{2c} \left(\mu \sqrt{K^2 + 4\sigma^2\tau^2} \right)^{1/2} \quad \textbf{(5.5.17)}$$

$$\gamma = \frac{\omega}{2c} \left(\mu \sqrt{K^2 + 4\sigma^2\tau^2} \right)^{1/2} \quad \textbf{(5.5.18)}$$

where $\tau = 2\pi/\omega$. Once again we see that the waves are dispersive. For a given value of k, we can solve Eq. (5.5.17) for ω and that ω would depend upon external parameters and the wave-number k.

The most important portion of this problem comes from noting that as the wave travels the distance $2\pi/\gamma$, its amplitude is reduced in the ratio of $e^{2\pi}$ to 1. Consequently this distance may be regarded as a convenient measure of the range or penetrating power of the wave. In the case of metals, $2\sigma\tau >> K$, so that $\gamma = k = \frac{\omega}{c}\sqrt{\mu\sigma\tau}$. For copper ($\sigma = 5.14 \times 10^{17}$, $\mu = 1$)

λ	1 cm	1 m	100 m	10 km
$2\pi/\gamma$	$2.4 \times 10^{-4} cm$	0.024 mm	0.24 mm	2.4 mm

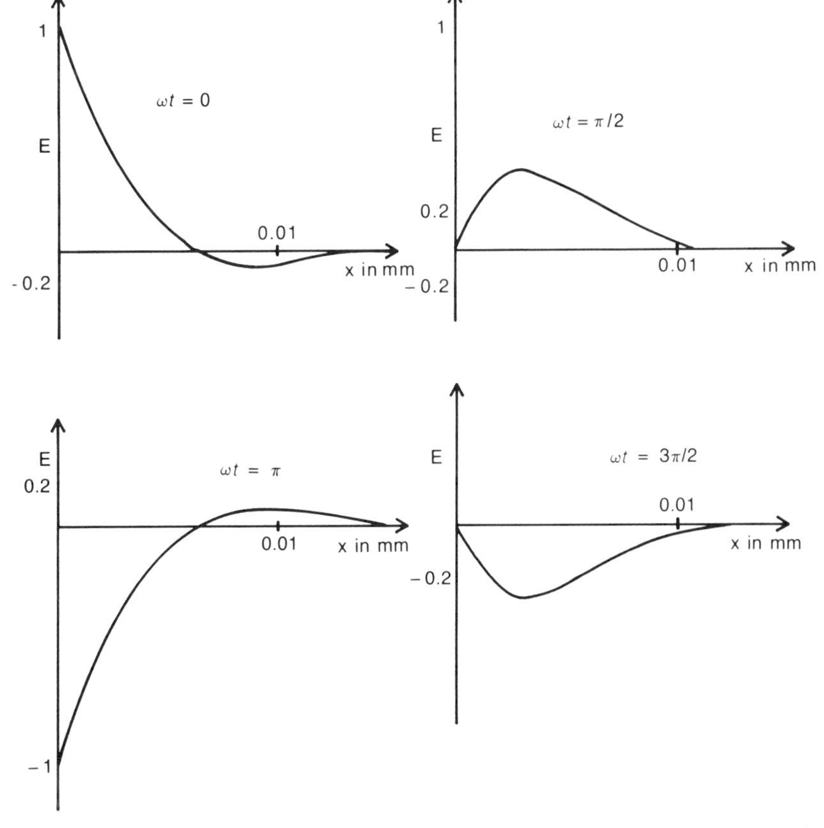

Fig. 5.5.2: The electric field within copper at various times when an electric wave of amplitude unity impinges on it.

As these numerical values show, electromagnetic waves cannot penetrate metals beyond a very shallow depth. This phenomena is known as the *skin effect*. This is illustrated in Fig. 5.5.2 for copper.

Exercises

1. The differential equation for the longitudinal vibrations of a rod within a viscous fluid is

$$\frac{\partial^2 u}{\partial t^2} + 2n \frac{\partial u}{\partial t} = c^2 \frac{\partial^2 u}{\partial t^2}$$

where c is the velocity of sound in the rod and n the damping coefficient. If the rod is fixed at $x = 0$ and allowed to oscillate freely at $x = L$, where L is the length of the rod, the boundary conditions are

$$u(0,t) = u_x(L,t) = 0.$$

Show that the vibrations are given by

$$u(x,t) = \frac{8L}{\pi^2} e^{-nt} \sum_{n=1}^{\infty} \frac{(-1)^n}{(2n+1)^2} \sin\left(\frac{(2n+1)\pi x}{2L}\right) \left[\cos(pt) + \frac{n}{p}\sin(pt)\right]$$

where

$$p = \left[\frac{c^2(2n+1)^2\pi^2}{4L^2} - n^2\right]^{1/2}$$

if the initial condition is given by $u(x,0) = x$ and $u_t(x,0) = 0$.

2. The pressure and velocity oscillations resulting from water hammer in a pipe with friction are given by

$$\frac{1}{c^2}\frac{\partial p}{\partial t} = -\varrho \frac{\partial u}{\partial x}$$

$$\frac{\partial u}{\partial t} = -\frac{1}{\varrho}\frac{\partial p}{\partial x} - \frac{fu}{\varrho}$$

where p denotes the pressure perturbation, u the velocity perturbation, c the speed of sound in water, ϱ the density of water and

f the frictional factor. (See Rich, G. R., 1945: Water-hammer analysis by the Laplace-Mellon transform. *Trans. ASME, 67*, 361-376.) These two first-order partial differential equations may be combined to yield

$$c^2 \frac{\partial^2 p}{\partial x^2} = \frac{\partial^2 p}{\partial t^2} + 2h \frac{\partial p}{\partial t}$$

where $2h = f/p$. Show that the solution to this partial differential equation is

$$p - p_0 = \frac{pU\pi^2 c^4}{4L^3} \sum_{n=1}^{\infty} (-1)^n (2n-1)^2 e^{-ht} \sin\left(\frac{(2n-1)\pi x}{2L}\right)$$

$$\times \frac{[4h\beta\cos(\beta t) + 2h^2\sin(\beta t) - 2\beta^2\sin(\beta t)]}{\beta \left\{\frac{(2n-1)^2\pi^2c^2}{4L^2} + \beta^2\right\}^{1/2}}$$

where $\beta = \left[\frac{(2n-1)^2\pi^2c^2}{4L^2} - h^2\right]^{1/2}$ if $p(0,t) = p_0$ and $u(L,t) = 0$.

5.6 THE RADIAL VIBRATIONS OF A GAS WITHIN A SPHERE

In the previous sections, we have confined ourselves to the wave equation in Cartesian coordinates. We now turn our attention to a problem in spherical coordinates. Once again we shall use the method of separation of variables to obtain the solution.

The small-amplitude vibration of a gas about its equilibrium position is given by

$$\frac{\partial^2 u}{\partial r^2} + \frac{2}{r}\frac{\partial u}{\partial r} = \frac{1}{c^2}\frac{\partial^2 u}{\partial t^2} \qquad \textbf{(5.6.1)}$$

in the case of radial vibrations. The variable u denotes the velocity potential (i.e., the radial velocity equals $-u_r$) and c is the speed of sound. (See Lamb, H., 1925: *The Dynamic Theory of Sound*. New York: Dover Publishing Co. Section 71 for the derivation of this equation.) Because the motion takes place within a sphere, we require the normal component of the velocity at the surface of the sphere $r = R$ to be zero or

$$u_r(R,t) = 0. \tag{5.6.2}$$

We seek a particular solution of the form

$$u(r,t) = W(r)T(t). \tag{5.6.3}$$

Substituting this expression into Eq. (5.6.1), we obtain

$$\frac{T''}{c^2 T} = \frac{W''(r) + \frac{2}{r} W'(r)}{W(r)} = -k^2 \tag{5.6.4}$$

or

$$T'' + k^2 c^2 T = 0 \tag{5.6.5}$$

and

$$W'' + \frac{2}{r} W' + k^2 W = 0. \tag{5.6.6}$$

The general solution to the $W(r)$ equation is

$$W(r) = C \frac{\sin(kr)}{r} + D \frac{\cos(kr)}{r} \tag{5.6.7}$$

where C and D are arbitrary constants. Because $u(r,t)$ and $W(r)$ must be finite at $r = 0$, we must set $D = 0$. We now use the condition that $W'(R) = 0$ and obtain

$$kR \cos(kR) - \sin(kR) = 0 \tag{5.6.8}$$

or

$$\tan(\mu) = \mu \tag{5.6.9}$$

if $kR = \mu$. The roots to Eq. (5.6.9) may be found either graphically or numerically. For sufficiently large n, the eigenvalues are given approximately by

$$\mu_n = \frac{(2n+1)\pi}{2} \tag{5.6.10}$$

An improved value may be found by setting

$$\mu_n = \frac{(2n+1)\pi}{2} - \epsilon_n. \tag{5.6.11}$$

Then Eq. (5.6.9) becomes

$$\cot(\epsilon_n) = \frac{(2n+1)\pi}{2} \tag{5.6.12}$$

Because

$$\cot(\epsilon_n) = \frac{1}{\epsilon_n} - \frac{1}{3}\epsilon_n - \frac{1}{45}\epsilon_n^2 + \ldots, \quad \epsilon_n << 1, \tag{5.6.13}$$

Eq. (5.6.12) becomes

$$\epsilon_n = \frac{2}{(2n+1)\pi} + \frac{4\epsilon_n^2}{3(2n+1)\pi} \tag{5.6.14}$$

or

$$\epsilon_n = \frac{4}{(2n+1)\pi} - \frac{32}{3(2n+1)^3\pi^3} \tag{5.6.15}$$

In Table 5.6.1, the first ten μ's are listed. In addition to these values, the approximate values given by Eq. (5.6.11) and Eq. (5.6.15) are listed.
Corresponding to each of these positive roots of Eq. (5.6.9), which we shall denote by $\mu_0, \mu_1, \mu_2, \ldots$, the eigenfunction is

$$W_n(r) = \frac{1}{r}\sin\left(\frac{\mu_n r}{R}\right) \tag{5.6.16}$$

Table 5.6.1. The first ten roots of Eq. (5.6.9) and their approximate value given by (5.6.11) and (5.6.15).

n	μ_n	approximate μ_n
0	4.493	3.78317
1	10.904	10.58390
2	17.221	17.02684
3	23.519	23.38104
4	29.811	29.70412
5	36.101	36.01282
6	42.388	42.31371
7	48.674	48.60989
8	54.960	54.90305
9	61.245	61.19409

So far we have found the eigenfunctions for a negative separation constant. However, in this problem we must include the eigenfunction for $k = 0$. Unlike the string problem the eigenfunction associated with $k = 0$ is nontrival and we must include it in our total solution. The solution for $k = 0$, which also satisfies the boundary conditions, is

$$W_0(r) = 1. \qquad \textbf{(5.6.17)}$$

The temporal portion of the problem is given by

$$T_n(t) = A_n \cos(\mu_n ct/R) + B_n \sin(\mu_n ct/R) \qquad \textbf{(5.6.18)}$$

and

$$T_0(t) = A_0 + B_0 t \qquad \textbf{(5.6.19)}$$

for $k = 0$. Consequently the general solution is given by the series

$$u(r,t) = A_0 + B_0 t + \sum_{n=0}^{\infty} [A_n \cos(\mu_n ct/R) + B_n \sin(\mu_n ct/R)] \times \frac{\sin(\mu_n r/R)}{r} \qquad \textbf{(5.6.20)}$$

For our initial conditions, let us assume that

$$u(r,0) = f(r) \text{ and } u_t(r,0) = g(r) \qquad \textbf{(5.6.21)}$$

Consequently

$$f(r) = \sum_{n=0}^{\infty} A_n \frac{\sin(\mu_n r/R)}{r} + A_0 \qquad \textbf{(5.6.22)}$$

$$g(r) = \sum_{n=0}^{\infty} \frac{\mu_n c}{R} B_n \frac{\sin(\mu_n r/R)}{r} + B_0 \qquad \textbf{(5.6.23)}$$

Assuming that Eq. (5.6.22) converges uniformly, we determine A_n by multiplying both sides of Eq. (5.6.22) by $r \sin(\mu_m r/R)\, dr$ and integrating from 0 to R. We obtain

$$\int_0^R r\, f(r) \sin(\mu_m r/R)\, dr = \sum_{n=0}^{\infty} A_n \int_0^R \sin(\mu_m r/R) \sin(\mu_n r/R)\, dr + ① \qquad \textbf{(5.6.24)}$$

$$\textcircled{1} = A_0 \int_0^R r \sin(\mu_m r/R)\, dr$$

Now

$$\int_0^R \sin(\mu_m r/R) \sin(\mu_n r/R)\, dr = \frac{R}{2}\left[\frac{\sin(\mu_m - \mu_n)}{\mu_m - \mu_n} - \frac{\sin(\mu_m + \mu_n)}{\mu_m + \mu_n}\right] \tag{5.6.25}$$

$$= \frac{R}{2}\, \frac{\cos(\mu_n) \sin(\mu_m) \left[\mu_n \tan(\mu_m) - \mu_m \tan(\mu_n)\right]}{\mu_m^2 - \mu_n^2} \tag{5.6.26}$$

Using Eq. (5.6.9) in Eq. (5.6.26) we obtain

$$\int_0^R \sin(\mu_m r/R) \sin(\mu_n r/R)\, dr = 0 \quad n \neq m \tag{5.6.27}$$

Similarly, we have

$$\int_0^R r \sin(\mu_m r/R)\, dr = \frac{R^2}{\mu_m^2} \sin(\mu_m r/R)\Big|_0^R - \frac{rR}{\mu_m} \cos(\mu_m r/R)\Big|_0^R \tag{5.6.28}$$

$$= \frac{R^2}{\mu_m^2} \sin(\mu_m) - \frac{R^2}{\mu_m} \cos(\mu_m) = 0 \tag{5.6.29}$$

However, if $n = m$

$$\int_0^R \sin^2(\mu_n r/R)\, dr = \frac{R}{2}\left[1 - \frac{\sin^2(\mu_n)}{\mu_n^2}\right] = \frac{R}{2}\left[1 - \frac{\tan^2(\mu_n)}{1 + \tan^2(\mu_n)}\right] = \frac{R}{2}\, \frac{\mu_n^2}{1 + \mu_n^2} \tag{5.6.30}$$

Consequently

$$A_n = \frac{2}{R}\left(1 + \frac{1}{\mu_n^2}\right) \int_0^R r\, f(r) \sin(\mu_n r/R)\, dr \tag{5.6.31}$$

In an analogous manner, we obtain

$$B_n = \frac{2}{c\mu_n}\left(1+\frac{1}{\mu_n^2}\right)\int_0^R r\,g(r)\sin(\mu_n r/R)\,dr. \qquad \textbf{(5.6.32)}$$

Finally, to determine C, we multiply both sides of Eq. (5.6.22) by $r^2\,dr$ and integrate from 0 to R:

$$\int_0^R f(r)\,r^2\,dr = \sum_{n=0}^{\infty} A_n \int_0^R r\sin(\mu_n r/R)\,dr + A_0\int_0^R r^2\,dr$$

(5.6.33)

From Eq. (5.6.29) we see that each term in the summation vanishes and we have

$$A_0 = \frac{3}{R^3}\int_0^R f(r)\,r^2 dr \qquad \textbf{(5.6.34)}$$

Similarly we find that

$$B_0 = \frac{3}{R^3}\int_0^R r^2 g(r) dr. \qquad \textbf{(5.6.35)}$$

To illustrate our solution, which combines Eq. (5.6.20), (5.6.31), (5.6.32), (5.6.34) and (5.6.35), we find the radial velocity field produced when the boundary separating a spherical region of high pressure ($0<r<R/2$) from lower pressure ($R/2<r<R$) is annihilated. Because there is initially no motion, $u(r,0)$ is constant everywhere, and may be taken as zero. On the other hand, the linearized, time-dependent Bernoulli equation (see Lamb, H., 1932: *Hydrodynamics*. New York: Dover Publications, Inc. Section 20.) states that the pressure field is given by

$$p(r,t) = \varrho u_t \qquad \textbf{(5.6.36)}$$

where ϱ is the density of the gas. Consequently, in our problem,

$$f(r) = 0 \qquad \textbf{(5.6.37)}$$

and

$$g(r) = \begin{cases} \dfrac{P}{\varrho} & 0<r<R/2 \\ 0 & R/2<r<R \end{cases} \qquad \textbf{(5.6.38)}$$

if we subtract off a reference pressure so that lower pressure is equal to zero.

With these initial conditions, we find that

$$A_n = A_0 = 0 \qquad \textbf{(5.6.39)}$$

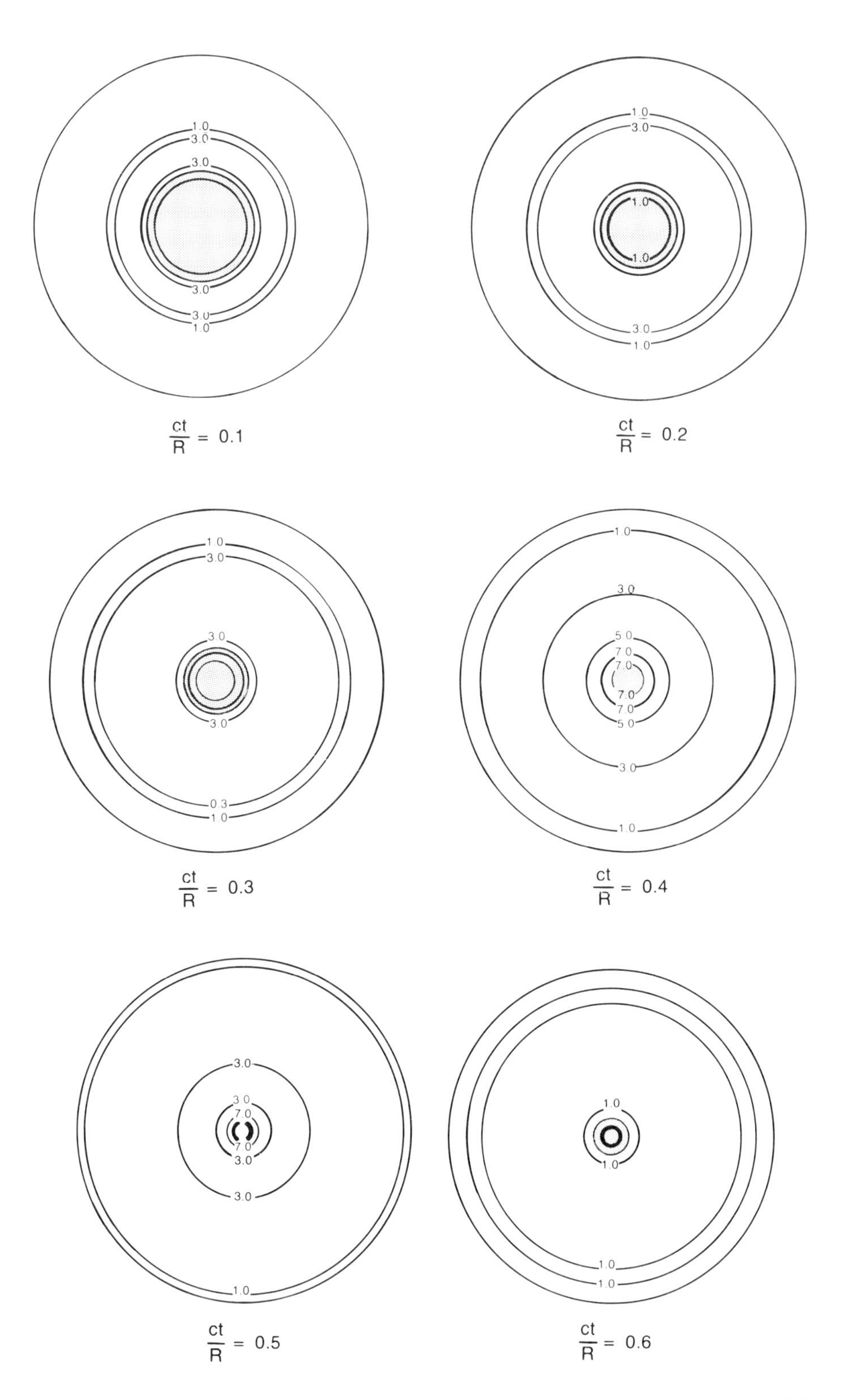
1.0
3.0
3.0
3.0
3.0
1.0
$\frac{ct}{R} = 0.1$
1.0
3.0
1.0
1.0
3.0
1.0
$\frac{ct}{R} = 0.2$
1.0
3.0
3.0
3.0
0.3
1.0
$\frac{ct}{R} = 0.3$
1.0
3.0
5.0
7.0
7.0
7.0
7.0
5.0
3.0
1.0
$\frac{ct}{R} = 0.4$
3.0
3.0
7.0
7.0
3.0
3.0
1.0
$\frac{ct}{R} = 0.5$
1.0
1.0
1.0
1.0
$\frac{ct}{R} = 0.6$

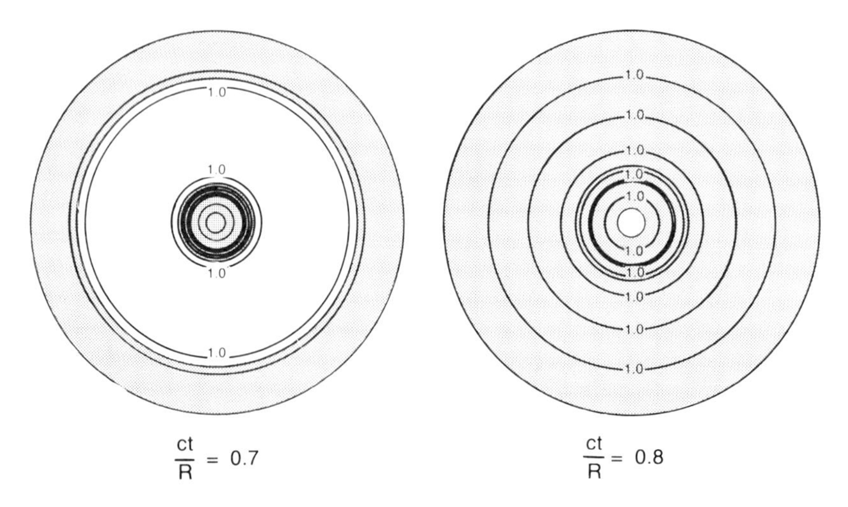

Fig. 5.6.1: The radial velocity within a spherical cavity for various times and positions. The shaded area denotes negative values. The initial conditions consist of a region of higher pressure separated from lower pressure by a boundary that dissolves at t = 0.

and

$$B_n = \frac{2}{c\mu_n}\left(1 + \frac{1}{\mu_n^2}\right)\frac{P}{\varrho}\int_0^{R/2} r\sin(\mu_n r/R)\,dr \qquad \textbf{(5.6.40)}$$

$$= \frac{2P}{c\mu_n\varrho}\left(1 + \frac{1}{\mu_n^2}\right)\left(\frac{R^2}{\mu_n^2}\sin(\mu_n r/R) - \frac{Rr}{\mu_n}\cos(\mu_n r/R)\right)\Bigg|_0^{R/2}$$

(5.6.41)

$$= \frac{2PR^2}{\varrho c\mu_n^3}\left(1 + \frac{1}{\mu_n^2}\right)\Big[(\sin(\mu_n/2) - \mu_n/2\cos(\mu_n/2)\Big]$$

5.6.42)

with $B_0 = P/8\varrho$ and the radial velocity is given by

$$-\frac{P}{\varrho}\sum_{n=0}^{\infty}\frac{2R^2}{c\mu_n^3}\left(1 + \frac{1}{\mu_n^2}\right)\ [\sin(\mu_n/2) - \mu_n/2\cos(\mu_n/2)]$$

$$+\sin\left(\frac{\mu_n ct}{R}\right)\left(\frac{\mu_n\cos(\mu_n r/R)}{rR} - \frac{\sin(\mu_n r/R)}{r^2}\right) \qquad \textbf{(5.6.43)}$$

In Fig. 5.6.1 the velocity field has been graphed at various times for various values of r/R in units of $0.1P/c\varrho$. From this figure we see that the air initially rushes from the region of higher pressure into the region of lower pressure; the velocity is directed outwards in the outer half and inwards in the inner half. Later on, after the pressure wave has reached the center and the discontinuity is completely eliminated ($ct/R = 0.5$), we have a simple oscillator where the radial velocity is all positive one moment (e.g., $ct/R = 0.5$) and then becomes all negative later on. The air sways from side to side in much the same manner as in a doubly closed pipe.

5.7 AXIALLY SYMMETRIC VIBRATIONS OF A CIRCULAR MEMBRANE

Vibrations of a circular membrane are governed by the two-dimensional wave equation

$$\frac{\partial^2 u}{\partial r^2} + \frac{1}{r}\frac{\partial u}{\partial r} + \frac{1}{r^2}\frac{\partial^2 u}{\partial \theta^2} = \frac{1}{c^2}\frac{\partial^2 u}{\partial t^2} \quad (5.7.1)$$

where u is the vertical displacement of the membrane, r the radius, θ the meridional angle and c the square root of the ratio of the tension of the membrane to its density. We shall solve this equation when the motion is axisymmetric and no θ dependence occurs. The initial conditions will be assumed to be that the membrane is initially at rest, $u(r,0) = 0$, when it is struck in a manner so that

$$u_t(r,0) = \begin{cases} \dfrac{P}{\pi\epsilon^2\varrho} & 0<r<\epsilon \\ 0 & \epsilon<r<a. \end{cases} \quad (5.7.2)$$

Because this problem can be solved by separation of variables, we assume that

$$u(r,t) = R(r)\ T(t). \quad (5.7.3)$$

The substitution of Eq. (5.7.3) into Eq. (5.7.1) and the separation of variables leads to

$$\frac{1}{rR}\frac{d}{dr}\left(r\frac{dR}{dr}\right) = \frac{1}{c^2T}\frac{d^2T}{dt^2} = -k^2 \quad (5.7.4)$$

or

$$\frac{1}{r}\frac{d}{dr}\left(r\frac{dR}{dr}\right) + k^2R = 0 \quad (5.7.5)$$

and

$$\frac{d^2T}{dt^2} + k^2c^2\, T = 0 \qquad \textbf{(5.7.6)}$$

The separation constant k must be positive in order that the solution remains bounded in the region $0 \leq r < a$ where a is the radius corresponding to the boundary of the membrane. At this boundary, $u(a,t) = R(a)T(t) = 0$ or $R(a) = 0$.

The solutions of Eqs. (5.7.5) – (5.7.6), subject to the boundary conditions, are

$$R_n(r) = J_0\left(\frac{\lambda_n r}{a}\right) \qquad \textbf{(5.7.7)}$$

and

$$T_n(t) = A_n \sin\left(\frac{\lambda_n ct}{a}\right) + B_n \cos\left(\frac{\lambda_n ct}{a}\right) \qquad \textbf{(5.7.8)}$$

where λ_n satisfies the equation $J_0(\lambda_n) = 0$. Because

$$u(r,0) = T_n(t) = 0,$$

we have $B_n = 0$. Consequently the product solution is

$$u(r,t) = \sum_{n=1}^{\infty} A_n J_0\left(\frac{\lambda_n r}{a}\right) \sin\left(\frac{\lambda_n ct}{a}\right) \qquad \textbf{(5.7.9)}$$

To determine A_n, we use the condition

$$u_t(r,0) = \sum_{n=1}^{\infty} \frac{\lambda_n c}{a} A_n J_0\left(\frac{\lambda_n r}{a}\right) = \begin{cases} \dfrac{P}{\pi\epsilon^2\varrho} & 0<r<\epsilon \\ 0 & \epsilon<r<a \end{cases} \qquad \textbf{(5.7.10)}$$

Equation (5.7.10) is a Fourier-Bessel expansion in the orthogonal function $J_0(\mu_n r/a)$ where A_n is given by

$$\frac{\lambda_n c}{a} A_n = \frac{2}{a^2 J_1(\lambda_n)^2} \int_0^{\epsilon} \frac{P}{\pi\epsilon^2\varrho}\, r\, J_0\left(\frac{\lambda_n r}{a}\right) dr \qquad \textbf{(5.7.11)}$$

from Eqs (1.7.58) and (1.7.66) in Section 1.7. Carrying out the integration, we have

$$A_n = \frac{2\, P\, J_1(\lambda_n \epsilon/a)}{c\pi\epsilon\varrho\, \lambda_n^{\,2}\, J_1\, (\lambda_n)^2} \qquad \textbf{(5.7.12)}$$

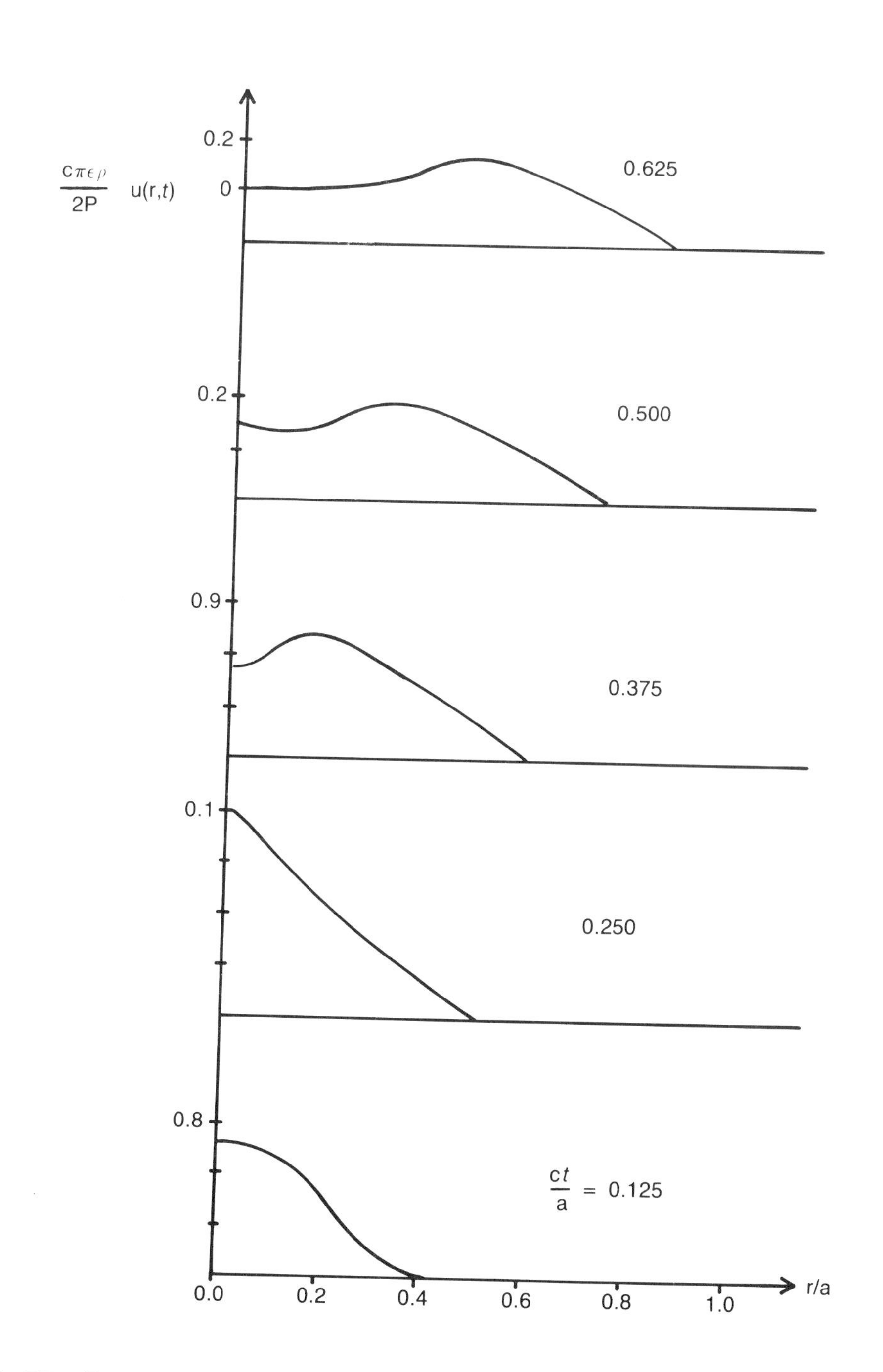

Fig. 5.7.1: The axially symmetric vibrations of a circular membrane when struck by a hammer of radius a/4.

Table 5.7.1. The First Ten Roots of $J_0(\lambda_n) = 0$ and the corresponding A_n for $\epsilon = a/4$.

n	λ_n	$\frac{c\pi\epsilon\varrho}{2P} A_n$
1	2.405	0.18429
2	5.520	0.15257
3	8.654	0.10165
4	11.792	0.04768
5	14.931	0.00424
6	18.071	−0.02049
7	21.212	−0.02561
8	24.352	−0.01667
9	27.493	−0.00246
10	30.635	0.00884

or

$$u(r,t) = \frac{2P}{c\pi\epsilon\varrho} \sum_{n=1}^{\infty} \frac{J_1(\lambda_n \epsilon/a)}{\lambda_n^2 J_1(\lambda_n)^2} J_0(\lambda_n \frac{r}{a}) \sin(\frac{\lambda_n ct}{a}) \quad \textbf{(5.7.13)}$$

In Fig. 5.7.1 the solution (5.7.13) is graphed for various times when $\epsilon = a/4$. In Table 5.7.1 the first ten roots of $J_0(\lambda_n) = 0$ are given along with the corresponding A_n.

5.8 THE SMALL-AMPLITUDE VIBRATIONS OF A ROTATING STRING

As an example of the application of Legendre polynomials, let us examine the problem of a vibrating, homogenous string of length L that is fixed at one end and that rotates freely around the point at which it is fixed. If we neglect gravity and air resistance, the equilibrium position of the string will be a straight line that is rotating with angular velocity ω in a plane passing through the fixed point.

In this problem the displacement $u(x,t)$ of the string from equilibrium is assumed parallel to the axis of rotation and perpendicular to the plane of rotation. Unlike our derivation of the wave equation in Section 5.1, the tension at the point x is not constant because the acceleration at each point equals $-\omega^2 x$. Consequently the tension is equal to the sum of these forces acting on all elements of the string from the point x to its free end:

$$\text{tension} = \int_x^L \varrho\omega^2 x \, dx = \varrho\omega^2 (L^2 - x^2)/2 \quad \textbf{(5.8.1)}$$

where ϱ is the density of the string. Therefore, if we modify our derivation of the wave equation, we obtain

$$u_{tt} = c^2 [(L^2 - x^2)u_x]_x \tag{5.8.2}$$

where $c^2 = \omega^2/2$, x the distance from the fixed point, and t time. The boundary conditions are

$$u(0,t) = 0, \; u(x,0) = f(x), \; u_t(x,0) = g(x). \tag{5.8.3}$$

Let us seek particular solutions to Eq. (5.8.2) of the form

$$u(x,t) = X(x)\, T(t). \tag{5.8.4}$$

Substituting this equation into Eq. (5.8.2), we obtain

$$\frac{T''}{c^2 T} = \frac{1}{X} [(L^2 - x^2)X']' = -k^2 \tag{5.8.5}$$

or

$$T'' + c^2 k^2 T = 0 \tag{5.8.6}$$

and

$$[(L^2 - x^2)X']' + k^2 X = 0. \tag{5.8.7}$$

We note that Eq. (5.8.7) is a form of Legendre's equation if we scale x by L.

As we saw in Section 1.6, the solution to Eq. (5.8.7) can remain finite only if $k^2 = n(n+1)$ where n is a positive integer and the eigenfunction is the Legendre polynomial $P_n(x/L)$. In order to satisfy the boundary condition $u(0,t) = 0$, $P_n(0) = 0$. This is only possible if n is an odd integer. Therefore

$$X_m(x) = P_{2m-1}(x/L) \tag{5.8.8}$$

with

$$k_m^{\;2} = 2m(2m-1) \tag{5.8.9}$$

and $m = 1,2,3, \ldots$

The temporal part is given by

$$T_m(t) = A_m \cos\left[\sqrt{2m(2m-1)}\; ct\right] + B_m \sin\left(\sqrt{2m(2m-1)}\; ct\right) \tag{5.8.10}$$

and the general solution, by the summation:

$$u(x,t) = \sum_{m=1}^{\infty} \left\{ A_m \cos [\sqrt{2m(2m-1)}ct] + B_m \sin [\sqrt{2m(2m-1)}ct] \right\} P_{2m-1}(x/L) \quad \textbf{(5.8.11)}$$

From the initial conditions we have

$$u(x,0) = \sum_{m=1}^{\infty} A_m P_{2m-1}(x/L) = f(x) \quad \textbf{(5.8.12)}$$

$$u_t(x,0) = \sum_{m=1}^{\infty} \sqrt{2m(2m-1)}\, c\, B_m P_{2m-1}(x/L) = g(x) \quad \textbf{(5.8.13)}$$

Assuming that the series (5.8.12) converges uniformly, we can determine the coefficient A_m by multiplying both sides of Eq. (5.8.12) by $P_{2n-1}(x/L)\, dx$ and integrating from 0 to L. Using the orthogonality condition for Legendre functions,

$$\int_0^L f(x) P_{2m-1}(x/L)\, dx = A_m \int_0^L P^2_{2m-1}(x/L)\, dx$$

$$= \frac{L}{2} A_m \int_{-1}^{1} P^2_{2m-1}(\nu)\, d\nu = \frac{L}{4m-1} A_m \quad \textbf{(5.8.14)}$$

Hence

$$A_m = \frac{4m-1}{L} \int_0^L f(x) P_{2m-1}(x/L)\, dx. \quad \textbf{(5.8.15)}$$

In an analogous way, we obtain

$$B_m = \frac{4m-1}{cL\sqrt{2m(2m-1)}} \int_0^L g(x) P_{2m-1}(x/L)\, dx. \quad \textbf{(5.8.16)}$$

We note that the frequency ω_m of the mth harmonic is given by $\sqrt{m(2m-1)}\omega$. Thus the frequencies of the vibration depend upon the angular velocity and not on the length of the string or its density (as long as the density is constant).

To illustrate the solution, which combines (5.8.11), (5.8.15) and (5.8.16), we find the solution when

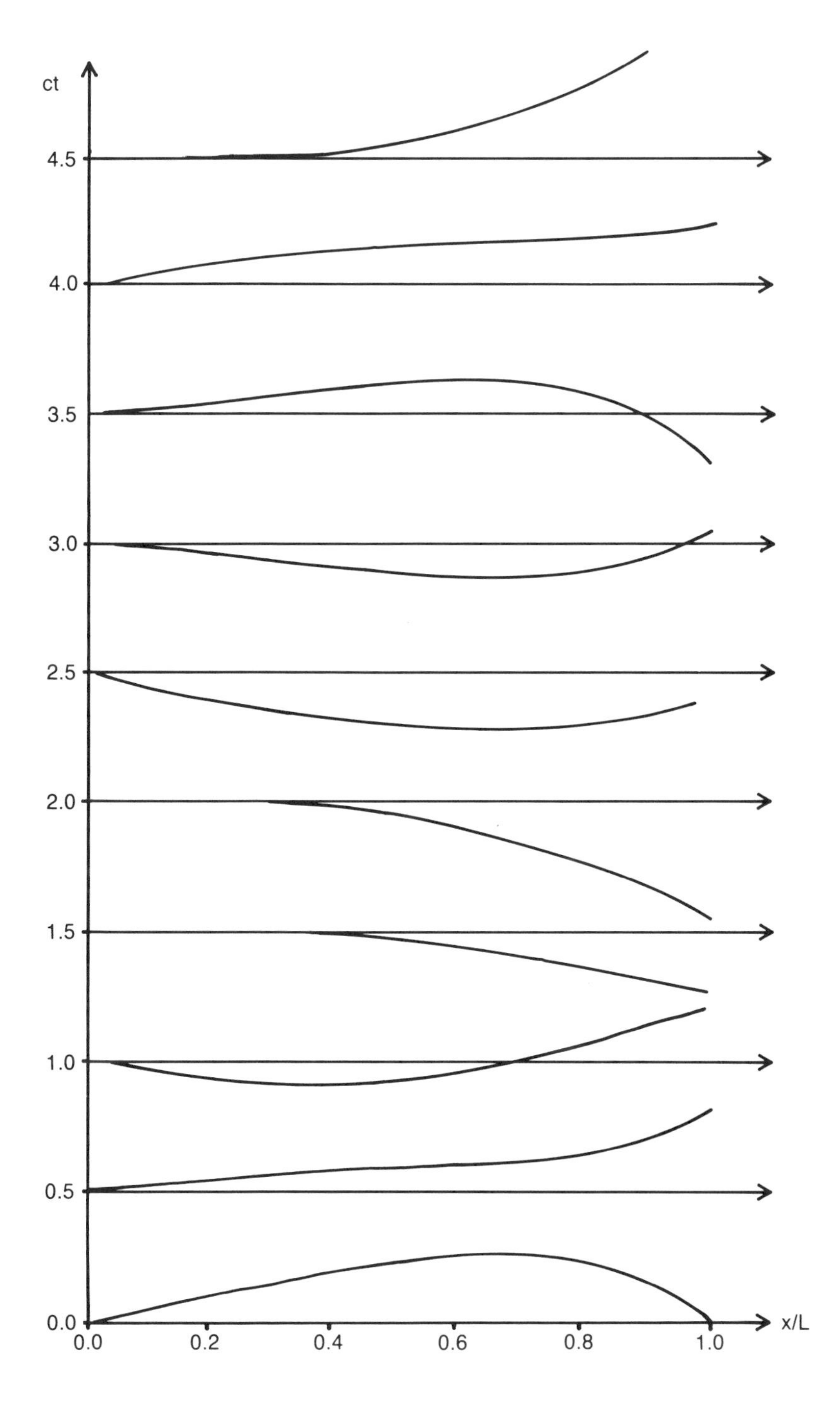

Fig 5.8.1: The vibration of a rotating string for various positions and times when $f(x) = x/L - (x/L)^5$.

$$f(x) = \frac{x}{L} - (\frac{x}{L})^5 \quad \textbf{(5.8.17)}$$

and

$$g(x) = 0. \quad \textbf{(5.8.18)}$$

The reexpression of $f(x)$ in terms of Legendre polynomials is most conveniently done by writing $f(x)$ as

$$f(x) = A_1 P_1 (\frac{x}{L}) + A_2 P_3 (\frac{x}{L}) + A_3 P_5 (\frac{x}{L}) \quad \textbf{(5.8.19)}$$

and computing A_1, A_2 and A_3 so that $f(x) = \frac{x}{L} - (\frac{x}{L})^5$ by direct substitution of the Legendre polynomials involved. When this is done, we find that

$$A_1 = 4/7, A_2 = -4/9 \text{ and } A_3 = -8/63.$$

In Fig. 5.8.1, the subsequent wave motion as a result of the initial displacement $f(x)$ is presented.

5.9 SPECTRAL METHODS: SEICHES INDUCED BY AN EARTHQUAKE

Forced or free oscillations of long water waves in an enclosed or semienclosed basin are called *seiches*. Among the causes of seiches are the horizontal displacement of the earth caused by major (even distant) earthquakes, passage of meteorological pressure fronts, impact of wind gusts, flood discharge, direct air-water coupling of atmospheric disturbances, etc. In this section we will demonstrate how spectral techniques may be used to solve the wave equation describing a seiche caused by a large earthquake. This analysis assumes that the natural period of a basin is close to the period of the seismic waves, the energy of the earthquake is above a certain threshold value, and that the epicenter of the earthquake is distant from the basin.

In the case of long waves, traveling along a straight canal of depth H, the equation governing the time integral of the velocity past the plane x = constant up to the time t is given by

$$\frac{\partial^2 \Theta}{\partial t^2} + k \frac{\partial \Theta}{\partial t} - gH \frac{\partial^2 \Theta}{\partial x^2} = F(x,t), \ k > 0 \quad \textbf{(5.9.1)}$$

where x axis is taken along the length of the canal. See McGarr, A., 1965: Excitation of seiches in channels by seismic waves. *J. Geophy. Res., 70*, 847-854. $F(x,t)$ represents the horizontal acceleration associated with the seismic waves propagating along the surface from the epicenter. The second term on the left-hand side of Eq. (5.9.1) gives a velocity-dependent dissipative force. The boundary conditions are

$$\Theta(0,t) = \Theta(L,t) = 0 \qquad \textbf{(5.9.2)}$$

and the initial conditions

$$\Theta(x,0) = \Theta_t(x,0) = 0 \qquad 0 \leq x \leq L \qquad \textbf{(5.9.3)}$$

where L is the canal's width.

To solve Eq. (5.9.1) we assume that the solution can be written as a sum of a complete set of eigenfunction on $0 \leq x \leq L$, namely

$$\Theta(x,t) = \sum_{n=1}^{\infty} A_n(t) \sin\left(\frac{n\pi x}{L}\right) \qquad \textbf{(5.9.3)}$$

This is certainly not revolutionary in light of our use of the principle of superposition and Fourier series in the previous sections. What is unique about this approach is that it may be generalized so that *any* infinite series of smooth and, preferably, orthogonal functions may be used to eliminate the physical space variable from the problem and reduce the solution of the partial differential equation to the solution (usually) of a set of ordinary differential equations in the other independent variable. Because of its close association with Fourier series, the expansion coefficients are referred to as spectra and this approach is called the *spectral method*. The functions used in the expansion are usually chosen so that they satisfy some of the boundary conditions.

In our particular problem, let us assume that the variation of $F(x,t)$ over the canal is small so that $F(x,t) = F(t)$. Consequently our expansion for $F(x,t)$ becomes

$$F(x,t) = F(t) \begin{cases} 1 & 0 < x/L < 1 \\ 0 & x = 0, L \end{cases} = \frac{4\,F(t)}{\pi} \sum_{n=0}^{\infty} \frac{\sin[(2n+1)\pi x/L]}{2n+1}$$

$$\textbf{(5.9.4)}$$

Upon substituting our expansion for Θ and F into Eq. (5.9.1) and equating the coefficients of $\sin[(2n+1)\pi x/L]$, we obtain a differential equation for the unknown coefficients:

$$\frac{d^2}{dt^2} A_{2n+1}(t) + k \frac{d}{dt} A_{2n+1}(t) + \frac{(2n+1)^2 \pi^2 c_0^2}{L^2} A_{2n+1}(t) = \frac{4}{\pi(2n+1)} F(t)$$

(5.9.5)

where $c_0 = \sqrt{gH}$, the velocity of free gravity waves in water of depth much less than the wavelength.

Equation (5.9.5) is a linear, second-order, nonhomogeneous ordinary differential equation. Because the expansion for the forcing contains only odd harmonics, then $A_{2n}(t) = 0$. A general solution depends upon the magnitude of k compared with $\pi c_0/L$. In most cases $k << 1$ so that the explicit solution is given approximately by

$$A_{2n+1}(t) = \frac{-4}{\pi(2n+1)} \int_0^t F(\tau)\, e^{-k(t-\tau)/2} \frac{\sin[\beta(t-\tau)]}{\beta} d\tau$$

(5.9.6)

where

$$\beta = \pi c_0 (2n+1)/L. \qquad \textbf{(5.9.7)}$$

To illustrate Eq. (5.9.6), suppose $F(\tau)$ is a short pulse of unit intensity, lasting the brief instant ϵ:

$$F(\tau) = \begin{matrix} 1 & 0<\tau<\epsilon \\ 0 & \text{afterwards} \end{matrix} \qquad \textbf{(5.9.8)}$$

Then Eq. (5.9.6) becomes

$$A_{2n+1}(t) = \frac{-4}{\pi(2n+1)} \int_0^\epsilon e^{-k(t-\tau)/2} \frac{\sin[\beta(t-\tau)]}{\beta} d\tau \qquad \textbf{(5.9.9)}$$

$$= \frac{e^{-k(t-\epsilon)/2}\{-k/2 \sin[\beta(t-\epsilon)] - \beta\cos[\beta(t-\epsilon)]\} - e^{-kt/2}[-k/2\sin(\beta t) - \beta\cos(\beta t)]}{\pi(2n+1)\,\beta\,(k^2/4 + \beta^2)/4}$$

(5.9.10)

$$A_{2n+1}(t) = -\frac{4\epsilon}{\pi(2n+1)\beta} \sin(\beta t)\, e^{kt/2} \qquad \textbf{(5.9.11)}$$

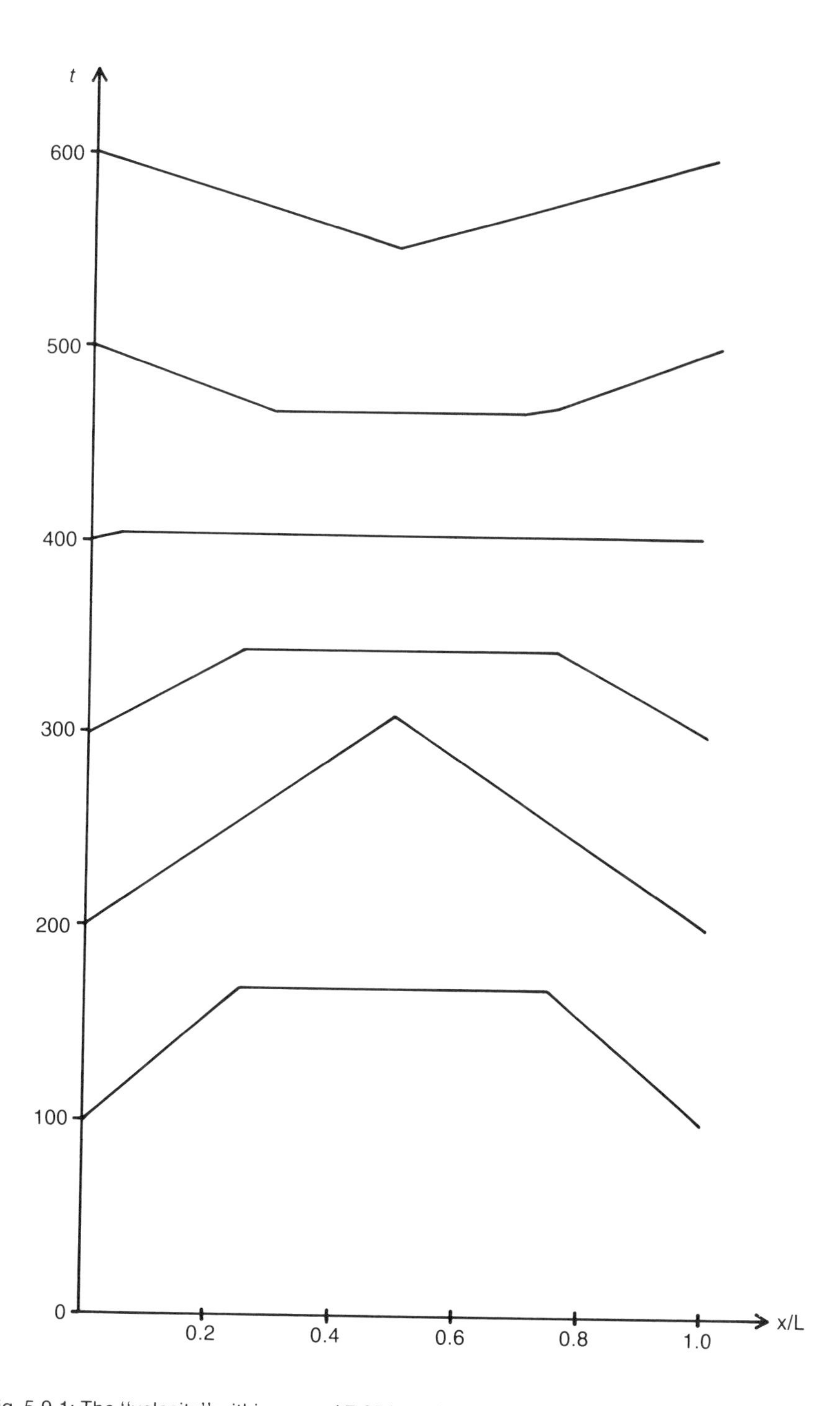

Fig. 5.9.1: The ''velocity'' within a canal 7.85 km wide resulting from an earthquake in a channel 40 m deep.

in the limit as $\epsilon \to 0$. Consequently we see that the temporal solution behaves like a damped oscillator.

In Fig. 5.9.1, the solution (5.9.3) along with Eq. (5.9.11) is graphed when $k = 4 \times 10^{-3}$/sec, $H = 40\ m$ ($c_0 = 20\ m$/sec) and $L = 7.85\ km$. From Fig. 5.9.1, we see that the solution is more strongly damped than dispersed.

5.10 THE VIBRATION OF A BAR WITH A LOAD AT THE END

At this point, the reader may have the impression that separation-of-variables solutions on a limited domain always leads to a Sturm-Liouville problem. To show that this not true, we consider the vibration of a bar with a load at its end. See Fig. 5.10.1. If the mass of the bar is small compared to the mass of the load at the end, it can be neglected and the longitudinal displacements from its equilibrium position $u(x,t)$ is given by

$$u_{tt} = c^2 u_{xx} \qquad \textbf{(5.10.1)}$$

where $c^2 = E/\varrho$, E denotes the modulus of elasticity, and ϱ the density of the material of the bar. At the top, the displacement is zero during the vibrations so that $u(0,t) = 0$. At the lower end, at which the load is attached, the tensile force in the bar must be equal to the inertia force of the oscillating load W and we have

$$AE\, u_x(L,t) = -\frac{W}{g}\, u_{tt}(L,t) \qquad \textbf{(5.10.2)}$$

where A denotes the cross-sectional area and g the gravitational attraction. See Timoshenko and D. H. Young, 1955: *Vibration Problems in Engineering*. 3rd Ed. Princeton (NJ): D. Van Nostrand Co., Inc., Chapter 5 for details.

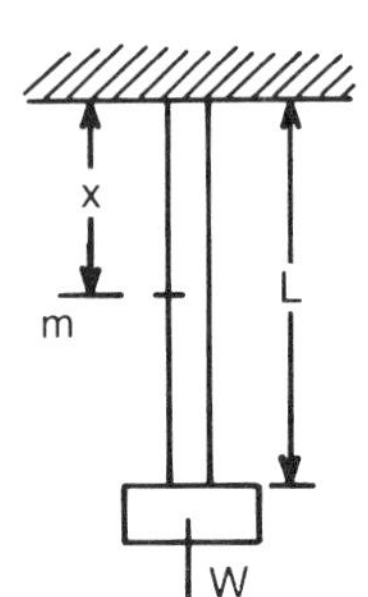

Fig. 5.10.1: Diagram of a bar with a load at the end.

From previous problems, we anticipate that $u(x,t)$ can be written as

$$u(x,t) = X(x) [A \cos(\omega t) + B \sin(\omega t)]. \qquad \textbf{(5.10.3)}$$

Substituting Eq. (5.10.3) into Eq. (5.10.1), we obtain

$$c^2 \frac{d^2X}{dx^2} + \omega^2 X = 0 \qquad \textbf{(5.10.4)}$$

whose solution is $X(x) = C \cos\left(\frac{\omega x}{c}\right) + D \sin\left(\frac{\omega x}{c}\right)$ **(5.10.5)**

where C and D are constants of integration. The boundary condition $u(0,t) = 0$ gives $C = 0$. The boundary condition (5.10.2) gives

$$AE \frac{\omega}{c} \cos\left(\frac{\omega L}{c}\right) = \frac{W}{g} \omega^2 \sin\left(\frac{\omega L}{c}\right). \qquad \textbf{(5.10.6)}$$

If we define $\alpha = g\varrho AL/W$, the ratio of the weight of the bar to the weight of the load W, and $\beta = \omega L/c$, we obtain

$$\alpha = \beta \tan(\beta). \qquad \textbf{(5.10.7)}$$

This is the frequency equation, the roots of which can be obtained graphically for a given α. Because the fundamental mode of vibration is usually the most important in practical applications, we first solve for the smallest root of Eq. (5.10.7), β_1. For various values of α, they are

α	0.01	0.10	0.50	1.00	5.00	10.0	20.0	100.0	∞
β_1	0.1	0.32	0.65	0.86	1.32	1.42	1.52	1.568	$\pi/2$

In the limit $\alpha << 1$, the roots β_1 will be small and Eq. (5.10.7) can be simplified by putting $\tan(\beta) = \beta$ so that

$$\beta^2 = \alpha = \frac{g\varrho AL}{W} \qquad \textbf{(5.10.8)}$$

or

$$\beta = \omega L/c = \sqrt{g\varrho AL/W} \text{ or } \omega = \frac{c}{L} \sqrt{g\varrho AL/W} \qquad \textbf{(5.10.9)}$$

Assuming that for a given c we calculate α, then the consecutive roots $\beta_1, \beta_2, \beta_3, \ldots$ of the frequency equation (5.10.7) can be calculated, and upon substituting $\beta_n c/L$ for ω, we obtain

$$u_n(x,t) = \sin(\beta_n x/L)[A_n \cos(\beta_n ct/L) + \beta_n \sin(\beta_n ct/L)]. \quad \textbf{(5.10.10)}$$

This solution represents a principal mode of vibration of our system.

By the principle of superposition, any vibration of the bar with a load at the end can be obtained in the form of a series

$$u(x,t) = \sum_{n=1}^{\infty} \sin(\beta_n x/L)\,[A_n \cos(\beta_n ct/L) + B_n \sin(\beta_n ct/L)]$$

(5.10.11)

where the constants A_n and B_n are determined from the initial conditions.

As an example, let us find the motion of a bar that is at rest under the action of a tensile force P applied at the lower end. At time $t = 0$, this force is suddenly removed. Because $u_t(x,0) = 0$, then $B_n = 0$. The coefficients A_n are given by the initial configuration of the system:

$$u(x,0) = \frac{Px}{AE} = \sum_{n=1}^{\infty} A_n \sin(\beta_n x/L). \quad \textbf{(5.10.12)}$$

From Eq. (5.10.12), it appears as if the initial condition is given by a generalized Fourier sine series. This is not true because the ordinary differential equation (5.10.4) along with the boundary condition (5.10.6) and $x(0)$ does not constitute a Sturm-Liouville problem. The eigenvalue ω^2/c^2 occurs not only in the differential equation but also in the boundary condition (5.10.12).

In order to obtain the coefficient A_n, we multiply both sides of Eq. (5.10.12) by $\sin(\beta_m x/L)\,dx$ and integrate from 0 to L. Then

$$\int_0^L \sin^2\left(\frac{\beta_n x}{L}\right) dx = \frac{L}{2}\left(1 - \frac{\sin(2\beta_n)}{2\beta_n}\right) \quad \textbf{(5.10.13)}$$

$$\frac{P}{AE}\int_0^L x \sin\left(\frac{\beta_n x}{L}\right) dx = \frac{PL^2}{AE}\left[-\frac{\cos(\beta_n)}{\beta_n} + \frac{\sin(\beta_n)}{\beta_n^{\,2}}\right]$$

(5.10.14)

$$\text{①} = \int_0^L \sin\left(\frac{L}{\beta_n x}\right) \sin\left(\frac{\beta_m x}{L}\right) dx$$

$$= \frac{L}{2}\,\frac{\cos(\beta_n)\cos(\beta_m)\,[\beta_m \tan(\beta_n) - \beta_n \tan\beta_m)]}{\beta_n^{\,2} - \beta_m^{\,2}} \quad \textbf{(5.10.15)}$$

$$(1) = -\frac{W}{A\varrho g} \sin(\beta_n)\sin(\beta_m) = -\frac{L}{\alpha} \sin(\beta_n)\sin(\beta_m) \qquad \textbf{(5.10.16)}$$

Note that if $n \neq m$, the right-hand side of Eq. (5.10.16) does not vanish proving that $\sin(\beta_n x/L)$ is not an eigenfunction. Therefore

$$\int_0^L \sin(\frac{\beta_m x}{L}) \sum_{n=1}^{\infty} A_n \sin(\frac{\beta_n x}{L})\, dx$$

$$= \frac{P}{AE} \int_0^L x \sin(\frac{\beta_n x}{L})\, dx \qquad \textbf{(5.10.17)}$$

or

$$A_n \frac{L}{2} (1 - \frac{sin(2\beta_n)}{2\beta_n}) - \frac{L}{\alpha} \sin(\beta_n) \sum_{(m=1,2,3,\ldots,n-1,n+1,\ldots)}^{\infty} A_m \sin(\beta_m)$$

$$= \frac{PL^2}{AE} (-\frac{\cos(\beta_n)}{\beta_n} + \frac{\sin(\beta_n)}{\beta_n^2}). \qquad \textbf{(5.10.18)}$$

However, because

$$\sum_{(m=1,2,3,\ldots,n-1,n+1,\ldots)}^{\infty} A_m \sin(\beta_m) = u(L,0) - A_n \sin(\beta_n) \qquad \textbf{(5.10.19)}$$

$$= \frac{PL}{AE} - A_n \sin(\beta_n), \qquad \textbf{(5.10.20)}$$

we obtain

$$\frac{A_n L}{2}\left[1 - \frac{\sin(2\beta_n)}{2\beta_n}\right] - \frac{L}{\alpha} \sin(\beta_n)\left[\frac{PL}{AE} - A_n \sin(\beta_n)\right]$$

$$= \frac{PL^2}{AE}\left[-\frac{\cos(\beta_n)}{\beta_n} + \frac{\sin(\beta_n)}{\beta_n^2}\right]. \qquad \textbf{(5.10.21)}$$

If we use Eq. (5.10.7)

$$\frac{L}{\alpha} \sin(\beta_n) = \frac{L \cos(\beta_n)}{\beta_n}, \qquad (5.10.22)$$

we obtain

$$A_n = \frac{4\, PL \sin(\beta_n)}{AE\, \beta_n\, [2\beta_n + \sin(2\beta_n)]}. \qquad (5.10.23)$$

Consequently, the initial displacement will be

$$u(x,0) = \frac{Px}{AE} = \frac{4PL}{AE} \sum_{n=1}^{\infty} \frac{\sin(\beta_n)\sin(\frac{\beta_n x}{L})}{\beta_n[(2\beta_n + \sin(2\beta_n)]} \qquad (5.10.24)$$

and the vibration of the bar is given by

$$u(x,t) = \frac{4PL}{AE} \sum_{n=1}^{\infty} \frac{\sin(\beta_n)\sin(\frac{\beta_n x}{L})\cos(\frac{\beta_n ct}{L})}{\beta_n\, [2\beta_n + \sin(2\beta_n\,]} \qquad (5.10.25)$$

In Table 5.10.1 the first eleven β_n when $\alpha = 1$ are given. The table also includes the first eleven A_n's divided by PL/AE. As the β_n's become larger, they approach a multiple of π.

In Fig. 5.10.2 the solution (5.10.25) is illustrated at various times.

Table 5.10.1: The First Eleven β_n's and A_n from Eqs. (5.10.7) and (5.10.23) for $\alpha = 1$.

n	β_n	$(n-1)\pi$	$\frac{4 \sin(\beta_n)}{\beta_n (2\beta_n + \sin(2\beta_n))}$
1	0.8603	0.0000	1.30085
2	3.4256	3.1416	−0.04428
3	6.4373	6.2832	0.00724
4	9.5293	9.4248	−0.00227
5	12.6453	12.5664	0.00098
6	15.7713	15.7008	−0.00051
7	18.9024	18.8496	0.00029
8	22.0365	21.9911	−0.00019
9	25.1725	25.1327	0.00013
10	28.3096	28.2743	−0.00009
11	31.4477	31.4159	0.00006

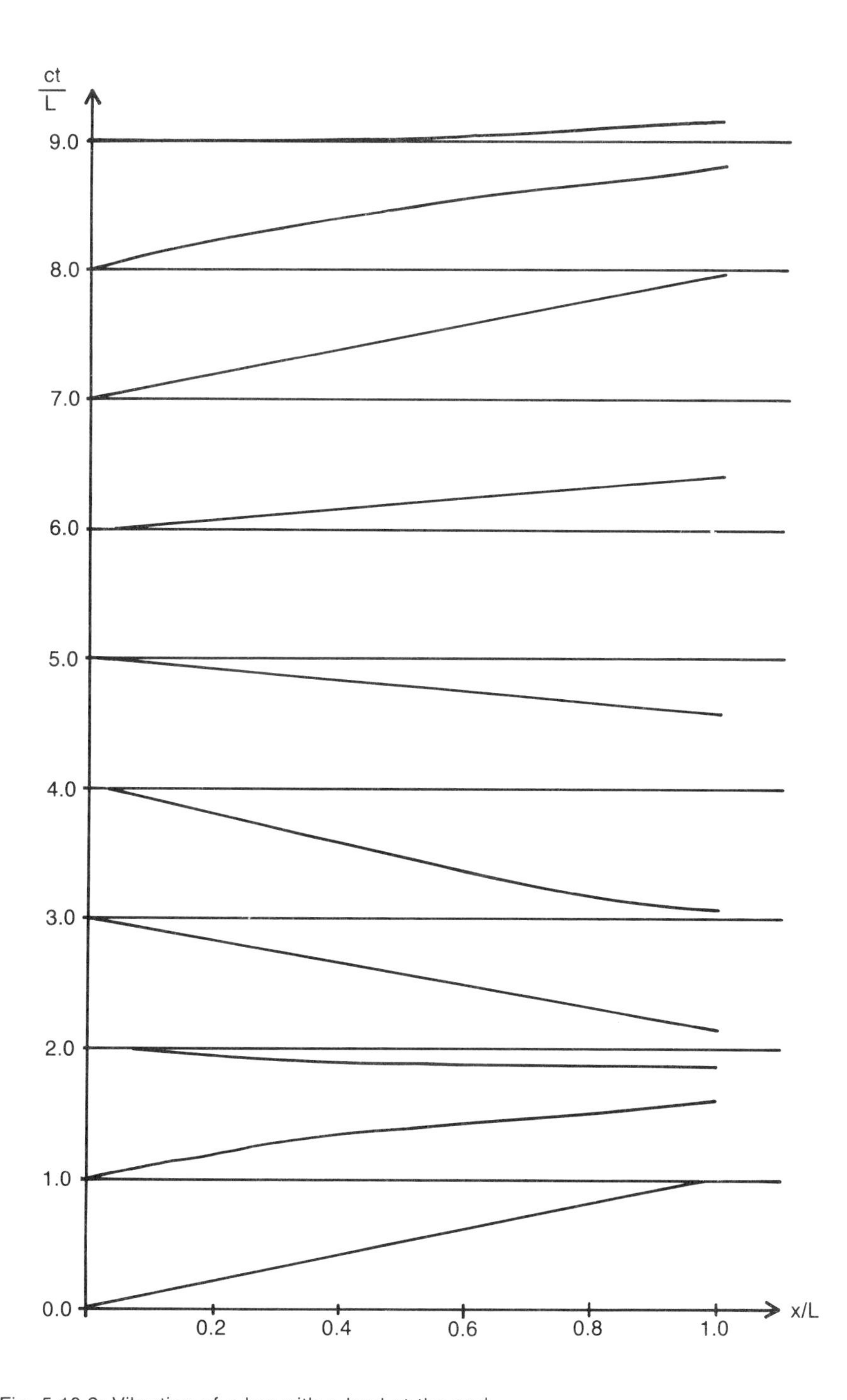

Fig. 5.10.2: Vibration of a bar with a load at the end.

Exercises

In problem 1-7, we refer to the problem

$$u_{tt} = c^2 u_{xx}$$

with the boundary conditions

$$u(0,t) = u(L,t) = 0 \qquad t>0$$

and the initial conditions

$$u(x,0) = f(x),\ u_t(x,0) = g(x) \qquad 0<x<L$$

Find the solution when

1. $f(x) = 0,\ g(x) = 1.$
2. $f(x) = 1,\ g(x) = 0.$
3. $f(x) = \begin{cases} 3hx/2L & 0<x<2L/3 \\ 3h(L-x)/L & 2L/3<x<L \end{cases} \qquad g(x) = 0$
4. $f(x) = \sin^3(\pi x/L) \qquad g(x) = 0$
   (Hint: $\sin^3(\pi x/L) = [3\sin(\pi x/L) - \sin(3\pi x/L)]/4$
5. $f(x) = \sin(\pi x/L) \qquad g(x) = \begin{cases} 0 & 0<x<L/4 \\ a & L/4<x<3L/4 \\ 0 & 3L/4<x<L \end{cases}$
   (Hint: $\cos(3n\pi/4) - \cos(n\pi/4) = -2\sin(n\pi/2)\sin(n\pi/4)$)
6. $f(x) = 0 \qquad g(x) = \begin{cases} ax/L & 0<x<L/2 \\ a(L-x)/L & L/2<x<L \end{cases}$
7. $f(x) = \begin{cases} x & 0<x<L/2 \\ L-x & L/2<x<L \end{cases} \qquad g(x) = 0$

8. Solve the equation $u_{tt} = c^2 u_{xx}$ $(0<x<\pi,\ t>0)$ subject to the boundary condition $u_x(0,t) = u_x(\pi, t) = 0$ $(t>0)$ and the initial condition $u(x,0) = 0$ and $u_t(x,0) = 1 + \cos^3(x)$ $(0\leq x\leq \pi)$. (Hint: You must include the separation constant $k = 0$.)

9. The displacement $u(x,t)$ of a uniform thin beam satisfies the equation

$$c^2 u_{xxxx} + u_{tt} = 0 \qquad 0<x<a,\ t>0$$

If the beam is simply supported at the ends, the boundary conditions are

$$u(0,t) = u(a,t) = u_{xx}(0,t) = u_{xx}(a,t) = 0.$$

Use separation of variables to solve this problem.

In problems 10-13, find the solution of the given problem by the method of separation of variables.

10. $u_{tt} - u_{xx} + u = 0$, $0<x<\pi$, $t>0$; $u(x,0) = 0$, $u_t(x,0) = \sin^3(x)$, $0\leq x\leq \pi$; $u(0,t) = 0$, $u(\pi,t) = 0$, $t\geq 0$.

11. $u_{tt} - u_{xx} + u = 0$, $0<x<\pi$, $t>0$; $u(x,0) = 0$, $u_t(x,0) = 1 + \cos^3(x)$, $0\leq x\leq \pi$; $u_x(0,t) = 0$, $u_x(\pi,t) = 0$, $t\geq 0$.

12. $u_{tt} - u_{xx} - 2u_x = 0$, $0<x<1$, $t>0$;
$u(x,0) = e^{-x}(2 \sin(2\pi x) - 3 \sin(5\pi x))$, $0\leq x\leq 1$;
$u_t(x,0) = 0$, $0\leq x\leq 1$; $u(0,t) = u(1,t) = 0$, $t>0$.

13. $u_{tt} + u_t - u_{xx} = 0$, $0 <x<\pi, t>0$; $u(x,0) = 0$, $u_t(x,0) = \sin^3(x/2)$, $0\leq x\leq \pi$; $u(0,t) = u_x(\pi,t) = 0$ $t>0$.

In problem 14-17, find all particular solutions of each of the following partial differential equations satisfying the given initial and boundary conditions.

14. $u_{tt} - x^2 u_{xx} - x u_x = 0$, $1<x<e$, $t>0$;
$u(x,0) = u(1,t) = u_x(e,t) = 0$.

15. $u_{tt} - u_{xx} + 2u_t - 2u_x + u = 0$, $0<x<\pi$, $t>0$;
$u_t(x,0) = u(0t) = u(\pi,t) = 0$.

16. $u_{tt} - x^2 u_{xx} = 0$, $1<x<2$, $t>0$; $u(x,0) = u(1,t) = 0$, $u(2,t) = 0$.

17. $u_{tt} - (x^2/(1+t)^2) u_{xx} = 0$, $1<x<2$, $t>0$;
$u(x,0) = u(1,t) = u(2,t) = 0$.

18. *Pipe problem.* A closed pipe of length L contains air whose density is slightly greater than that of the outside air in the ratio of $1+s_0$ to 1. Everything being at rest, the disk closing one end of the pipe is suddenly drawn aside. We want to determine what happens *inside* the pipe after the disk is removed.

As the air rushes outside, sound waves are generated within the pipe. These waves are governed by the wave equation

$$u_{tt} = c^2 u_{xx}$$

where c is the speed of sound and u is the velocity potential. Without going into the fluid mechanics of the problem, it can be shown that the boundary conditions are

1) No flow through the closed end.

$$u_x(0,t) = 0$$

2) No infinite accelerations at the open end.

$$u_{xx}(L,t) = 0$$

3) Air is initially at rest.

$$u_x(x,0) = 0$$

4) Air initially has a density greater than the surrounding air by the amount s_0

$$u_t(x,0) = -c^2 s_0$$

Show that the solution is

$$u(x,t) = -\frac{8Lcs_0}{\pi^2} \sum_{n=0}^{\infty} \frac{(-1)^n}{(2n+1)^2} \cos\left(\frac{(2n+1)\pi x}{2L}\right)$$

$$\times \sin\left(\frac{(2n+1)\pi ct}{2L}\right).$$

19. *Violin problem.* One of the classic applications of the wave equation has been the explanation of the acoustical properties of string instruments. Usually a string is excited in one of three ways: by plucking (as in the harp, zether, etc.), by striking with a hammer (piano), or by bowing (violin, violoncello, etc.). In all these cases, the governing partial differential equation is $u_{tt} = c^2 u_{xx}$ with the boundary conditions $u(0,t) = u(L,t) = 0$. For each of the following methods of exciting a string instrument, find the complete solution to the problem.

(a) Plucked string

$$u(x,0) = \begin{cases} \beta x/a & 0 \le x \le a \\ \beta(L-x)/(L-a) & a \le x \le L \end{cases}$$

$$u_t(x,0) = 0$$

Show that the solution is

$$u(x,t) = \frac{2\beta L^2}{\pi^2 a(L-a)} \sum_{n=1}^{\infty} \frac{1}{n^2} \sin\left(\frac{n\pi a}{L}\right) \sin\left(\frac{n\pi x}{L}\right) \cos\left(\frac{n\pi ct}{L}\right)$$

We note that harmonics of order n will be altogether absent if $\sin(n\pi a/L) = 0$. Thus, if the string is plucked at the center, all the harmonics of even order will be absent. Furthermore, the intensities of the successive harmonics will vary as n^{-2}. The higher harmonics (overtones) are therefore relatively feeble compared to the $n = 1$ term (the fundamental).

(b) String excited by impact

The effect of the impact of a hammer depends upon the manner and duration of the contact, and is more difficult to estimate. However, as a first estimate, let $u(x,0) = 0$ and

$$u_t(x,0) = \begin{cases} \mu & a-\epsilon < x < a+\epsilon \\ 0 & \text{otherwise} \end{cases} \qquad \epsilon << 1$$

and show that the solution is

$$u(x,t) = \frac{4\mu L}{\pi^2 c} \sum_{n=1}^{\infty} \frac{1}{n^2} \sin(\frac{n\pi\epsilon}{L}) \sin(\frac{n\pi a}{L}) \sin(\frac{n\pi x}{L}) \sin(\frac{n\pi ct}{L}).$$

As in part (a), the nth mode is absent if the origin is at a node. The intensity of the overtones are now of the same order of magnitude; higher harmonics (overtones) are relatively more in evidence than in part (a).

(c) Bowed violin string

The theory of the vibration of a string when excited by bowing is somewhat difficult. The mode of action of the bow appears to be that it drags the string with it for a time by friction until the string springs back; after a further interval the string is carried forward again, and so on. It can be shown (see Lamb's *Dynamical Theory of Sound*, Sec. 27) that the proper boundary conditions are $u(x,0) = 0$ and $u_t(x,0) = 4\beta_0 c(L-x)/L^2$ where β_0 is the maximum displacement. Show that the solution is

$$u(x,t) = \frac{8\beta_0}{\pi^2} \sum_{n=1}^{\infty} \frac{1}{n^2} \sin(\frac{n\pi x}{L}) \sin(\frac{n\pi ct}{L}).$$

20. Show that the solution to

$$\frac{\partial^2 u}{\partial t^2} = \frac{\partial^2 u}{\partial r^2} + \frac{1}{r}\frac{\partial u}{\partial r} + \frac{\partial^2 u}{\partial z^2} - \frac{u}{r^2}$$

with

$$\begin{aligned} u(r,z.0) &= u_t(r,z,0), = 0 \qquad (1)\\ u(r,0,t) &= 0, u_z(r,L,t) = re^{-t} \qquad (2)\\ u_r(a,z,t) &- u(a,z,t)/a = 0 \qquad (3) \end{aligned}$$

is

$$u(r,z,t) = \frac{\sinh(z)}{\cosh(L)} r e^{-t} + 8Lr \sum_{n=0}^{\infty} \frac{(-1)^n}{4L^2 + (2n+1)^2\pi^2)}$$

$$\times \sin\left[\frac{(2n+1)\pi z}{2L}\right] \left[-\cos(\alpha_n t) + \frac{\sin(\alpha_n t)}{\alpha_n}\right]$$

where $\alpha_n = (2n+1)\pi/(2L)$.

Hint: break $u(r,z,t)$ into two parts, u_1 and u_2. The solution u_1 is the particular solution that satisfies the boundary conditions (2) – (3). Then find u_2 which satisfies the boundary condition $u_z(r,L,t) = 0$, the other boundary conditions and the initial condition by guessing that

$$u_2(r,z,t) = r \sum_{n=0}^{\infty} \sin(kz) \; [A \cos(kt) + B \sin(kt)].$$

5.11 D'ALEMBERT'S FORMULA

In the previous sections we have been seeking solutions to the homogeneous wave equation in the form of a product $X(x)T(t)$. For the one-dimensional wave equation, there is an elegant method of constructing the solution without restricting ourselves to product solutions. It was given by d'Alembert in 1747.

We want to determine a solution of the homogeneous wave equation

$$u_{tt} = c^2 u_{xx}, \quad -\infty < x < \infty, \; t > 0 \tag{5.11.1}$$

which satisfies the initial conditions

$$u(x,0) = f(x) \text{ and } u_t(x,0) = g(x) \quad -\infty < x < \infty. \tag{5.11.2}$$

We begin our analysis by introducing the new variables ξ, η defined by

$$\xi = x + ct \tag{5.11.3}$$

$$\eta = x - ct \tag{5.11.4}$$

and set $u(x,t) = w(\xi,\eta)$. The variables ξ and η are called the *characteristics* of the wave equation. Using the chain rule, we find

$$\frac{\partial}{\partial x} = \frac{\partial \xi}{\partial x} \frac{\partial}{\partial \xi} + \frac{\partial \eta}{\partial x} \frac{\partial}{\partial \eta} = \frac{\partial}{\partial \xi} + \frac{\partial}{\partial \eta} \tag{5.11.5}$$

$$\frac{\partial}{\partial t} = \frac{\partial \xi}{\partial t} \frac{\partial}{\partial \xi} + \frac{\partial \eta}{\partial t} \frac{\partial}{\partial \eta} = c \frac{\partial}{\partial \xi} - c \frac{\partial}{\partial \eta} \tag{5.11.6}$$

$$\frac{\partial^2}{\partial x^2} = \frac{\partial \xi}{\partial x} \frac{\partial}{\partial \xi} \left(\frac{\partial}{\partial \xi} + \frac{\partial}{\partial \eta} \right) + \frac{\partial \eta}{\partial x} \frac{\partial}{\partial \eta} \left(\frac{\partial}{\partial \xi} + \frac{\partial}{\partial \eta} \right) \tag{5.11.7}$$

$$= \frac{\partial^2}{\partial \xi^2} + 2 \frac{\partial^2}{\partial \xi \partial \eta} + \frac{\partial^2}{\partial \eta^2} \tag{5.11.8}$$

and similarly

$$\frac{\partial^2}{\partial t^2} = c^2\left(\frac{\partial^2}{\partial \xi^2} - 2\frac{\partial^2}{\partial \xi \partial \eta} + \frac{\partial^2}{\partial \eta^2} \right) \qquad \textbf{(5.11.9)}$$

so that the wave equation becomes

$$\frac{\partial^2 w}{\partial \xi \partial \eta} = 0. \qquad \textbf{(5.11.10)}$$

The general solution is

$$w(\xi, \eta) = F(\xi) + G(\eta). \qquad \textbf{(5.11.11)}$$

Thus, the general solution of Eq. (5.11.11) is of the form

$$u(x,t) = F(x+ct) + G(x-ct) \qquad \textbf{(5.11.12)}$$

where F and G are arbitrary functions of one variable and are twice differentiable. Setting $t = 0$ in Eq. (5.11.12) and using the value of $u(x,0)$ from Eq. (5.11.2), we have

$$F(x) + G(x) = f(x). \qquad \textbf{(5.11.13)}$$

The differentiation of Eq. (5.11.12) with respect to t yields

$$u_t(x,t) = c\,F'(x+ct) - c\,G'(x-ct). \qquad \textbf{(5.11.14)}$$

Here the primes indicate differentiation with respect to the argument of the function. If we set $t = 0$ and use the value of $u_t(x,0)$ from Eq. (5.11.2), we find

$$c\,F'(x) - c\,G'(x) = g(x). \qquad \textbf{(5.11.15)}$$

The integration of Eq. (5.11.5) from 0 to any point x gives

$$F(x) - G(x) = \frac{1}{c}\int_0^x g(s)\,ds + C \qquad \textbf{(5.11.16)}$$

where C is a constant of integration. Combining this result with Eq. (5.11.15), we obtain

$$F(x) = f(x)/2 + \frac{1}{2c}\int_0^x g(s)\,ds + C/2 \qquad \textbf{(5.11.17)}$$

and

$$G(x) = f(x)/2 - \frac{1}{2c}\int_0^x g(s)\,ds - C/2. \qquad \textbf{(5.11.18)}$$

If we replace the variable x in the expression for F and G by $x + ct$ and $x - ct$, respectively, and substitute the results into Eq. (5.11.12), we finally arrive at the formula

$$u(x,t) = \frac{f(x+ct)+f(x-ct)}{2} + \frac{1}{2c}\int_{x-ct}^{x+ct} g(s)\,ds \qquad \textbf{(5.11.19)}$$

This is known as *d'Alembert's formula* for the solution of the wave equation Eq. (5.11.1) subject to the initial conditions Eq. (5.11.2).

To illustrate the usefulness of d'Alembert's formula, let us find the solution of the wave equation (5.11.1) satisfying the initial conditions $u(x,0) = \sin(x)$ and $u_t(x,0) = 0$, $-\infty < x < \infty$. By d'Alembert's formula (5.11.19), we have

$$u(x,t) = [\sin(x-ct)+\sin(x+ct)]/2 = \sin(x)\cos(ct) \qquad \textbf{(5.11.20)}$$

If we wish to find the solution of the wave equation (5.11.1) when $u(x,0) = 0$ and $u_t(x,0) = \sin(2x)$, then by d'Alembert's formula, the solution is

$$u(x,t) = \frac{1}{2c}\int_{x-ct}^{x+ct} \sin(2s)\,ds = \frac{\sin(2x)\sin(2ct).}{2}. \qquad \textbf{(5.11.21)}$$

Let us examine the solution Eq. (5.11.19) in more detail to gain a clearer understanding of the physics behind this solution of the wave equation. Suppose that the string is released with zero velocity after being given an initial displacement defined by $f(x)$. According to Eq. (5.11.19), the displacement of a point x at any time t is

$$u(x,t) = [f(x-ct)+f(x+ct)]/2. \qquad \textbf{(5.11.22)}$$

Consider the function $f(x-ct)$. We observe that the graph of $f(x-ct)$ is the same as the graph of $f(x)$ translated to the right by a distance equal to ct. (See Fig. 5.11.1.) This means that as time increases, $f(x-ct)$ represents a wave of the form $f(x)$ traveling to the right with the velocity c. We call the wave represented by $f(x-ct)$ a forward wave. Similarly, the function $f(x+ct)$ can be interpreted as representing a wave with the shape $f(x)$ traveling to the left with the velocity c. This wave is called a backward wave. With this interpretation, we see that the solution Eq. (5.11.19) is a superposition of forward

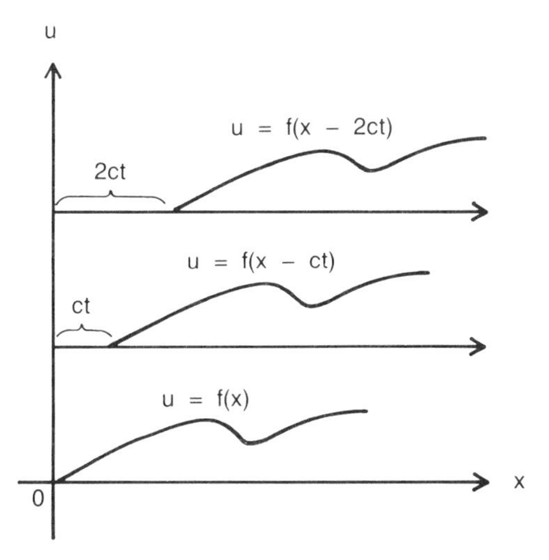

Fig. 5.11.1: Traveling wave form.

and backward waves traveling with the same velocity c and having the shape of the initial profile $f(x)$ with half of the amplitude. Clearly the characteristics $x + ct$ and $x - ct$ give the propagation paths along which the wave form $f(x)$ propagates.

To illustrate this, suppose that the string has an initial displacement defined by (see Fig. 5.11.2 (a))

$$f(x) = \begin{cases} a - |x| & -a \leq x \leq a \\ 0 & \text{otherwise.} \end{cases} \tag{(5.11.23)}$$

The forward and backward waves indicated by the dotted curve in Fig. 5.11.2 coincide at $t = 0$. At $t = a/2c$, both waves have moved in opposite directions through a distance $a/2$, resulting in the shape of the string shown in Fig. 5.11.2(b). At $t = a/c$, the forward backward waves are on the verge of separating from each other. For $t > a/c$, the motion of the string consists of the forward and backward waves traveling toward the ends of the string at the same velocity c, Figs. 5.11.2(d) and 5.11.2(e). It is seen from the figure that each point of the string returns to its original position of rest after the passage of each wave.

In the case when $u(x,0) = 0$ and $u_t(x,0) = g(x)$, the displacement is given by

$$u(x,t) = \frac{1}{2c} \int_{x-ct}^{x+ct} g(s)\, ds. \tag{5.11.24}$$

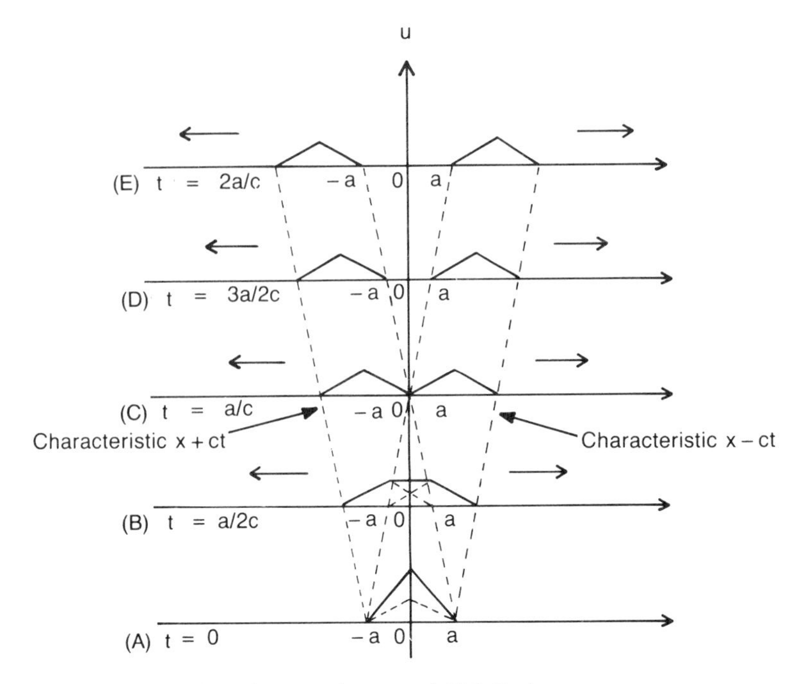

Fig. 5.11.2: The propagation of waves due to an initial displacement.

Let us define

$$\phi(x) = \frac{1}{2c} \int_0^x g(s)\, ds, \qquad \textbf{(5.11.25)}$$

then Eq. (5.11.24) can be written as

$$u(x,t) = -\phi(x-ct) + \phi(x+ct) \qquad \textbf{(5.11.26)}$$

which again shows that the solution is a superposition of a forward wave $-\phi(x-ct)$ and a backward wave $\phi(x+ct)$ traveling with the same velocity c. The forms of these waves are determined by the function ϕ, which is related to the initial velocity function $g(x)$ through the integral (5.11.25).

So far we have applied d'Alembert's formula to the one-dimensional wave equation when the domain extends indefinitely in space. However, in Section 5.3, we showed that the product solution for the domain $(0,L)$ could be written in a form similar to that of d'Alembert's formula. This similarity suggests that we could use d'Alembert's formula if we were very careful in our choice of F and G in Eq. (5.11.19).

Let us reconsider the problem of the vibrating string problem

$$u_{tt} = c^2\, u_{xx},\ 0<x<L,\ t>0 \qquad \textbf{(5.11.27)}$$

with

$$u(0,t) = u(L,t) = 0,\ t>0 \qquad \textbf{(5.11.28)}$$

and

$$u(x,0) = f(x),\ u_t(x,0) = g(x),\ 0<x<L. \qquad \textbf{(5.11.29)}$$

In principle we already know how to find $u(x,t)$. The problem is to choose F and G in such a way that the initial and boundary conditions are satisfied. We assume then that

$$u(x,t) = \psi(x+ct) + \phi(x-ct) \qquad \textbf{(5.11.30)}$$

as we did in our original derivation of d'Alembert's formula. the initial conditions give

$$\psi(x) + \phi(x) = f(x) \qquad 0<x<L \qquad \textbf{(5.11.31)}$$

$$c\psi'(x) - c\phi'(x) = g(x) \quad 0<x<L. \qquad \textbf{(5.11.32)}$$

If we divide through the second equation by c and integrate, it becomes

$$\psi(x) - \phi(x) = G(x) + C \quad 0<x<L \qquad \textbf{(5.11.33)}$$

where

$$G(x) = \frac{1}{c}\int_0^x g(u)\,du \qquad \textbf{(5.11.34)}$$

and C is an arbitrary constant of integration. Eq. (5.11.31) and Eq. (5.11.33) can now be solved simultaneously to determine

$$\psi(x) = (f(x) + G(x) + C)/2 \qquad \textbf{(5.11.35)}$$

$$0<x<L$$

$$\phi(x) = (f(x) - G(x) - C)/2. \qquad \textbf{(5.11.36)}$$

These equations give ψ and ϕ only for values of the argument between zero and L. But $x \pm ct$ may take on any value, so we must extend these functions to define them for arbitrary values of their argument, and in such a way that the boundary conditions are satisfied. The boundary conditions are

$$u(0,t) = \psi(ct) + \phi(-ct) = 0 \qquad \textbf{(5.11.37)}$$

$$t>0$$

$$u(L,t) = \psi(L+ct) + \phi(L-ct) = 0. \qquad \textbf{(5.11.38)}$$

Therefore

$$\bar{f}(ct) + \bar{G}(ct) + C + \bar{f}(-ct) - \bar{G}(-ct) - C = 0 \qquad \textbf{(5.11.39)}$$
$$\bar{f}(ct) + \bar{f}(-ct) + \bar{G}(ct) - \bar{G}(-ct) = 0 \qquad \textbf{(5.11.40)}$$

($\bar{f}$ and $\bar{G}$ denote the extensions of f and G that we are seeking.) Because these equations must be true for arbitrary f and G, we must have

$$\bar{f}(ct) = -\bar{f}(-ct) \text{ and } \bar{G}(ct) = \bar{G}(-ct). \qquad \textbf{(5.11.41)}$$

That is, $\bar{f}$ is an odd function and $\bar{G}$ is an even function.

At the second end point, a similar calculation shows that

$$\bar{f}(L+ct) + \bar{f}(L-ct) + \bar{G}(L+ct) - \bar{G}(L-ct) = 0 \qquad \textbf{(5.11.42)}$$

Once again, the independence of $\bar{f}$ and $\bar{G}$ implies that

$$\bar{f}(L+ct) = -\bar{f}(L-ct) \text{ and } \bar{G}(L+ct) = \bar{G}(L-ct). \qquad \textbf{(5.11.43)}$$

The oddness of $\bar{f}$ and evenness of $\bar{G}$ can be used to transform the right-hand side. Then

$$\bar{f}(L+ct) = \bar{f}(-L-ct) \text{ and } \bar{G}(L+ct) = \bar{G}(-L+ct). \qquad \textbf{(5.11.44)}$$

These equations say that f and G are both periodic with period $2L$. Consequently

$$\psi(x+ct) = (\bar{f}(x+ct) + \bar{G}(x+ct) + C)/2 \qquad \textbf{(5.11.45)}$$
$$\phi(x-ct) = (\bar{f}(x-ct) - \bar{G}(x-ct) - C)/2 \qquad \textbf{(5.11.46)}$$

where $\bar{f}$ is the odd periodic extension of f (with period $2L$) and $\bar{G}$ is the even periodic extension of G (with period $2L$). Finally, we arrive at an expression for the solution $u(x,t)$:

$$\begin{aligned} u(x,t) = & [\bar{f}(x+ct) + \bar{f}(x-ct)]/2 \\ & + [\bar{G}(x+ct) - \bar{G}(x-ct)]/2 \end{aligned} \qquad \textbf{(5.11.47)}$$

Equation (5.11.47) may be used to find $u(x,t)$, given $f(x)$ and $g(x)$, in two different methods.

The first method is graphical. We calculated $G(x)$ and then sketch our $\bar{f}(x+ct)$, $\bar{f}(x-ct)$, $\bar{G}(x+ct)$ and $\bar{G}(x-ct)$ at any desired time t as a function of x. We then take the appropriate summing, differencing and averaging.

The alternative method is to replace $f(x)$ and $G(x)$ by their equivalent Fourier sine and cosine series. Consequently d'Alembert's solution is merely another path to the use of Fourier series to obtain solutions for the wave equation on a closed interval.

5.12 AN EXAMPLE OF d'ALEMBERT THE FORMULA: VIBRATION OF A MOVING THREADLINE

The characterization and analysis of the oscillations of a string or yarn has an important application in the textile industry because it describes the way that yarn is wound on a bobbin. (See Fig. 5.12.1.) As shown in Section 5.1, the governing equation, the "threadline equation," is

$$u_{tt} + \alpha u_{xt} + \beta u_{xx} = 0 \tag{5.12.1}$$

where

$$\alpha = 2V, \tag{5.12.2}$$

$$\beta = V^2 - gT/\varrho, \tag{5.12.3}$$

V is the windup velocity, g the gravitational attraction, T the tension in the string and ϱ the density of the string. We now introduce the characteristics

$$\xi = x + \lambda_1 t \tag{5.12.4}$$
$$\eta = x + \lambda_2 t. \tag{5.12.5}$$

Upon substituting the new variables into Eq. (5.12.1), we have

$$(\lambda_1{}^2 + 2V\lambda_1 + V^2 - gT/\varrho)u_{\xi\xi} + (\lambda_2{}^2 + 2V\lambda_2 + V^2 - gT/\varrho)u_{\eta\eta}$$
$$+ (2V^2 - 2gT/\varrho + 2\text{V}\ (\lambda_2 + \lambda_2) + 2\lambda_1\lambda_2]u_{\xi n} = 0 \tag{5.12.6}$$

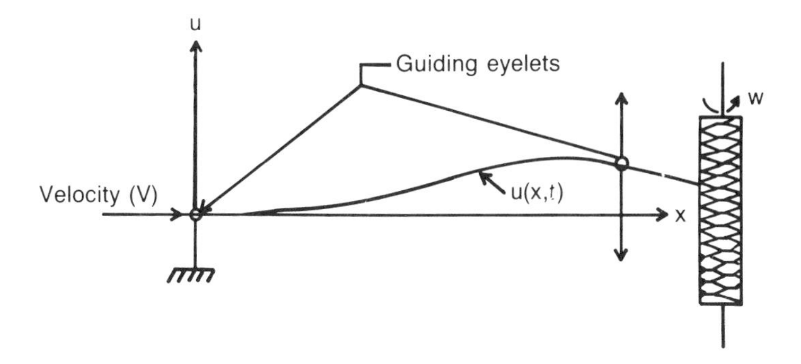

Fig. 5.12.1: Vibrations of a moving threadline. Reprinted with permission from *Journal of the Franklin Institute, 275*, R. D. Swope and W. F. Ames, Vibration of a moving threadline, Copyright 1963, Pergamon Press, Ltd.

If we choose λ_1 and λ_2 to be the roots of the equation

$$\lambda^2 + 2V\lambda + V^2 - gT/\varrho = 0. \tag{5.12.7}$$

Eq. (5.12.6) reduces to the simple form

$$u_{\xi\eta} = 0 \tag{5.12.8}$$

which has for its general solution

$$u = F(\xi) + G(\eta) \tag{5.12.9}$$

$$= F(x+\lambda_1 t) + G(x+\lambda_2 t) \tag{5.12.10}$$

Solving Eq. (5.12.7) gives

$$\lambda_1 = c - V \text{ and } \lambda_2 = -c - V \tag{5.12.11}$$

where $c = \sqrt{gT/\varrho}$. If the initial conditions are given by

$$u(x,0) = f(x) \tag{5.12.12}$$

and

$$u_t(x,0) = g(x), \tag{5.12.13}$$

then

$$u(x,t) = \frac{1}{2c}\left[\lambda_1 f(x+\lambda_2 t) - \lambda_2 f(x+\lambda_1 t) + \int_{x+\lambda_2 t}^{x+\lambda_1 t} g(s)\, ds\right] \tag{5.12.14}$$

Because λ_1 is not in general equal to λ_2, the two waves that constitute the motion of the string move with different magnitudes of velocity and have different shapes or forms. For example, if

$$f(x) = \frac{1}{1+x^2} \tag{5.12.15}$$

and

$$g(x) = 0, \tag{5.12.16}$$

we have

$$u(x,t) = \frac{1}{2c}\left\{\frac{c - V}{1 + [x-(c+V)t]^2} + \frac{c + V}{1 + [x+(c-V)t]^2}\right\} \tag{5.12.17}$$

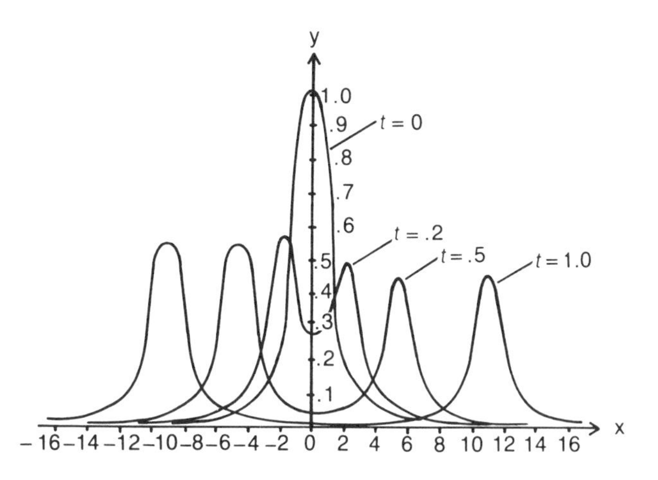

Fig. 5.12.2: Displacement of an infinite, moving threadline when c = 10 and V = 1. Reprinted with permission from *Journal of the Franklin Institute, 275*, R. D. Swope and W. F. Ames, Vibration of a moving threadline, Copyright 1963, Pergamon Press, Ltd.

Figures 5.12.2 and 5.12.3 show several diagrams from a paper by Swope and Ames (Swope, R.D. and W.F. Ames, 1963: Vibrations of a moving threadline. *J. of the Franklin Institute. 275*. 36-55.) In Fig. 5.12.2, $c = 10$ and $V = 1$; in Fig. 5.12.3, $c = 11$ and $V = 10$.

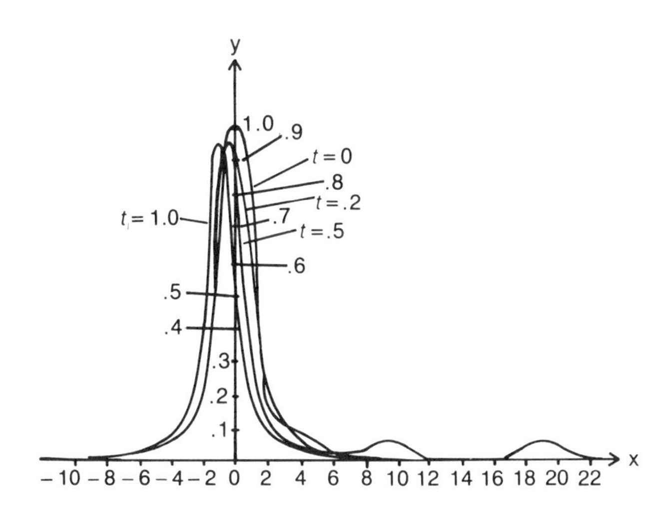

Fig. 5.12.3: Displacement of an infinite, moving threadline when c = 11 and V = 10. Reprinted with permission from *Journal of the Franklin Institute, 275*, R. D. Swope and W. F. Ames, Vibration of a moving threadline, Copyright 1963, Pergamon Press, Ltd.

5.13 SPHERICAL WAVES: THE BURSTING OF A BALLOON

In this section, we shall demonstrate how d'Alembert's method can be used to study oscillations that propagate radially away from a given source. In particular, we will determine the sound waves generated when a balloon of radius R bursts.

Because of the complete symmetry of the problem, all dependent variables will be a function of radius r and time t only. We assume that the density of the air within the balloon is uniform and exceeds the density of the surrounding air ϱ_0 by the amount s_0. At the instant $t = 0$, the balloon's surface dissolves and the sound waves generated by the density discontinuity are governed by the wave equation

$$(rs)_{tt} = c^2 (rs)_{rr} \tag{5.13.1}$$

where

$$s = \frac{\varrho - \varrho_0}{\varrho_0}, \tag{5.13.2}$$

$$c^2 = \frac{C_p p_0}{C_v \varrho_0}, \tag{5.13.3}$$

C_p the specific heat of air at constant pressure, C_v the specific heat of air at constant volume and p_0 the pressure of the surrounding air. See Lamb, H., 1925: *The Dynamical Theory of Sound.* New York: Dover Publishing Co. Section 71.

The easiest way to solve this problem is to introduce a new dependent variable $\psi = rs$ and transform the wave equation into $\psi_{tt} = c^2 \psi_{rr}$. Consequently the general solution of Eq. (5.13.1) will be of the form

$$s(r,t) = \frac{f(r-ct)}{r} + \frac{g(r+ct)}{r}. \tag{5.13.4}$$

The first term of this expression is a spherical wave that is propagated radially outward with velocity c from the region of the initial disturbance; the second term is a spherical wave that is propagated with the same velocity radially inward.

From our initial conditions, the initial velocities are zero. From the linearized equation of conservation of mass in spherical coordinates

$$r^2 s_t = (r^2 u)_r \tag{5.13.5}$$

where $u(r,t)$ is the velocity in the radial direction, this condition implies that

$$s_t(r,0) = 0 \tag{5.13.6}$$

for all r while the initial density is constant within the sphere and is equal to zero outside of the balloon:

$$s(r,0) = \phi(r) = \begin{array}{ll} s_0 & 0 \leq r < R \\ 0 & r > R. \end{array} \tag{5.13.7}$$

To determine the form of the functions f and g, we use the initial conditions Eqs. (5.13.6) and (5.13.7). From these, we have

$$f(r) + g(r) = r\,\phi\,(r) \tag{5.13.8}$$

and

$$f'(r) - g'(r) = 0. \tag{5.13.9}$$

Integrating the latter equation,

$$f(r) - g(r) = K. \tag{5.13.10}$$

Combining Eqs. (5.13.8) and (5.13.10),

$$f(r) = [r\,\phi(r) + K/2\,]/2 \tag{5.13.11}$$

and

$$g(r) = [r\,\phi(r) - K/2\,]/2. \tag{5.13.12}$$

Substituting these relations back into the equation for $s(r,t)$,

$$s(r,t) = 1/2 \left[\frac{r-ct}{r}\,\phi(r-ct) + \frac{r+ct}{r}\,\phi(r+ct) \right]. \tag{5.13.13}$$

Consequently the final solution is known once we have $\phi(r-ct)$ and $\phi(r+ct)$. The easiest method of determining these solutions is the graphical picture given in Fig. 5.13.1.

In Fig. 5.13.1 the first quadrant has been divided into four regions. They are defined as follows:

Region I	$0 < ct < R - r$
Region II	$r + R < ct < \infty$
Region III	$r - R < ct < r + R$
Region IV	$0 < ct < r - R$

These regions were selected because along each of the characteris-

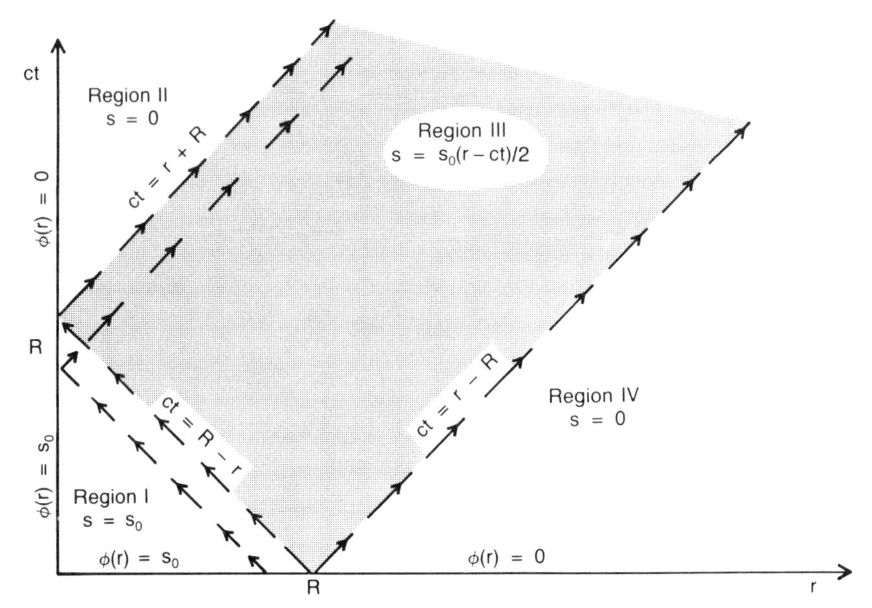

Fig. 5.13.1: Graphical solution of the bursting balloon problem.

tics $r - ct$ and $r + ct$, the function $\phi(r)$ remains constant. Consequently if we follow along any characteristic to either one of the axises, then we can find $s(r,t)$ for that particular point. The value of $\phi(r)$ along the r axis is given by the initial condition. The value along the time axis are determined by the leftward propagating characteristic $r + ct$ emitted from the r axis. Consequently they too are determined by the initial data. This technique of following the propagation paths of the characteristics to construct the solution is called the *method of characteristics*.

In our problem both the rightward $(r - ct)$ and leftward $(r + ct)$ characteristics emit into Region IV from the part of the r axis where $\phi(r) = 0$. Therefore $s = 0$. Similarly, in Region II, the characteristics in *both* directions emit from data where $\phi(r) = 0$ and $s = 0$. In Region I, however, the data along both of the characteristics is s_0. Consequently,

$$s = 1/2 \left(\frac{r - ct}{r} s_0 + \frac{r + ct}{r} s_0 \right) = s_0. \qquad \textbf{(5.13.14)}$$

Finally, in Region III, the data going to the right equals s_0 while the data from the left is equal to zero. Therefore,

$$s = 1/2 \left(\frac{r - ct}{r} s_0 + \frac{r + ct}{r} 0 \right) \qquad \textbf{(5.13.15)}$$

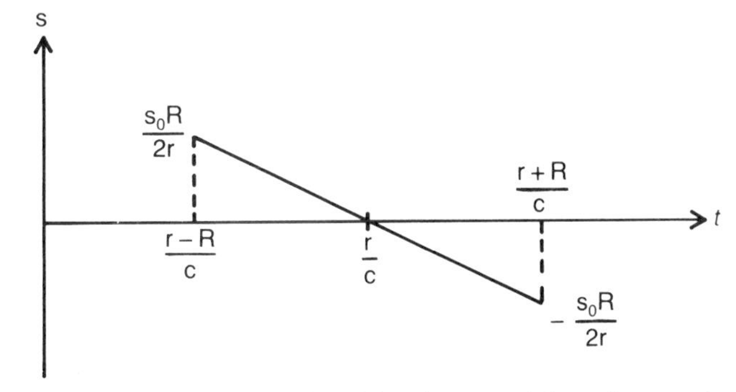

Fig. 5.13.2: The temporal evolution of density at a point r as the sound wave propagates by.

$$s = s_0 \frac{r-ct}{2r} . \qquad \textbf{(5.13.16)}$$

In Fig. 5.13.1 the results are summarized. In Fig. 5.13.2 the solution is graphed for a particular point r as the sound wave propagates by. We see from this figure that as the sound wave passes a point, compression takes place first followed by a decrease in density until rarefaction occurs. Finally s becomes zero.

In Fig. 5.13.3 a snap shot of the density field in the $x-z$ plane is presented. These so-called "N-wave" (The name comes from the

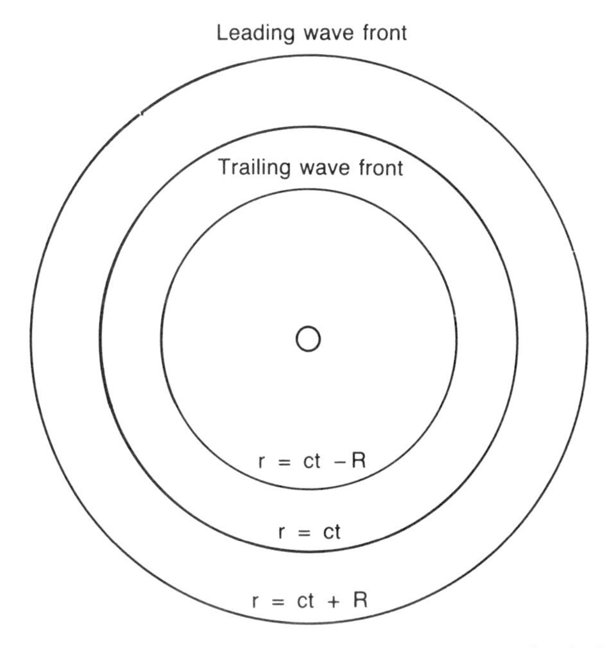

Fig. 5.13.3: At a given instant, the density field in the x-z plane after the balloon has burst.

density profile shown in Fig. 5.13.2.) may arise from the pressure waves generated by supersonic transport aircraft or explosions in the air. Because we have neglected heat conduction, viscosity and real-gas effects, the observed waves differ somewhat from our solution. Improved results can be obtained by numerical solution of the governing equations which include viscosity, heat conduction and vibrational relaxation. See Honma, H. and I. I. Glass, 1984: Weak spherical shock-wave transitions of N-wave in air with vibrational excitation. *Proc. R. Soc. Lond. A, 391*, 55-83.

Exercises

Use d'Alembert's formula to solve the wave equation for the following initial condition defined for $|x| < \infty$.

1. $u(x,0) = 2\sin(x)\cos(x)$ $\quad u_t(x,0) = \cos(x)$
2. $u(x,0) = x\sin(x)$ $\quad u_t(x,0) = \cos(2x)$
3. $u(x,0) = 1/(1+x^2)$ $\quad u_t(x,0) = e^x$
4. $u(x,0) = e^{-x}$ $\quad u_t(x,0) = 1/(1+x^2)$
5. $u(x,0) = \cos(\pi x/2)$ $\quad u_t(x,0) = \sinh(ax)$
6. $u(x,0) = \sin(3x)$ $\quad u_t(x,0) = \sin(2x) - \sin(x)$

5.14 THE SOLUTION OF PARTIAL DIFFERENTIAL EQUATIONS BY LAPLACE TRANSFORMS

The solution of linear partial differential equations by Laplace transforms is the most commonly employed technique after the method of separation of variables. Laplace transforms are used in problems where one of the variables, t say, varies in the interval $t \geq 0$ while the other independent variable x varies in a finite interval $0 \leq x \leq L$ or the infinite interval $-\infty < x < \infty$. Because the transformation denotes an integration with respect to time alone, we apply the transform only to time and time derivatives while the other independent variable acts like a constant. For each fixed value of x, we obtain a different transform which varies both in x and s:

$$\overline{u}(x,s) = \int_0^\infty e^{-st}\, u(x,t)\, dt. \qquad \textbf{(5.14.1)}$$

Consequently, from the rules involving Laplace transforms,

$$\mathcal{L}(u_t(x,t)) = s\,\overline{u}(x,s) - u(x,0+) \qquad \textbf{(5.14.2)}$$

$$\mathcal{L}(u_{tt}(x,t)) = s^2\,\overline{u}(x,s) - s\,u(x,0+) - u_t(x,0+). \qquad \textbf{(5.14.3)}$$

On the other hand, derivatives involving x become

$$\mathscr{L}(u_x(x,t)) = \frac{\partial}{\partial x}\left[\mathscr{L}(u(x,t))\right] = \frac{\partial \overline{u}(x,s)}{\partial x} \quad \textbf{(5.14.4)}$$

and

$$\mathscr{L}[u_{xt}(x,t)] = \frac{\partial}{\partial x}[\mathscr{L}(u_t)] = \frac{\partial}{\partial x}[s\,\overline{u}(x,s) - u(x,0+)]. \quad \textbf{(5.14.5)}$$

The values $u(x,0+)$ and $u_t(x,0+)$ are given by the initial conditions and must be specified for the problem to be well posed.

Because the time variable has been eliminated by the transformation, only $\overline{u}(x,s)$ and its derivatives remain in the equation. Consequently, we have transformed the partial differential equation into a boundary-value problem for an ordinary differential equation. Because such an equation is generally easier to solve than a partial differential equation, the original problem is considerably simplified through the use of Laplace transforms. This explains why many problems, which cannot be treated at all by other methods, or at best, in a complicated way, can be handled by this method.

Summing up this method, we obtain the following scheme:

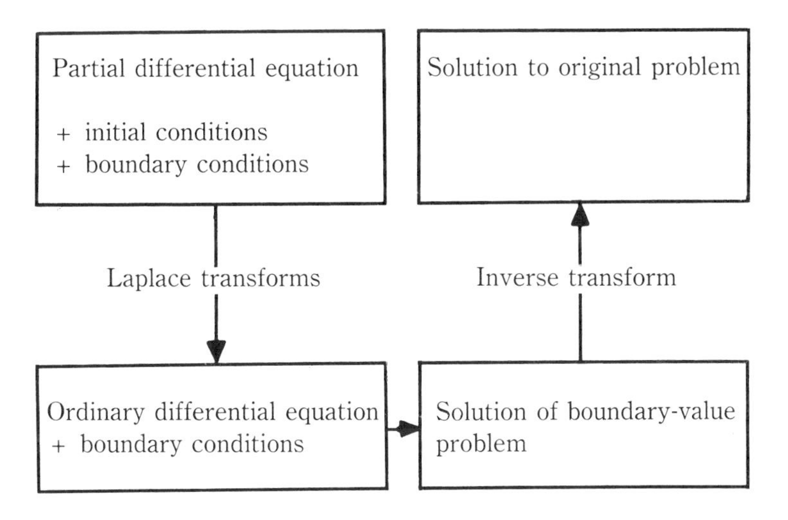

To illustrate the use of Laplace transforms, let us find the sound waves that arise when a sphere of radius a begins to pulsate at time $t = 0$. The wave equation in spherical coordinates is

$$\frac{1}{r}\frac{\partial^2}{\partial r^2}(ru) = \frac{1}{c^2}\frac{\partial^2 u}{\partial t^2} \quad \textbf{(5.14.6)}$$

where c is the speed of sound and $u(r,t)$ the velocity potential (i.e.,

$-\partial u/\partial r$ gives the velocity of the parcel of air). See Lamb, H., 1925: *The Dynamic Theory of Sound*. New York: Dover Publications, Inc. Section 71. If we assume that the air is initially at rest, $u(r,0) = u_t(r,0) = 0$ and the Laplace transform of Eq. (5.14.6) is

$$\frac{d^2}{dr^2}(r\,\bar{u}) - \frac{s^2}{c^2}\,r\,\bar{u} = 0 \qquad \textbf{(5.14.7)}$$

where $\bar{u}$ denotes the Laplace transform of $u(r,t)$. The solution of Eq. (5.14.7) is

$$r\,\bar{u} = A\exp(-sr/c). \qquad \textbf{(5.14.8)}$$

We have discarded the $\exp(sr/c)$ solution because it represents an inwardly propagating wave. Because there is no source of energy at infinity, we can only have outwardly propagating (divergent) waves.

At the surface of the sphere, the radial velocity must equal the velocity of the pulsating sphere

$$-\frac{\partial u}{\partial r} = \frac{d\xi}{dt} \qquad \textbf{(5.14.9)}$$

at $r = a$ where ξ is the displacement of the surface of the pulsating sphere. If we assume that $\xi = B\sin(\omega t)$, $t>0$, then

$$-\frac{d}{dr}\left(A\frac{e^{-rs/c}}{r}\right) = \frac{\omega B s}{s^2+\omega^2} = A\,e^{-as/c}\left(\frac{1}{a^2} + \frac{s}{ac}\right). \qquad \textbf{(5.14.10)}$$

Therefore, $r = a$

$$r\,\bar{u} = \frac{\omega B a^2 c s}{(s^2+\omega^2)(as+c)}\;e^{-s(r-a)/c} \qquad \textbf{(5.14.11)}$$

$$= \frac{\omega B a^2 c}{a^2\omega^2 + c^2}\left\{\frac{cs+\omega^2 a}{s^2+\omega^2} - \frac{c}{s+c/a}\right\}e^{-s(r-a)/c}. \qquad \textbf{(5.14.12)}$$

Using the second shifting theorem, the inversion of Eq. (5.14.12) follows directly:

$$r\,u(r,t) = \frac{\omega B a^2 c^2}{a^2\,\omega^2 + c^2}\left\{\cos\left[\omega\left(t - \frac{r-a}{c}\right)\right] + \frac{\omega a}{c}\sin\left[\omega\left(t - \frac{r-a}{c}\right)\right]\right.$$

$$\left. - \exp\left[-\frac{c}{a}\,t - \left(\frac{r-a}{c}\right)\right]\right\}H\left(t - \frac{r-a}{c}\right). \qquad \textbf{(5.14.13)}$$

5.15 THE SOLUTION OF THE UNIFORM TRANSMISSION LINE PROBLEM BY LAPLACE TRANSFORMS

In this section we shall employ Laplace transforms to solve the partial differential equations associated with a uniform transmission line. We suppose the line to have resistance R, inductance L, capacity C, and leakage conductance G, per unit length. Also let I be the current in the direction of positive x and let V be the voltage drop across the line at the point x. I and V will be functions of both distance x along the line and time t.

To derive the differential equations satisfied by the current and voltage in the line, consider the points A at x and B at $x + \delta x$ in Fig. 5.15.1. The current and voltage at A will be $I(x,t)$ and $V(x,t)$; at B,

$I + \frac{\partial I}{\partial x}\delta x$ and $V + \frac{\partial V}{\partial x}\delta x$. Here the voltage drop from A to B is

$-\frac{\partial V}{\partial x}\delta x$ and the current in it is $I + \frac{\partial I}{\partial x}\delta x$. Neglecting terms of

$O(\delta x^2)$, we have

$$\left(L\frac{\partial I}{\partial t} + RI\right)\delta x = -\frac{\partial V}{\partial x}\delta x. \tag{5.15.1}$$

Also the voltage drop over the parallel portion HK of the line is

V, and the current in it is $-\frac{\partial I}{\partial x}\delta x$. Thus,

$$\left(C\frac{\partial V}{\partial t} + GV\right)\delta x = -\frac{\partial I}{\partial x}\delta x. \tag{5.15.2}$$

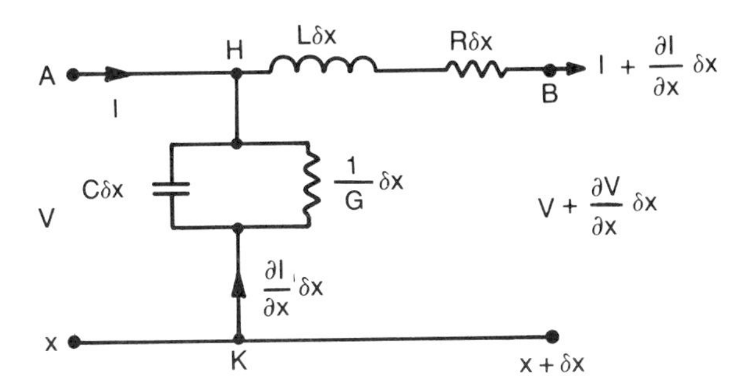

Fig. 5.15.1: Schematic of the uniform transmission line.

From Eqs. (5.15.1) and (5.15.2) the differential equations for I and V are

$$L \frac{\partial I}{\partial t} + RI = - \frac{\partial V}{\partial x} \qquad \textbf{(5.15.3)}$$

$$C \frac{\partial V}{\partial t} + GV = - \frac{\partial I}{\partial x}. \qquad \textbf{(5.15.4)}$$

We wish to solve these simultaneous partial differential equations with the initial conditions

$$I(x,0) = I_0(x) \qquad \textbf{(5.15.5)}$$

and $\qquad t>0$

$$V(x,0) = V_0(x). \qquad \textbf{(5.15.6)}$$

There are also boundary conditions to be satisfied at the ends of the line; these depend on the particular problem. For example, if the line is short-circuited at $x = a$, we must have $V = 0$ at $x = a$, $t>0$; if there is an open circuit as $x = a$, the condition is $I = 0$ for $x = a$ and all $t>0$.

To solve Eqs. (5.15.3) and (5.15.4) by Laplace transforms, we take the Laplace transforms of both sides of Eqs. (5.15.3) and (5.15.4). We obtain

$$(Ls + R)\, \bar{I}(x,s) = - \frac{d\bar{V}(x,s)}{dx} + L\, I_0\,(x) \qquad \textbf{(5.15.7)}$$

and

$$(Cs + G)\, \bar{V}(x,s) = - \frac{d\bar{I}(x,s)}{dx} + C\, V_0(x). \qquad \textbf{(5.15.8)}$$

Eliminating $\bar{I}$ gives the ordinary differential equation for $\bar{V}$,

$$\frac{d^2\bar{V}}{dx^2} - q^2\, \bar{V} = L \frac{dI_0(x)}{dx} - C(Ls + R)\, V_0(x) \qquad \textbf{(5.15.9)}$$

where

$$q^2 = (Ls + R)\,(Cs + G). \qquad \textbf{(5.15.10)}$$

When $\bar{V}$ has been found, $\bar{I}$ may be found from

$$\bar{I} = -\frac{1}{Ls + R}\frac{d\bar{V}}{dx} \quad \frac{L\, I_0(x)}{Ls + R}. \tag{5.15.11}$$

At this point, we treat several classic cases.

(A) The Semi-Infinite Line

We consider the problem of a semi-infinite line $x > 0$ with zero initial current and charge. The end $x = 0$ is maintained at constant voltage E for $t > 0$.

In this case

$$\frac{d^2\bar{V}}{dx^2} - q^2\,\bar{V} = 0,\ x > 0 \tag{5.15.12}$$

The boundary conditions at the ends of the line are

$$V = E \text{ at } x = 0,\ t > 0 \text{ and } V \text{ is finite as } x \to \infty. \tag{5.15.13}$$

The transform of the boundary conditions is

$$\bar{V}(0,s) = E/s \text{ and } \lim_{x \to \infty} \bar{V}(x,s) \text{ is finite.} \tag{5.15.14}$$

The general solution of Eq. (5.15.12) is

$$\bar{V} = A\, e^{-qx} + B\, e^{qx}. \tag{5.15.15}$$

The requirement that $\bar{V}$ remains finite as $x \to \infty$ forces $B = 0$. The boundary condition at $x = 0$ gives $A = E/s$. Thus,

$$\bar{V} = \frac{E}{s} \exp\left[-\sqrt{(Ls + R)\,(Cs + G)}\; x\right]. \tag{5.15.16}$$

In general, the inverse Laplace transform of Eq. (5.15.16) is a complicated integral involving Bessel functions. However, for the so-called "lossless" line where $R = G = 0$, we have

$$\bar{V} = \frac{E}{s} \exp(-sx/c) \tag{5.15.17}$$

where $c = (LC)^{-1/2}$. Consequently,

$$V(x,t) = E\, H(t - \frac{x}{c}) \tag{5.15.18}$$

where $H(t)$ is the Heaviside step function. The physical interpretation of this solution is that $V(x,t)$ is zero up to the time x/c at which time a wave travelling with speed c from the origin would arrive at the point x, and V has the constant value E afterwards.

For the so-called "distortionless" line, $R/L = G/C = \varrho$, we have

$$V(x,t) = E\, e^{-\varrho x/c} H\left(t - \frac{x}{c}\right). \qquad \textbf{(5.15.19)}$$

In this case the disturbance is propagated with velocity c, but there is attenuation as we move along the line.

Suppose, now, that instead of applying a constant voltage E at $x = 0$, we apply $f(x)$. The only modification is that in place of Eq. (5.15.16) we have

$$\overline{V} = \overline{f}(s)\, e^{-qx}. \qquad \textbf{(5.15.20)}$$

In the case of the distortionless line, $q = (s + \varrho)/c$, this becomes

$$\overline{V} = \overline{f}(s)\, e^{-(s+\varrho)x/c}, \qquad \textbf{(5.15.21)}$$

and

$$V(x,t) = e^{-\varrho x/c} f\left(t - \frac{x}{c}\right) H\left(t - \frac{x}{c}\right). \qquad \textbf{(5.15.22)}$$

This may be interpreted as stating that the voltage at x is zero up to the time x/c, and subsequently to that time it follows the voltage at $x = 0$ with a time lag of x/c, and a reduction in magnitude by the factor $e^{-\varrho x/c}$.

(B) The Travelling Wave Solution for the Finite Transmission Line

We now consider the problem of the finite transmission line $0<x<\ell$, with zero initial current and charge. The end $x = 0$ is grounded, and the end $x = \ell$ maintained at constant voltage E for $t>0$.

The transformed partial differential equations become in this case

$$\frac{d^2\overline{V}}{dx^2} - q^2\, \overline{V} = 0 \qquad 0<x<\ell \qquad \textbf{(5.15.23)}$$

The boundary conditions are

$$V(0,t) = 0 \text{ and } V(\ell,t) = \text{E} \qquad \text{for } t>0. \qquad \textbf{(5.15.24)}$$

The transform of these boundary conditions is

$$\overline{V}(0,s) = 0 \text{ and } \overline{V}(\ell,s) = E/s. \tag{5.15.25}$$

The solution to Eq. (5.15.23) which satisfies the boundary conditions is

$$\overline{V}(x,s) = \frac{E}{s}\frac{\sinh(qx)}{\sinh(q\ell)}. \tag{5.15.26}$$

Once again, $\overline{V}$ is a complicated function of s. If we write Eq. (5.15.26) in a form involving negative exponentials, and expand the denominator by the binomial theorem, this gives

$$\overline{V} = \frac{E}{s} e^{-qL} \frac{[1 - \exp(-2qx)]}{[1 - \exp(-2q\ell)]} \tag{5.15.27}$$

$$= \frac{E}{s} e^{-q(\ell-x)} (1 - e^{-2qx}) (1 + e^{-2q\ell} + e^{-4q\ell} + \ldots) \tag{5.15.28}$$

$$= \frac{E}{s} (e^{-q(\ell-x)} - e^{-q(\ell+x)} + e^{-q(3\ell-x)} - e^{-q(3\ell+x)} + \ldots) \tag{5.15.29}$$

In the special case of the lossless line, for which $q = s/c$, we have

$$V = \frac{E}{s} [e^{-s(\ell-x)/c} - e^{-s(\ell+x)/c} + e^{-s(3\ell-x)/c} - \ldots] \tag{5.15.30}$$

or

$$V(x,t) = E\,[H(t - \frac{\ell-x}{c}) - H(t - \frac{\ell+x}{c}) + H(t - \frac{3\ell-x}{c}) - \ldots \tag{5.15.31}$$

The graph of the above function is shown in Fig. 5.15.2. The voltage at x is zero up to the time $(\ell-x)/c$, at which time a wave travelling directly from the end $x = \ell$ would reach the point x. The voltage then has the constant value E up to the time $(\ell+x)/c$, at which time a wave travelling from the end $x = \ell$ and reflected back from the end $x = 0$ would arrive. From this time up to the time of arrival of a twice reflected wave, it has the value zero, and so on.

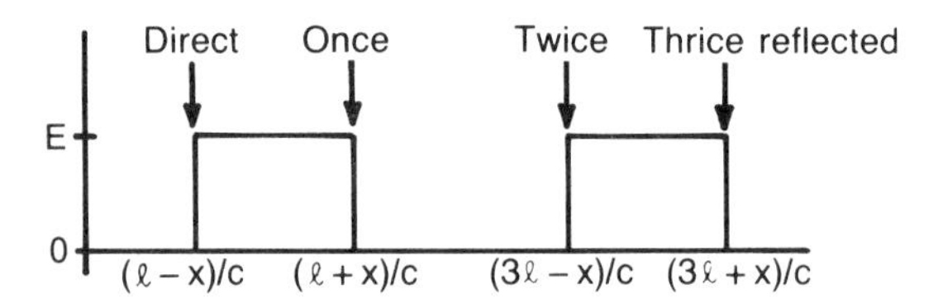

Fig. 5.15.2: A solution to the finite transmission line.

(C) Finite Line with Terminal Impedances

We now consider a lossless line ($R = G = 0$) line of length ℓ, insulated at $x = \ell$, and suppose that at $t = 0$ a condenser of capacity C_0, charged to voltage E, is discharged into the line at $x = 0$.

The Laplace transform of the partial differential equations for the line is

$$\frac{d^2\overline{V}}{dx^2} - \frac{s^2}{c^2}\overline{V} = 0 \qquad 0<x<\ell \qquad \textbf{(5.15.32)}$$

with

$$\overline{I} = \frac{d\overline{V}}{dx} = 0 \text{ at } x = \ell \qquad \textbf{(5.15.33)}$$

and $c = (LC)^{-1/2}$.

If V_0 and I_0 are the voltage and current at the point $x = 0$ of the line, the transform of the equation for the condensor C_0 may be shown to be

$$-\frac{1}{sC_0}\overline{I}_0 = \overline{V}_0 - \frac{E}{s} \qquad \textbf{(5.15.34)}$$

or

$$\frac{1}{LC_0s^2}\frac{d\overline{V}}{dx} = \overline{V} - \frac{E}{s} \text{ at } x = 0 \qquad \textbf{(5.15.35)}$$

after using Eq. (5.15.11). This is the boundary condition to be satisfied at $x = 0$.

A solution of Eq. (5.15.32) which also satisfies Eq. (5.15.33) is

$$\overline{V}(x,s) = A \cosh(s(\ell - x)/c) \qquad \textbf{(5.15.36)}$$

Substituting this into Eq. (5.15.35) gives

$$A\left[\frac{1}{LC_0cs}\sinh\left(\frac{s\ell}{c}\right)+\cosh\left(\frac{s\ell}{c}\right)\right]=\frac{E}{s}. \qquad \textbf{(5.15.37)}$$

Thus,

$$\overline{V}(x,s)=\frac{EC_0}{Cc}\,\frac{\cosh[s(\ell-x)/c]}{\sinh(s\ell/c)+(sC_0/Cc)\cosh(s\ell/c)}\,. \qquad \textbf{(5.15.38)}$$

In the application of Laplace transforms to the solution of ordinary differential equations, the transform of the solution is often in the form

$$\overline{\phi}(s)=\frac{f(s)}{g(s)} \qquad \textbf{(5.15.39)}$$

where $f(s)$ and $g(s)$ are polynomials of s. A powerful method of finding the solution $\phi(t)$ is through the use of the so-called "Heaviside's expansion theorem." (See Section 3.5.) This technique is essentially the use of partial fractions and states that the solution is given by

$$\phi(t)=\sum_{n=1}^{N}\frac{f(a_n)}{g'(a_n)}\exp(a_nt) \qquad \textbf{(5.15.40)}$$

where a_n is the nth zero of $g(s)$ and each zero must be different. Consequently, we must first find the zeros of $g(s)$ or $\sinh(s\ell/c)+(sC_0/Cc)\cosh(s\ell/c)$. They occur for $s=0$ and

$$s=\pm ic\alpha_n/\ell$$

where $n=1,2,3,4,\ldots$, α_n is the positive root of

$$\sin(\alpha)+k\alpha\cos(\alpha)=0, \qquad \textbf{(5.15.41)}$$

and

$$k=C_0/\ell C \qquad \textbf{(5.15.42)}$$

is the ratio of the terminal capacity C_0 to the capacity of the whole line. They may be found either numerically or through graphical methods. It can be easily shown that there are an infinite number of distinct roots.

The use of Heaviside's expansion theorem for an infinite number of zeros has not been proven although it might be suggested as an extension of the theorem for a finite number of zeros. This is precisely the approach that was used by Heaviside. The formal justification requires an investigation of the inverse integral for Laplace transforms *(the Bromwich integral)* and complex variables. The net result of this analysis is that we may use Heaviside's expansion theorem for a finite transmission line but not for a semi-infinite line.

To find $V(x,t)$ from Eq. (5.15.18) by the expansion theorem formula (5.15.40) we need

$$\frac{d}{ds}\left[\sinh\left(\frac{s\ell}{c}\right) + \frac{sC_0}{Cc}\cosh\left(\frac{s\ell}{c}\right)\right]_{s=\frac{ic\alpha_n}{\ell}}$$

$$= \left\{\left[\frac{C_0}{cC} + \frac{\ell}{c}\right]\cosh\left(\frac{s\ell}{c}\right) + \frac{s\ell C_0}{C\,c}\sinh\left(\frac{s\ell}{c}\right)\right\}_{s=\frac{ic\alpha_n}{\ell}} \quad \textbf{(5.15.43)}$$

$$= \left(\frac{C_0}{cC} + \frac{\ell}{c}\right)\cos(\alpha_n) - \frac{C_0\alpha_n}{cC}\sin(\alpha_n) \quad \textbf{(5.15.44)}$$

$$= \frac{\ell}{c}\,(k + 1 + k^2\,\alpha_n^{\,2})\cos(\alpha_n). \quad \textbf{(5.15.45)}$$

For the root $s = 0$, we have

$$\frac{d}{ds}\left[\sinh\left(\frac{s\ell}{c}\right) + \frac{sC_0}{Cc}\cosh\left(\frac{s\ell}{c}\right)\right]_{s\,=\,0} = \frac{\ell(1+k)}{c}. \quad \textbf{(5.15.46)}$$

Using these results in conjunction with Eq. (5.15.40) gives

$$V(x,t) = \frac{kE}{1+k} + 2kE\sum_{n=1}^{\infty}\frac{\cos[\alpha_n(\ell-x)/\ell]\cos(\alpha_n ct/\ell)}{(1+k+k^2\alpha^2_n)\cos(\alpha_n)}. \quad \textbf{(5.15.47)}$$

(D) The Semi-Infinite Line Reconsidered

In subsection (A) we showed that the transform of the solution for the semi-infinite line is

$$\overline{V}(x,s) = \frac{E}{s}\,e^{-qx} \quad \textbf{(5.15.48)}$$

where

$$q^2 = (Ls + R)(Cs + G). \quad \textbf{(5.15.49)}$$

In the case of a lossless line ($R = G = 0$), we found travelling wave solutions.

In this section, we shall study the case of a submarine cable where $L = G = 0$. In this special case, we have

$$\overline{V}(x,s) = \frac{E}{s}\,e^{-x(s/\varkappa)^{1/2}} \quad \textbf{(5.15.50)}$$

where $\varkappa = 1/(RC)$. Then it follows from Eq. (3.7.1) that

$$V(x,t) = E \operatorname{erfc}\left(\frac{x}{2\sqrt{\varkappa t}}\right). \quad \textbf{(5.15.51)}$$

Equation (5.15.51) is presented in Fig. 5.15.3 for various values of $\varkappa t$.

(E) The General Solution of the Equation of Telegraphy

In this section we drop all restrictions on the parameters L, R, C and G and find the general solution to the equation of telegraphy. Our analysis begins by eliminating I from Eqs. (5.15.3) and (5.15.4):

$$\left[CL \frac{\partial^2}{\partial t^2} + (GL + RC) \frac{\partial}{\partial t} + RG\right] V = \frac{\partial^2 V}{\partial x^2}. \quad \textbf{(5.15.52)}$$

Eq. (5.15.52) can be further simplified by substituting

$$V(x,t) = \exp\left(-\frac{GL + RC}{2LC} t\right) v(x,t) \quad \textbf{(5.15.53)}$$

to give

$$\frac{\partial^2 v}{\partial t^2} = c^2 \frac{\partial^2 v}{\partial x^2} + k^2 v \quad \textbf{(5.15.54)}$$

where

$$c^2 = \frac{1}{LC} \quad \textbf{(5.15.55)}$$

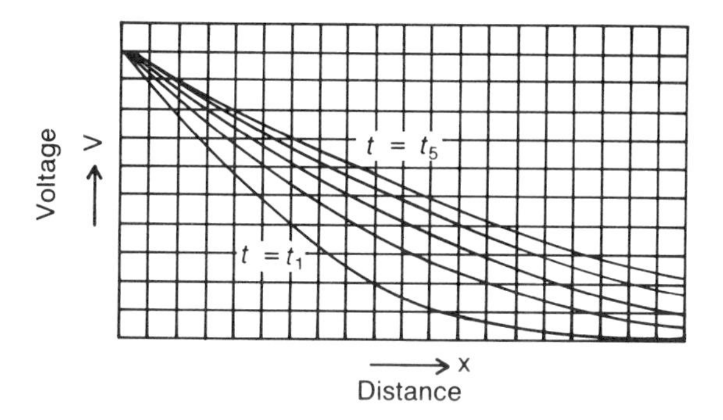

Fig. 5.15.3: Solution for submarine cable.

and

$$k = \frac{|LG - RC|}{2LC} \tag{5.15.56}$$

Equation (5.15.54) is best solved through the use of Fourier integrals. Assuming that $v(x,t)$ has a Fourier transform,

$$v(x,t) = \frac{1}{2\pi}\int_{-\infty}^{\infty} \overline{v}(\omega,t)\, e^{i\omega x}\, d\omega, \tag{5.15.57}$$

the Fourier transform of Eq. (5.15.54) is

$$\frac{d^2\overline{v}}{dt^2} = (k^2 - c^2\omega^2)\overline{v} \tag{5.15.58}$$

which has the general solution

$$\overline{v}(\omega,t) = A(\omega)\ \cos\ [(\omega^2c^2 - k^2)^{1/2}t] + B(\omega)\ \sin\ [(\omega^2c^2 - k^2)^{1/2}t]. \tag{5.15.59}$$

If we assume that the initial conditions can be written

$$v(x,0) = f(x) \text{ and } v_t\,(x,0) = g(x), \tag{5.15.60}$$

then

$$v(x,t) = v_1 + v_2 = \frac{1}{2\pi}\iint_{-\infty}^{\infty} f(\lambda)\cos((\omega^2c^2 - k^2)^{1/2}t)\, e^{i\omega(x-\lambda)}\, d\lambda\, d\omega$$

$$+ \frac{1}{2\pi}\iint_{-\infty}^{\infty} g(\lambda)\, \frac{\sin[(\omega^2c^2 - k^2)^{1/2}t]}{\sqrt{\omega^2c^2 - k^2}}\, e^{i\omega(x-\lambda)}\, d\lambda\, d\omega \tag{5.15.61}$$

We must now integrate Eq. (5.15.61) with respect to ω. Before we can, we must derive some intermediate results.

Consider now the integral

$$\frac{1}{2\pi}\int_{-\pi}^{\pi} \exp(ix\cos(y))\, dy$$

$$= \frac{1}{2\pi}\int_{-\pi}^{\pi} \sum_{n=0}^{\infty} \frac{i^n x^n \cos^n(y)}{n!}\, dy \tag{5.15.62}$$

$$= \sum_{n=0}^{\infty} \frac{i^n x^n}{n!}\, \frac{1}{2\pi}\int_{-\pi}^{\pi} \cos^n(y)\, dy \tag{5.15.63}$$

$$\frac{1}{2\pi}\int_{-\pi}^{\pi} \exp\,(ix\cos(y))dy = \sum_{m=0}^{\infty} \frac{(-1)^m}{m!}\left(\frac{x}{2}\right)^{2m} = J_0(x) \quad \textbf{(5.15.64)}$$

We now apply this result to evaluate the integral

$$\int_0^{\pi} J_0\,(r\sin(\phi)\sin(\theta))\exp(ir\cos(\phi)\cos(\theta))\sin(\theta)\,d\theta$$

$$= \frac{1}{2\pi}\int_0^{\pi}\int_{-\pi}^{\pi} \exp[ir(\cos(\phi)\cos(\theta) + \sin(\phi)\sin(\theta)\cos(\omega)]\sin(\theta)\,d\omega\,d\theta$$

(5.15.65)

Equation (5.15.65) is a surface integration over a sphere of unit radius. To carry out this integration, spherical trigonometry is employed. From Fig. 5.15.4 we find that

$$\cos(\Theta) = \cos(\theta)\cos(\phi) + \sin(\theta)\sin(\phi)\cos(\omega) \quad \textbf{(5.15.66)}$$

and

$$\sin(\theta)\,d\theta\,d\omega = dS, \quad \textbf{(5.15.67)}$$

where dS is the element of area on the surface of the sphere. Because the integration is over the entire sphere, we can chose the pole to be at Ω rather than ω. Then

$$dS = \sin(\Theta)\,d\Theta\,d\Omega. \quad \textbf{(5.15.68)}$$

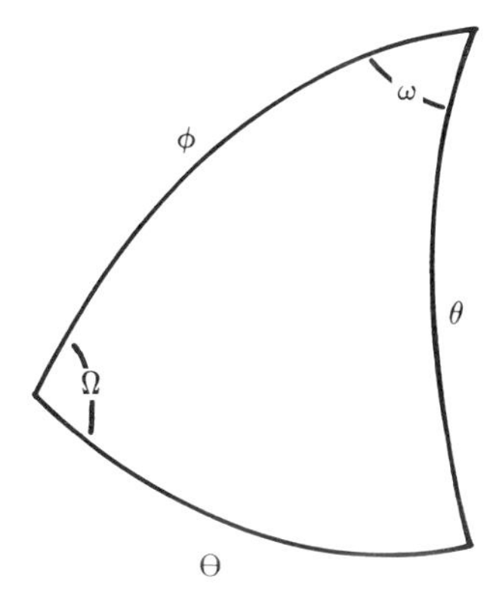

Fig. 5.15.4: Spherical triangle used in the integration of Eq. (5.15.65).

Therefore,

$$\int_0^{\pi} J_0\,[r\sin(\phi)\sin(\theta)]\,\exp[ir\cos(\phi)\cos(\theta)]\,\sin(\theta)\,d\theta$$

$$= \frac{1}{2\pi}\int_0^{\pi}\int_{-\pi}^{\pi} \exp[ir\cos(\Theta)]\,\sin(\Theta)\,d\Omega\,d\Theta \quad \textbf{(5.15.69)}$$

$$= \frac{e^{ir} - e^{-ir}}{ir} = \frac{2\sin(r)}{r} \quad \textbf{(5.15.70)}$$

Consequently,

$$\frac{\sin(r)}{r} = 1/2\int_0^{\pi} J_0\,(r\sin(\phi)\sin(\theta))\,e^{ir\cos(\phi)\cos(\theta)}\sin(\theta)\,d\theta. \quad \textbf{(5.15.71)}$$

Let us now substitute

$$r\cos(\phi) = -\omega ct \quad \textbf{(5.15.72)}$$

$$r\sin(\phi) = ikt \quad \textbf{(5.15.73)}$$

$$r^2 = t^2(c^2\omega^2 - k^2) \quad \textbf{(5.15.74)}$$

$$\cos(\theta) = \frac{\beta}{ct},\; -\sin(\theta)\,d\theta = \frac{d\beta}{ct}, \quad \textbf{(5.15.75)}$$

we have

$$\frac{\sin[t(\omega^2c^2 - k^2)^{1/2}]}{t(\omega^2c^2 - k^2)^{1/2}} = 1/2\int_{-ct}^{ct} J_0[ikt(1 - \frac{\beta^2}{c^2t^2})^{1/2}]\,e^{-i\omega\beta}\,\frac{d\beta}{ct}$$

$$\textbf{(5.15.76)}$$

$$= \frac{1}{2c}\int_{-ct}^{ct} I_0\,[\frac{k}{c}\,(c^2t^2 - \beta^2)^{1/2}]\,e^{-i\omega\beta}\,d\beta$$

$$\textbf{(5.15.77)}$$

where I_0 is the modified Bessel function of the first kind, of order zero. It is given by the power series

$$I_0(x) = \sum_{k=0}^{\infty} \frac{1}{(k!)^2} \left(\frac{x}{2}\right)^2. \qquad \textbf{(5.15.78)}$$

Therefore,

$$\frac{1}{2\pi} \int \int_{-\infty}^{\infty} g(\lambda) \frac{\sin[(c^2\omega^2 - k^2)^{1/2}t]}{\sqrt{c^2\omega^2 - k^2}} e^{i\omega(x-\lambda)}\, d\omega\, d\lambda$$

$$= \frac{1}{4\pi c} \int_{-\infty}^{\infty} d\omega \int_{-\infty}^{\infty} g(\lambda)\, d\lambda$$

$$\int_{-ct}^{ct} I_0 \left[\frac{k}{c} (c^2t^2 - \beta^2)^{1/2} \right] e^{i\omega(x-\lambda-\beta)}\, d\beta \qquad \textbf{(5.15.79)}$$

$$= \frac{1}{4\pi c} \int_{-\infty}^{\infty} e^{-i\omega\lambda} g(\lambda)\, d\lambda \int_{-\infty}^{\infty} d\omega$$

$$\int_{-ct}^{ct} I_0 \left[\frac{k}{c} (c^2t^2 - \beta^2)^{1/2} \right] e^{i\omega(x-\beta}\, d\beta \qquad \textbf{(5.15.80)}$$

The two inner integrals define a Fourier transform

$$\phi(x) = \frac{1}{2\pi} \int_{-\infty}^{\infty} d\omega \int_{-\infty}^{\infty} \phi(\beta)\, e^{i\omega(x-\beta)}\, d\beta \qquad \textbf{(5.15.81)}$$

if we define

$$\phi(x) = \begin{cases} 0 & \text{when } |x| > c|t| \\ I_0 \left[\frac{k}{c} (c^2t^2 - x^2)^{1/2} \right] & \text{when } |x| < c|t|. \end{cases} \qquad \textbf{(5.15.82)}$$

Therefore, using Eq. (5.15.81) in conjunction with the shifting theorem, we have

$$v_2 = \frac{1}{2c} \int_{-\infty}^{\infty} g(\lambda)\, \phi(x - \lambda)\, d\lambda \qquad \textbf{(5.15.83)}$$

$$= \frac{1}{2c} \int_{x-ct}^{x+ct} g(\lambda)\, I_0 \left\{ \frac{k}{c} \left[c^2t^2 - (x - \lambda)^2 \right]^{1/2} \right\} d\lambda \qquad \textbf{(5.15.84)}$$

The v_1 term immediately follows by using Leibnitz's rule on Eq. (5.15.80) to give

$$v_1 = \frac{1}{2c}\frac{\partial}{\partial t}\int_{x-ct}^{x+ct} f(\lambda)\, I_0\left\{\frac{k}{c}\,[c^2t^2 - (x-\lambda)^2]^{1/2}\right\} d\lambda \quad \textbf{(5.15.85)}$$

$$= [f(x+ct) + f(x-ct)]/2$$

$$+ \frac{1}{2c}\int_{x-ct}^{x+ct} f(\lambda)\frac{\partial}{\partial t} I_0\left\{\frac{k}{c}\,[c^2t^2 - (x-\lambda)^2]^{1/2}\right\} d\lambda \quad \textbf{(5.15.86)}$$

because $I(0) = 1$.

There only remains the task of back substitution for $V(x,t)$ and writing the initial conditions in terms of $V(x,0)$ and $V_t(x,0)$:

$$V(x,0) = F(x) \text{ and } V_t(x,0) = G(x). \quad \textbf{(5.15.87)}$$

When this is done, we find

$$V(x,t) = 1/2\,\exp\left(-\frac{GL+RC}{2LC}\,t\right)\Bigg(F(x+ct) + F(x-ct)$$

$$+ \frac{1}{c}\int_{x-ct}^{x+ct} F(\lambda)\frac{\partial}{\partial t} I_0\left\{\frac{k}{c}\,[c^2t^2 - (x-\lambda)^2]^{1/2}\right\} d\lambda$$

$$+ \frac{1}{c}\int_{x-ct}^{x+ct} \left[G(\lambda) + \frac{GL+RC}{2LC}\,F(\lambda)\right]$$

$$\times I_0\left\{\frac{k}{c}\,[c^2t^2 - (x-\lambda)^2]^{1/2}\right\} d\lambda\Bigg) \quad \textbf{(5.15.88)}$$

$$V(x,t) = 1/2\,\exp\left(-\frac{GL+RC}{2LC}\,t\right)\Bigg[F(x+ct) + F(x-ct)$$

$$+ \frac{k^2}{c}\,t\int_{x-ct}^{x+ct} F(\lambda)\,\frac{I_1\left\{\frac{k}{c}\,[c^2t^2 - (x-\lambda)^2]^{1/2}\right\}}{\sqrt{c^2t^2 - (x-\lambda)^2}}\, d\lambda$$

$$+ \frac{1}{c}\int_{x-ct}^{x+ct} \left[G(\lambda) + \frac{GL+RC}{2LC}\,F(\lambda)\right]$$

$$\times I_0\left\{\frac{k}{c}\,[c^2t^2 - (x-\lambda)^2]^{1/2}\right\} d\lambda\Bigg] \quad \textbf{(5.15.89)}$$

The physical interpretation of the first two terms in Eq.

(5.15.89) is straightforward. They represent damped progressive waves. One is propagating to the right; the other, to the left. There is, however, an entirely new phenomenon because even after the wave has passed the two integrals show that there is an effect from all points where originally F and G are not zero within a distance ct from the point in question. This effect persists through all time,

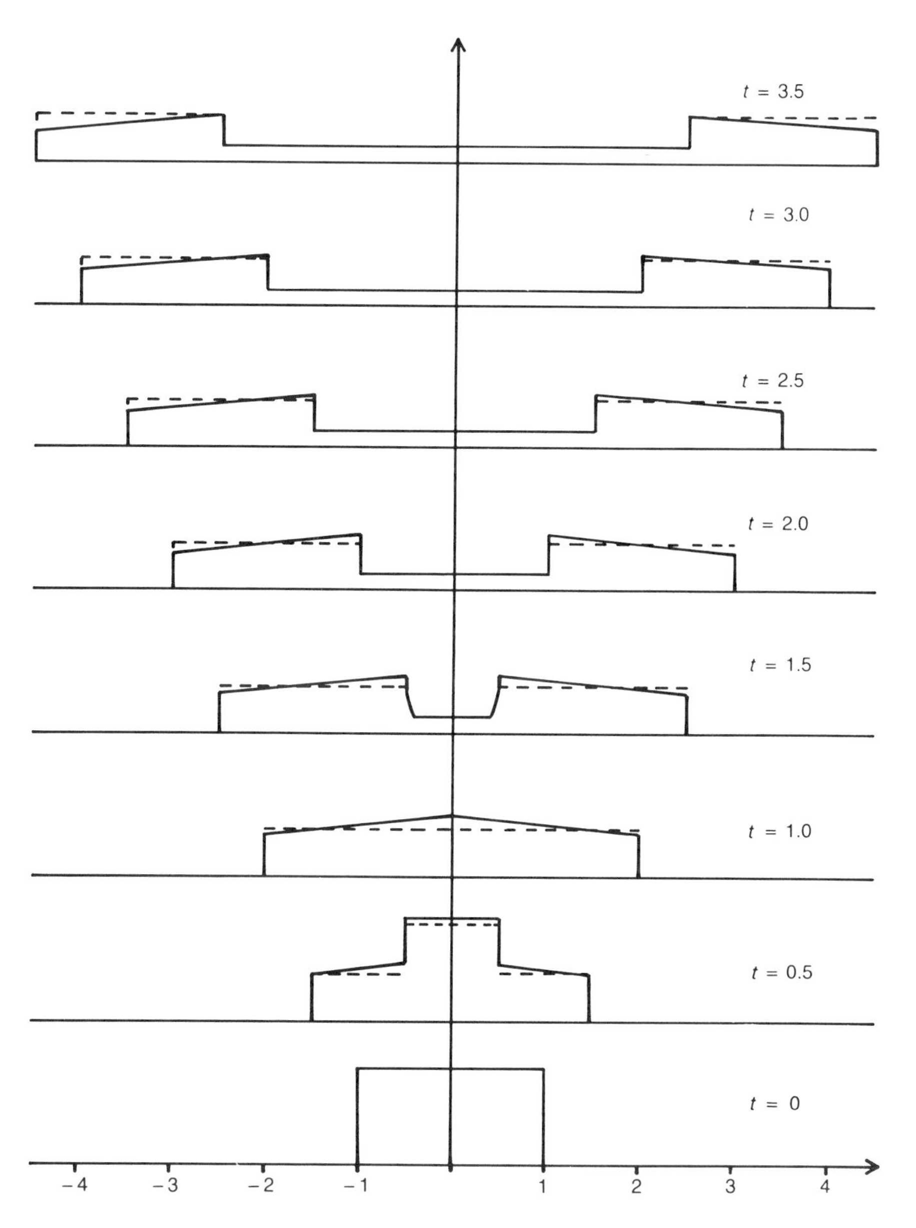

Fig. 5.15.5: The evolution of the voltage with time given by the general equation of telegraphy when the initial voltage is shown in t = 0.

although dying away, and constitutes a residue or tail. This is illustrated in Fig. 5.15.5 for

$$c = 1,\ k = 0.2 \text{ and } (GL + RC)/2LC = 0.1.$$

The integrals were evaluated by numerical integration with Simpson's rule. If $k = 0$ (the lossless case), an initial voltage

$$V(x,0) = \begin{cases} 1 & |x| < 1 \\ 0 & \text{otherwise} \end{cases} \qquad \textbf{(5.15.90)}$$

(with $V_t(x,0) = 0$) would propagate to the right and left pulses shown by the dotted line in Fig. 5.15.5. However, with the addition of resistance or leakage, the solution leaves a residue after the leading edge of the wave passes. This differs from the submarine cable problem where inductance and leakage is neglected because it takes time for the disturbance to arrive.

5.16 THE PROPAGATION OF SOUND WAVES INTO A VISCOUS FLUID

The propagation of small-amplitude sound waves into a viscous fluid is governed by the equation

$$\upsilon\, u_{x^*x^*t^*} + c^2 u_{x^*x^*} = u_{t^*t^*} \qquad \textbf{(5.16.1)}$$

where $t*$ denotes time, $x*$ the distance along the axis of propagation and c the speed of sound in the absence of viscosity. The constant υ equals four thirds of the kinematic viscosity. See Lamb, H., 1932: *Hydrodynamics*. New York: Dover Publication, Inc. for details. We wish to find the motion that arises when an impulse is applied to the fluid at $t* = 0$. In this situation the boundary condition may be written

$$u(0,t*) = \delta(t*) \text{ and } u(\infty,t*) = 0 \qquad \textbf{(5.16.2)}$$

where $\delta(t*)$ is the Dirac delta function. For simplicity we choose the initial conditions to be

$$u(x*,0) = u_{t*}(x*,0) = 0. \qquad \textbf{(5.16.3)}$$

This problem will be solved by Laplace transforms. We have chosen this particular problem because the inversion of the transform of the solution involves a branch point and cut.

To simplify the analysis, we nondimensionalize Eq. (5.16.1) by introducing the nondimensional time and position

$$t = \frac{c^2}{\upsilon} t* \text{ and } x = \frac{c}{\upsilon} x*$$

so that Eq. (5.16.2) becomes

$$u_{xxt} + u_{xx} = u_{tt} \tag{5.16.4}$$

with

$$u(0,t) = \delta(t), \; u(\infty,t) = 0$$

and

$$u(x,0) = u_t(x,0) = 0.$$

The Laplace transform of Eq. (5.16.4) is

$$(1 + s)\overline{u}'' + s^2 \overline{u} = 0 \tag{5.16.5}$$

after applying the initial conditions. The general solution of Eq. (5.16.5) which satisfies the boundary conditions is

$$\overline{u}(x,s) = \exp\left(-\frac{sx}{(1+s)^{1/2}}\right). \tag{5.16.6}$$

Consequently, the inverse of $\overline{u}(x,s)$ is given by the Bromwich integral

$$u(x,t) = \frac{1}{2\pi i}\int_{c-\infty i}^{c+\infty i} \exp\left(st - \frac{sx}{\sqrt{1+s}}\right) ds. \tag{5.16.7}$$

The exact contour depends upon the nature of the integrand. Because of the square root in the exponential, it is a multivalued function. Its branch point is located at $s = -1$. Because our contour must run parallel to the imaginary axis, we take the branch cut of this multivalued function to lie along the negative real axis so that the contour will not cross the branch cut.

It is convenient to shift the branch point to the origin. This is done by letting $s = \sigma - 1$, so that Eq. (5.16.7) becomes

$$u(x,t) = \frac{e^{-t}}{2\pi i}\int_{c-\infty i}^{c+\infty i} e^{\sigma t} \exp\left(\left(\frac{1}{\sqrt{\sigma}} - \sqrt{\sigma}\right)x\right) d\sigma \tag{5.16.8}$$

We now convert the contour from $c - \infty i$ to $c + \infty i$ into a closed contour. There are an infinite number of ways of doing this; however, the new contour must not cross the negative real axis (i.e., the branch cut). Hamin (Hamin, M. 1957: Propagation of an aperiodic wave in a compressible viscous medium. *J. Math. and Phys.*, *36*, 234-249.) found that a particularly useful contour consists of an infinite semicircle to the left of $c > 1$ with an indentation along the negative real axis to the unit circle with its center at $\sigma = 0$. See Fig. 5.16.1. The branch cut is denoted by the wavy line.

Because the integrand enclosed by the contour is analytic, the value of the integral along the entire contour is zero. It is readily shown that the contribution from the arcs at infinity is zero. The contribution from the unit circle is

$$-\frac{e^{-t}}{2\pi i}\int_{C_2} e^{\sigma t}\exp[x(\sigma^{-1/2} - \sigma^{1/2})]d\sigma$$

$$= -\frac{e^{-t}}{2\pi}\int_{\pi}^{-\pi} \exp[e^{i\theta} + x(e^{-i\theta/2} - e^{i\theta/2})]\,e^{i\theta}\,d\theta \qquad \textbf{(5.16.9)}$$

$$= \frac{e^{-t}}{2\pi}\int_{-\pi}^{\pi} \exp[t\cos(\theta) + it\sin(\theta) - 2xi\sin\left(\frac{\theta}{2}\right) + i\theta]d\theta$$

(5.16.10)

$$= \frac{e^{-t}}{2\pi}\int_{-\pi}^{\pi} e^{t\cos(\theta)}\cos\left[t\sin(\theta) - 2x\sin\left(\frac{\theta}{2}\right) + \theta\right]d\theta$$

(5.16.11)

$$= \frac{e^{-t}}{\pi}\int_{0}^{\pi} e^{t\cos(\theta)}\cos[t\sin(\theta) - 2x\sin(\theta/2)]\cos(\theta)d\theta$$

$$-\frac{e^{-t}}{\pi}\int_{0}^{\pi} e^{t\cos(\theta)}\sin[t\sin(\theta) - 2x\sin(\theta/2)]\sin(\theta)\,d\theta$$

(5.16.12)

$$= \frac{1}{\pi}\int_0^1 \exp(-2t\,\omega^2)\cos[2t\omega(1-\omega^2)^{1/2} - 2x\omega]\,(1-2\omega^2)\frac{2\,d\omega}{\sqrt{1-\omega^2}}$$

$$-\frac{1}{\pi}\int_0^1 \exp(-2t\,\omega^2)\sin[2t\omega(1-\omega^2)^{1/2} - 2x\omega]\,4\omega\,d\omega$$

(5.16.13)

where

$$\sigma = e^{i\theta},\ \sin(\theta) = (e^{i\theta} - e^{-i\theta})/2i,\ \omega = \sin(\theta/2),$$
$$\cos(\theta) = 1-2\omega^2,\ \sin(\theta) = 2\omega(1-\omega^2)^{1/2}.$$

The integral of $\exp[t\cos(\theta)]\sin[t\sin(\theta) - 2x\sin(\theta/2) + \theta]$ integrates to zero over the symmetric interval; the integral of $\exp[t\cos(\theta)]\cos[t\sin(\theta) - 2x\sin(\theta/2) + \theta]$ equals twice the

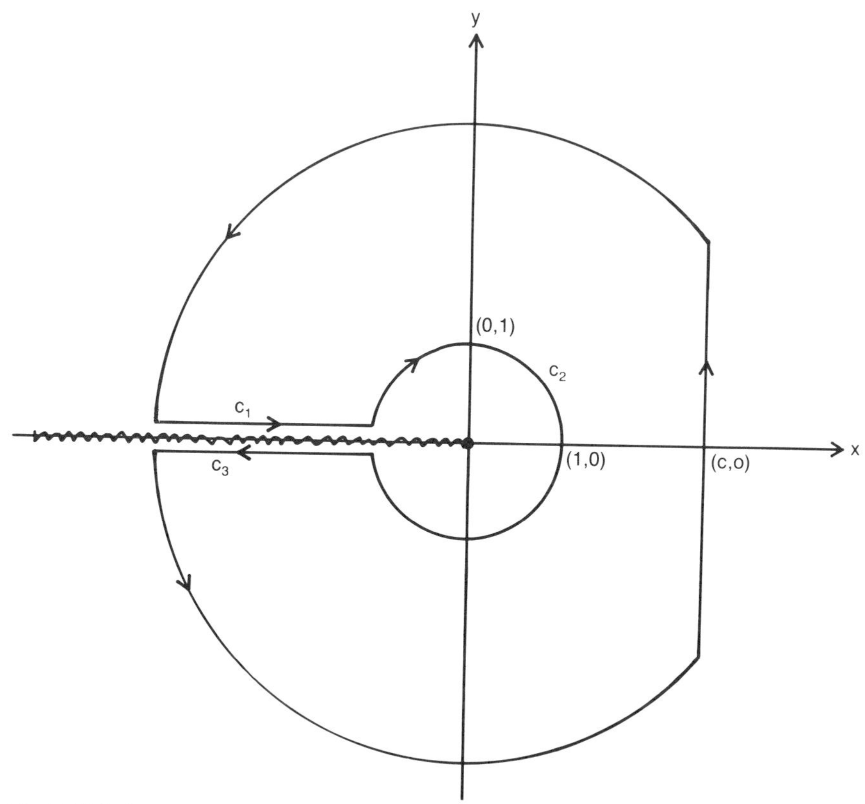

Fig. 5.16.1: Contour used in finding the sound waves that propagate into a viscous fluid from a unit impulse.

value of the integral from 0 to π.

Because

$$\frac{1}{t}\frac{d}{d\omega}\{\exp(-2t\,\omega^2)\sin[2t\omega(1-\omega^2)^{1/2} - 2x\omega$$

$$= -4\omega\exp(-2t\,\omega^2)\sin\,[2t\omega(1-\omega^2)^{1/2} - 2x\omega] \qquad \textbf{(5.16.14)}$$

$$+\ 2\left(\frac{1-2\omega^2}{\sqrt{1-\omega^2}} - \frac{x}{t}\right)\exp(-2t\omega^2)\cos\,[2t\omega(1-\omega^2)^{1/2} - 2x\omega]\,,$$

we have

$$① = -\frac{e^{-t}}{2\pi i}\int_{C_2} e^{\sigma t}\exp(x(\sigma^{-1/2} - \sigma^{1/2}))\,d\sigma$$

$$① = \frac{2x}{\pi t} \int_0^1 \exp(-2t\omega^2) \cos[2t\omega(1-\omega^2)^{1/2} - 2x\omega]\, d\omega$$

$$+ \frac{1}{\pi t} \exp(-2t\omega^2) \sin[2t\omega(1-\omega^2)^{1/2} - 2x\omega] \Big|_0^1 \tag{5.16.15}$$

$$= \frac{2x}{\pi t} \int_0^1 \exp(-2t\omega^2) \cos[2t\omega(1-\omega^2)^{1/2} - 2x\omega]\, d\omega$$

$$- \frac{1}{\pi t} \exp(-2t) \sin(2x). \tag{5.16.16}$$

Along C_1, $\sigma = \lambda e^{i\pi}$ and $\sigma^{1/2} = \lambda^{1/2} e^{i\pi/2} = i\lambda^{1/2}$,

$$-\frac{e^{-t}}{2\pi i} \int_{C_1} e^{\sigma t} \exp[x(\sigma^{-1/2} - \sigma^{1/2})]\, d\sigma$$

$$= -\frac{e^{-t}}{2\pi i} \int_{\infty}^{1} e^{-\lambda t} \exp[x(\frac{-i}{\sqrt{\lambda}} - i\sqrt{\lambda})](-d\lambda) \tag{5.16.17}$$

$$= \frac{e^{-t}}{2\pi i} \int_1^{\infty} e^{-\lambda t} \exp[ix(\frac{\lambda+1}{\sqrt{\lambda}})]\, d\lambda \tag{5.16.18}$$

Along C_3, $\sigma = \lambda e^{-i\pi}$ and $\sigma^{1/2} = \lambda^{1/2} e^{-i\pi/2} = -i\lambda^{1/2}$,

$$-\frac{e^{-t}}{2\pi i} \int_{C_3} e^{\sigma t} \exp[x(\sigma^{-1/2} - \sigma^{1/2})]\, d\sigma$$

$$= -\frac{e^{-t}}{2\pi i} \int_1^{\infty} e^{-\lambda t} \exp[x(\frac{i}{\sqrt{\lambda}} + i\sqrt{\lambda})](-d\lambda) \tag{5.16.19}$$

$$= \frac{e^{-t}}{2\pi i} \int_1^{\infty} e^{-\lambda t} \exp[ix(\frac{\lambda+i}{\sqrt{\lambda}})]\, d\lambda \tag{5.16.20}$$

Finally, because the integration from $c - i\infty$ to $c + i\infty$ equals the sum of the integrals (5.16.16), (5.16.18) and (5.16.19) minus the negligible contribution from the arc at infinity, we have

$$\frac{e^{-t}}{2\pi i} \int_{c-\infty i}^{c+\infty i} e^{\sigma t} \exp(x(\sigma^{-1/2} - \sigma^{1/2}))\, d\sigma = ①$$

$$① = \frac{e^{-t}}{\pi} \int_1^\infty e^{-\lambda t} \sin(x \frac{\lambda + 1}{\sqrt{\lambda}})\, d\lambda - \frac{1}{\pi t} e^{-2t} \sin(2x)$$

$$+ \frac{2x}{\pi t} \int_0^1 \exp(-2t\omega^2) \cos[2t\omega(1-\omega^2)^{1/2} - 2x\omega]\, d\omega \quad \mathbf{(5.16.21)}$$

$$= \frac{2x}{\pi t} \int_0^1 \exp(-2t\omega^2) \cos[2t\omega(1-\omega^2)^{1/2} - 2x\omega]\, d\omega$$

$$+ \frac{e^{-2t}}{\pi} \int_0^\infty e^{-t\omega} [\sin(\frac{2+\omega}{\sqrt{1+\omega}} x) - \sin(2x)]\, d\omega. \quad \mathbf{(5.16.22)}$$

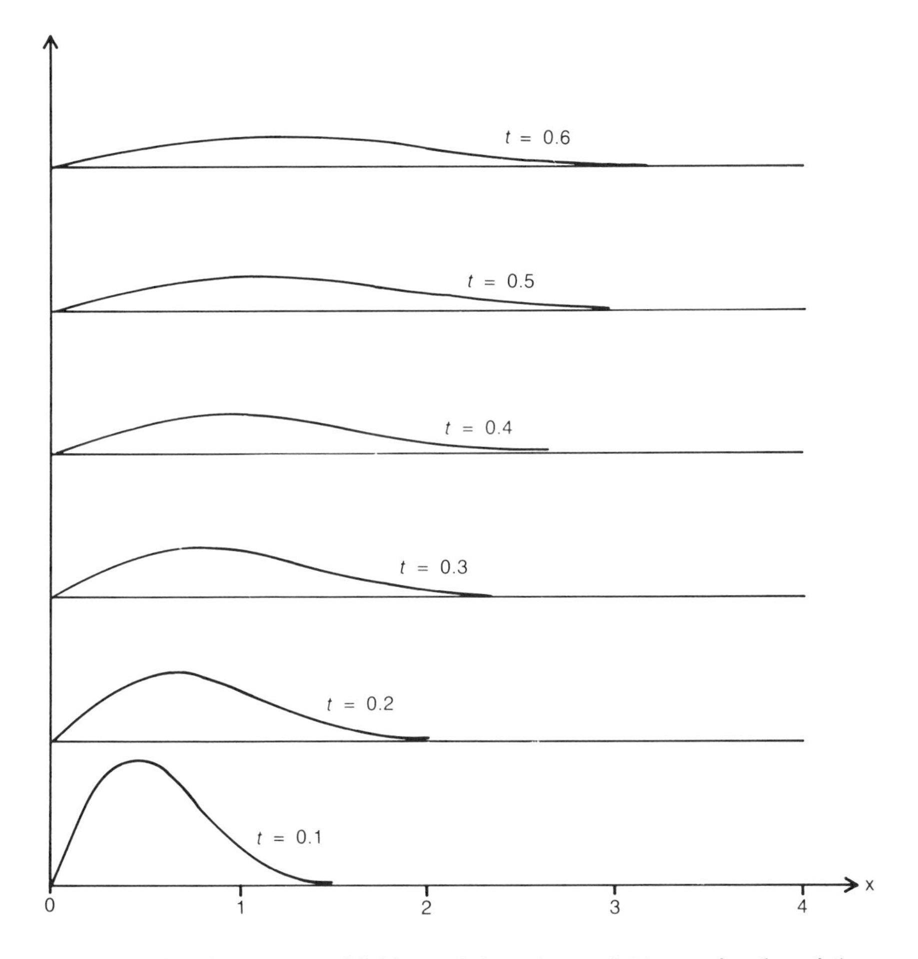

Fig. 5.16.2: The displacement of fluid parcels in a viscous fluid as a function of time and position.

The integrals in Eq. (5.16.22) must be evaluated numerically. However, because of their particular form, they converge very rapidly. In Fig. 5.16.2 the solution is shown for various nondimensional times. The effect of viscosity is to reduce the sharpness of the wave and smooth out the initially imposed discontinuity. The wave is thereby dispersed by continual spreading of the region of the main disturbance.

Solutions to this problem with more complicated boundary conditions are very difficult to find. Blackstock (Blackstock, D. T., 1967: Transient solution for sound radiated into a viscous fluid. *Acoust. Soc. Am. J.*, *41*, 1312-1319) has found the approximate solution for small viscosity when the forcing is given by $u(0,t) = u_0 \sin(\omega t) H(t)$.

Exercises

In problems 1-3, solve the wave equation $u_{tt} = u_{xx}$, $0<x<1$, $t>0$, using transform methods for the following set of initial and boundary conditions.

1. $u(0,t) = u(1,t) = 0$, $u(x,0) = \sin(\pi x)$, $u_t(x,0) = -\sin(\pi x)$.
2. $u(0,t) = u_x(1,t) = 0$, $u(x,0) = 0$, $u_t(x,0) = x$.
3. $u(0,t) = u(1,t) = 0$, $u(x,0) = 0$, $u_t(x,0) = 1$.

4. Solve $u_{tt} - u_{xx} = te^{-x}$ with the conditions $u(x,0) = 0$, $u_t(x,0) = x$, and $u(0,t) = 1 - e^{-t}$, $x>0$.

5. Solve $u_{tt} - u_{xx} = xe^{-t}$ with the conditions $u(x,0) = 1$, $u_t(x,0) = 0$, and $u(0,t) = \cos(t)$, $x>0$.

6. Solve $2u_{tt} + 3u_{xt} + u_{xx} = 0$ with the conditions $u(x,0) = u_t(x,0) = u(0,t) = 0$ and $u_x(0,t) = 1$.

7. The function $u(x,t)$ satisfies the wave equation $u_{tt} = c^2 u_{xx}$ for $0<x<a$, $t>0$. Given that $u(x,0) = u_t(x,0) = u(a,t) = 0$ and that $u(0,t) = \sin(\omega t)$, find the series expansion for $u(x,t)$ using transform methods. Consider separately the case in which $\omega a/c$ is an integral multiple of π.

8. A heavy spring has a mass ϱ per unit length when free of tension, and is attached at one end. At $t = 0$, when the spring is hanging vertically in equilibrium, a mass m is attached to the free end $x = L$ of the spring and released from rest. If the additional extension $z(x,t)$ satisfies the equation

$$z_{tt} = \frac{k}{\varrho} z_{xx}$$

with $z(x,0) = z_t(x,0) = 0$ and

$$z(0,t) = 0 \text{ and } z_{tt}(L,t) + \frac{k}{m} z_x(L,t) = g.$$

Show that the motion of the mass m is given by

$$z(L,t) = \frac{mgL}{k}\left(1 - \sum_{n=1}^{\infty} \frac{2\cos(\alpha_n ct/L)}{\alpha_n^2(\alpha_n^2\lambda^2 + \lambda + 1)}\right.$$

where $c^2 = k/\varrho$, $\lambda = m/(\varrho L)$, and $\alpha_1, \alpha_2, \alpha_3, \alpha_4, \ldots$ are the positive roots of $\cot(\alpha) = \lambda\alpha$.

9. Constant voltage E is applied at $t = 0$ at the end $x = 0$ of a lossless line of length ℓ, the end $x = \ell$ being insulated $(d\overline{V}/dx = 0)$. If the initial current and charge are zero, show that the voltage at any point of the line is

$$E \sum_{n=0}^{\infty} (-1)^n H(t - \frac{2n\ell + x}{c}) + H(t - \frac{(2n+1)\ell - x}{c}).$$

10. A lossless transmission line extends from $x = -\ell$ to $x = a$, the end $x = a$ being insulated, and the end $x = -\ell$ being grounded. At $t = 0$ there is no current in the line, the portion $0<x<a$ is charged to the voltage E, and the portion $-\ell<x<0$ is uncharged. Show that the current at $x = -\ell$ at time t is

$$-E\sqrt{C/L} \sum_{n=0}^{\infty} (-1)^n H\left[t - \frac{2n(\ell + a) + \ell}{c}\right]$$

$$- H\left[t - \frac{(2n+1)(\ell + a) + a}{c}\right]$$

where $c = (LC)^{-1/2}$

11. Constant voltage E is applied at $t = 0$ through an inductive resistance L_0, R_0 to the end $x = 0$ of a lossless line of length ℓ, the end $x = \ell$ being insulated, and there being zero initial current and charge. Show that the current at $x = 0$ is

$$\frac{E}{R_0 + R_1}(1 - e^{-\alpha t}) - \frac{2ER_1}{(R_0 + R_1)^2}\left[1 - (\alpha t + 1 - \frac{2\alpha\ell}{c}) e^{-\alpha(t - 2\ell/c)} H(t - \frac{2\ell}{c})\right] + \ldots$$

where $c = (LC)^{-1/2}$, $R_1 = (L/C)^{1/2}$, and $\alpha = (R_0 + R_1)/L_0$.

12. A submarine cable ($L = G = 0$ and $\varkappa = 1/(RC)$) of length ℓ has zero initial current and charge. The end $x = 0$ is insulated, and a constant voltage E is applied at $x = \ell$. Show that the voltage at any point is

$$E + \frac{4E}{\pi} \sum_{n=1}^{\infty} \frac{(-1)^n}{(2n-1)} e^{-\varkappa(2n-1)^2\pi^2 t/4\ell^2} \cos\left[\frac{(2n-1)\pi x}{2\ell}\right].$$

13. A lossless line of length ℓ has zero initial current and charge, and is insulated at $x = \ell$. Voltage $E \sin(\omega t)$ is applied at $x = 0$ for $t>0$. Show that the voltage at any point is

$$\frac{E \cos(\omega(\ell - x)/c)\sin(\omega t)}{\cos(\omega\ell/c)}$$

$$+ 8E\omega\ell c \sum_{n=0}^{\infty} \frac{\sin[(2n+1)\pi ct/2\ell]\sin[(2n+1)\pi x/2\ell]}{4\omega^2\ell^2 - (2n+1)^2\pi^2 c^2}$$

where $c = (LC)^{-1/2}$, and ω is not to be equal to $(2n+1)\pi c/2\ell$ for any n.

14. A submarine cable of length ℓ has the end $x = \ell$ grounded, and constant voltage E is applied at $x = 0$ with zero initial conditions. Show that the current at $x = 0$ is

$$\frac{E}{R\sqrt{\pi\varkappa t}} \left(1 + 2 \sum_{n=1}^{\infty} \exp\left(-n^2\ell^2/(\varkappa t)\right) \right).$$

5.17 THE WAVE EQUATION IN AN UNBOUNDED REGION: THE FOURIER TRANSFORM AND GROUP VELOCITY

In Section 5.11 we showed how we could solve the wave equation in an unbounded region using d'Alembert's formula. The derivation of this formula required the introduction of two new independent variables $\xi = x + ct$ and $\eta = x - ct$.

In the case of a more complicated wave equation, the choice of such a "trick" substitution becomes much more problematical. Consequently the question arises of whether we might be able to apply the product solution technique on an unbounded domain and avoid quessing the new variables that are necessary to simplify the problem. The answer is that we can but we must use the Fourier integral which was introduced in Section 2.13 rather than Fourier series.

Consider the problem

$$u_{tt} = c^2 u_{xx} \qquad -\infty < x < \infty,\ t>0 \tag{5.17.1}$$

with

$$u(x,0) = f(x),\ u_t(x,0) = 0 \qquad -\infty < x < \infty \tag{5.17.2}$$

and

$$u(0,t) = 0 \qquad t>0. \tag{5.17.3}$$

We also require that the solution $u(x,t)$ be bounded as $|x| \rightarrow \infty$. Transforming the wave equation with respect of x, we have

$$\mathfrak{F}(u_{xx}) = (ik)^2 \, \overline{V}(ik,t) \tag{5.17.4}$$

where $\overline{V}(ik,t)$ is the Fourier transform of $u(x,t)$. In most aspects, we are performing the same operations as we did with Laplace transforms in the previous section. In this case, however, we are applying transform techniques to the x domain from $-\infty$ to ∞.

Turning to the time domain

$$\mathfrak{F}(u_{tt}) = \int_{-\infty}^{\infty} u_{tt}(x,t) \, e^{-ikx} \, dx = \frac{d^2 \overline{V}(ik,t)}{d^2 t}, \tag{5.17.5}$$

$$\mathfrak{F}(u(x,0)) = \overline{V}(ik,0) = \mathfrak{F}(f(x)) = \overline{F}(ik), \tag{5.17.6}$$

and

$$\mathfrak{F}(u_t(x,0)) = \frac{d}{dt} (\mathfrak{F}(u(x,0))) = \frac{d\overline{V}(ik,0)}{dt}. \tag{5.17.7}$$

From these equations, we obtain the ordinary differential equation

$$\frac{d^2 \overline{V}(ik,t)}{dt^2} = c^2 (ik)^2 \, \overline{V}(ik,t) \tag{5.17.8}$$

with

$$\overline{V}(ik,0) = \overline{F}(ik) \text{ and } \frac{d\overline{V}(ik,0)}{dt} = 0. \tag{5.17.9}$$

The general solution is

$$\overline{V}(ik,t) = A(ik) \, e^{+ikct} + B(ik) \, e^{-ikct}. \tag{5.17.10}$$

The constants of integration A and B are evaluated by the boundary conditions:

$$\overline{V}(ik,0) = A(ik) + B(ik) = \overline{F}(ik) \tag{5.17.11}$$

$$\frac{d\overline{V}(ik,0)}{dt} = ikc \, A(ik) - ikc \, B(ik) = 0. \tag{5.17.12}$$

Consequently

$$A(ik) = B(ik) = \overline{F}(ik)/2 \tag{5.17.13}$$

and

$$\overline{V}(ik,t) = \overline{F}(ik)\,(e^{ikct} + e^{-ikct})/2. \qquad \textbf{(5-17-14)}$$

The desired function $u(x,t)$ is the inverse Fourier transform of $\overline{V}(ik,t)$, namely,

$$u(x,t) = \frac{1}{2\pi}\int_{-\infty}^{\infty} \overline{V}(ik,t)\, e^{ikx}\, dk \qquad \textbf{(5.17.15)}$$

$$= \frac{1}{4\pi}\int_{-\infty}^{\infty} \overline{F}(ik)\,(e^{ikct} + e^{-ikct})\, e^{ikx}\, dk \qquad \textbf{(5.17.16)}$$

Because

$$\mathfrak{F}^{-1}(\overline{F}(ik)e^{ikct}) = f(x+ct) \qquad \textbf{(5.17.17)}$$

and

$$\mathfrak{F}^{-1}(\overline{F}(ik)e^{-ikct}) = f(x-ct), \qquad \textbf{(5.17.18)}$$

then

$$u(x,t) = \{\mathfrak{F}^{-1}[\overline{F}(ik)e^{ikct}] + \mathfrak{F}^{-1}[\overline{F}(ik)e^{-ikct}]\}/2 \qquad \textbf{(5.17.19)}$$

$$= [\,f(x + ct) + f(x - ct)]/2 \qquad \textbf{(5.17.20)}$$

which is the same result that we would have obtained using d'Alembert's formula.

The usefulness of this technique is not in finding solutions to the simple wave equation on an unbounded domain. It is far easier to use d'Alembert formula. The importance of this technique lies with the ability to write the solution of any wave equation as a Fourier integral once the Fourier transform of the initial condition is known.

Another important result follows from our use of Fourier integrals. We already have

$$u(x,t) = \frac{1}{4\pi}\int_{-\infty}^{\infty} \overline{F}(ik)\,(e^{ikx-i\omega t} + e^{ikx+i\omega t})\, dk \qquad \textbf{(5.17.21)}$$

where $\omega = kc$. Suppose now that the initial condition is given by

$$\overline{F}(ik) = \begin{cases} F_0 & |k-k_0| \leq \Delta k/2 \\ 0 & \text{otherwise} \end{cases} \qquad \textbf{(5.17.22)}$$

where $k << k_0$. Then

$$u(x,t) = \frac{F_0}{4\pi} \int_{k_0 - \Delta k/2}^{k_0 + \Delta k/2} (e^{ikx - i\omega t} + e^{ikx + i\omega t})\, dk. \quad \textbf{(5.17.23)}$$

In this equation not only does the wave number vary but also the frequency because $\omega = \omega(k)$ in the case of dispersive waves. Because the wave number lies within a narrow band, we can expand ω in a Taylor expansion about k_0:

$$\omega(k) = \omega(k_0) + (k - k_0) \frac{d\omega}{dk}\Big|_{k = k_0} + \ldots \quad \textbf{(5.17.24)}$$

The quantity $\omega(k_0) = \omega_0$ is generally called the *carrier frequency*. Consequently,

$$u(x,t) = \frac{F_0}{4\pi}\left\{\int_{k_0 - \Delta k/2}^{k_0 + \Delta k/2} \exp[i(k - k_0)(x - \frac{d\omega}{dk}\Big|_{k=k_0} t]dk\right\} e^{i(k_0 x - \omega_0 t)}$$

$$+ \frac{F_0}{4\pi}\left\{\int_{k_0 - \Delta k/2}^{k_0 + \Delta k/2} \exp[i(k - k_0)(x + \frac{d\omega}{dk}\Big|_{k=k_0} t)]\, dk\right\} e^{i(k_0 x + \omega_0 t)}$$

$$\textbf{(5.17.25)}$$

If we now introduce the new variable $\xi = k - k_0$, we have

$$u(x,t) = \frac{F_0}{4\pi} e^{i(k_0 x - \omega_0 t)} \int_{-\Delta k/2}^{\Delta k/2} \exp\left[i\xi(x - \frac{d\omega}{dk}\Big|_{k=k_0} t)\right] d\xi$$

$$+ \frac{F_0}{4\pi} e^{i(k_0 x + \omega_0 t)} \int_{-\Delta k/2}^{\Delta k/2} \exp\left[i\xi(x + \frac{d\omega}{dk}_{k=k_0} t)\right] d\xi$$

$$\textbf{(5.17.26)}$$

$$= \frac{F_0}{4\pi}\left[\frac{2 \sin\left[\Delta k/2(x - \frac{d\omega}{dk} t)\right]}{x - \frac{d\omega}{dk} t} + \frac{2 \sin\left[(\Delta k/2(x + \frac{d\omega}{dk} t)\right]}{x + \frac{d\omega}{dk} t}\right]$$

$$\textbf{(5.17.27)}$$

Let us now examine the expression that we have obtained. We shall only explain the first expression because the same thing holds true for the second term.

The term $\frac{F_0}{4\pi} e^{i(k_0 x + \omega_0 t)}$ represents a traveling wave, homogeneous in space, with a mean "carrier" frequency ω_0. However the amplitude of the resultant wave is no longer constant in space because of the second factor

$$\frac{2 \sin \left[\Delta k/2 \left(x - \left.\frac{d\omega}{dk}\right|_{k=k_0} t\right) \right]}{x - \left.\frac{d\omega}{dk}\right|_{k=k_0} t}. \qquad \textbf{(5.17.28)}$$

This factor has its maximum at $x = \frac{d\omega}{dk} t$ where the argument of the sine and the denominator are equal to zero. This maximum is not stationary but moves in space with the velocity $\frac{d\omega}{dk}$. A disturbance of this kind concentrated in space is called a *wave packet* because the wave packet represents the superimposing of a group of waves (in our case, with wave numbers from $k_0 - \Delta k/2$ to $k_0 + \Delta k/2$. The speed at which the wave packet moves, namely $\frac{d\omega}{dk}$, is given the special name of *group velocity*. Unlike a monochromatic wave with its constant amplitude, a wave packet displaces its maximum with time and thereby can move a concentrated electromagnetic disturbance, for example, from one location to another. Because this concentrated disturbance is generally associated with a signal, the use of Fourier integrals and the concept of group velocity is of fundamental importance in the theory of signal transmission.

As a second example of the Fourier integral, we examine the work of Sharpe (Sharpe, J. A., 1942: The production of elastic waves by explosion pressures. I. Theory and empirical field observations. *Geophysics, 7,* 144-154.) who used Fourier transforms to find the wave motion resulting when a pressure $p(t)$ is applied to the interior surface of a spherical cavity of radius a in an ideal elastic medium. If we assume that all motions are radially symmetric and irrotational, the motion of the medium is governed by the scalar wave equation

$$\frac{\partial^2}{\partial t^2}(r\phi) = c^2 \frac{\partial^2}{\partial r^2}(r\phi) \qquad \textbf{(5.17.29)}$$

where ϕ is the velocity potential and c the velocity of propagation of compressional waves. The radial displacement $u(r,t)$ of the elastic wave is given by ϕ_r.

If $\phi(r,t)$ has a Fourier transform, Fourier's theorem states that it can be written as

$$\phi(r,t) = \frac{1}{2\pi}\int_{-\infty}^{\infty} \overline{\phi}(r,\omega)\, e^{i\omega t}\, d\omega \qquad \textbf{(5.17.30)}$$

Substituting Eq. (5.17.30) into Eq. (5.17.29), we have

$$c^2 \frac{d^2}{dr^2}(r\overline{\phi}) = -\,\omega^2 r\overline{\phi} \qquad \textbf{(5.17.31)}$$

which has the general solution

$$\overline{\phi}(r,\omega) = \frac{A(\omega)}{r} \exp\left[-\frac{i\omega}{c}(r-a)\right]. \qquad \textbf{(5.17.32)}$$

The solution is written in this particular format for computational ease later on. We have discarded the $\exp[\frac{i\omega}{c}(r-a)]$ solution because we can only have outwardly propagating waves; energy must propagate away from the source of the disturbance. The combination of Eq. (5.17.30) and (5.17.32) yields

$$\phi(r,t) = \frac{1}{2\pi r}\int_{-\infty}^{\infty} A(\omega) \exp\left[i\omega t - \frac{i\omega}{c}(r-a)\right] d\omega. \qquad \textbf{(5.17.33)}$$

$A(\omega)$ is determined by the boundary condition that the radial component of stress in the medium equals the pressure inside the cavity

$$-[(\lambda + 2\mu)u_r + 2\lambda u/r] = p(t) \qquad \textbf{(5.17.34)}$$

at $r = a$. The parameters λ and μ are Lamé elastic constants. Because we are interested in this problem from its mathematical, rather than physical, aspect, we now take $\lambda = \mu$. The general solution for $\lambda \neq \mu$ has been given by Blake, F. G., 1952: Spherical wave propagation in solid media. *J. Acoust. Soc. Amer.*, *24*, 211-215. Therefore, we have

$$-\,\varrho c^2 \left(u_r + \frac{2u}{3r}\right) = p(t) \qquad \textbf{(5.17.35)}$$

because $c = [(\lambda + 2\mu)/\varrho]^{1/2} = (3\mu/\varrho)^{1/2}$ where ϱ denotes the density.

If $p(t)$ has a Fourier transform, $p(t)$ can be expressed as

$$p(t) = \frac{1}{2\pi}\int_{-\infty}^{\infty}\left(\int_{-\infty}^{\infty} p(x)\, e^{-i\omega x}\, dx\right) e^{i\omega t}\, d\omega \qquad \textbf{(5.17.36)}$$

$$= \frac{1}{2\pi}\int_{-\infty}^{\infty}\int_{-\infty}^{\infty} p(x)\, e^{i\omega t - i\omega x}\, dx\, d\omega \qquad \textbf{(5.17.37)}$$

Upon substituting Eqs. (5.17.37) and (5.17.33) into Eq. (5.17.35), we have

$$-A(\omega)\ \varrho c^2\left(\frac{4}{3a^2}+\frac{4i\omega}{3ca}-\frac{\omega^2}{a^2}\right)=\int_{-\infty}^{\infty} p(x)\ e^{-i\omega x}\,dx \qquad \textbf{(5.17.38)}$$

at $r = a$ because $u = \phi_r$. Eqs. (5.17.33) and (5.17.38) can be combined to yield

$$\phi(r,t) = \frac{a}{2\pi r\varrho}\int_{-\infty}^{\infty}\int_{-\infty}^{\infty}\frac{p(x)\ \exp\left[-i\omega x + i\omega t - \frac{i\omega}{c}(r-a)\right]}{\omega^2-\frac{4i\omega c}{3a}-\frac{4c^2}{3a^2}}\,dx\,d\omega$$

(5.17.39)

To illustrate Eq. (5.17.39), let us take

$$p(t) = p_0\ e^{-\alpha t}\ H(t) \qquad \textbf{(5.17.40)}$$

where $H(t)$ is the Heaviside step function, p_0 the initial and highest pressure attained and α a positive constant. The Fourier transform of $p(t)$ is

$$\mathfrak{F}\,(p(t)) = \frac{p_o}{i\omega+\alpha} \qquad \textbf{(5.17.41)}$$

so that Eq. (5.17.39) becomes

$$\phi(r,t) = \frac{ap_o}{2\pi r\varrho}\int_{-\infty}^{\infty}\frac{\exp\left[\ i\omega t - \frac{i\omega}{c}(r-a)\right]}{(i\omega+\alpha)\ (\omega^2-\frac{4i\omega c}{3a}-\frac{4c^2}{3a^2})}\,d\omega \qquad \textbf{(5.17.42)}$$

$$= \frac{ap_o}{2\pi r\varrho}\int_{-\infty}^{\infty}\frac{\exp\left[i\omega t - \frac{i\omega}{c}(r-a)\right]}{(i\omega+\alpha)\left[(\omega+\frac{2c}{3a}(-i+\sqrt{2})\right]\left[(\omega+\frac{2c}{3a}(-i-\sqrt{2})\right]}\,d\omega$$

(5.17.43)

The integrand has three simple poles, all of them located in the upper half-plane. Consequently Eq. (5.17.43) can be evaluated by the residue theorem (see Section 8 of Appendix B) where the contour runs along the x-axis from $-\infty$ to ∞ and then along the infinite semi-circle in the upper half-plane. Applying the residue theorem, we have

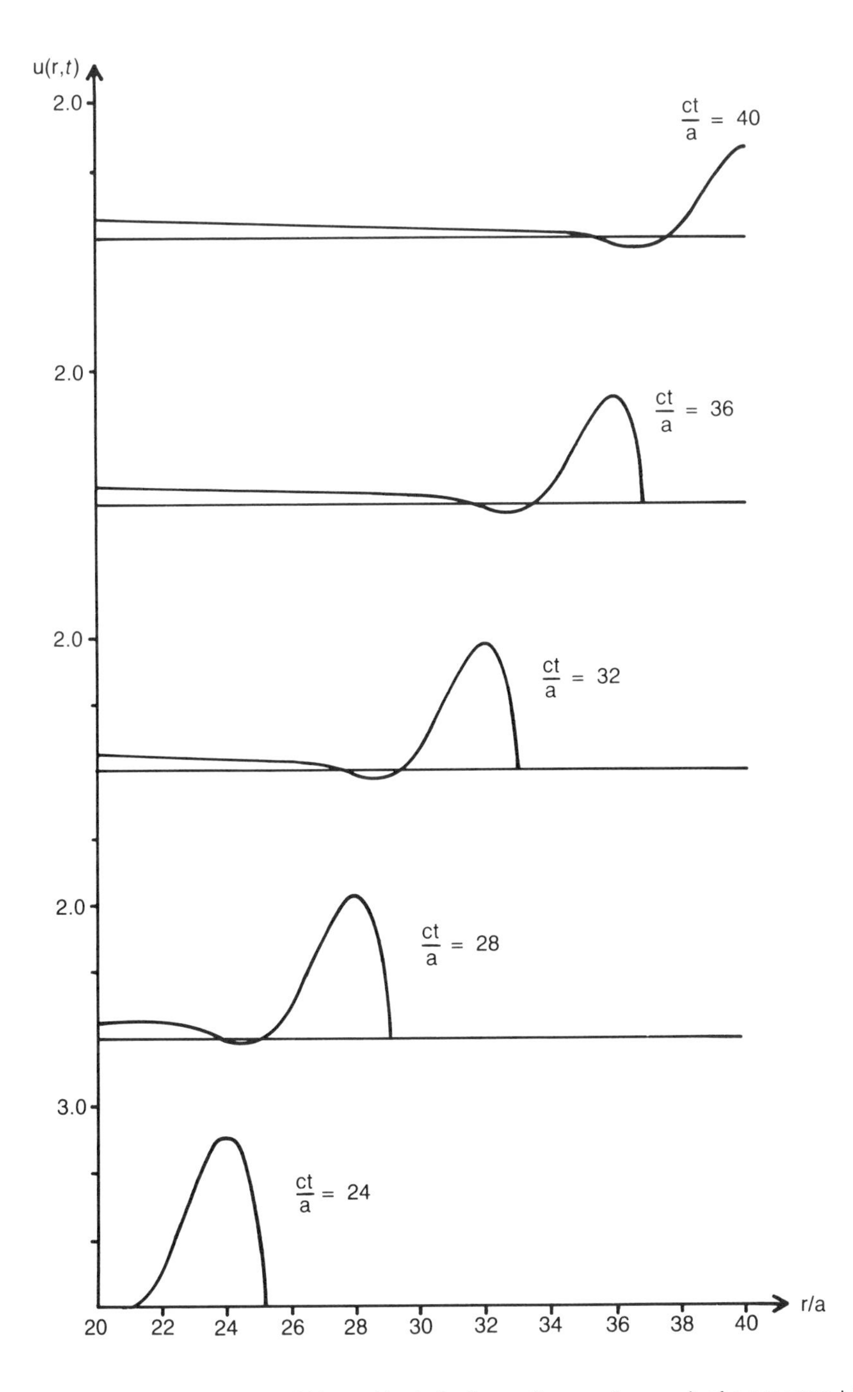

Fig. 5.17.1: The displacement within an ideal elastic medium as the result of a pressure blast within a spherical cavity of a radius a at time t = 0.

$$\phi(r,t) = \frac{ap_0/\varrho r}{(\beta/\sqrt{2} - \alpha)^2 + \beta^2} \left\{ -e^{-\alpha\tau} + e^{-\alpha\tau/\sqrt{2}} \left[\left(\frac{1}{\sqrt{2}} - \frac{\alpha}{\beta}\right)\sin(\beta\tau) + \cos(\beta\tau) \right] \right\} H(\tau)$$

(5.17.44)

where

$$\tau = t - \frac{r-a}{c}$$

and

$$\beta = \frac{2\sqrt{2}\,c}{3a}$$

and

$$u(r,t) = -\frac{\phi(r,t)}{r} + \frac{p_0/\varrho r}{(\beta/\sqrt{2} - \alpha)^2 + \beta^2} \left\{ -\frac{a\alpha}{c} e^{-\alpha\tau} + e^{-\alpha\tau/\sqrt{2}} \left[\left(\sqrt{2} - \frac{\alpha a}{\sqrt{2}\,c}\right) \sin(\beta\tau) + \frac{a\alpha}{c}\right) \cos(\beta\tau) \right] \right\}$$

(5.17.45)

In Fig. 5.17.1 the displacement $u(r,t)$ far away from the spherical cavity is presented at various times as a function of r/a. The limit of $\alpha \to 0$ (a step function) has been selected when $ap_0/4\mu = 100\ m$.

5.18 THE TWO-DIMENSIONAL WAVE EQUATION: THE DYNAMIC WATER PRESSURE ON THE FACE OF A DAM DURING AN EARTHQUAKE

Since the earliest days of dam construction, designing dams to withstand the abnormal stresses placed on them during an earthquake has been of utmost importance to civil engineers. The complexity of the coupled dam-water-soil system forced engineers in the first half of the century to consider separately the vibrations generated within a dam during an earthquake from those within the reservoir.

In 1933, Westergaard (Westergaard, E. M., 1933: Water pressure on dams during earthquakes. *Trans. Amer. Soc. Civ. Eng.*, *98*, 418-423.) published the first paper to consider the pressure on the face of a dam caused by waves generated within a reservoir during an earthquake. Although this paper is essentially of academic interest today, it provided, in conjunction with later papers, guidance in designing dams for several decades. During the mathematical analysis of this problem, it was necessary to solve the two-dimensional wave equation. Consequently we shall use this problem to illustrate the mathematical techniques used in solving the two-dimensional wave equation on a finite domain. This analysis was first done by Werner, P. M. and K. J. Sundquist, 1949: On hydrodynamic earthquake effects. *Trans. Amer. Geophys., 30,* 636-657.

The derivation of the governing equation begins with the linearized momentum equations for a frictionless fluid in the x and y directions:

$$\frac{\partial u}{\partial t} = -\frac{1}{\varrho_0}\frac{\partial p}{\partial x} \tag{5.18.1}$$

$$\frac{\partial v}{\partial t} = -\frac{1}{\varrho_0}\frac{\partial p}{\partial y} \tag{5.18.2}$$

where x points along the horizon and y downwards. The corresponding Eulerian velocities are u and v. The constant ϱ_0 denotes of the fluid's density when at rest: p the portion of the pressure field that arises from the oscillations.

To complete the set of equations, we combine the continuity and equation of state for water together to give

$$\frac{\partial u}{\partial x} + \frac{\partial v}{\partial y} = -\frac{1}{gE}\frac{\partial p}{\partial t} \tag{5.18.3}$$

where E is the compressibility of water (2.1×10^8 kg/m^2) and g the gravitational constant. Equations (5.18.1)-(5.18.3) can be combined together to yield

$$c^2\left(\frac{\partial^2 p}{\partial x^2} + \frac{\partial^2 p}{\partial y^2}\right) = \frac{\partial^2 p}{\partial t^2} \tag{5.18.4}$$

where $c^2 = gE/\varrho_0 = 2 \times 10^6$ m^2/sec^2. Equation (5.18.4) is the two-dimensional wave equation.

Turning to the boundary conditions, we must have

$$p(x,0,t) = 0 \tag{5.18.5}$$

along the free surface (the surface dividing the water from air). The pressure must be continuous across the free surface because Eq. (5.18.1) and (5.18.2) would predict an infinite acceleration otherwise. Because the pressure perturbation within the air is zero since there is no motion occurring there, $p(x,0,t)$ must be zero.

At the bottom, the velocity must be zero

$$v(x,h,t) = 0 \tag{5.18.6}$$

or

$$p_y(x,h,t) = 0 \tag{5.18.7}$$

from Eq. (5.18.2). In the horizontal direction, the reservoir is forced at the left wall by an acceleration

$$u_t(0,y,t) = \alpha g \cos(\omega t) \quad \textbf{(5.18.8)}$$

or

$$p_x(0,y,t) = -\alpha g \cos(\omega t) \quad \textbf{(5.18.9)}$$

caused by an earthquake. At the other end, we assume that a fault line has separated the actual reservoir from the right wall so that the right wall does not move. Consequently, the boundary condition at $x = L$ is

$$u_t(L,y,t) = 0 \quad \textbf{(5.18.10)}$$

or

$$p_x(L,y,t) = 0 \quad \textbf{(5.18.11)}$$

The technique that we shall use to solve Eq. (5.18.4) with boundary conditions (5.18.5), (5.18.7), (5.18.9) and (5.18.11) is separation of variables. Because of the boundary condition (5.18.9), we look for solutions to the wave equation of form

$$p(x,y,t) = F(x,y) \cos(\omega t). \quad \textbf{(5.18.12)}$$

Then Eq. (5.18.4) becomes

$$\frac{\partial^2 F}{\partial x^2} + \frac{\partial^2 F}{\partial y^2} + \frac{\omega^2}{c^2} F = 0 \quad \textbf{(5.18.13)}$$

where Eq. (5.18.13) is often called the two-dimensional *Helmholtz equation* or the reduced wave equation. The boundary conditions then become

$$F(x,0) = 0 \quad \textbf{(5.18.14)}$$

$$F(x,h) = 0 \quad \textbf{(5.18.15)}$$

$$F_x(L,y) = 0 \quad \textbf{(5.18.16)}$$

and

$$F_x(0,y) = -\alpha g \varrho_0. \quad \textbf{(5.18.17)}$$

To solve Helmholtz's equation, we again apply separation of variables

$$F(x,y) = X(x)Y(y). \tag{5.18.18}$$

Substituting into Eq. (5.18.13), we have

$$\frac{X''}{X} + \frac{Y''}{Y} + \frac{\omega^2}{c^2} = 0 \tag{5.18.19}$$

with the boundary conditions

$$Y(0) = Y'(h) = 0 \tag{5.18.20}$$

and

$$X'(L) = 0. \tag{5.18.21}$$

As in the case of the one-dimensional wave equation, the ratio X''/X and Y''/Y must be constant

$$\frac{X''}{X} = -k^2 \tag{5.18.22}$$

and

$$\frac{Y''}{Y} = -\ell^2 \tag{5.18.23}$$

Turning to Eq. (5.18.23) first, the solution to Eq. (5.18.23) with the boundary conditions (5.18.20) is

$$Y(y) = \sin\left[\frac{(2n+1)\pi y}{2h}\right] \tag{5.18.24}$$

with $n = 0, 1, 2, 3, \ldots$

For the solution in the x direction, we assume that

$$k_n^2 = \frac{\omega^2}{c^2} - \frac{(2n+1)^2\pi^2}{4h^2} > 0. \tag{5.18.25}$$

For most parameters of practical interest, this is true. The solution under this assumption of Eq. (5.18.22) is

$$X_n(x) = A_n \cos[k_n(L-x)] + B_n \sin[k_n (L-x)]. \tag{5.18.26}$$

Equation (5.18.26) is written in this particular form for ease of future computations.

From boundary condition (5.18.21), we have $B_n = 0$ and

$$F(x,y) = \sum_{n=0}^{\infty} A_n \cos[k_n(L-x)] \sin\left[\frac{(2n+1)\pi y}{2h}\right]. \quad (5.18.27)$$

To determine A_n, we note that

$$F_x(0,y) = -\alpha g \varrho_0 = -\sum_{n=0}^{\infty} A_n k_n \sin(k_n L) \sin\left[\frac{(2n+1)\pi y}{2h}\right]. \quad (5.18.28)$$

Equation (5.18.28) is an eigenfunction expansion in the eigenfunctions $\sin\left[\frac{(2n+1)\pi y}{2h}\right]$. Therefore, the coefficient A_n equals

$$k_n \sin(k_n L)\, A_n = \frac{\int_0^h \alpha g \varrho_0 \sin\left(\frac{(2n+1)\pi y}{2h}\right) dy}{\int_0^h \sin^2\left(\frac{(2n+1)\pi y}{2h}\right) dy} \quad (5.18.29)$$

$$= \frac{4\alpha g \varrho_0}{\pi\,(2n+1)} \quad (5.18.30)$$

Therefore,

$$p(x,y,t) = \frac{4\alpha g \pi \varrho_o}{\pi} \cos(\omega t) \sum_{n=0}^{\infty} \frac{\cos[k_n(L-x)] \sin\left[\frac{(2n+1)\pi y}{2h}\right]}{k_n \sin(k_n L)\,(2n+1)}. \quad (5.18.31)$$

or

$$p(x,y,t) = \frac{\alpha g \varrho_0}{2} \sum_{n=0}^{\infty} \frac{1}{k_n \sin(k_n L)\,(2n+1)} (①+②+③+④)$$

$$① = \sin\left[k_n(x-L) + \frac{(2n+1)\pi y}{2h} + \omega t\right]$$

$$② = -\sin\left[k_n(x-L) + \frac{(2n+1)\pi y}{2h} - \omega t\right]$$

$$③ = -\sin\left[k_n(x-L) - \frac{(2n+1)\pi y}{2h} + \omega t\right]$$

$$④ = \sin\left[k_n(x-L) - \frac{(2n+1)\pi y}{2h} - \omega t\right] \qquad \textbf{(5.18.32)}$$

The analogy between Eq. (5.18.32) and Eq. (5.3.28) is very striking. However, from Eq. (5.18.32) we see that there are waves traveling in many more than just two directions. These waves travel in several directions at once, changing their shape as they propagate.

A particularly interesting case occurs when $\sin(k_n L) = 0$. This condition implies resonance and it occurs whenever

$$k_n L = m\pi, \qquad \textbf{(5.18.33)}$$

$m = 0, 1, 2, 3, \ldots$ and resonance will occur if the reservoir has certain critical depths:

$$h_{n,m} = \frac{(2n+1)\pi}{2\left(\frac{\omega^2}{c^2} - \frac{m^2\pi^2}{L^2}\right)^{1/2}} \qquad \textbf{(5.18.34)}$$

For typical values of the parameters in Eq. (5.18.34), the depth of the reservoir must be fairly deep.

In our analysis of the hydrodynamic pressures due to earthquake motions, we have assumed that the dam is rigid. However, the infinite pressures on the upstream face of the dam that we predict at resonance will cause the dam to deform, which in turn will affect the hydrodynamic pressures. When the response of the dam is included (Chakrabarti, P. and A. K. Chopra, 1973: Earthquake analysis of gravity dams including hydrodynamic interactions. *Earthquake Engineering and Structural Dynamics, 2,* 143-160.), resonance is still found to occur although the infinite pressure predicted earlier does not. More recently, however, it has been shown that the unbounded resonant peaks for hydrodynamic forces on a rigid dam are eliminated by including the absorbency of the reservoir bottom (see Fenves, G. and A. K. Chopra, 1983: Effects of reservoir bottom absorption. *Earthquake Engineering and Structural Dynamics, 11,* 807-829.).

In this section we have only found the forced modes driven by an earthquake. The complete transient solution from a state of rest for an infinite reservoir was given by Chopra, A. K., 1967:

Hydrodynamic pressures on dams during earthquakes. *J. Engng. Mech. Div., ASCE, 93,* 205-223.

Exercises

1. Solve the Helmholtz equation

$$u_{xx} + u_{yy} + u = 0$$

when

$$u(0,y) = u(a,y) = u(x,0) = u(x,b) = 0.$$

2. The three-dimensional wave equation is

$$u_{tt} = c^2(u_{xx} + u_{yy} + u_{zz})$$

where c is the constant phase speed. In the case of sound wave in a cubical cavity of edge length d, $u(x,y,z,t)$ represents density. Assuming that the solution can be written as

$$u(x,y,z,t) = X(x)Y(y)Z(z)\cos(\omega t),$$

find the solution if the gradient of u normal to a fixed boundary must be zero.

5.19 SOMMERFELD'S RADIATION CONDITION

In the problems that we have worked in this chapter, we have always chosen solutions where the waves propagate away from their source of generation. In many complicated problems in an unbounded domain, such as electromagnetic scattering problems, choosing this physical solution from several possible solutions is often difficult. To formulate a method to circumvent this difficulty, consider two independent and radially symmetric solutions of the wave equation:

$$u_1 = \cos[\omega(t + \frac{r}{c})]/r \qquad \textbf{(5.19.1)}$$

and

$$u_2 = \cos[\omega(t - \frac{r}{c})]/r \qquad \textbf{(5.19.2)}$$

Clearly u_1 represents a disturbance propagating in the direction of decreasing r while u_2 represents a disturbance propagating outwards. Although both solutions might be necessary in the domain near the origin to satisfy the initial condition (as in the case of the bursting balloon problem), only solutions which propagate outwards can be present as $r \to \infty$. There is no source of energy at infinity to generate inward propagating disturbances.

These physical considerations suggest that the unique solution could be found by examining the behavior of the general solution as $r \to \infty$ and then choosing that solution which corresponds to the divergent wave. This was indeed the method used by Sommerfeld (Sommerfeld, A., 1909: Über die Ausbreitung der Wellen in der draftlosen Telegraphie. *Ann. Physik, 28,* 665-736.) who first annunciated this condition in his classic paper on electromagnetic waves generated by a Hertzian dipole near a flat, conducting earth.

Although Sommerfeld used this technique successfully, applying it to other problems may be more difficult. A far better approach is based upon the concept of group velocity and the fact that energy must propagate away from its source of generation. To derive this criteria, we consider the solutions to the two-dimensional wave equation:

$$(u_{xx} + u_{zz})_{tt} + N^2 u_{xx} = 0. \quad \textbf{(5.19.3)}$$

This equation arises in the study of small-amplitude waves in a continuous stratified, incompressible fluid. (See Gill, A. E., 1982: *Atmosphere-Ocean Dynamics*. New York: Academic Press. Section 6.4 for its derivation.) and the resulting wave motions are referred to as internal-gravity waves. We wish to find the solutions when the fluid is forced at $z = 0$ in such a manner that

$$u(x,0,t) = \sin(kx) \sin(\omega t). \quad \textbf{(5.19.4)}$$

Because we are seeking outward propagating waves, we look for solutions of the form

$$u(x,z,t) = A \sin(mz - \omega t) \sin(kx). \quad \textbf{(5.19.5)}$$

Substituting Eq. (5.19.5) into Eq. (5.19.3), we see that for every ω, there are two possible m's:

$$m = \pm k \left(\frac{N^2}{\omega^2} - 1\right)^{1/2} \quad \textbf{(5.19.6)}$$

if $N^2 > \omega^2$. (If $N^2 < \omega^2$, then the solution in the vertical direction is a decaying exponential and we are not interested in these *evanescent waves*.) Consequently, the most general solution is

$$u(x,z,t) = (A \sin(mz - \omega t) + B \sin(mz + \omega t)) \sin(kx). \quad \textbf{(5.19.7)}$$

The first term in Eq. (5.19.7) corresponds to the positive root of Eq. (5.19.7) while the latter one, to the negative root. We can also express Eq. (5.19.7) as

$$u(x,z,t) = - A/2 \cos(mz + kx - \omega t) + A/2 \cos(mz - kx - \omega t) - B/2 \cos(mz + kx + \omega t) + B/2 \cos(mz - kx + \omega t) \quad \textbf{(5.19.8)}$$

In this particular formulation the first two terms represent outward propagating waves while the remaining terms are inward propagating. However, these waves are not propagating at angles perpendicular to the x axis but at some angle between the x and z axis. This can be illustrated graphically by a wave-number vector, which points in the direction that the ridges and troughs are propagating and we have done so for the outwarding propagating waves in Fig. 5.19.1. In the case of one-dimensional waves, we introduced the concept of phase speed

$$c = k\omega \quad \textbf{(5.19.9)}$$

where ω is frequency. We can extend this concept by defining *phase velocity*

$$\mathbf{c} = \mathbf{k}\omega \quad \textbf{(5.19.10)}$$

where k is the wave-number vector (k,ℓ,m). Consequently, phase velocity points in the direction of propagation of the ridges and troughs.

In Section 5.17 we showed that in dispersive wave systems the energy of a group of waves does not necessarily propagate with the phase velocity of the average wave-number but with the group velocity. If the analysis given in that section is generalized to include waves in three dimensions, it is found that the direction in which the energy of a group of three-dimensional waves propagates

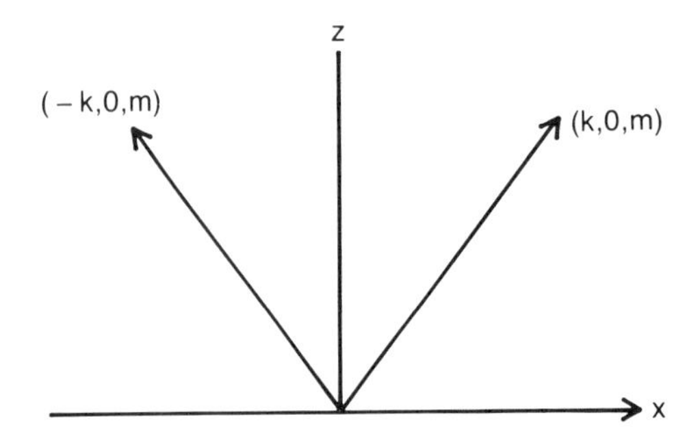

Fig. 5.19.1: Wave number diagram.

is given by the vectorial group velocity:

$$\mathbf{c_g} = \left(\frac{\partial \omega}{\partial k}, \frac{\partial \omega}{\partial \ell}, \frac{\partial \omega}{\partial m}\right) \tag{5.19.11}$$

In general, the direction of the phase velocity and group velocity are not the same. Consequently, Sommerfeld's radiation condition requires that the group velocity must point outward regardless of the direction of the phase velocity.

In our case the vertical component of the group velocity is

$$\frac{\partial \omega}{\partial m} = \frac{\partial}{\partial m}\left(\frac{Nk}{\sqrt{k^2 + m^2}}\right) = -\frac{Nkm}{\sqrt{(k^2 + m^2)^3}}. \tag{5.19.12}$$

Because k is positive by definition, we come to the startling conclusion that we must take the solution corresponding to negative m in order for the wave energy to propagate outward. Consequently, $A = 0$ in order for the radiation boundary condition to be satisfied.

Exercise

Given the partial differential equation

$$(u_{xx} + u_{zz})_t + (u_{xx} + u_{zz})_x + u_x = 0$$

show that the frequency is given by

$$\omega = k - \frac{1}{m^2 + k^2}$$

if

$$u(x,z,t) = Re[U(z) \exp(ikx - i\omega t)]$$

$$k > \omega$$

$$U(z) = \exp(\pm\ imz)$$

Then show that the radiation condition is satisfied (i.e., the group velocity points upward) if the positive value of m is taken.

Answers

P. 254

1. $u(x,t) = \dfrac{4L}{c\pi^2} \displaystyle\sum_{n=1}^{\infty} \frac{1}{(2n-1)^2} \sin\left[\frac{(2n-1)\pi x}{L}\right] \sin\left[\frac{(2n-1)\pi ct}{L}\right]$

2. $u(x,t) = \frac{4}{\pi} \sum_{n=1}^{\infty} \frac{1}{(2n-1)} \sin\left[\frac{(2n-1)\pi x}{L}\right] \cos\left[\frac{(2n-1)\pi ct}{L}\right]$

3. $u(x,t) = \frac{9h}{\pi^2} \sum_{n=1}^{\infty} \frac{1}{n^2} \sin\left(\frac{2n\pi}{3}\right)\sin\left(\frac{n\pi x}{L}\right)\cos\left(\frac{n\pi ct}{L}\right)$

4. $u(x,t) = \frac{3}{4} \sin\left(\frac{\pi x}{L}\right)\cos\left(\frac{\pi ct}{L}\right) - \frac{1}{4} \sin\left(\frac{3\pi x}{L}\right)\cos\left(\frac{3\pi ct}{L}\right)$

5. $u(x,t) = \sin\left(\frac{\pi x}{L}\right) \cos\left(\frac{\pi ct}{L}\right)$

$$+ \frac{4aL}{\pi^2 c} \sum_{n=1}^{\infty} \frac{(-1)^n}{(2n-1)^2} \sin\left[\frac{(2n-1)\pi}{4}\right] \sin\left(\frac{n\pi x}{L}\right) \sin\left(\frac{n\pi ct}{L}\right)$$

6. $u(x,t) = \frac{4L}{\pi^3} \sum_{n=0}^{\infty} \frac{(-1)^n}{(2n+1)^3} \sin\left[\frac{(2n+1)\pi x}{L}\right] \sin\left[\frac{(2n+1)\pi ct}{L}\right]$

7. $u(x,t) = \frac{4L}{\pi^2} \sum_{n=0}^{\infty} \frac{(-1)^n}{(2n+1)^2} \sin\left[\frac{(2n+1)\pi x}{L}\right] \cos\left[\frac{(2n+1)\pi ct}{L}\right]$

8. $u(x,t) = 1 + 1/4\left[\frac{3}{c}\cos(x)\sin(ct) + \frac{1}{3c}\cos(3x)\sin(3ct)\right]$

9. $u(x,t) = \sum_{n=1}^{\infty}\left[A_n \cos\left(\frac{n\pi ct}{a}\right) + B_n \sin\left(\frac{n\pi ct}{a}\right)\right] \sin\left(\frac{n\pi x}{a}\right)$

10. $u(x,t) = \frac{3\ 2^{1/2}}{8} \sin(x)\sin(2^{1/2}t) - \frac{10^{1/2}}{40} \sin(3x)\sin(10^{1/2}\ t)$

11. $u(x,t) = \sin(t) + \frac{3\ 2^{1/2}}{8}\cos(x)\sin(2^{1/2}t)$

$$+ \frac{10^{1/2}}{40}\cos(3x)\sin(10^{1/2}\ t)$$

12. $u(x,t) = e^{-x}\left[2\sin(2\pi x)\cos[(4\pi^2+1)^{1/2}t]\right.$

$\left.- 3\sin(5\pi x)\cos[(25\pi^2+1)^{1/2}t)]\right]$

13. $u(x,t) = \frac{3}{4}\, t\, e^{-t/2}\sin\left(\frac{x}{2}\right) - \frac{2^{1/2}}{8}\, e^{-t/2}\sin(2^{1/2}t)\sin\left(\frac{3x}{2}\right)$

14. $u_n(x,t) = \sin[(2n-1)\pi\, \ell n(x)/2]\sin[(2n-1)\pi t/2]$

15. $u_n(x,t) = \exp(-x-t)\cos[(n^2+1)^{1/2}t]\sin(nx)$

16. $u_n(x,t) = x^{1/2}\sin[n\pi\, \ell n(x)/\ell n(2)]\sin[(\frac{n^2\pi^2}{\ell n(x)^2} + 1/4)^{1/2}t]$

17. $u_n(x,t) = (1 + t^{1/2})\sin[n\pi\, \ell n(1+t)/\ell n(2)]$

$\times \sin[n\pi\, \ell n(x)/\ell n(2)]$

P. 274

1. $u(x,t) = \sin(2x)\cos(4ct) + \frac{1}{2c}\cos(x)\sin(2ct)$

2. $u(x,t) = x\sin(x)\cos(3ct) + 3ct\cos(x)\sin(3ct)$

$+ \frac{1}{6c}\cos(2x)\sin(6ct)$

3. $u(x,t) = \frac{1 + x^2 + (ct)^2}{[1 + (x+ct)^2][1 + (x-ct)^2]} + \frac{1}{c}e^x\sinh(ct)$

4. $u(x,t) = e^{-x}\cosh(ct) + \frac{1}{2c}[\tan^{-1}(x+ct) - \tan^{-1}(x-ct)]$

5. $u(x,t) = \cos(\pi x/2)\cos(\pi ct/2) + \frac{1}{ac}\sinh(ax)\sinh(act)$

6. $u(x,t) = \sin(3x)\cos(3ct) + \frac{1}{2c}\sin(2x)\sin(2ct)$

$-\frac{1}{c}\sin(x)\sin(ct)$

P. 297

1. $u(x,t) = \sin(\pi x)\cos(\pi t) - \frac{1}{\pi}\sin(\pi x)\sin(\pi t)$

2. $u(x,t) = \frac{16}{\pi^3}\sum_{n=1}^{\infty}\frac{(-1)^n}{(2n-1)^3}\sin\left[\frac{(2n-1)\pi x}{2}\right]\sin\left[\frac{(2n-1)\pi t}{2}\right]$

3. $u(x,t) = 4\sum_{n=1}^{\infty}\frac{1}{(2n-1)^2}\sin\,[(2n-1)x]\sin[(2n-1)t]$

4. $u(x,t) = xt + [\sinh(t) - t]e^{-x}$
$+ H(t-x)\,[t - x - 1 - \sinh(t)e^{-x} - e^{-t}\cosh(x)]$

5. $u(x,t) = 1 + x\,(t - 1 + e^{-t}) + H\,(t-x)\,(\cos(t-x) - 1)$

6. $u(x,t) = H(t-x)\,(t-x) - (t-2x)\,H(t-2x)$

7. $u(x,t) = \frac{\sin(\omega(a-x)/c)}{\sin(\omega a/c)}\sin(\omega t)$
$+ \frac{2\omega a}{c}\sum_{n=1}^{\infty}\frac{\sin(n\pi x/a)\sin(n\pi ct/a)}{(\omega a/c)^2 - n^2\pi^2}$

or

$$= \frac{2k}{\pi}\sum_{\substack{n=1\\ n\neq k}}^{\infty}\frac{\sin(n\pi x/a)\,\sin(n\pi ct/a)}{k^2 - n^2}$$

$$+ \left[\left(\frac{1}{2k\pi} - \frac{ct}{a}\right)\sin\left(\frac{n\pi x}{a}\right) + \left(1 - \frac{x}{a}\right)\cos\left(\frac{n\pi x}{a}\right)\right]\sin\left(\frac{k\pi ct}{a}\right)$$

when $\omega a/c = k\pi$.

P. 314

1. $u(x,y) = \sum_{n=0}^{\infty}\sum_{m=0}^{\infty}A_{nm}\sin\left(\frac{n\pi x}{a}\right)\sin\left(\frac{n\pi y}{b}\right)$

with the condition that

$$\frac{n^2\pi^2}{a^2} + \frac{m^2\pi^2}{b^2} = 1.$$

2. $$u(x,y,z,t) = \sum_{k=0}^{\infty} \sum_{n=0}^{\infty} \sum_{m=0}^{\infty} A_{knm} \cos\left(\frac{k\pi x}{d}\right) \times \cos\left(\frac{n\pi x}{d}\right) \cos\left(\frac{m\pi z}{d}\right) \cos(\omega t)$$

with the condition

$$\frac{\omega^2}{c^2} = \frac{k^2\pi^2}{d^2} + \frac{n^2\pi^2}{d^2} + \frac{m^2\pi^2}{d^2}.$$

Chapter 6

The Heat Equation

IN THIS CHAPTER WE SHALL BE CONCERNED WITH SOME problems dealing with the linear parabolic partial differential equation

$$u_t - a^2 u_{xx} = F(x,t)$$

in two independent variables x and t. This equation, known as the one-dimensional heat equation, arises in the conduction of heat in solids as well as in a variety of diffusive phenomena. The heat equation is similar to the wave equation in that one of the variables in a practical context is usually time. Consequently, both equations are equations of evolution. However, unlike the wave equation, the heat equation is not "conservative" because if the sign of t is reversed, we obtain a different sort of solution. This reflects the fact that entropy is continually increasing as time goes on, whereas for the wave equation the energy remains constant (unless friction is included).

6.1 THE DERIVATION OF THE HEAT EQUATION

To derive the heat equation, consider a heat-conducting homogeneous rod, extending from $x = 0$ to $x = L$ along the x-axis (See Fig. 6.1.1). The rod has uniform cross section A and constant

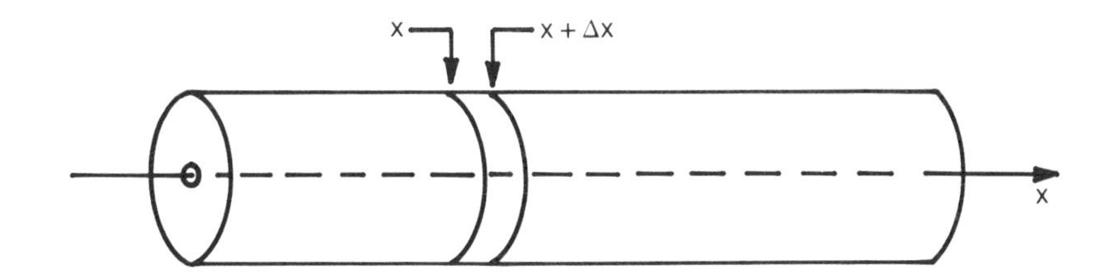

Fig. 6.1.1: Heat conduction in a thin bar.

density ϱ. Furthermore, the rod is insulated laterally so that heat flows only in the x-direction. The rod is sufficiently thin so that the temperature at all points on a cross section is constant. Let $u(x,t)$ denote the temperature of the cross section at the point x at any instant of time t, and let c denote the specific heat of the rod (that is, the amount of heat required to raise the temperature of a unit mass of the rod by one degree). In the segment of the rod between the cross section at x and the cross section at $x + \Delta x$, the amount of heat in this segment is

$$Q(t) = \int_{x}^{x+\Delta x} c\, \varrho A\; u(s,t)\; ds. \qquad \textbf{(6.1.1)}$$

On the other hand, the rate at which heat flows into the segment across the cross section at x is known to be proportional to the cross section and the gradient of the temperature of the cross section (Fourier's law of heat conduction); that is, it is equal to

$$-\; \varkappa A\; u_x(x,t) \qquad \textbf{(6.1.2)}$$

where $\varkappa$ denotes the thermal conductivity of the rod. The negative sign in Eq. (6.1.2) indicates that heat flows in the direction of decreasing temperature. Similarly, the rate at which heat flows out of the segment through the cross section at $x + \Delta x$ is given by

$$-\; \varkappa A\; u_x(x + \Delta x, t) \qquad \textbf{(6.1.3)}$$

The difference between the amount of heat that flows in through the cross section at x and the amount of heat that flows out through the cross section at $x + \Delta x$ must equal the change in the heat content of the segment $x \leq s \leq x + \Delta x$. Hence, by subtracting Eq. (6.1.3) from (6.1.2) and equating the result to the time derivative of Eq. (6.1.1), we obtain

$$Q_t = \int_x^{x+\Delta x} c\varrho A\, u_t(s,t)\, ds \tag{6.1.4}$$

$$= \varkappa A\, [u_x(x+\Delta x,t) - u_x(x,t)] \tag{6.1.5}$$

We assume that the integrand in Eq. (6.1.4) is a continuous function of s. Then, by the mean value theorem for integrals, we have

$$\int_x^{x+\Delta x} u_t(s,t)\, ds = u_t(\xi,t)\Delta x \quad x<\xi<x+\Delta x \tag{6.1.6}$$

so that Eq. (6.1.5) becomes

$$c\varrho\, u_t(\xi,t)\, \Delta x = \varkappa[\, u_x(x+\Delta x,t) - u_x(x,t)\,]. \tag{6.1.7}$$

Dividing both sides of this equation by $c\varrho\, \Delta x$ and taking the limit as $\Delta x \to 0$, we finally obtain

$$u_t(x,t) = a^2\, u_{xx}(x,t) \tag{6.1.8}$$

or

$$u_t(x,t) - a^2\, u_{xx}(x,t) = 0 \tag{6.1.9}$$

with $a^2 = \varkappa/(c\varrho)$. Equation (6.1.7) is called the one-dimensional heat equation. The constant a^2 is called *diffusivity*.

If heat is supplied to the rod from an external source at a rate $f(x,t)$ per unit volume per unit time, then the term $\int_x^{x+\Delta x} f(s,t)\, ds$ must be added to the right-hand side of Eq. (6.1.4). Thus, in passing to the limit $\Delta x \to 0$, we get

$$u_t(x,t) - a^2\, u_{xx}(x,t) = F(x,t) \tag{6.1.10}$$

where $F(x,t) = f(x,t)/\,(c\varrho)$ is the source density. This equation is called the nonhomogeneous heat equation.

6.2 INITIAL AND BOUNDARY CONDITIONS

In the case of heat conduction in a thin rod, the temperature function $u(x,t)$ must satisfy the heat equation (6.1.9) or (6.1.10). From physical considerations, we know that the differential equation alone cannot determine the temperature distribution in the rod at any subsequent time. We need to have additional information regarding the initial temperature of the rod and the conditions imposed at its two ends. We must specify $u(x,t)$ at some initial time (say, at time $t = 0$) and describe how the two ends of the rod exchange heat energy with the surrounding medium. If (1) there is no heat source, (2) the temperature of the rod at $t = 0$ is described by the function

$f(x)$, $0 \leq x \leq L$, and (3) the two ends are maintained at zero temperature for all time, then the temperature distribution $u(x,t)$ in the rod at any later time $t>0$ is found by solving the differential equation

$$u_t = a^2 u_{xx} \qquad 0<x<L,\ t>0 \tag{6.2.1}$$

subject to the conditions

$$u(x,0) = f(x) \qquad 0 \leq x \leq L \tag{6.2.2}$$

and

$$u(0,t) = u(L,t) = 0 \qquad t \geq 0 \tag{6.2.3}$$

This problem is called an initial-boundary value problem for the heat equation. The auxiliary conditions (6.2.2) and (6.2.3) are called the initial condition and boundary conditions, respectively.

We notice in the problem (6.2.1)-(6.2.3) that only the value of $u(x,t)$ is prescribed initially and not both $u(x,t)$ and $u_t(x,t)$, as in the case of the wave equation. This is, of course, dictated by the physical problem being considered. But even from a mathematical standpoint, u_t cannot be prescribed arbitrarily, because it is related to u_{xx} through the differential equation.

The condition (6.2.3) is an example of a "Dirichlet condition" or "condition of the first kind." Dirichlet boundary conditions give the value of the solution (which is not necessarily equal to zero) along a boundary. If both ends of the rod are insulated so that no heat flows from the ends, then according to Eq. (6.1.2) the boundary condition assumes the form

$$u_x(0,t) = u_x(L,t) = 0. \qquad t \geq 0 \tag{6.2.4}$$

This is an example of a "Neumann condition" or "condition of the second kind." A Neumann condition gives the value of the normal derivative (which may not be equal to zero) of the solution along the boundary. Finally, if there is radiation of heat from the ends of the rod into the surrounding medium, we shall show that the boundary condition is of the form

$$u_x(0,t) - h\, u(0,t) = \text{a constant} \tag{6.2.5}$$

$$u_x(L,t) + h\, u(L,t) = \text{another constant} \tag{6.2.6}$$

for $t \geq 0$, where h is a constant. This is an example of a "condition of the third kind" or "Robin's condition" and is a linear mixture of a Dirichlet and Neumann conditions. Notice that all these boundary conditions are linear.

Exercises

1. Verify that each of the given functions satisfies the heat equation (6.1.8) for $0<x<\pi$ and the accompanying boundary and initial conditions.

(a) $u(x,t) = \exp(-a^2t)\sin(x);\ u(x,0) = \sin(x),$
$u(0,t) = u(\pi,t) = 0.$

(b) $u(x,t) = \exp(-a^2t)\cos(x);\ u(x,0) = \cos(x),$
$u_x(0,t) = u_x(\pi,t) = 0.$

(c) $u(x,t) = 1/2 + 1/2\exp(-4a^2t)\cos(2x);\ u(x,0) = \cos^2(x),$
$u_x(0,t) = u_x(\pi,0) = 0.$

2. Verify that the function

$$u(x,t) = t^{-1/2}\exp(-x^2/4t)$$

satisfies the heat equation (5.5) for $a = 1$, $t>0$ and that

$$\lim_{t\to 0+} u(x,t) = 0$$

provided $x \neq 0$.

6.3 THE SOLUTION OF THE HEAT EQUATION BY SEPARATION OF VARIABLES

We now consider the problem of finding a solution of the homogeneous heat equation

$$u_t = a^2 u_{xx} \qquad 0\le x\le L,\ t>0 \tag{6.3.1}$$

which satisfies the initial condition

$$u(x,0) = f(x) \qquad 0\le x\le L \tag{6.3.2}$$

and the boundary conditions

$$u(0,t) = u(L,t) = 0 \quad t\ge 0. \tag{6.3.3}$$

We shall solve this problem by the method of separation of variables presented in Section 5.3. Accordingly we seek particular solutions of Eq. (6.3.1) in the form

$$u(x,t) = X(x)\,T(t) \tag{6.3.4}$$

which satisfy the boundary conditions (6.3.3). Because

$$u_t = X(x)\,T'(t) \tag{6.3.5}$$

and

$$u_{xx} = T(t)\, X''(x), \tag{6.3.6}$$

Equation (6.3.1) becomes

$$T'(t)\, X(x) = a^2\, X''(x) T(t). \tag{6.3.7}$$

Separating the variables yields

$$\frac{T'}{a^2 T} = \frac{X''}{X} = -\, k^2 \tag{6.3.8}$$

This leads to the two ordinary differential equations

$$X'' + k^2 X = 0 \tag{6.3.9}$$

and

$$T' + k^2 a^2\, T = 0 \tag{6.3.10}$$

for the functions $X(x)$ and $T(t)$, respectively. For the present moment, let us assume that $k^2 > 0$.

Because $u(0,t) = X(0)T(t) = 0$ and $u(L,t) = X(L)\, T(t) = 0$, we must have $X(0) = X(L) = 0$ if $T(t) \neq 0$. Thus, the function $X(x)$ is determined by solving the regular Sturm-Liouville problem

$$X'' + k^2 X = 0 \qquad 0 \leq x \leq L \tag{6.3.11}$$

with

$$X(0) = X(L) = 0.$$

This is the same problem that we encountered in Chapter 5. We found that

$$k_n = \frac{n\pi}{L},\; n = 1,2,3,4,\; .\;.\;. \tag{6.3.12}$$

and the eigenfunctions were

$$X_n(x) = \sin\left(\frac{n\pi x}{L}\right) \tag{6.3.13}$$

Now, for each k_n, $n \geq 1$, a solution of the equation (6.3.10) is given by

$$T_n(t) = B_n \exp(-a^2 k_n^2 t). \qquad \textbf{(6.3.14)}$$

Thus, the functions

$$u_n(x,t) = B_n \exp(-a^2 k_n^2 t) \sin\left(\frac{n\pi x}{L}\right), \quad n = 1,2,3, \ldots \qquad \textbf{(6.3.15)}$$

with k_n given by Eq. (6.3.12), are all particular solutions of Eq. (6.3.1) and satisfy the homogeneous boundary conditions (6.3.3).

To obtain the total solution of our problem, we take a linear sum of these particular solutions:

$$u(x,t) = \sum_{n=1}^{\infty} B_n \exp(-a^2 k_n^2 t) \sin\left(\frac{n\pi x}{L}\right) \qquad \textbf{(6.3.16)}$$

and determine the coefficients B_n so as to satisfy the initial condition (6.3.2). Thus, setting $t = 0$ in Eq. (6.3.16) we see from (6.3.2) that the coefficients B_n must satisfy the relation

$$f(x) = \sum_{n=1}^{\infty} B_n \sin\left(\frac{n\pi x}{L}\right). \quad 0 \le x \le L \qquad \textbf{(6.3.17)}$$

This is precisely the Fourier half-range sine series for $f(x)$ on the interval $[0,L]$. Therefore, the coefficients B_n are given by the formula

$$B_n = \frac{2}{L} \int_0^L f(x) \sin\left(\frac{n\pi x}{L}\right) dx \quad n = 1,2, \ldots \qquad \textbf{(6.3.18)}$$

For example, if L $= \pi$ and $u(x,0) = x(\pi - x)$, then

$$B_n = \frac{2}{\pi} \int_0^{\pi} x(\pi - x) \sin(nx)\, dx \qquad \textbf{(6.3.19)}$$

$$= 2 \int_0^{\pi} x \sin(nx)\, dx - \frac{2}{\pi} \int_0^{\pi} x^2 \sin(nx)\, dx \qquad \textbf{(6.3.20)}$$

$$= 4 \; \frac{1 - (-1)^n}{\pi n^3} \qquad \textbf{(6.3.21)}$$

Hence, we have

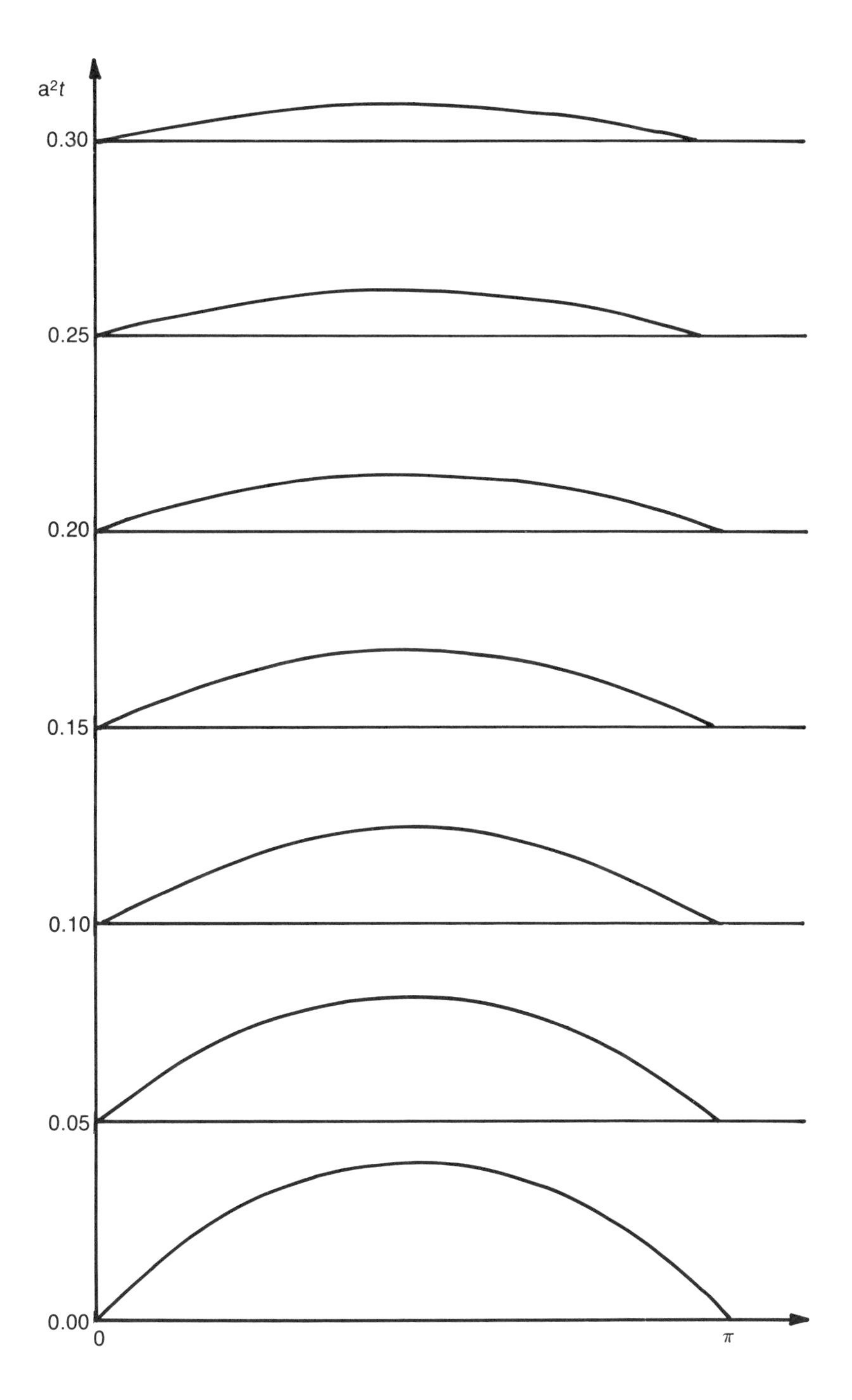

Fig. 6.3.1: Heat conduction in a rod when both ends are maintained at zero and the initial temperature is given by $x(\pi - x)$.

$$u(x,t) = \frac{8}{\pi} \sum_{n=1}^{\infty} \frac{e^{-(2n-1)^2 a^2 t}}{(2n-1)^3} \sin((2n-1)x). \qquad \textbf{(6.3.22)}$$

Equation (6.3.22) is graphed for various times in Fig. 6.3.1.

As a second example, let us solve the heat equation

$$u_t = a^2 u_{xx} \qquad \textbf{(6.3.23)}$$

subject to the boundary conditions

$$u_x(0,t) = u(L,t) = 0 \qquad \textbf{(6.3.24)}$$

and the initial condition

$$u(x,0) = x. \qquad \textbf{(6.3.25)}$$

The boundary condition $u_x(0,t) = 0$ states that no heat flow is allowed through the left boundary (insulated end condition).

Once again we employ separation of variables. For $X(x)$ we find

$$X'' + k^2 X = 0 \qquad \textbf{(6.3.26)}$$

with

$$X'(0) = X(L) = 0 \qquad \textbf{(6.3.27)}$$

because $u_x(0,t) = X'(0)\, T(t) = 0$ and $u(L,t) = X(L)\, T(t) = 0$. This regular Sturm-Liouville problem has the solution

$$X_n(x) = \cos\left(\frac{(2n-1)\pi x}{2L}\right), \; n = 1,2,3, \ldots \qquad \textbf{(6.3.28)}$$

The temporal solution then becomes

$$T_n(t) = B_n \exp\left(-\frac{(2n-1)^2 \pi^2 a^2 t}{4L^2}\right). \qquad \textbf{(6.3.29)}$$

Consequently the total solution is given by a linear superposition of the particular solutions:

$$u(x,t) = \sum_{n=1}^{\infty} B_n \cos\left(\frac{(2n-1)\pi x}{2L}\right) \exp\left(-\frac{(2n-1)^2 \pi^2 a^2 t}{4L^2}\right). \qquad \textbf{(6.3.30)}$$

Our final task remains to find the B_n's. Evaluating Eq. (6.3.30) at $t = 0$, we have

$$f(x) = x = \sum_{n=1}^{\infty} B_n \cos\left(\frac{(2n-1)\pi x}{2L}\right). \qquad \textbf{(6.3.31)}$$

Equation (6.3.31) is not a half-range cosine expansion; it is an expansion in the orthogonal functions $\cos\left(\frac{(2n-1)\pi x}{2L}\right)$ corresponding to the regular Sturm-Liouville problem (6.3.26)-(6.3.27). Consequently, B_n is given by Eq. (1.4.4) with $r(x) = 1$ as

$$B_n = \frac{\int_0^L x \cos\left(\frac{(2n-1)\pi x}{2L}\right) dx}{\int_0^L \cos^2\left(\frac{(2n-1)\pi x}{2L}\right) dx} \qquad \textbf{(6.3.32)}$$

$$= \frac{\frac{4L^2}{(2n-1)^2\pi^2} \cos\left[\frac{(2n-1)\pi x}{2L}\right]\Bigg|_0^L + \frac{(2n-1)\pi x}{2L} \sin\left[\frac{(2n-1)\pi x}{2L}\right]\Bigg|_0^L}{\frac{1}{2}\left(x + \frac{L}{(2n-1)\pi} \sin\left[\frac{(2n-1)\pi x}{L}\right]\right)\Bigg|_0^L} \qquad \textbf{(6.3.33)}$$

$$= \frac{4L^2}{(2n-1)^2\pi^2}\left(\cos\left[\frac{(2n-1)\pi}{2}\right] - 1\right) + \frac{\frac{(2n-1)\pi}{2} \sin\left[\frac{(2n-1)\pi}{2}\right]}{L/2} \qquad \textbf{(6.3.34)}$$

$$= \frac{8L}{(2n-1)^2\pi^2}\left[\frac{(2n-1)\pi}{2}(-1)^{n+1} - 1\right] \qquad \textbf{(6.3.35)}$$

because $\cos((2n-1)\pi/2) = 0$ and $\sin((2n-)\pi/2) = (-1)^{n+1}$. Consequently the final solution can be written

$$u(x,t) = \frac{8L}{\pi^2} \sum_{n=1}^{\infty} \frac{1}{(2n-1)^2}\left(\frac{(2n-1)\pi}{2}(-1)^{n+1} - 1\right)$$

$$\times \cos\left[\frac{(2n-1)\pi x}{2L}\right] \exp\left(-\frac{(2n-1)^2\pi^2 a^2 t}{4L^2}\right) \qquad \textbf{(6.3.36)}$$

Equation (6.3.36) is graphed for various times in Fig. 6.3.2.

A slight variation on our original problem is

$$u_t = a^2 u_{xx} \tag{6.3.37}$$

with

$$u(x,0) = u(0,t) = 0 \text{ and } u(L,t) = \theta. \tag{6.3.38}$$

We now blindly employ the technique of separation of variables to the problem. Once again we obtain the ordinary differential equation (6.3.9) and (6.3.10). The initial and boundary conditions become, however,

$$X(0) = T(0) = 0 \tag{6.3.39}$$

and

$$X(L)\ T(t) = \theta. \tag{6.3.40}$$

Although Eq. (6.3.39) is all right, Eq. (6.3.40) gives us an impossible condition because $T(t)$ cannot be a constant. If it were, it would have to equal zero by Eq. (6.3.39).

To find a way around this difficulty, suppose we wanted the solution to our problem at a time long after $t = 0$. We know from experience that after a long time heat conduction with time-independent boundary conditions results in an evolution from the initial condition to some time-independent (*steady-state*) equilibrium. If we denote this steady-state solution by $w(x)$, it must satisfy the heat equation

$$a^2 w''(x) = 0 \tag{6.3.41}$$

and the boundary conditions

$$w(0) = 0 \text{ and } w(L) = \theta. \tag{6.3.42}$$

Equation (6.3.41) can be integrated immediately to give

$$w(x) = A + Bx \tag{6.3.43}$$

and the boundary conditions (6.3.42) result in

$$w(x) = \frac{\theta x}{L}. \tag{6.3.44}$$

Clearly Eq. (6.3.44) cannot hope to satisfy the initial condition. It was never designed to. However, if we add a time-varying *(transient)* solution $v(x,t)$ to $w(x)$ so that

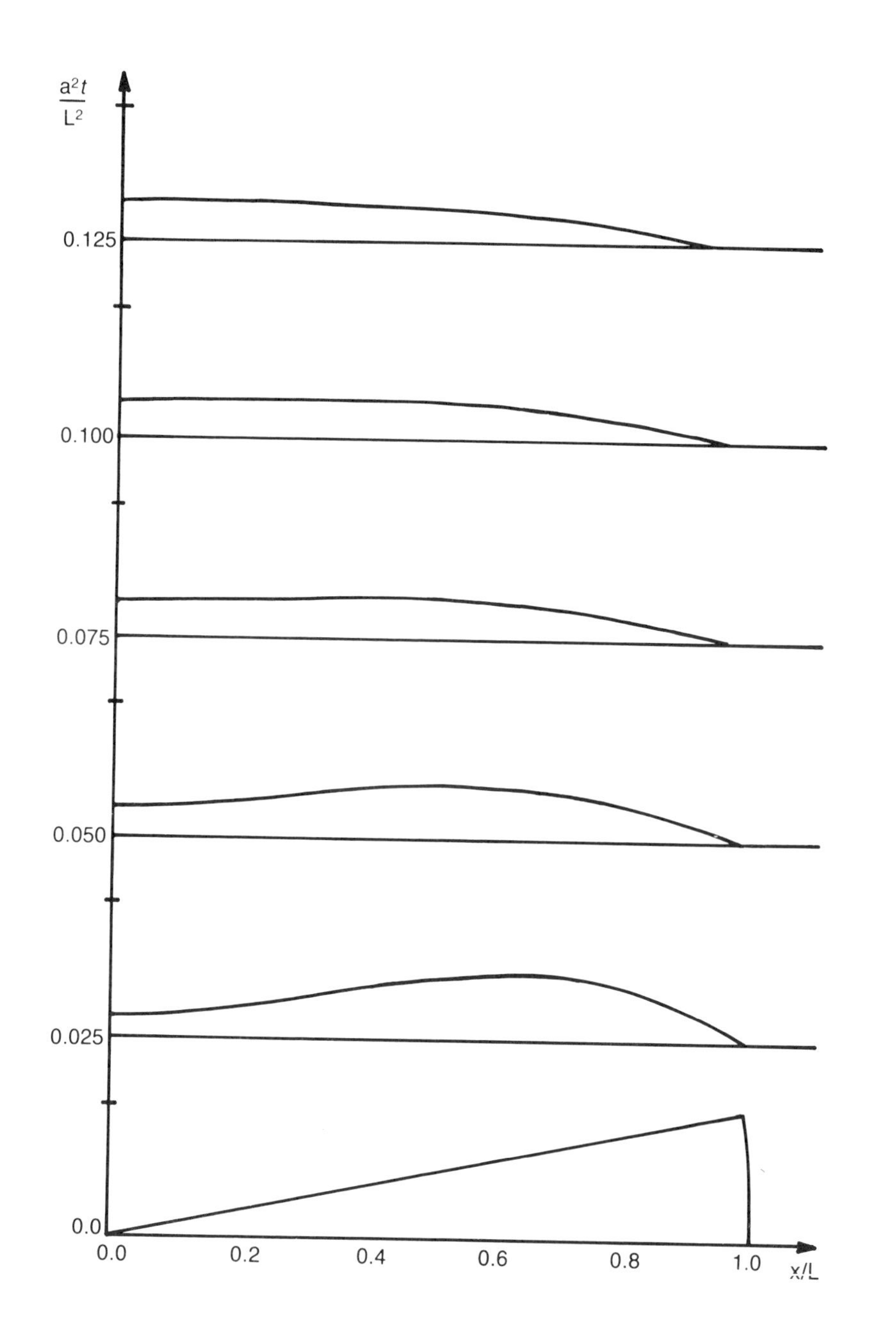

Fig. 6.3.2: Heat conduction in a rod when the left end is insulated, the right end held at the temperature of zero and the initial temperature is given by u(x,0) = x.

$$u(x,t) = w(x) + v(x,t), \tag{6.3.45}$$

we could satisfy the initial condition if

$$v(x,0) = u(x,0) - w(x) \tag{6.3.46}$$

and $v(x,t)$ tends to zero at $t \to \infty$. Furthermore, because

$w''(x) = w(0) = 0$ and $w(L) = \theta$, we have

$$v_t = a^2 v_{xx} \tag{6.3.47}$$

with the boundary conditions

$$v(0,t) = 0 \text{ and } v(L,t) = 0. \tag{6.3.48}$$

We know that we can solve Eqs. (6.3.46), (6.3.47) and (6.3.48) by separation of variables. We did it in the first problem in this section. However, in place of $f(x)$, we now have $u(x,0) - w(x)$ or $-w(x)$ because $u(x,0) = 0$. Therefore, the solution for $v(x,t)$ which dies away as $t \to \infty$ is

$$v(x,t) = \sum_{n=1}^{\infty} B_n \sin\left(\frac{n\pi x}{L}\right) \exp\left(-\frac{a^2 n^2 \pi^2 t}{L^2}\right) \tag{6.3.49}$$

with

$$B_n = \frac{2}{L} \int_0^L -w(x) \sin\left(\frac{n\pi x}{L}\right) dx \tag{6.3.50}$$

$$= \frac{2}{L} \int_0^L -\frac{\theta x}{L} \sin\left(\frac{n\pi x}{L}\right) dx \tag{6.3.51}$$

$$= -\frac{2\theta}{L^2} \left[-\frac{xL}{n\pi} \cos\left(\frac{n\pi x}{L}\right) + \frac{L^2}{n^2\pi^2} \sin\left(\frac{n\pi x}{L}\right) \right] \Bigg|_0^L \tag{6.3.52}$$

$$= \frac{2\theta}{L^2} \frac{L^2}{n\pi} \cos(n\pi) = (-1)^n \frac{2\theta}{n\pi}. \tag{6.3.53}$$

Thus, the entire solution is

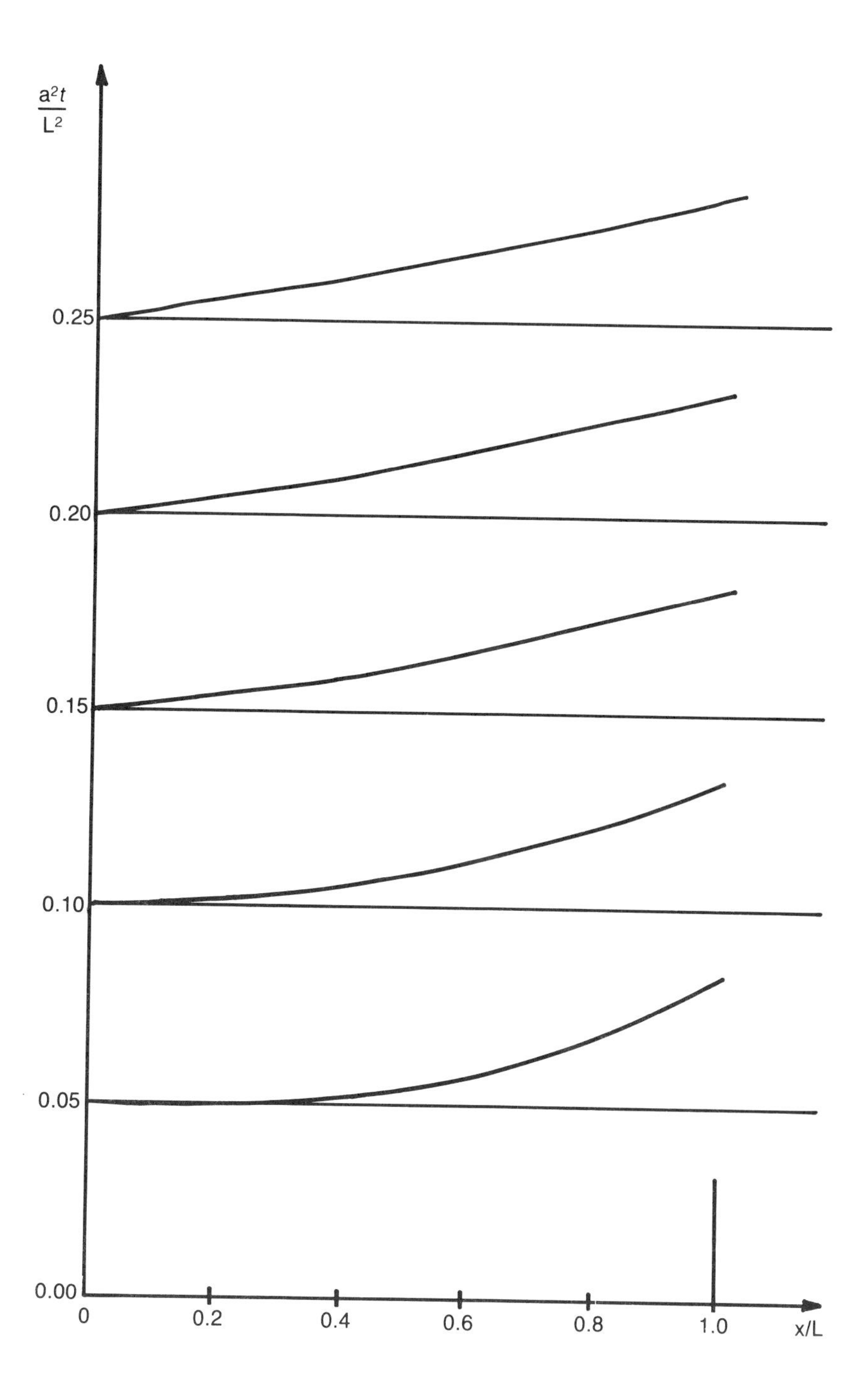

Fig. 6.3.3: Solution of the heat equation with $u(0,t) = 0$ and $u(L,t) = \theta$ and $u(x,0) = 0$.

$$u(x,t) = \frac{\theta x}{L} + \frac{2\theta}{\pi} \sum_{n=1}^{\infty} \frac{(-1)^n}{n} \sin\left(\frac{n\pi x}{L}\right) \exp\left(-\frac{n^2\pi^2 a^2 t}{L^2}\right). \quad (6.3.54)$$

The quantity a^2t/L^2 is often called the *Fourier number*. Our results have been illustrated in Fig. 6.3.3.

Finally, let us consider the heat equation

$$u_t = a^2 u_{xx} \quad (6.3.55)$$

subject to the Neumann boundary conditions

$$u_x(0,t) = u_x(L,t) = 0 \quad (6.3.56)$$

and the initial condition that

$$u(x,0) = x. \quad (6.3.57)$$

Assuming that $u(x,t) = X(x)T(t)$, we have

$$\frac{X''}{X} = \frac{T'}{a^2T} = -k^2 \quad (6.3.58)$$

where we have presently assumed that the separation constant is negative. The Neumann conditions give $u_x(0,t) = X'(0)T(t) = 0$ and $u_x(L,t) = X'(L)T(t) = 0$ so that $X'(0) = X'(L) = 0$. The x dependence is therefore given by the Sturm-Liouville problem

$$X'' + k^2X = 0 \quad (6.3.59)$$

and

$$X'(0) = X'(L) = 0 \quad (6.3.60)$$

which has the eigenfuction solution of

$$X_n(x) = \cos\left(\frac{n\pi x}{L}\right) \quad (6.3.61)$$

where $k_n = n\pi/L$ and $n = 1,2,3,4, \ldots$

The corresponding temporal part is given by the solution of

$$T_n' + a^2k_n^2T_n = T_n' + \frac{a^2n^2\pi^2}{L^2} T_n = 0 \quad (6.3.62)$$

which is

$$T_n(t) = A_n \exp(-\frac{a^2n^2\pi^2t}{L^2}). \qquad \textbf{(6.3.63)}$$

The product solution given by a negative separation constant is given by

$$u_n(x,t) = X_n(x)T_n(t)$$

$$= B_n \cos(\frac{n\pi x}{L}) \exp(-\frac{a^2n^2\pi^2t}{L^2}). \qquad \textbf{(6.3.64)}$$

Unlike our previous problems, there is a nontrival solution for $k = 0$. In this instance, the x-dependence is given by

$$X(x) = Ax + B. \qquad \textbf{(6.3.65)}$$

The boundary condition $X'(0) = X'(L) = 0$ forces A to be zero but B is completely free. Consequently, the eigenfunction in this particular case is

$$X_0(x) = 1. \qquad \textbf{(6.3.66)}$$

Because $T_0' = 0$, the temporal part equals a constant which we shall take to be $A_0/2$. Therefore, the product solution corresponding to $k = 0$ is

$$u_0(x,t) = X_0(x)T_0(t) = A_0/2. \qquad \textbf{(6.3.67)}$$

The most general solution to our problem equals the sum of all of the possible solutions:

$$u(x,t) = A_0/2 + \sum_{n=1}^{\infty} A_n \cos(\frac{n\pi x}{L}) \exp(-\frac{a^2n^2\pi^2t}{L^2}). \qquad \textbf{(6.3.68)}$$

Upon substituting $t = 0$ into Eq. (6.3.68), we can determine A_n because

$$u(x,0) = x = A_0/2 + \sum_{n=1}^{\infty} A_n \cos(\frac{n\pi x}{L}) \qquad \textbf{(6.3.69)}$$

is merely a half-range Fourier cosine expansion of the function x over the interval $[0,L]$. From Eqs. (2.6.9)-(2.6.10), we have

$$A_0 = \frac{2}{L}\int_0^L x\,dx = L \qquad \textbf{(6.3.70)}$$

and

$$A_n = \frac{2}{L}\int_0^L x \cos\left(\frac{n\pi x}{L}\right) dx \qquad (6.3.71)$$

$$= \frac{2}{L}\left\{ \frac{L^2}{n^2\pi^2} \cos\left(\frac{n\pi x}{L}\right) + \frac{Lx}{n\pi} \sin\left(\frac{n\pi x}{L}\right) \right\}\Big|_0^L \qquad (6.3.72)$$

$$= \frac{2L}{n^2\pi^2}\left((-1)^n - 1\right). \qquad (6.3.73)$$

The solution may finally be written as

$$u(x,t) = \frac{L}{2} - \frac{4L}{\pi^2}\sum_{m=1}^{\infty} \frac{1}{(2m-1)^2} \cos\left(\frac{(2m-1)\pi x}{L}\right) \exp\left(-\frac{a^2(2m-1)^2\pi^2 t}{L^2}\right)$$

$$(6.3.74)$$

because all the even harmonics vanish and the odd harmonics may be rewritten with $n = 2m-1$, $m = 1,2,3, \ldots$

Exercises

In problems 1 through 13, find the solution of the initial-boundary value problem for the heat equation (6.1.8), $0<x<L$, $t>0$, satisfying the given initial and boundary conditions.

1. $u(x,0) = A$, a constant; $u(0,t) = u(\pi,t) = 0$, $t>0$.

2. $u(x,0) = \sin^3(x)$; $u(0,t) = u(\pi,t) = 0$, $t>0$. Hint: $\sin^3(x) = [3\sin(x) - \sin(3x)]/4$

3. $u(x,0) = \cos^2(x)$; $u_x(0,t) = u_x(\pi,t) = 0$, $t>0$. Hint: $\cos^2(x) = [1+\cos(2x)]/2$

4. $u(x,0) = x(\pi - x)$; $u(0,t) = u(\pi,t) = 0$, $t>0$.

5. $u(x,0) = \begin{cases} x & 0\le x\le \pi/2 \\ \pi - x & \pi/2 \le x \le \pi \end{cases}$; $u(0,t) = u(\pi,t) = 0$, $t>0$.

6. $u(x,0) = x^2 - \pi^2$; $u_x(0,t) = u(\pi,t) = 0$, $t>0$.

7. $u(x,0) = x$; $u_x(0,t) = u_x(\pi,t) = 0$, $t>0$.

(Hint: Include the separation constant $k = 0$.)

8. $u(x,0) = \sin(x);\ u(0,t) = u(\pi,t) = 0,\ t>0.$

9. $u(x,0) = T_1;\ u(0,t) = u(L,t) = T_0,\ t>0.$

10. $u(x,0) = T_1 x/L;\ u(0,t) = T_0,\ u_x(L,t) = 0,\ t>0.$

11. $u(x,0) = T_0;\ u_x(0,t) = u_x(L,t) = \Delta T/L,\ t>0.$

12. $u(x,0) = T_1;\ u(0,t) = T_0,\ u_x(L,t) = 0,\ t>0.$

13. $u(x,0) = \begin{matrix} T_0 & 0<x<L/2 \\ T_1 & L/2<x<L \end{matrix};\ u_x(0,t) = u_x(L,t) = 0,\ t>0.$

14. *The warming of walls.* It is well known that a room with masonry walls is often very difficult to heat. In this problem we shall explain why this is so and see what parameters determine the indoor temperature of a wall of thickness d, conductivity $\varkappa$, and diffusivity a^2 which is heated at a constant rate H. The temperature of the outside face of the wall is assumed constant at T_0 and the entire wall initially had the uniform temperature T_0.

The equation governing for this problem is

$$T_t = a^2\, T_{xx}$$

with the boundary conditions

$$T(d,t) = T_0,\ T(x,0) = T_0,\ T_x(0,t) = -\frac{H}{\varkappa}\,.$$

Show that the temperature anywhere in the wall is given by

$$T(x,t) = T_0 + \frac{Hd}{\varkappa}\left[1 - \frac{x}{d} - \frac{8}{\pi^2}\sum_{n=0}^{\infty}\frac{1}{(2n+1)^2}\right.$$

$$\left.\times \cos\left(\frac{(2n+1)\pi x}{2d}\right)\exp-\left(\frac{(2n+1)^2\pi^2\, a^2 t}{4d^2}\right)\right].$$

Consequently the rise of temperature at the indoor wall $x = 0$ is

$$T_0 + \frac{Hd}{\varkappa}\left[1 - \frac{8}{\pi^2}\sum_{n=0}^{\infty}\frac{1}{(2n+1)^2}\exp\left(-\frac{(2n+1)^2\pi^2\, a^2 t}{4d^2}\right)\right].$$

This series can be approximated by $T_0 + (5a^2t)^{1/2}H/(2\varkappa)$. We thus see that the temperature will rise as the square root of the time and diffusivity and inversely with the conductivity. For an average rock, $\varkappa = 0.0042$ g/cm sec and $a^2 = 0.0118$ cm²/sec while for wood (Spruce) $\varkappa = 0.0003$ gm/cm sec and $a^2 = 0.0024$ cm²/sec.

15. Show that the solution of

$$u_t = u_{xx} - u$$

with the boundary conditions

$$u(0,t) = 1 \text{ and } u(L,t) = 0$$

and the initial condition

$$u(x,0) = 0$$

is

$$u(x,t) = \frac{2\pi}{L^2} \sum_{n=1}^{\infty} \frac{n}{1 + \frac{n^2\pi^2}{L^2}} \sin\left(\frac{n\pi x}{L}\right)$$

$$\times \left\{ \exp\left[-\left(1 + \frac{n^2\pi^2}{L^2}\right) t \right] - 1 \right\}.$$

Hint: Show that $w(x) = \sinh(L-x)/\sinh(L)$ and reexpress $w(x)$ in a Fourier half-range expansion over the interval 0 to L.

16. *Aquifer response to sudden change in reservior level.* The height of the water table $u(x,t)$ above some reference point is governed by Boussinesq's equation

$$u_t = a^2 u_{xx}$$

where a^2 is the product of the storage coefficient times the hydraulic coefficient divided by the aquifer thickness. A typical value of a^2 is 10 m²/min. Consider the problem of a strip of land of width L that separates two reserviors of depth h_1. Consequently, the height of the water table would initially be h_1. Suddenly the reservior on the right $(x = L)$ is lowered to a depth h_2 $(u(0,t) = h_1$, $u(L,t) = h_2$, and $u(x,0) = h_1)$ show that the water table is given by

$$u(x,t) = h_1 + (h_2 - h_1)\frac{x}{L} + \frac{2}{\pi}\sum_{n=1}^{\infty}\frac{(h_2 - h_1)(-1)^n}{n}\sin\left(\frac{n\pi x}{L}\right)\exp\left(-\frac{a^2n^2\pi^2 t}{L^2}\right).$$

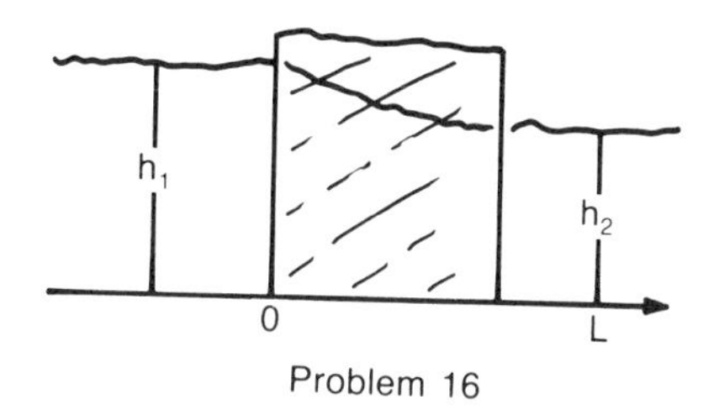

Problem 16

17. *Land drainage.* The height of water table $u(x,t)$ is governed by the equation (see Problem 16)

$$u_t = a^2 u_{xx}.$$

Consider the problem of a piece of land that suddenly has two drains placed at the points $x = 0$ and $x = L$ so that $u(0,t) = u(L,t) = 0$. If the water table initially has the profile

$$u(x,0) = \frac{8H}{L^4}(L^3x - 3L^2x^2 + 4Lx^3 - 2x^4),$$

show that the water table is subsequently given by

$$u(x,t) = \frac{192H}{\pi^5}\sum_{n=0}^{\infty}\frac{(2n+1)^2\pi^2 - 8}{(2n+1)^5}$$

$$\times \sin\left(\frac{(2n+1)\pi x}{L}\right)\exp\left(-\frac{(2n+1)^2\pi^2a^2t}{L^2}\right)$$

Problem 17

18. *Flow in a confined aquifer.* We want to find the rise of the water table of an aquifer which is sandwiched between a canal and impervious rocks if the water level in the canal is suddenly raised h_0 units above its initial elevation and maintained constant. The level of the water table is governed by the Boussinesq equation (see Problem 16)

$$u_t = a^2 u_{xx}$$

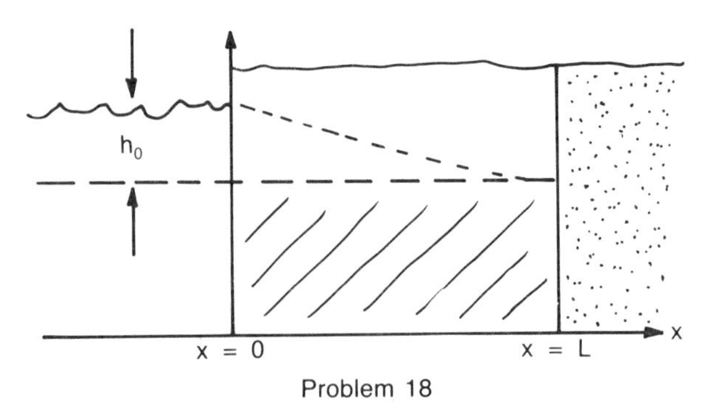

Problem 18

with the boundary conditions

$$u(0,t) = h_0 \text{ and } u_x(L,t) = 0$$

and the initial condition

$$u(x,0) = 0.$$

Show that the solution is

$$u(x,t) = h_0 - \frac{4h_0}{\pi} \sum_{n=0}^{\infty} \frac{1}{2n+1} \times \sin\left[\frac{(2n+1)\pi x}{L}\right] \exp\left[-\frac{a^2(2n+1)^2\pi^2 t}{L^2}\right]$$

6.4 COOLING OF AN INFINITELY LONG CIRCULAR CYLINDER

In this section we illustrate how separation of variables may be employed in solving the axisymmetric heat equation in an infinitely long cylinder. In circular coordinates the heat equation is

$$\frac{\partial u}{\partial t} = a^2 \left(\frac{\partial^2 u}{\partial r^2} + \frac{1}{r}\frac{\partial u}{\partial r}\right) \quad \textbf{(6.4.1)}$$

where r denotes the radial distance and a^2 the thermal diffusivity. Let us assume that this cylinder of radius b has been heated to the temperature T_0 and then allowed to cool by having its surface held at the temperature zero starting from the time $t = 0$.

We begin our analysis by assuming that the solution is of the form

$$u(r,t) = R(r)\, T(t) \quad \textbf{(6.4.2)}$$

so that

$$\frac{1}{R}\left(\frac{d^2R}{dr^2} + \frac{1}{r}\frac{dR}{dr}\right) = \frac{1}{a^2T}\frac{dT}{dt} = -\frac{k^2}{b^2} \quad \textbf{(6.4.3)}$$

The only values of the separation constant that yields finite solutions as $r \to 0$ are negative. These solutions are

$$R(r) = J_0(k \frac{r}{b}) \quad \textbf{(6.4.4)}$$

where J_0 is the Bessel function of the first kind, of order zero. The separation constant $k = 0$ gives $R(r) = \ln(r)$, which becomes infinite at the origin. Positive separation constants yield the modified Bessel function $I_0(\frac{k}{b} r)$. It can be shown that this solution cannot satisfy the boundary condition that $u(b,t) = R(b)\ T(t) = 0$ or $R(b) = 0$.

The boundary condition that $R(b) = 0$ requires that $J_0(k) = 0$. This transcendental equation yields an infinite number of k_n. For each of these k_n's, the temporal part of the solution is given by

$$\frac{dT_n}{dt} + \frac{k_n^2 a^2 T}{b^2} = 0 \quad \textbf{(6.4.5)}$$

which has the solution

$$T_n(t) = A_n \exp(-\frac{k_n^2 a^2 t}{b^2}). \quad \textbf{(6.4.6)}$$

Consequently, the product solution is

$$u_n(r,t) = A_n J_0(k_n \frac{r}{b}) \exp(-\frac{k_n^2 a^2 t}{b^2}). \quad \textbf{(6.4.7)}$$

The total solution is a linear superposition of all of the particular solutions:

$$u(r,t) = \sum_{n=1}^{\infty} A_n J_0(k_n \frac{r}{b}) \exp(-\frac{a^2 k_n^2 t}{b^2}). \quad \textbf{(6.4.8)}$$

Our final task remains to determine A_n. From the initial condition that $u(r,0) = T_0$, we have

$$u(r,0) = T_0 = \sum_{n=1}^{\infty} A_n J_0(k_n \frac{r}{b}). \quad \textbf{(6.4.9)}$$

From Eq. (1.7.58) and (1.7.66),

$$A_n = \frac{2T_0}{J_1(k_n)^2 b^2} \int_0^b r\, J_0(k_n \frac{r}{b})\, dr \quad \textbf{(6.4.10)}$$

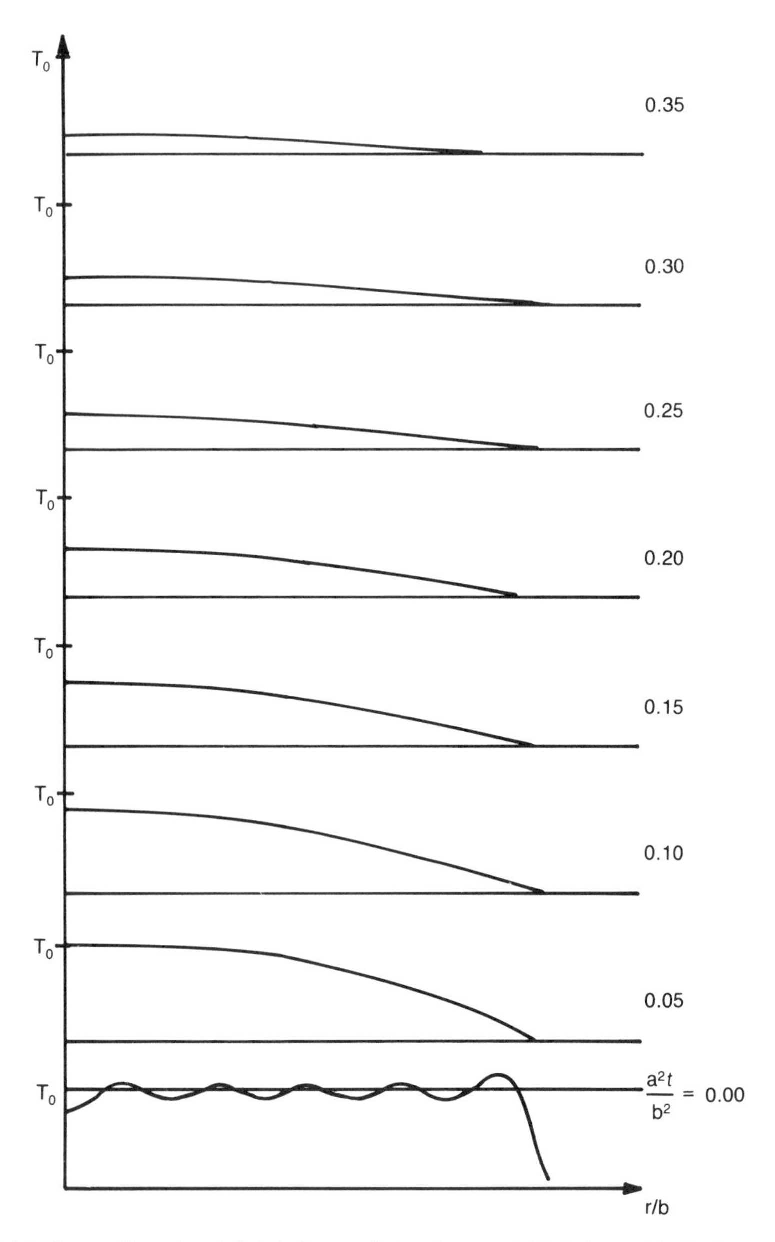

Fig. 6.4.1: The cooling of an infinitely long cylinder that was initially heated to the temperature T_0 and then allowed to cool with u(b,t) = 0.

$$A_n = \frac{2T_0}{J_1(k_n)^2 k_n^{\,2}} \left(\frac{r}{b}\, k_n \right) J_1\!\left(k_n \frac{r}{b}\right) \Bigg|_0^b \qquad \textbf{(6.4.11)}$$

$$= \frac{2\ T_0}{J_1(k_n)\ k_n} \qquad \textbf{(6.4.12)}$$

from Eq. (1.7.18). Thus, the final solution is

$$u(r,t) = 2\, T_0 \sum_{n=1}^{\infty} \frac{J_0\left(k_n \frac{r}{b}\right) \exp\left(-\frac{k_n{}^2 a^2 t}{b^2}\right).}{k_n\, J_1(k_n)} \tag{6.4.13}$$

In Fig. 6.4.1, the solution (6.4.13) has been graphed for various Fourier numbers a^2t/b^2. Only the first ten terms of Eq. (6.4.13) have been used.

6.5 THE SOLUTION OF THE NONHOMOGENEOUS HEAT EQUATION: THE PROBLEM OF REFRIGERATING APPLES

Some decades ago, ship loads of apples, going from Australia to England, deteriorated from a disease, called "brown heart," which occurred under insufficient cooling conditions. Apples, when placed on shipboard, are usually warm and have to be cooled down to be carried in cold storage. They also generate heat by their respiration. It was suspected that this heat generation effectively counteracted the refrigeration of the apples, resulting in the "brown heart."

This was the problem which induced Awberry (Awberry, J.H., 1927: Flow of heat in a body generating heat. *Philosophical Magazine. Series 7. 4,* 629-638.) to study the heat distribution within a sphere in which heat is being generated. Awberry first assumed that the apples are initially at a uniform temperature. We can take this initial temperature to be zero by the appropriate choice of temperature scale. At time $t = 0$, the skins of the apples assume the temperature θ immediately when they are introduced into the hold.

Because of the spherical geometry, the nonhomogeneous heat equation becomes

$$\frac{1}{a^2}\frac{\partial u}{\partial t} = \frac{1}{r^2}\frac{\partial}{\partial r}\left(r^2 \frac{\partial u}{\partial r}\right) + \frac{G}{\varkappa} \tag{6.5.1}$$

where a^2 is the thermal diffusivity, $\varkappa$ the thermal conductivity, and G the heat rate (per unit time per unit volume). If we try to use separation of variables on Eq. (6.5.1), we find that it does not work because of the $G/\varkappa$ term. To circumvent this difficulty, we ask the simpler question of what happens after a very long time. We anticipate that a balance will eventually be established where the heat produced within the apple is conducted to the surface of the apple where it is removed. Consequently just as we were forced to introduce a steady-state solution in the previous section, we again anticipate a steady-state solution $w(r)$ where the heat conduction removes the heat generated within the apple:

$$\frac{1}{r^2}\frac{d}{dr}(r^2\frac{dw}{dr}) = -\frac{G}{\varkappa}. \quad (6.5.2)$$

Furthermore, just as we were forced to introduce a transient solution in the previous section which allowed our solution to satisfy the initial condition, we must also have one here:

$$\frac{\partial v}{\partial t} = \frac{a^2}{r^2}\frac{\partial}{\partial r}(r^2\frac{\partial v}{\partial r}). \quad (6.5.3)$$

Solving Eq. (6.5.2) first, we find

$$w(r) = C + \frac{D}{r} - \frac{Gr^2}{6\varkappa}. \quad (6.5.4)$$

We must take $D = 0$ because the solution must be finite at $r = 0$. Because the steady-state solution must satisfy the boundary condition $w(b) = \theta$.

$$C = \theta + \frac{Gb^2}{6\varkappa}. \quad (6.5.5)$$

Turning to the transient problem, we introduce a new dependent variable $y(r,t) = r\,v(r,t)$. This new dependent variable allows us to replace Eq. (6.5.3) with

$$y_t = a^2 y_{rr} \quad (6.5.6)$$

which we know how to solve. If we assume that $y(r,t) = R(r)T(t)$, the $R(r)$ equation becomes

$$\frac{d^2R}{dr^2} + k^2 R = 0 \quad (6.5.7)$$

which has the solution

$$R(r) = A\cos(kr) + B\sin(kr). \quad (6.5.8)$$

We must take $A = 0$ because the solution (6.5.8) must go to zero at $r = 0$ in order that $v(0,t)$ remains finite. However because

$$\theta = w(b) + v(b,t) \quad (6.5.9)$$

for all time, and

$$v(R,t) = \frac{R(b)T(t)}{b} = 0 \quad (6.5.10)$$

or

$$R(b) = 0. \quad \textbf{(6.5.11)}$$

Consequently

$$k_n = \frac{n\pi}{b} \quad \textbf{(6.5.12)}$$

and

$$v_n(r,t) = B_n \frac{\sin(\frac{n\pi r}{b})}{r} \exp(-\frac{n^2\pi^2a^2t}{b^2}). \quad \textbf{(6.5.13)}$$

The total solution is given by the principle of superposition:

$$u(r,t) = \theta + \frac{G}{6\varkappa}(b^2 - r^2) + \sum_{n=1}^{\infty} B_n \frac{\sin(\frac{n\pi r}{b})}{r} \times \exp(-\frac{n^2\pi^2a^2t}{b^2}). \quad \textbf{(6.5.14)}$$

Finally we determine the B_n's by the initial condition that
$u(r,0) = 0.$

Therefore

$$B_n = -\frac{2}{b}\int_0^b r\left[\theta + \frac{G}{6\varkappa}(b^2 - r^2)\right] \sin(\frac{n\pi r}{b})\, dr. \quad \textbf{(6.5.15)}$$

$$= \frac{2\theta b}{n\pi}(-1)^n + \frac{2G}{n}(\frac{b}{n\pi})^3 (-1)^n \quad \textbf{(6.5.16)}$$

The final solution may therefore be written as

$$u(r,t) = \theta + \frac{2\theta b}{\pi}\sum_{n=1}^{\infty} \frac{(-1)^n}{n} \frac{\sin(\frac{n\pi r}{b})}{r} \exp(-\frac{n^2\pi^2a^2t}{b^2})$$

$$+ \frac{G}{6\varkappa}(b^2 - r^2) + \frac{2Gb^3}{\varkappa\pi^3}\sum_{n=1}^{\infty}\frac{(-1)^n}{n^3}\frac{\sin(\frac{n\pi r}{b})}{r^3} \times \exp(-\frac{n^2\pi^2a^2t}{b^2}) \quad \textbf{(6.5.17)}$$

The first line of Eq. (6.5.17) gives the temperature distribution due to the imposition of the temperature θ on the surface of the apple while the second line gives the rise in the temperature due to the

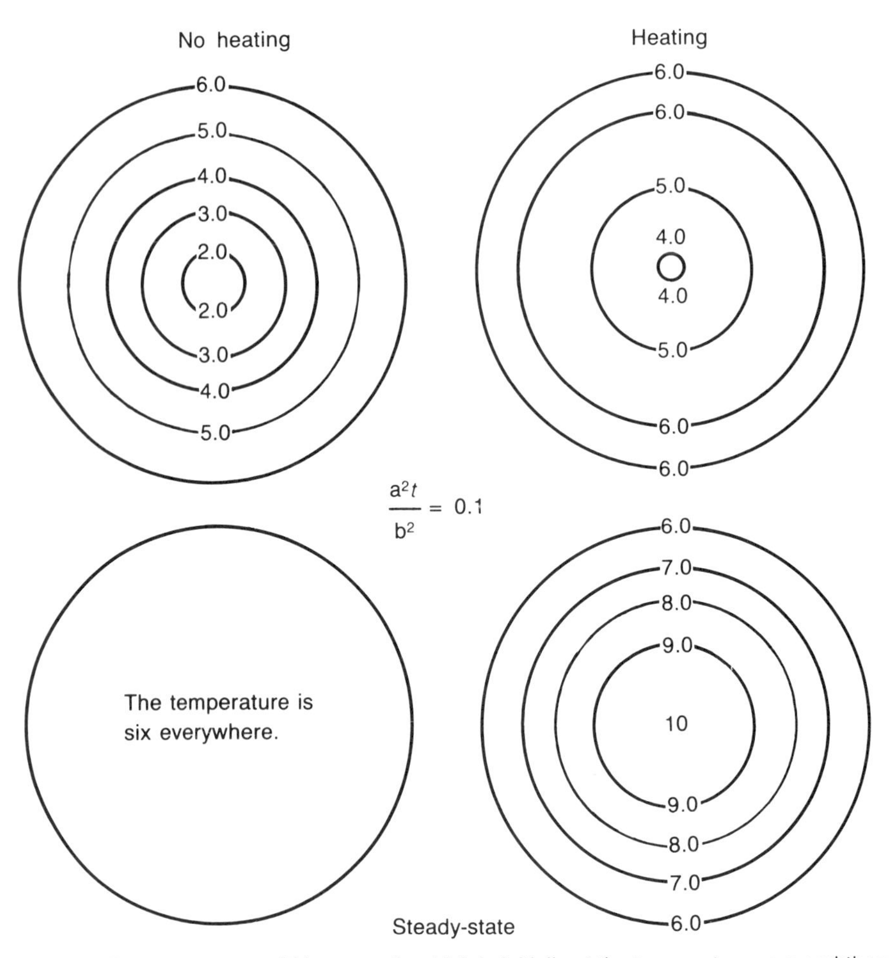

Fig. 6.5.1: The temperature within an apple which is initially at the temperature zero and then its surface is subjected to a temperature of six degrees. Two cases are presented. The illustrations on the left are for the case of no interior heating. The ones on the right are for the case when $Gb^2/6\varkappa = 4$.

interior heating. In Fig. 6.5.1, this solution is illustrated when $\theta = 6°$ and $Gb^2/6\varkappa = 4$ or 0 for various times.

Returning to our original problem of whether the interior heating is strong enough to counteract the cooling by the refrigeration, we merely have to use the second line of Eq. (6.5.17) to find how much the temperature deviates from what we normally expect. Because the center of each apple is at the highest temperature, the value there is the only one of interest in this practical problem. Assuming $b = 4$ cm as the radius of an apple, $G = 1.33 \times 10^{-5}$ °C/sec and $\varkappa = 1.55 \times 10^{-3}$ cm²/sec, the temperature effect of the heat generation is very small, only 0.0232 °C when, after about 2 hours, equilibrium is reached. For more powerful reactions than those occurring in an apple, the temperature rise can be quite significant.

Exercises

1. Find the solution of the problem $u_t - u_{xx} = F(x,t)$ with the conditions that $u(x,0) = 0$, $u(0,t) = u(\pi,t) = 0$ where $F(x,t) = x$ when $0 \leq x \leq \pi/2$ and $\pi - x$ when $\pi/2 \leq x \leq \pi$.

2. Solve the problem $u_t - a^2 u_{xx} = e^{-x}$ with the conditions that $u(x,0) = f(x)$ and $u(0,t) = u_x(\pi,t) = 0$.

3. Solve the problem $u_t - a^2 u_{xx} = A \cos(\omega t)$ where A is a constant and the initial-boundary conditions are $u(x,0) = f(x)$ and $u_x(0,t) = u_x(\pi,t) = 0$. (Hint: Observe that any function of t satisfies the boundary condition.)

4. *Dielectric heating.* A uniform, conducting rod of length L and thermometric conductivity (or diffusivity) a^2 is initially at temperature zero. Heat is supplied uniformly throughout the rod in such a manner that the heat conduction equation is $a^2 u_{xx} = u_t - P$, where P is rate at which the temperature would rise if there was no conduction. The outside is maintained at the temperature zero. Find the temperature at any subsequent time.

5. Show that the solution to the equation

$$u_t = u_{xx} - 1 \qquad 0 \leq x \leq 1, \ t \geq 0$$

with

$$u_x(0,t) = u_x(1,t) = 0$$

and

$$u(x,0) = (1 - x^2)/2 \qquad 0 < x < 1$$

is

$$u(x,t) = -t + \frac{1}{6} + \frac{2}{\pi^2} \sum_{n=1}^{\infty} \frac{\cos(n\pi x)}{n^2} \exp(-n^2\pi^2 t)$$

6. Solve $r^2 u_t = a^2(r^2 u_r)_r$ subject to the boundary conditions that $u(u,t)$ is finite and $u(1,t) = 0$ and the initial condition that $u(r,0) = 1$.

6.6 THE TEMPERATURE FIELD WITHIN A MULTICONDUCTOR ELECTRIC CABLE

In the design of cable installations it is necessary to know the temperature reached within an electrical cable as a function of current and other parameters. For us this problem provides an example of solving the nonhomogeneous heat equation in cylindrical coordinates with the radiation boundary condition. The solution was published by Iskenderian, H.P., and W.J. Horvath, 1946: Determination of the temperature rise and the maximum safe current through multiconductor electric cables. *J. Appl. Phys., 17,* 255-262.

The derivation of the heat equation follows from the conservation of energy:

$$\text{heat generated} = \text{heat dissipated} + \text{heat stored.} \tag{6.6.1}$$

$$I^2RN\,dt = -\varkappa\left(2\pi r \left.\frac{\partial u}{\partial r}\right|_r - 2\pi(r+\Delta r)\left.\frac{\partial u}{\partial r}\right|_{r+\Delta r}\right) dt$$

$$+ 2\pi r\Delta r\, c\varrho\, du \qquad \textbf{(6.6.2)}$$

where I is the current through each wire, R the resistance of each conductor, N the number of conductors in the shell between radii r and $r + \Delta r = \dfrac{2\pi r\Delta r}{\pi b^2} n$, b the radius of the cable, n the number of conductors, $\varkappa$ the thermal conductivity, ϱ the density, c the average specific heat, and u the temperature. In the limit of $\Delta r \to 0$, Eq. (6.6.2) becomes

$$\frac{\partial u}{\partial t} = A + a^2 \frac{1}{r}\frac{\partial}{\partial r}\left(r\frac{\partial u}{\partial r}\right) \qquad \textbf{(6.6.3)}$$

where

$$A = \frac{I^2Rn}{\pi b^2 c\varrho} \qquad \textbf{(6.6.4)}$$

and

$$a^2 = \frac{\varkappa}{\varrho c}. \qquad \textbf{(6.6.5)}$$

Equation (6.6.3) is the nonhomogeneous heat equation within an infinite, axisymmetric cylinder. The easiest method of solving it is by separation of variables. From our earlier study of the nonhomogeneous heat equation in Section 6.5, we know that we must write the temperature as the sum of a steady-state and transient solution: $u(r,t) = w(r) + v(r,t)$.

The steady-state solution $w(r)$ satisfies

$$\frac{1}{r}\frac{d}{dr}\left(r\frac{dw}{dr}\right) = -\frac{A}{a^2} \qquad \textbf{(6.6.6)}$$

or

$$w(r) = T_c - \frac{A}{4a^2} r^2 \qquad \textbf{(6.6.7)}$$

where T_c is the (yet unknown) temperature in the center of the cable.

The transient equation is

$$\frac{\partial v}{\partial t} = a^2 \frac{1}{r}\frac{\partial}{\partial r}\left(r\frac{\partial v}{\partial r}\right) \qquad \textbf{(6.6.8)}$$

with the initial condition that $u(r,0) = T_c - \frac{Ar^2}{4a^2} + v(0,t) = 0$. At the surface we assume that the cable is radiating to free space so that the boundary condition is

$$\frac{\partial u}{\partial r} = -hu \quad \textbf{(6.6.9)}$$

where h is the surface conductance. Because the steady-state temperature must be true when all transient effects die away, it must satisfy this radiation boundary condition (6.6.9) regardless of the transient solution. This requires that

$$T_c = \frac{A}{a^2}\left(\frac{b^2}{4} + \frac{b}{2h}\right). \quad \textbf{(6.6.10)}$$

Therefore, $v(b,t)$ must satisfy

$$\frac{\partial v(b,t)}{\partial r} = -h\, v(b,t). \quad \textbf{(6.6.11)}$$

The transient solution $v(r,t)$ is found by separation of variables $v(r,t) = R(r)T(t)$. Substituting into Eq. (6.6.8), we have

$$\frac{1}{rR}\frac{d}{dr}\left(r\frac{dR}{dr}\right) = \frac{1}{a^2T}\frac{dT}{dt} = -k^2 \quad \textbf{(6.6.12)}$$

or

$$\frac{dT}{dt} + k^2a^2\, T = 0 \quad \textbf{(6.6.13)}$$

and

$$\frac{d}{dr}\left(r\frac{dR}{dr}\right) + k^2r\, R = 0 \quad \textbf{(6.6.14)}$$

with

$$\frac{dR}{dr} = -h\, R \quad \textbf{(6.6.15)}$$

at $r = b$. The only solution of Eq. (6.6.14) which remains finite at $r = 0$ is $R(r) = C\, J_0(kr)$ where J_0 is the zero-order Bessel function of the first kind. See Section 1.7.

Substituting $C\, J_0(kr)$ into the boundary condition (6.6.15), we have the transcendental equation

$$kb\, J_1(kb) + hb\, J_0(kb) = 0. \quad \textbf{(6.6.16)}$$

For a given value of h and b, Eq. (6.6.16) yields an infinite number of unique zeros k_n. The temporal solution to the problem is

$$T_n(t) = A_n \exp(-a^2 k_n^2 t) \qquad \textbf{(6.6.17)}$$

so that the complete product solution is

$$v(r,t) = \sum_{n=1}^{\infty} A_n J_0(k_n r) \exp(-a^2 k_n^2 t). \qquad \textbf{(6.6.18)}$$

Our final task remains to compute A_n. This is done by evaluating Eq. (6.6.18) at $t = 0$,

$$v(r,0) = \frac{Ar^2}{4a} - T_c = \sum_{n=1}^{\infty} A_n J_0(k_n r) \qquad \textbf{(6.6.19)}$$

which is a Fourier-Bessel series in $J_0(k_n r)$. In Section 1.7, we showed that the coefficient of a Fourier-Bessel series with the orthogonal function $J_0(k_n r)$ and the boundary condition (6.6.16) is given by

$$A_n = \frac{2\, k_n^2}{(k_n^2 b^2 + h^2 b^2)(J_0(k_n b))^2} \int_0^b r(\frac{Ar^2}{4a^2} - T_c) J_0(k_n r)\, dr$$

(6.6.20)

from Eqs. (1.7.58) and (1.7.68). Carrying out the indicated integration,

$$A_n = \frac{2\, b^2}{[(k_n^2 + h^2)(J_0(k_n b)]^2} \left[(\frac{A k_n b}{4a^2} - \frac{A}{k_n b a^2} - \frac{T_c k_n}{b^2}) J_1(k_n b) + \frac{A}{2a^2} J_0(k_n b) \right]$$

(6.6.21)

Equation (6.6.21) was obtained by using Eq. (1.7.34) and integration by parts in a similar manner as was done in Eqs. (1.7.79)-(1.7.84).

To illustrate this solution, let us compute the solution for the typical parameters $b = 4$ cm, $hb = 1$, $a^2 = 1.14$ cm^2/sec, $A = 2.2747$ C/sec and $T_c = 23.94$ C. The value of A corresponds to 37 wires of #6 AWG copper wire within a cable carrying a current of 22 amp. In Fig. 6.6.1 the solution is plotted as a function of radius at various times.

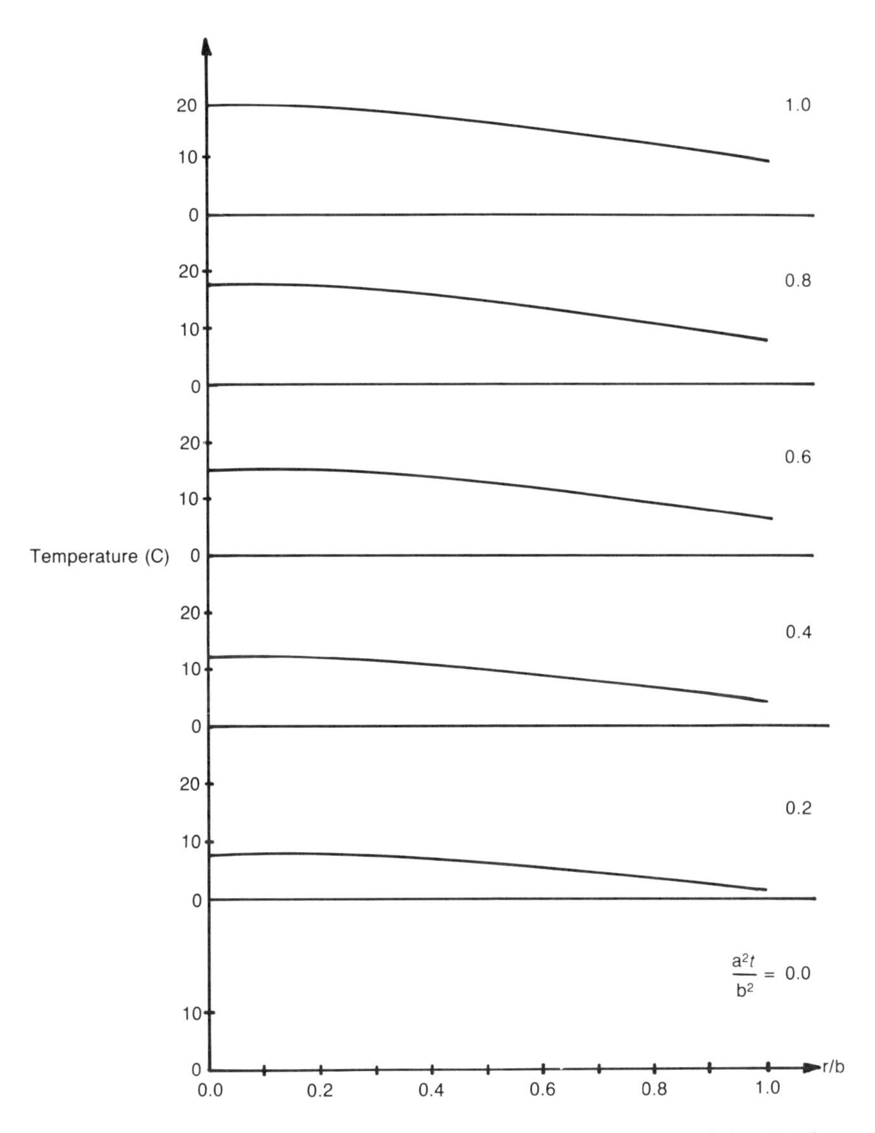

Fig. 6.6.1: The temperature field within an electric copper cable containing 37 wires when 22 amps. are flowing.

Exercises

1. The equation

$$\frac{\partial u}{\partial t} = \frac{G}{\varrho} + \upsilon\left(\frac{\partial^2 u}{\partial r^2} + \frac{1}{r}\frac{\partial u}{\partial r} \right)$$

governs the velocity along the pipe $u(r,t)$ of an incompressible fluid of density ϱ and kinematic viscosity υ flowing in a long circular pipe of radius a with an imposed, constant pressure gradient, $-G$.

If we assume that the fluid is initially at rest $u(r,0) = 0$ and there is no slip at the wall $u(a,t) = 0$, show that

$$u(r,t) = \frac{G}{4\varrho}(a^2 - r^2) - \frac{2Ga^2}{\mu}\sum_{n=1}^{\infty}\frac{J_0(k_n\frac{r}{a})}{k_n^3 J_1(k_n)}\exp\left(-\frac{k_n^2 \upsilon t}{a^2}\right)$$

where k_n the positive roots of $J_0(k_n) = 0$ and μ the dynamic viscosity. (Szymanski, F, 1932: Quelques solutions exactes des équations de l'hydrodynamique de fluide visqueux dans le cas d'un tube cylindreque. *J. de math. pures et appliquées,* Series 9, *11*, 67-107.)

2. The flow of water within a sloping unconfined aquifer is approximately governed by the linear Dupuit-Forchheimer equation

$$H\frac{\partial^2 u}{\partial x^2} + I\frac{\partial u}{\partial x} = \frac{\eta}{K}\frac{\partial u}{\partial t} - \frac{R}{K}$$

where $u(x,t)$ denotes the deviation of the hydraulic head from its average value H, I the slope of the substratum, η the aquifer porosity, K the hydraulic conductivity and R the areal accretion. (See Bear, J., 1972: *Dynamics of Fluids in Porous Medium*. New York: American Elsevier Publ. Co., p. 377.) If $u(0,t) = h_0$ while $u(L,t) = h_L$, show that the solution is

$$u(x,t) = h_0 + \frac{(h_L - h_0)x}{L} + \sum_{n=1}^{\infty}\frac{A_n}{\beta_n^2}\left(\exp\left(-\frac{\beta_n t}{L}\right) - 1\right)$$

$$\times \sin\left(\frac{n\pi x}{L}\right)\exp\left(-\frac{Ix}{2H}\right)$$

if the initial condition is

$$h(x,0) = h_0 + \frac{(h_L - h_0)x}{L}$$

where $\beta_n^2 = Hn^2\pi^2/L^2 + I^2/(4H)$, $c = \eta/K$ and

$$A_n = -\frac{2n\pi[R/K + I(h_L - h_0)/L]}{(IL/2H)^2 + n^2\pi^2}[1 - (-1)^n \exp(IL/2H)].$$

6.7 THE LASER HEATING OF AN OPAQUE SOLID WITHOUT VAPORIZATION

When a sufficiently high rate of heat (100 kW/cm^2, for example) is supplied to any material, vaporization or sublimation will eventually occur. Lasers provide a means of realizing such heat fluxes on a targeted surface for relatively long durations. The process of laser-beam heating and burn-through can be modeled as a one-dimensional, nonlinear heat transfer problem and has been recently resolved by Harrach (Harrach, R. J., 1977: Analytical solutions for laser heating and burnthrough of opaque solid slabs. *J. Appl. Phys., 48,* 2370-2383.)

The general problem is too difficult for us to solve here. However, a portion of the problem, namely, the laser heating of the slab until it begins to vaporize, provides a useful example in how to solve the heat equation when the heat fluxes are specified at both ends (i.e., Neumann boundary conditions at both ends). The evolution of temperature within the slab is given by

$$u_t = a^2 u_{xx}. \tag{6.7.1}$$

We assume that the slab initially has the temperature of zero, i.e., $u(x,0) = 0$. At the end $x = L$, we irradiate the slab with laser light of constant surface heat flux αI, where α is the absorptivity. At the origin we assume no heat flux. Consequently

$$u_x(0,t) = 0 \text{ and } \varkappa u_x(L,t) = \alpha I. \tag{6.7.2}$$

Our solution will be valid only until vaporization begins to occur.

The difficulty in solving this problem by separation of variables is the nonhomogeneous boundary condition $\varkappa u_x = \alpha I$ at $x = L$. In Section 6.3 we were able to find a steady-state solution $w(x)$ which removed those terms in the boundary condition that blocked our use of separation of variables. In the present case energy is continuously being supplied to the system so that no steady state exists.

Although we do not have a steady-state solution, we can still use the concept of finding a simple solution $w(x,t)$ which satisfies the heat equation and boundary conditions. We could then find a solution $v(x,t)$ by separation of variables so that $v(x,0) + w(x,0)$ satisfies the initial conditions.

We begin our search by guessing $w(x,t) = Ax^2$ because

$$w_x(0,t) = 0 \text{ and } w_x(L,t) = 2AL = \alpha I/\varkappa.$$

The only difficulty with this solution is the fact that $w_{xx}(x,t) \neq 0$ and the heat equation is no longer satisfied. However, if we modify $w(x,t)$ to be

$$w(x,t) = \frac{\alpha I x^2}{2L\varkappa} + \frac{\alpha I a^2 t}{\varkappa L}, \tag{6.7.3}$$

then $w(x,t)$ satisfies the heat equation and the boundary conditions. Consequently, if the solid initially has the temperature of zero, the total solution is

$$u(x,t) = \frac{\alpha I x^2}{2L\varkappa} + \frac{\alpha I a^2 t}{\varkappa L} + v(x,t) \qquad \textbf{(6.7.4)}$$

where the function $v(x,t)$ satisfies the equation

$$v_t = a^2 v_{xx} \qquad \textbf{(6.7.5)}$$

and

$$v_x(0,t) = v_x(L,t) = 0 \text{ and } v(x,0) = -\frac{\alpha I x^2}{2L\varkappa}. \qquad \textbf{(6.7.6)}$$

To solve $v(x,t)$ we use separation of variables $v(x,t) = X(x)T(t)$. Then

$$\frac{1}{X}\frac{d^2X}{dx^2} = \frac{1}{T}\frac{dT}{dt} = -k^2. \qquad \textbf{(6.7.7)}$$

The eigenfunctions which satisfy the boundary conditions are

$$\cos(\frac{n\pi x}{L}),\ n = 0,1,2,3,\ \ldots \qquad \textbf{(6.7.8)}$$

and the temporal part is

$$A_n \exp(-\frac{n^2\pi^2 a^2 t}{L^2}). \qquad \textbf{(6.7.9)}$$

Consequently, the total solution is given by

$$u(x,t) = \frac{\alpha I}{\varkappa}(\frac{x^2}{2L} + \frac{a^2 t}{L})$$

$$+ \sum_{n=0}^{\infty} A_n \cos(\frac{n\pi x}{L}) \exp(-\frac{n^2\pi^2 a^2 t}{L^2}). \qquad \textbf{(6.7.10)}$$

The coefficient A_n is given by

$$A_0 = \frac{1}{L}\int_0^L -\frac{\alpha I}{\varkappa}\frac{x^2}{2L}\,dx = -\frac{\alpha I}{2L^2}\frac{x^3}{3}\Big|_0^L$$

$$= -\frac{\alpha I}{6L^2\varkappa}L^3 = -\frac{\alpha I L}{6\varkappa}. \qquad \textbf{(6.7.11)}$$

$$A_n = \frac{2}{L}\int_0^L -\frac{\alpha I}{\varkappa}\frac{x^2}{2L}\cos(\frac{n\pi x}{L})\,dx \qquad \textbf{(6.7.12)}$$

$$= -\frac{\alpha I}{\varkappa L^2}\left[\frac{2x\cos(\frac{n\pi x}{L})}{(\frac{n\pi}{L})^2} + \frac{\frac{n^2\pi^2x^2}{L^2}-2}{(\frac{n\pi}{L})^3}\sin(\frac{n\pi x}{L})\right]\Bigg|_0^L \qquad \textbf{(6.7.13)}$$

$$= -\frac{\alpha I}{L^2}\frac{2L(-1)^n}{\varkappa(\frac{n\pi}{L})^2} = -\frac{2\alpha IL}{\varkappa}\frac{(-1)^n}{(n\pi)^2} \qquad \textbf{(6.7.14)}$$

$$n = 1,2,3,\ldots$$

The final solution may be written as

$$\frac{u(x,t)}{T_v} = \frac{\alpha IL}{T_v\varkappa\pi^2}\left[\frac{a^2\pi^2t}{L^2} + \frac{\pi^2}{2}(\frac{x}{L})^2 - \frac{\pi^2}{6} + \sum_{n=1}^{\infty}\frac{2(-1)^{n+1}}{n^2}\cos(\frac{n\pi x}{L})\exp(-\frac{n^2\pi^2a^2t}{L^2})\right].$$

$$\textbf{(6.7.15)}$$

where T_v is the temperature of vaporization.

As stressed earlier, this solution is valid until vaporization occurs at time t_v. Because the highest temperature occurs at $x = L$, vaporization will occur at $x = L$ when $x(L,t_v) = T_v$ or

$$\frac{T_v\varkappa\pi^2}{\alpha IL} = \tau_v + \frac{\pi^2}{3} - 2\sum_{n=1}^{\infty}\frac{1}{n^2}e^{-n^2\tau_v} \qquad \textbf{(6.7.16)}$$

where

$$\tau_v = \frac{\pi^2a^2t_v}{L^2} \qquad \textbf{(6.7.17)}$$

In Fig. 6.7.1, this nondimensional time constant has been graphed. For example, if we wanted to find out how long it would take a piece of aluminum of 10 cm thickness to reach the vaporization temperature when irradiated with a megawatt laser (α = 0.012), Eq. (6.7.16) and (6.7.17) predict the time as 6.9 sec. In Fig. 6.7.2, the temperature field has been illustrated within the slab at various times when $T_v\varkappa\pi^2/\alpha IL = 8$.

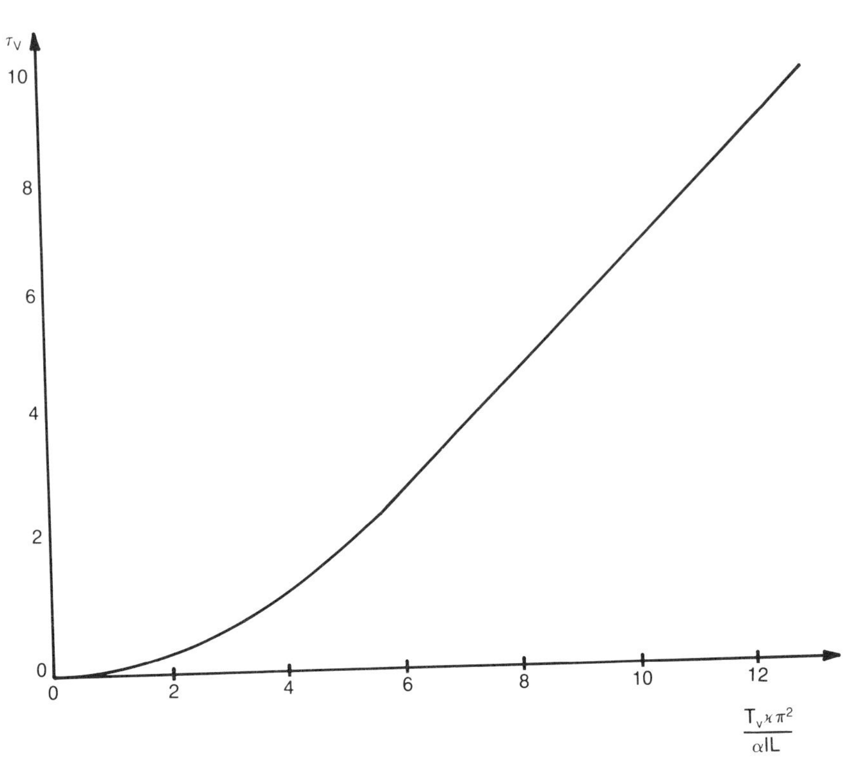

Fig. 6.7.1: Graphical illustration of Eq. (6.5.16) which gives the (nondimensional) time needed for a laser of intensity I to raise the temperature of a solid of thickness L to the vaporization temperature T_v.

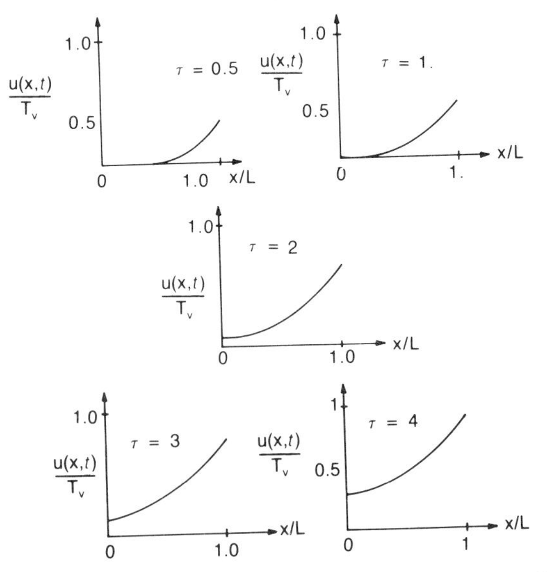

Fig. 6.7.2: Temperature field within the slab at various τ when $\frac{T_v \varkappa \pi^2}{\alpha I L} = 8.$

Exercise

1. The drawdown of finite, circular aquifer with constant well discharge is governed by the equation

$$\frac{\partial^2 u}{\partial r^2} + \frac{1}{r}\frac{\partial u}{\partial r} = \frac{S}{T}\frac{\partial u}{\partial t}$$

where r is the radial distance from the discharging well, $u(r,t)$ the groundwater head, t time, S the storage coefficient of the aquifer and T the transmissivity of the aquifer. If the rate of discharge of the well is Q, the boundary and initial conditions are

$$\frac{\partial u(b,t)}{\partial r} = 0, \ \lim_{r\to 0} r\frac{\partial u}{\partial r} = \frac{Q}{2\pi T} \text{ and } u(r,0) = 0$$

where b is the radius of the aquifer. Show that $u(r,t)$ equals

$$\frac{2\pi T}{Q}u(r,t) = \frac{3}{4} + \ln(\frac{r}{b}) - \{(\frac{r}{b})^2 + \frac{4Tt}{Sb}\}/2$$

$$+ 2\sum_{m=1}^{\infty}\frac{J_0(\alpha_m r)}{\alpha_m^2 b^2 J_0(\alpha_m b)^2}\exp(-\frac{\alpha_m^2 Tt}{S})$$

where α_m is the mth root of $J_0'(\alpha_m b) = 0$. See Kuiper, L. K., 1972: Drawdown in a finite circular aquifer with constant well discharge. *Water Resources Res., 8,* 734-736.

2. Show that the solution of

$$\frac{\partial u}{\partial t} = a^2\frac{1}{r}\frac{\partial}{\partial r}(r\frac{\partial u}{\partial r})$$

subject to the conditions that $u(0,t)$ is finite, $u_t(R,t) = h$ and $u(r,0) = u_0$ is

$$u(r,t) = u_0 + ht$$

$$-\frac{2hR^2}{a^2}\sum_{n=1}^{\infty}\frac{J_0(\lambda_n\frac{r}{R})[1-\exp(-\frac{a^2\lambda_n^2 t}{R^2})]}{\lambda_n^3 J_1(\lambda_n)}$$

where $J_0(\lambda_n) = 0$.

[Hint: Show that $u(r,t) = v(r,t) + u_0 + ht - \frac{hR^2}{4a^2}(1-(\frac{r}{R})^2)$ with $v_t(R,T) = 0$.]

6.8 THE SOLUTION OF THE HEAT EQUATION WITH RADIATION AT ONE END

So far we have dealt with problems where the temperature or flux of heat has been specified at the ends of the rod. In many physical applications, one or both of the ends are allowed to radiate to free space at temperature u_0. According to *Stefan's law*, the amount of heat radiated from a given area dA in a given time interval dt is

$$\sigma(u^4 - u_0^4)\, dA\, dt \tag{6.8.1}$$

where σ is called the Stefan-Bolzmann constant. On the other hand, the quantity of heat that may come to the surface by conduction from the interior of the body, assuming that we are at the right end of the bar, is given by

$$-\varkappa\, u_x\, dA\, dt \tag{6.8.2}$$

where $\varkappa$ is the thermal conductivity. Because these quantities must be equal, we obtain

$$-\varkappa\, u_x = \sigma(u^4 - u_0^4) \tag{6.8.3}$$

$$= \sigma(u - u_0)(u^3 + u^2 u_0 + u\, u_0^2 + u_0^3). \tag{6.8.4}$$

If u and u_0 are very nearly equal, the second bracketed term on the right-hand side of Eq. (6.8.4) may be approximated as $4u_0^3$. We write this approximate form of Eq. (6.8.4) as

$$-u_x = h(u - u_0) \tag{6.8.5}$$

where $h = 4\sigma u_0^3/\varkappa$. Equation (6.8.5) is often called the "radiation" boundary condition or "Newton's Law" because Newton's Law of Cooling states that cooling of a body by forced convection is given by Eq. (6.8.5). The constant h is called the *surface conductance* or the *coefficient of surface heat transfer*.

Let us now solve the problem of a rod that was initially heated to the temperature 100. We then allow the rod to cool by maintaining the temperature at zero at $x = 0$ and radiatively cooling to the surrounding air at the temperature of zero at $x = L$. We may restate the problem as

$$u_t = a^2\, u_{xx} \tag{6.8.6}$$

with

$$u(x,0) = 100, \tag{6.8.7}$$

$$u(0,t) = 0 \tag{6.8.8}$$

and

$$u_x(L,t) + h\, u(L,t) = 0. \tag{6.8.9}$$

Once again we assume a product solution $u(x,t) = X(x)T(t)$ so that

$$\frac{X''}{X} = \frac{T'}{a^2 T} = -k^2. \tag{6.8.10}$$

We obtain

$$X'' + k^2 X = 0 \tag{6.8.11}$$

but the boundary conditions are now

$$X(0) = 0 \text{ and } X'(L) + hX(L) = 0 \tag{6.8.12}$$

The most general solution is

$$X(x) = A \cos(kx) + B \sin(kx). \tag{6.8.13}$$

However, $A = 0$ because $X(0) = 0$. On the other hand, we obtain

$$k \cos(kL) + h \sin(kL) = kL \cos(kL) + hL \sin(kL) = 0 \tag{6.8.14}$$

if $B \neq 0$. The nondimensional number hL is often called the *Biot number* and is completely dependent upon the physical characteristics of the rod.

In Chapter 1, we saw how to find the roots of the transcendental equation

$$\alpha + hL \tan(\alpha) = 0 \tag{6.8.15}$$

where $\alpha = kL$. Consequently if α_n is the nth root of Eq. (6.8.15), then

$$X_n(x) = \sin(\alpha_n x/L) \tag{6.8.16}$$

In Table 6.8.1, the first ten roots of Eq. (6.8.15) are listed for $hL = 1$.

In general, Eq. (6.8.15) must be solved either numerically or graphically. If α is large, however, we can find approximate values by noting that

Table 6.8.1: The First Roots of Eq. (6.8.15) and C_n for hL = 1.

n	α_n	Approximate α_n	C_n
1	2.0288	2.2074	118.9193
2	4.9132	4.9246	31.3402
3	7.9787	7.9813	27.7554
4	11.0856	11.0865	16.2878
5	14.2075	14.2079	14.9923
6	17.3364	17.3366	10.8359
7	23.6044	23.6043	8.0998
8	26.7410	26.7409	7.7478
9	29.8786	29.8776	6.4625
10	33.0170	33.0170	6.2351

$$\cot(\alpha) = -\frac{hL}{\alpha} \simeq 0 \quad (6.8.17)$$

or

$$\alpha_n = \frac{(2n-1)\pi}{2} \quad (6.8.18)$$

where $n = 1,2,3, \ldots$ A better approximation may be obtained by

$$\alpha_n = \frac{(2n-1)\pi}{2} - \epsilon_n \quad (6.8.19)$$

where $\epsilon_n << 1$. Substituting into Eq. (6.8.15),

$$[(2n-1)\pi/2 - \epsilon_n] \cot[(2n-1)\pi/2 - \epsilon_n] + hL = 0. \quad (6.8.20)$$

We can simplify Eq. (6.8.20) to

$$\epsilon_n^2 + (2n-1)\pi\epsilon_n/2 + hL = 0 \quad (6.8.21)$$

because $\cot[(2n-1)\pi/2 - \theta] = \tan(\theta)$ and $\tan(\theta) \simeq \theta =$ for $\theta << 1$. Solving for ϵ_n, we find

$$\epsilon_n \simeq \frac{-2hL}{(2n-1)\pi} \quad (6.8.22)$$

and

$$\alpha_n \simeq \frac{(2n-1)\pi}{2} + \frac{2hL}{(2n-1)\pi}. \quad (6.8.23)$$

In Table 6.8.1, the approximate roots given by Eq. (6.8.23) are compared with the actual roots.

The temporal part is

$$T_n(t) = C_n \exp(-k_n^2 a^2 t) = C_n \exp(-\frac{\alpha_n^2 a^2 t}{L^2}). \quad \textbf{(6.8.24)}$$

Consequently the general solution is given by

$$u(x,t) = \sum_{n=1}^{\infty} C_n \sin(\alpha_n \frac{x}{L}) \exp(-\frac{\alpha_n^2 a^2 t}{L^2}) \quad \textbf{(6.8.25)}$$

where α_n is the nth root of Eq. (6.8.15).

To determine C_n, we use the initial condition Eq. (6.8.7):

$$100 = \sum_{n=1}^{\infty} C_n \sin(\alpha_n x/L). \quad \textbf{(6.8.26)}$$

Equation (6.8.26) is an eigenfunction expansion of 100 for the Sturm-Liouville problem

$$X'' + k^2 X = 0 \quad \textbf{(6.8.27)}$$

with

$$X(0) = X'(L) + h\, X(L) = 0. \quad \textbf{(6.8.28)}$$

Consequently the coefficient C_n is given in Eq. (1.4.4):

$$C_n = \frac{\int_0^L 100 \sin(\alpha_n x/L)\, dx}{\int_0^L \sin^2(\alpha_n x/L)\, dx} \quad \textbf{(6.8.29)}$$

because $r(x) = 1$. Peforming the integrations, we find

$$C_n = \frac{100 \frac{L}{\alpha_n} [1 - \cos(\alpha_n)]}{[L - \frac{L}{2\alpha_n} \sin(2\alpha_n)/2\alpha_n]/2} \quad \textbf{(6.8.30)}$$

$$= \frac{200}{\alpha_n} \frac{1 - \cos(\alpha_n)}{1 + \cos^2(\alpha_n)/hL} \quad \textbf{(6.8.31)}$$

because $\sin(2\alpha_n) = 2\cos(\alpha_n)\sin(\alpha_n)$ and $\alpha_n = -hL\tan(\alpha_n)$.

The final solution may be written

$$u(x,t) = \sum_{n=1}^{\infty} \frac{200}{\alpha_n} \frac{1 - \cos(\alpha_n)}{1 + \cos^2(\alpha_n)/hL}$$

$$\times \sin(\alpha_n x/L) \exp(- \frac{a^2\alpha_n{}^2 t}{L^2}). \quad \textbf{(6.8.32)}$$

This solution has been graphed for $hL = 1$ at various times and positions in Fig. 6.8.1. Only ten terms of the series (6.8.32) were used.

6.9 COOLING OF AN INFINITELY LONG CYLINDER BY RADIATION

In this section we find the evolution of the temperature field within a cylinder of radius b as it radiatively cools from an initial temperature T_0. The heat equation is

$$\frac{\partial u}{\partial t} = a^2(\frac{\partial^2 u}{\partial r^2} + \frac{1}{r} \frac{\partial u}{\partial r}) \quad \textbf{(6.9.1)}$$

which we shall solve by separation of variables

$$u(r,t) = R(r)\ T(t). \quad \textbf{(6.9.2)}$$

Therefore

$$\frac{1}{R} (\frac{d^2R}{dr^2} + \frac{1}{r} \frac{dR}{dr}) = \frac{1}{a^2T} \frac{dT}{dt} = - \frac{k^2}{b^2} \quad \textbf{(6.9.3)}$$

because only a negative separation constant yields a $R(r)$ which is finite at the origin. This solution is

$$R(r) = J_0(k \frac{r}{b}) \quad \textbf{(6.9.4)}$$

where J_0 is the Bessel function of the first kind of order zero.

The radiative boundary condition may be expressed as

$$\frac{\partial u(b,t)}{\partial r} + h\ u(b,t) = T(t)\ [\frac{dR(b)}{dr} + h\ R(b)] = 0. \quad \textbf{(6.9.5)}$$

Because $T(t)$ cannot equal to zero, we have

$$k\ J_0'(k) + hb\ J_0(k) = -\ k\ J_1(k) + hb\ J_0(k) = 0 \quad \textbf{(6.9.6)}$$

where the product hb is called the Biot number. The solution of

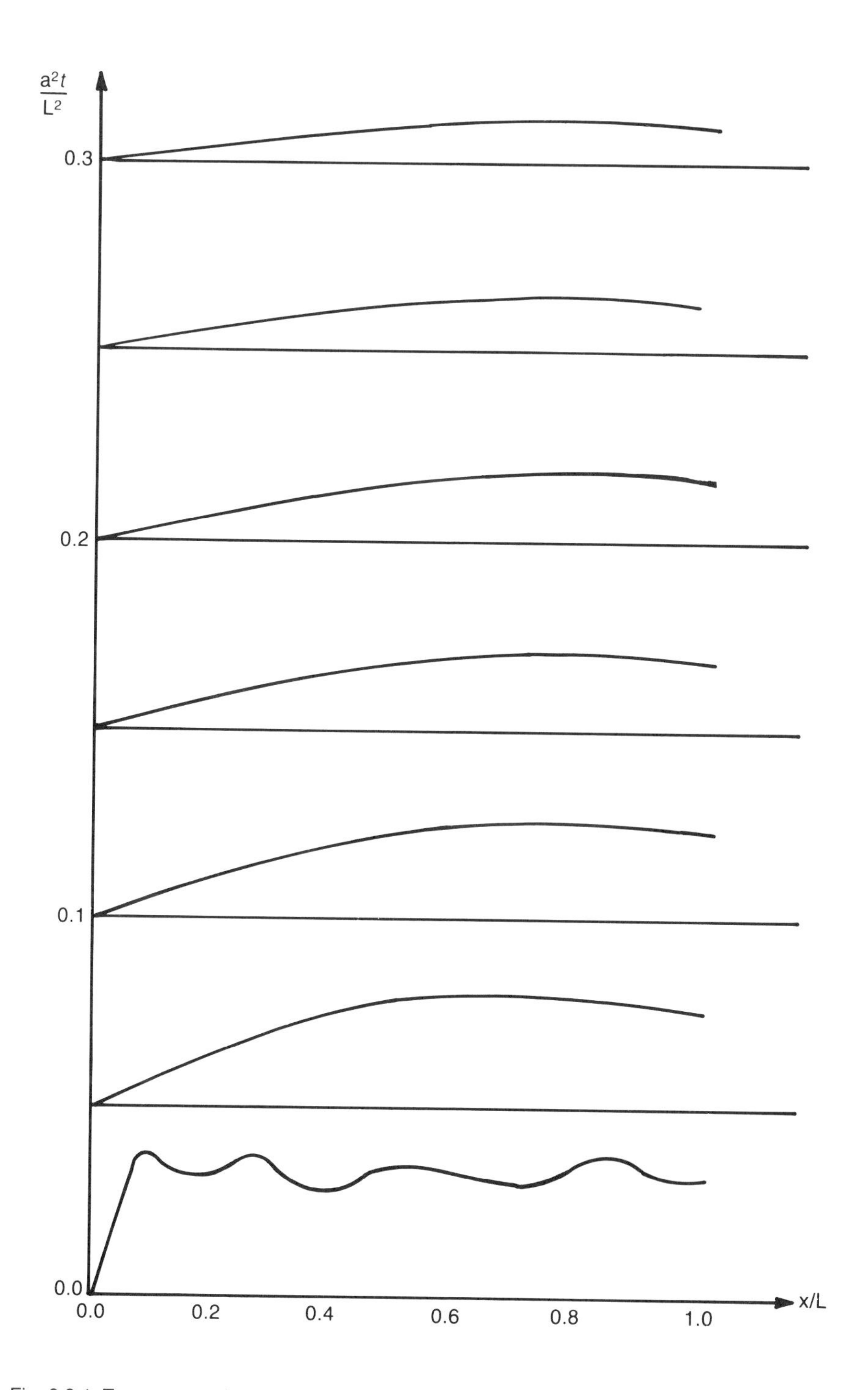

Fig. 6.8.1: Temperature field in a bar initially at 100 degrees and then allowed to radiatively cool at $x = L$ while the temperature is zero at $x = 0$.

the transcendental equation (6.9.6) yields an infinite number of distinct k_n. For each of these k_n's, the temporal part of the problem is given by

$$\frac{dT_n}{dt} + \frac{k_n^2 a^2}{b^2} T_n = 0 \qquad \textbf{(6.9.7)}$$

which has the solution

$$T_n(t) = A_n \exp(-\frac{k_n^2 a^2 t}{b^2}). \qquad \textbf{(6.9.8)}$$

The product solution is, therefore,

$$u_n(r,t) = A_n J_0(k_n \frac{r}{b}) \exp(-\frac{k_n^2 a^2 t}{b^2}) \qquad \textbf{(6.9.9)}$$

and the most general solution is a sum of these product solutions

$$u(r,t) = \sum_{n=1}^{\infty} A_n J_0(k_n \frac{r}{b}) \exp(-\frac{k_n^2 a^2 t}{b^2}). \qquad \textbf{(9.9.10)}$$

Finally, we must determine A_n. From the initial condition that $u(r,0) = T_0$, we have

$$T_0 = \sum_{n=1}^{\infty} A_n J_0(k_n \frac{r}{b}) \qquad \textbf{(6.9.11)}$$

where

$$A_n = \frac{2 k_n^2 T_0}{b^2(k_n^2 + b^2h^2)J_0(k_n)^2} \int_0^b r J_0(k_n \frac{r}{b}) dr \qquad \textbf{(6.9.12)}$$

$$= \frac{2 T_0}{(k_n^2 + b^2h^2)J_0(k_n)^2} (k_n \frac{r}{b}) J_1(k_n \frac{r}{b}) \Big|_0^b \qquad \textbf{(6.9.13)}$$

$$= \frac{2 T_0 k_n J_1(k_n)}{(k_n^2 + b^2h^2)J_0(k_n)^2} \qquad \textbf{(6.9.14)}$$

$$= \frac{2 T_0 k_n J_1(k_n)}{k_n^2 J_0(k_n)^2 + b^2h^2 J_0(k_n)^2} \qquad \textbf{(6.9.15)}$$

$$= \frac{2 T_0 k_n J_1(k_n)}{k_n^2 J_0(k_n)^2 + k_n^2 J_1(k_n)^2} \qquad \textbf{(6.9.16)}$$

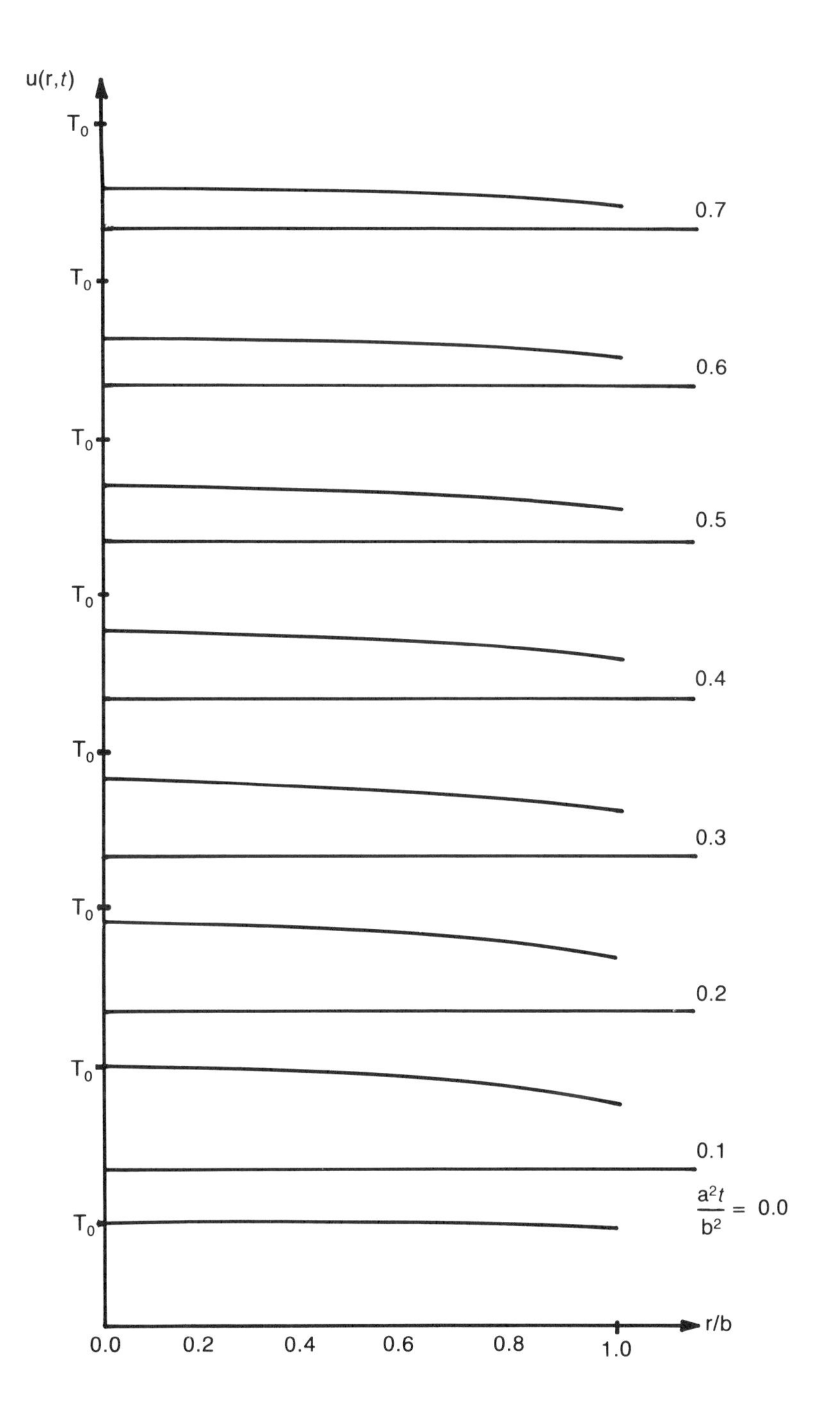

Fig. 6.9.1: The temperature within an infinitely long cylinder as it radiates to space after being heated to the temperature T_0.

$$A_n = \frac{2\,T_0}{J_0(k_n)^2 + J_1(k_n)^2} \; \frac{J_1(k_n)}{k_n} \qquad \textbf{(6.9.17)}$$

from Eqs. (1.7.18), (1.7.32), (1.7.42) and (6.9.6).

Consequently, the final solution is

$$u(r,t) = 2T_0 \sum_{n=1}^{\infty} \frac{J_1(k_n) J_0(k_n r/b) \exp(-k_n{}^2 a^2 t/b^2)}{(J_0(k_n)^2 + J_1(k_n)^2) \;\; k_n} \qquad \textbf{(6.9.18)}$$

This solution is graphed in Fig. 6.9.1 for the Biot number $hb = 1$ at various Fourier numbers a^2t/b^2. Only the first ten terms of the expansion were used.

6.10 THE SOLUTION OF THE HEAT EQUATION WITH THERMAL FLUXES AT THE BOUNDARIES: THE TEMPERATURE REACHED IN DISC BRAKES

The importance of heat transfer in the proper design of automobiles has long been recognized. In this section we determine the transient temperatures reached in a disc brake during a single brake application. It will involve solving a *Robin's* (third kind) *boundary condition* at both ends.

Disc brakes consist of two blocks of frictional material known as pads which are pressed against each side of a rotating annulus usually made of a ferrous material. See Fig. 6.10.1. If in a single brake application a constant torque acts on each pad and produces a uniform disc deceleration, then the rate of heat generation from the friction surfaces decreases linearly with time:

$$N(1 - Mt) \qquad \textbf{(6.10.1)}$$

where M and N are constants determined from the rate of evolution of heat during breaking. As Newcomb has shown (Newcomb, T. P., 1960: Temperature reached in disc brakes. *J. Mech. Eng. Sci.*, *2,* 167-177) for conventional pad materials, the heat generated flows wholly into the brake disc and the problem can be treated as linear flow of heat through an infinite slab bounded by parallel planes $(x = \pm L)$ where the thermal flux through the two plane boundaries decreases linearly with time.

In this problem the heat equation is given by

$$u_t = a^2\, u_{xx} \qquad \textbf{(6.10.2)}$$

where a^2 is the diffusivity of the disc and the origin is at the midpoint of the brake disc. At the planes $x = \pm L$ there is a uniformly

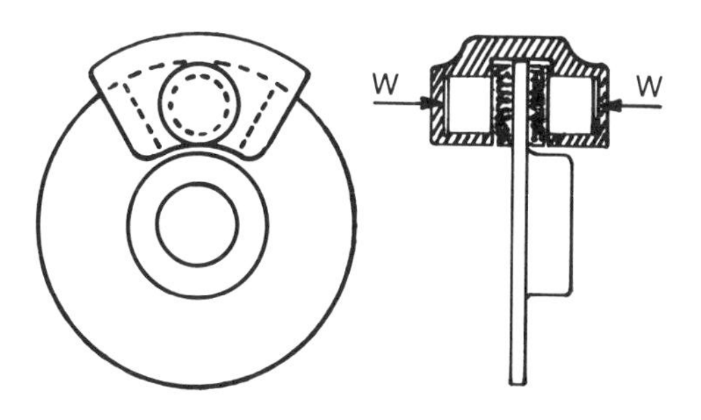

Fig. 6.10.1: Diagram of a disc brake.

distributed, thermal flux $N(1 - Mt)$ together with an average loss of heat due to convection which is proportional to the temperature. That is,

$$\varkappa u_x = N(1 - Mt) - hu \qquad \text{at } x = L \qquad \textbf{(6.10.3)}$$

and

$$\varkappa u_x = -N(1 - Mt) + hu \qquad \text{at } x = -L \qquad \textbf{(6.10.4)}$$

where $\varkappa$ is the thermal conductivity and h the heat transfer coefficient.

If we attempt to use separation of variables in this problem, we immediately run into difficulty from the boundary conditions. In Section 6.7 we introduced a simple function of x and t that satisfied the heat equation and boundary conditions but not necessarily the initial condition. Because the problem is linear, we found a second solution $v(x,t)$ by separation of variables that allowed the sum of $w(x,t) + v(x,t)$ to satisfy the heat equation and boundary conditions as well as the initial condition. We will again employ this technique here although in most instances it will not work and Laplace transforms must be used (see Section 6.14).

If we guess the solution $w(x,t) = \frac{N}{h}(1 - Mt)$, we satisfy the boundary conditions because $w_x = 0$. On the other hand, we do not satisfy the heat equation. If we backtrack and guess

$$w(x,t) = -\frac{NM}{2a^2h}x^2 + Bx + C + \frac{N}{h}(1 - Mt) \qquad \textbf{(6.10.5)}$$

then we satisfy the heat equation. The arbitrary constants B and C are chosen so that we satify the boundary conditions:

$$\varkappa\left[-\frac{NML}{a^2h}+B\right]+h\left[-\frac{NML^2}{2a^2h}+BL+C\right]=0 \quad \textbf{(6.10.6)}$$

and

$$\varkappa\left[\frac{NML}{a^2h}+B\right]-h\left[-\frac{NML^2}{2a^2h}-BL+C\right]=0 \quad \textbf{(6.10.7)}$$

Solving Eq. (6.10.6)-(6.10.7), we find that

$$B = 0 \quad \textbf{(6.10.8)}$$

and

$$C = \frac{NM}{a^2}\left(\frac{L^2}{2h}+\frac{\varkappa L}{h^2}\right). \quad \textbf{(6.10.9)}$$

Having found this particular solution to the problem, we can write

$$u(x,t) = w(x,t) + v(x,t) \quad \textbf{(6.10.10)}$$

where

$$w(x,t) = \frac{NM}{a^2}\left[\frac{L^2-x^2}{2h}+\frac{\varkappa L}{h^2}\right]+\frac{N}{h}(1-Mt) \quad \textbf{(6.10.11)}$$

so that

$$v_t = a^2\, v_{xx} \quad \textbf{(6.10.12)}$$

with

$$\varkappa\, v_x \pm h\, v = 0 \quad \text{at } x = \pm L \quad \textbf{(6.10.13)}$$

We recognize the boundary condition (6.10.13) as a Robin's boundary condition. Assuming that $v(x,t)$ can be written as $X(x)T(t)$, we obtain for $X(x)$

$$\frac{d^2X}{dx^2} + k^2X = 0 \quad \textbf{(6.10.14)}$$

with

$$\varkappa\, X'(L) + h\, X(L) = 0 \quad \textbf{(6.10.15)}$$

and

$$\varkappa X'(-L) - h X(-L) = 0. \qquad \textbf{(6.10.16)}$$

The only nontrival solution to Eq. (6.10.14) is

$$X(x) = A \cos(kx) + B \sin(kx). \qquad \textbf{(6.10.17)}$$

Substituting into Eqs. (6.10.15)-(6.10.16)

$$\varkappa (-kA \sin(kL) + kB \cos(kL)) + h [A \cos(kL) + B \sin(kL)] = 0 \qquad \textbf{(6.10.18)}$$

$$\varkappa (kA \sin(kL) + kB \cos(kL)) - h [A \cos(kL) - B \sin(kL)] = 0 \qquad \textbf{(6.10.19)}$$

which leads to either

$$B = 0 \text{ and } (kL) \tan(kL) = \frac{hL}{\varkappa} \qquad \textbf{(6.10.20)}$$

or

$$A = 0 \text{ and } \tan(kL) = \frac{\varkappa}{hL} (kL). \qquad \textbf{(6.10.21)}$$

At present the solution $u(x,t)$ is made up of two infinite series. One corresponds to terms from the roots of Eq. (6.10.20) and the other from Eq. (6.10.21). However, the initial condition

$$u(x,0) = w(x,0) + u(x,0) = 0$$

shows that we must discard the sine terms because $w(x,0)$ is an even function and only cosine terms can construct an even function over the interval $[-L,L]$.

Therefore

$$u(x,t) = \frac{NM}{a^2} \left(\frac{L^2 - x^2}{2h} + \frac{\varkappa L}{h^2} \right) + \frac{N}{h} (1 - Mt) + \sum_{n=0}^{\infty} A_n \cos(k_n x)\, e^{-k_n^2 a^2 t} \qquad \textbf{(6.10.22)}$$

where k_n are the positive roots of Eq. (6.10.20). If $w(x,0)$ had contained both even and odd functions, then we would have used both sine and cosine terms. In our present problem A_n is given by

$$A_n = \frac{-\int_{-L}^{L} \left[\frac{NM}{a^2} \left(\frac{L^2 - x^2}{2h} + \frac{\varkappa L}{h^2} \right) + \frac{N}{h} \right] \cos(k_n x)\, dx}{\int_{-L}^{L} \cos^2(k_n x)\, dx} \qquad \textbf{(6.10.23)}$$

$$A_n = \frac{\left[\frac{NM}{a^2}\left(\frac{L^2}{2h} + \frac{\varkappa L}{h^2}\right) + \frac{N}{h}\right]\frac{2\sin(k_nL)}{k_n}}{L + \frac{\sin(k_nL)}{2k_n}}$$

$$+ \frac{\frac{2NM}{a^2}\left[\frac{2L\cos(k_nL)}{2hk_n^2} + \frac{k_n^2L^2 - 2}{2hk_n^3}\sin(k_nL)\right]}{L + \frac{\sin(k_nL)}{2k_n}} \quad \textbf{(6.10.24)}$$

$$= \frac{-\frac{2}{k_n}\sin(k_nL)\left(\frac{NM\varkappa L}{a^2h^2} + \frac{N}{h}\right)}{L + \frac{\sin(k_nL)\cos(k_nL)}{k_n}}$$

$$- \frac{\frac{2NM}{a^2hk_n^2}\left(\frac{L\varkappa}{h}k_n\sin(k_nL) - \frac{\sin(k_nL)}{k_n}\right)}{L + \frac{\sin(k_nL)\cos(k_nL)}{k_n}} \quad \textbf{(6.10.25)}$$

because $\cos(k_nL) = \varkappa k_n \sin(k_nL)/h$.

$$A_n = \frac{-2N\sin(k_nL)\left(1 + \frac{M}{a^2k_n^2}\right)}{hk_n\left(L + \frac{\varkappa}{h}\sin^2(k_nL)\right)} \quad \textbf{(6.10.26)}$$

$$= \frac{-2N\varkappa}{h^2}\,\frac{\sin^2(k_nL)}{\cos(k_nL)}\,\frac{1 + \frac{M}{a^2k_n^2}}{L + \frac{\varkappa}{h}\sin^2(k_nL)} \quad \textbf{(6.10.27)}$$

$$= \frac{-2N\varkappa}{h^2\cos(k_nL)}\,\frac{1 + \frac{M}{a^2k_n^2}}{L\csc^2(k_nL) + \varkappa/h} \quad \textbf{(6.10.28)}$$

$$= \frac{-2N/h}{\cos(k_nL)}\,\frac{1 + \frac{M}{a^2k_n^2}}{1 + hL/\varkappa + L\varkappa k_n^2/h} \quad \textbf{(6.10.29)}$$

The value of k_n have been computed from Eq. (6.10.20) by Newton's method when h = 0.0013 cal/cm^2 °C sec, $\varkappa$ = 0.115 cal/cm sec °C and L = 0.635 cm and given the results in Table

Table 6.10.1: Eigenvalues and Spectral Coefficients for the Five Gravest Modes from the Solution of the Temperature Distribution with a Disc Brake as a Racing Car Stops from 150 mph to Rest in 10 Seconds.

n	k_n (cm^{-1})	$n\pi/L$ (cm^{-1})	A_n (°C)
0	0.13327	0.00000	−1,952,634
1	4.95098	4.94739	63
2	9.89658	9.89478	−15
3	14.84337	14.84217	7
4	19.79045	19.78956	−4

6.10.1. The eigenfunctions have also been compared for $n = 0$ and 1 with the corresponding eigenfunctions in a Fourier cosine series in Fig. 6.10.2 for various $hL/\varkappa$.

In most practical cases involving disc brakes, h is small compared to $\varkappa$ and the eigenvalues are very nearly given by $n\pi/L$. Furthermore, as Table 6.10.1 shows, the coefficient A_0 is so large that all the remaining terms may be neglected except for times near $t = 0$. Consequently the value of k_0 must be computed very accurately.

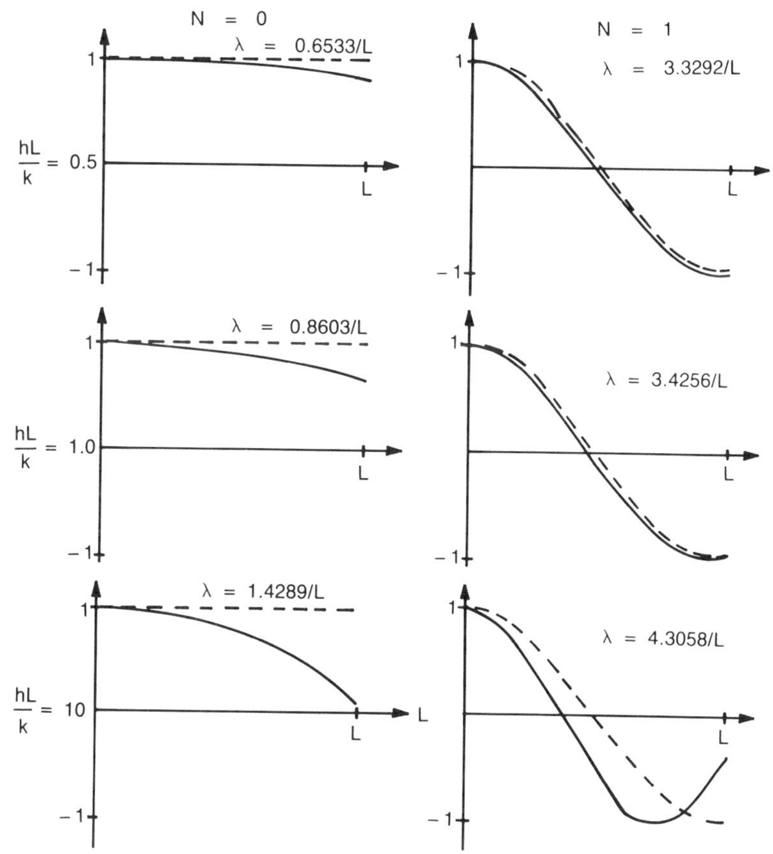

Fig. 6.10.2: Plots of the two gravest modes (n = 0 and n = 1) for the disc brake problem (solid line). Also plotted are the first two modes from a Fourier cosine series (dashed line).

In Fig. 6.10.3 the solutions have been graphed when

$$a^2 = 0.123 \text{ cm}^2/\text{sec},$$

$$h = 0.0013 \text{ cal/cm}^2 \text{ °C sec}, \ \varkappa = 0.115 \text{ cal/cm sec °}C,$$

$$L = 0.635 \text{ cm}, \ N = 54.2 \text{ cal/cm}^2 \text{ sec}$$

(corresponding to a braking from 150 mph to rest in 10 sec) and

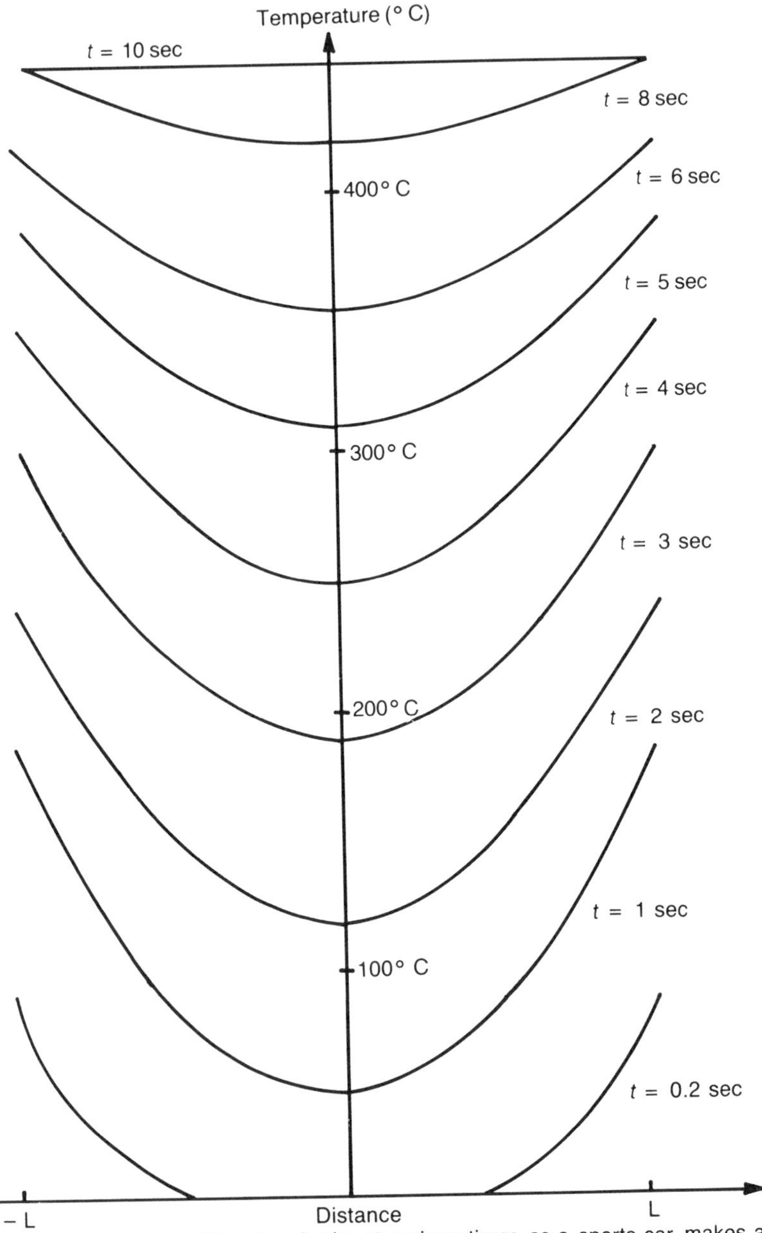

Fig. 6.10.3: Temperature with a disc brake at various times as a sports car makes a panic stop from 150 mph to rest in 10 sec.

$M = 1/\tau$ where τ is the time of a single brake application (10 sec). Because the diffusivity of the lining is so small compared to the disc ($a^2 = 10^{-3}$ cm²/sec), all the heat flow occurs within the disc. From Fig. 6.10.3 we see that the temperature with a disc brake can climb over 400 °C in the panic stop of a sports car.

Exercises

1. Find the temperature within a rod of length L with insulated lateral boundaries when the end $x = 0$ is maintained at $u(0,t) = T_1$ and heat escapes through the end $x = L$ according to Newton's law of cooling

$$hL\ u_x(L,t) + u(L,t) - T_0 = 0$$

where T_0 is the temperature of the surrounding medium and h is a constant. The initial temperature distribution along the rod is prescribed as $u(x,0) = f(x)$.

2. Find the temperature $u(x,t)$ in a slender rod of length L whose curved surface and left-end are perfectly insulated $u_x(0,t) = 0$ and whose right-hand end radiates into air of constant temperature 0 °C (that is, $u_x(L,t) = -h\ u(L,t)$). The rod is initially at the temperature of 100 °C throughout.

3. Rework problem 2 with the only difference being

$$u_x(L,t) = -h(u(L,t) - 70).$$

(Hint: Let $U(x,t) = u(x,t) - 70$ be the new dependent variable.)

4. Rework problem 2 if the only change is $u(0,t) = 100$.

6.11 THE PERIODIC VARIATION IN THE EARTH'S CRUSTAL TEMPERATURE

Observations of the temperature at points near the surface of the earth have been carried out at a large number of meteorological stations in different parts of the world for many years. These results have established that the variations of the surface temperature from the heat by day to the cold at night do not affect the temperature of points at a depth of more than three to four feet while annual changes from winter to summer may be observed up to a depth of 60 to 70 feet. Below that depth the temperature remains essentially constant from day to day and is not subject to alterations due to changes at the surface. In other words, the heat waves due to the changes at the surface die away before they penetrate to the depth of more than 60 to 70 feet. In this section we will explain this phenomena through the study of the heat equation subject to a periodic forcing at one of the boundaries.

Our analysis of this problem starts with the heat equation

$$u_t = a^2 u_{xx} \qquad \textbf{(6.11.1)}$$

where x increases downward into the ground. The boundary conditions at the earth's surface is $u(0,t) = T_0 \cos(\omega t)$. We also require that the temperature remains finite as x tends to infinity. We now guess a solution of the form:

$$u(x,t) = w(x) \cos(\omega t) + v(x) \sin(\omega t). \qquad \textbf{(6.11.2)}$$

Substituting into Eq. (6.11.1) we have the system of ordinary differential equations

$$a^2 w''(x) - \omega v(x) = 0 \qquad \textbf{(6.11.3)}$$
$$a^2 v''(x) + \omega w(x) = 0 \qquad \textbf{(6.11.4)}$$

where $w(0) = T_0$, $v(0) = 0$, $w(\infty)$ and $v(\infty)$ are finite. These equations may be combined to give

$$a^4 w^{(IV)} + \omega^2 w(x) = 0. \qquad \textbf{(6.11.5)}$$

The general solution of Eq. (6.11.5) is

$$\begin{aligned} w(x) &= e^{-\gamma x}[A \cos(\gamma x) + B \sin(\gamma x)] \\ &+ e^{\gamma x}[C \cos(\gamma x) + D \sin(\gamma x)] \end{aligned} \qquad \textbf{(6.11.6)}$$

where $\gamma = (\omega/2a^2)^{1/2}$. In order for the solution to remain finite as x goes to infinity, we must take $C = D = 0$. The other two conditions that we must satisfy are $w(0) = T_0$ and

$$a^2 w''(0) = \omega v(0) = 0$$

which leads immediately to $A = T_0$ and $B = 0$. From Eq. (6.11.3) we find that $v(x) = T_0 e^{-\gamma x} \sin(\gamma x)$.
Consequently

$$u(x,t) = T_0 e^{-\gamma x}[\cos(\gamma x) \cos(\omega t) + \sin(\gamma x) \sin(\omega t)] \qquad \textbf{(6.11.7)}$$

$$= T_0 e^{-\gamma x} \cos(\omega t - \gamma x) \qquad \textbf{(6.11.8)}$$

Equation (6.11.8) is a wave propagating down in the earth (with a wavelength $\lambda = 2\pi/\gamma$) which decays with depth. For typical rock material with $a^2 = 0.01$ cm^2/sec, the wavelength is about 2.7 cm for a frequency of 1 cycle/min., 1 m for 1 cycle/day, and 20 m for 1 cycle/year. For a metallic conductor with $a^2 = 1$ cm^2/sec, the wavelength is 3.5 cm at 1 cycle/sec, and 27 cm at 1 cycle/min. while for metals at temperatures near absolute zero, where a^2 is of the order 10^4 cm^2/sec, the wavelength is 11 cm at 1000 cycles/sec.

The time at which a maximum or minimum of temperature will occur at any point is that for which

$$\omega t - \gamma x = (2n+1)\pi/2 \qquad \textbf{(6.11.9)}$$

or

$$t = \frac{\gamma x + (2n+1)\pi/2}{\omega} \tag{6.11.10}$$

where odd values of n gives minima, and even, maxima. Fixing our attention on the minimum that occurs at the surface when, say, $\omega t = 3\pi/2$, we see that if both x and t increase so that

$$\omega t - \gamma x = \frac{3}{2}\pi \tag{6.11.11}$$

we may think of this particular minimum being propagated into the medium and reaching any point x at the time given by this equation. This is later than its occurrence at the surface by an amount

$$t_1 = \frac{x}{\sqrt{2a^2\omega}} \tag{6.11.12}$$

which may be called the "lag" of the temperature wave. The same reasoning holds for the maximum, or zero, or any other phase.

In Fig.6.11.1 the diurnal temperature wave is plotted at the moment when the surface temperature is equal to zero. It was assumed that the daily temperature varied from −4 °C to 16 °C. The temperature lag in this particular case is 9.7 hours.

The lag between the time that a surface is heated and the time that it takes for that information to reach a point in the interior plays an important role in the design of electrical appliances, such as ovens, when the heating element is considerably separated in distance from the thermostat. Because a finite amount of time is

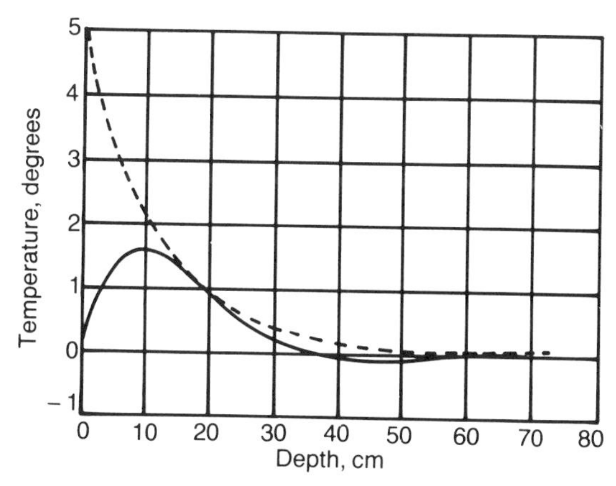

Fig. 6.11.1: Curves showing the penetration of the diurnal temperature wave in soil of diffusivity of 0.0049 cgs. Solid line is curve of the temperature in the early evening. Broken line is the curve when the temperature at the surface is 5° C. Taken from Ingersoll, L. R., O. J. Zobel and A. C. Ingersoll, 1954: *Heat conduction with engineering, geological and other applications*. Madison, WI: The University of Wisconsin Press, p. 51. (Permission of the University of Wisconsin Press.)

required for the effects of the heating to reach the thermostat, an excessive amount of heating will take place near the heating element. Eventually the temperature at the thermostat will rise to the point where the thermostat will turn off the heater and cooling will begin to occur. As the extra amount of heat moves towards and by the thermostat, the heater will remain off even though the area around the heater may drop below the desired quantity. Eventually this information will reach the thermostat, the heater will be turned on and the entire cycle will repeat.

To illustrate this effect quantitatively, let us solve the heat equation

$$u_t = a^2 u_{xx} \tag{6.11.13}$$

where a^2 is the thermal diffusivity when a heater is placed at $x = L$ and the thermostat at $x = 0$. At $x = 0$, the wall is insulated while at $x = L$ heat is supplied to the wall at a rate proportional to the departure of the temperature from the desired temperature T_0:

$$-\varkappa A\, u_x(L,t) = Q(u(0,t) - T_0). \tag{6.11.14}$$

The constant $\varkappa$ is the thermal conductivity, A the area of the end of the bar and Q the heating rate.

Because we anticipate that the solution will be periodic in time, we guess

$$u(x,t) = T_0 + w(x) \cos(\omega t) + v(x) \sin(\omega t). \tag{6.11.15}$$

We have already found the general solution for $u(x)$, namely Eq. (6.11.6). However, it proves very useful to write this solution using complex variables

$$w(x) = c_1 \cosh[(1+i)\gamma x] + c_2 \sinh[(1+i)\gamma x] \tag{6.11.16}$$

$$v(w) = 2ic_1 \cosh[(1+i)\gamma x] + 2ic_2 \sinh[(1+i)\gamma x] \tag{6.11.17}$$

where ω, and hence γ, is still unknown. Because

$$\cosh[(1+i)\gamma x] = \cosh(\gamma x) \cos(\gamma x) + i \sinh(\gamma x) \sin(\gamma x) \tag{6.11.18}$$

$$\sinh[(1+i)\gamma x] = \sinh(\gamma x) \cos(\gamma x) + i \cosh(\gamma x) \sin(\gamma x) \tag{6.11.19}$$

$$\cosh(\gamma x) = (e^{\gamma x} + e^{-\gamma x})/2 \tag{6.11.20}$$

$$\sinh(\gamma x) = (e^{\gamma x} - e^{-\gamma x})/2 \tag{6.11.21}$$

it is readily shown that Eq. (6.11.16) and Eq. (6.11.6) are identical if the complex constants c_1 and c_2 are chosen in the appropriate manner. The usefulness of employing complex numbers is immediately seen in applying the boundary conditions that

$$u_x(0,t) = w'(0) = v'(0) = 0.$$

Because sinh(0) = 0 and cosh(0) = 1, $c_2 = 0$, the boundary condition at $x = L$ becomes

$$A\ w'(L) = -\ Q\ w(0) \qquad \textbf{(6.11.22)}$$

$$A\ v'(L) = -\ Q\ v(0) \qquad \textbf{(6.11.23)}$$

In either case, we find

$$\varkappa A\ (1+i)\gamma\ \sinh[(1+i)\gamma L] = Q. \qquad \textbf{(6.11.24)}$$

The real and imaginary parts of Eq. (6.11.24) are

$$Q = \varkappa A\gamma[\cosh(\gamma L)\sin(\gamma L) - \sinh(\gamma L)\cos(\gamma L)] \qquad \textbf{(6.11.25)}$$

and

$$\tanh(\gamma L) = -\ \tan(\gamma L). \qquad \textbf{(6.11.26)}$$

The roots of the transcendental equation (6.11.26) are accurately approximated by $3\pi/4$, $7\pi/4$, $11\pi/4$, etc., because the hyperbolic tangent rapidly equals one as γL increases. However, only $3\pi/4$, $11\pi/4$, etc., are physically meaningful because they give a positive Q. From Eq. (6.11.26), Eq. (6.11.25) simplifies to

$$Q = 2\varkappa A\gamma\ \sin(\gamma L)\ \cosh(\gamma L). \qquad \textbf{(6.11.27)}$$

In Table 6.11.1 the lowest five γL that give a positive Q and the corresponding Q are listed. The lowest value of γL is the most important because it corresponds to the lowest possible Q where the oscillation will be observed.

These oscillatory solutions have been observed experimentally. The interested reader is referred to a paper by Turner (Turner, L. B., 1936: Self-oscillations in a retroacting thermal conductor. *Proc. Camb. Phil. Soc., 32,* 663-675.) for experimental results as well as further refinements on this problem.

Table 6.11.1: The Lowest Five γL's Which Would Give an Oscillation in the Temperature Field if Q Equalled a Critical Value.

Approximate value of γL	Actual γL	QL/$\varkappa$A
2.356	2.365	1.7799 10^1
8.639	8.639	3.4514 10^4
14.923	14.923	3.1928 10^7
21.206	21.206	2.4293 10^{10}
27.489	27.206	1.6863 10^{13}

Exercises

1. The temperature within soil is governed by the equation

$$u_t = a^2 u_{xx}$$

where a^2 is the soil's diffusivity. If the temperature at the surface is given by the balance between the incoming solar radiation and the heat loss at the surface:

$$I_0 + I_1 \sin(\omega t - \phi_I)$$
$$= u_x(0,t) + h[u(0,t) - T_0 - T_1 \sin(\omega t - \phi_a)],$$

show that the temperature within the soil after the transients die away is

$$u(x,t) = T_0 + \frac{I_0}{h} + \frac{I_1}{D}[(h-\omega')\sin(\omega t - \phi_I - \omega' x)$$
$$+ \omega' \cos(\omega t - \phi_I - \omega' x)]e^{-\omega' x}$$
$$+ \frac{hT_1}{D}[(h-\omega')\sin(\omega t - \phi_a - \omega' x)$$
$$+ \omega' \cos(\omega t - \phi_a - \omega' x)]e^{-\omega' x}$$
$$= T_0 + \frac{I_0}{h} + \frac{e^{-\omega' x}}{\sqrt{D}}[I_1 \sin(\omega t - \phi_I - \omega' x - \delta)$$
$$+ hT_1 \quad \sin(\omega t - \phi_a - \omega' x - \delta)]$$

where $D = (h-\omega')^2 + \omega'^2$, $\omega' = \sqrt{\omega/2a^2}$ and $\delta = \tan^{-1}[\omega'/(h^2+\omega'^2).]$

6.12 UNSTEADY HEAT TRANSFER IN ENGINES

The temperature within the cylinder of a steam or combustion engine varies periodically with time. Consequently, along the interior surface of the cylinder the temperature is given by the Fourier series:

$$u(x,0) = A_0 + \sum_{n=1}^{\infty} A_n \cos(n\omega t) + B_n \sin(n\omega t) \qquad \textbf{(6.12.1)}$$

where ω is the angular frequency associated with the movement of the piston. From observations it is found that the temperature outside of the cylinder (which is usually surrounded by some coolant such as water or oil) does not fluctuate with time. Thus the temporal variations within the cylinder wall must damp out just as they do in the case of thermal waves within the solid earth. If we exploit this analogy, then we can guess the solution for a cylinder's wall of thickness L to be

$$u(x,t) = T_i - (T_i - T_0)\frac{x}{L} + \sum_{n=1}^{\infty} e^{-\gamma x}[(A_n \cos(n\omega t - \gamma x) + B_n \sin(n\omega t - \gamma x)] \quad \mathbf{(6.12.2)}$$

where

$$\gamma = (\frac{n\omega}{2a^2})^{1/2} \quad \mathbf{(6.12.3)}$$

and T_i is the mean temperature in the combustion chamber and T_0 is the temperature on the exterior of the cylinder. We obtain this solution through the use of the principle of superposition where we replace our complex problem by the simpler one of periodic forcing at the surface $x = 0$ by the harmonic function $\cos(n\omega t)$ or $\sin(n\omega t)$. For each forcing we can use the techniques and solutions developed in the previous section. The final answer is given by the sum of all these solutions.

An important quantity is the heat flux at the cylinder's wall $x = 0$. At that point, the heat flux equals

$$\frac{\varkappa}{L}(T_i - T_0) + \varkappa \sum_{n=1}^{\infty} \gamma(B_n - A_n) \sin(n\omega t) + \gamma(B_n + A_n) \cos(n\omega t) \quad \mathbf{(6.12.4)}$$

This formula shows that considerable heat loss and decrease in efficiency will occur if γ is large. This will occur if ω is particularly large or the cylinder's wall has a large constant of diffusivity.

Equations (6.12.2) and (6.12.4) have been applied to several studies of the heat flux through the cylinder walls in internal combustion engines. In Fig. 6.12.1 the observed temperature difference between the inside and outside walls of a cylinder liner is presented from a paper by Dahl (Dahl, O. G. C., 1924: Temperature and stress distribution in hollow cylinders. *ASME Trans.*, *46*, 161-208.) From this data the first eleven harmonics were computed from Eq. (6.12.2) in terms of amplitudes and phases by rewriting Eq. (2.9.2) in terms of cosine series. See Eq. (2.8.2).

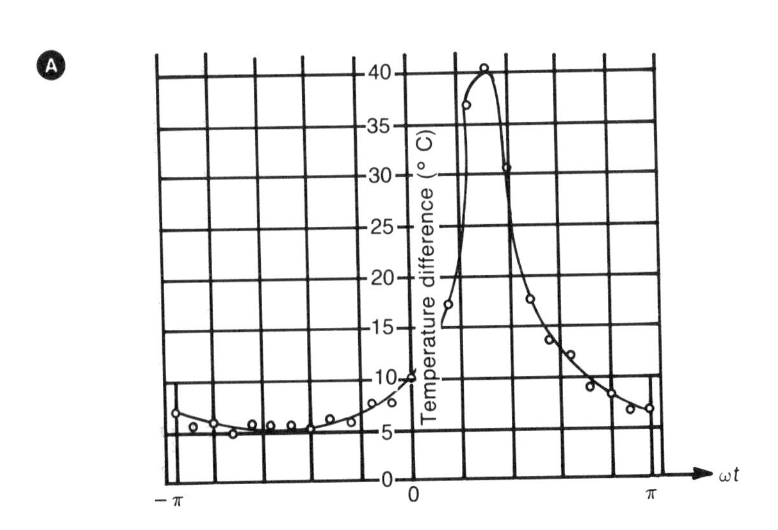

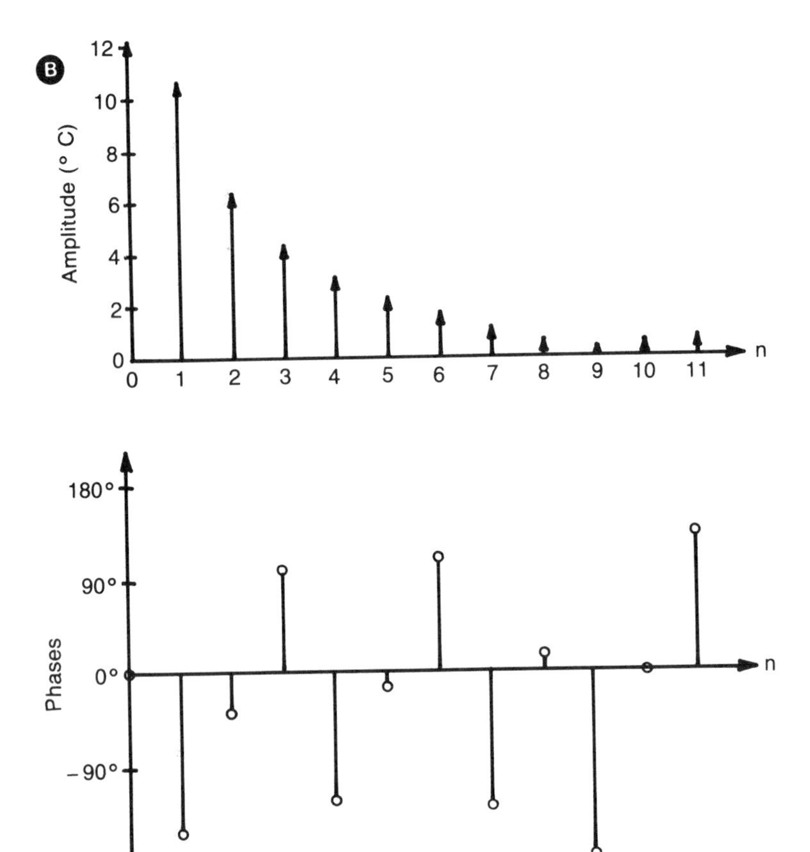

Fig. 6.12.1: The temperature difference between the inner and outer walls of a cylinder of an internal combustion engine as given (A) by an experiment by Dahl, O. G. C., 1924: Temperature and stress in hollow cylinders. *ASME. Trans., 46,* p. 185 and (B) Eq. (6.9.2) written as a sine series with amplitudes and phases.

In Fig. 6.12.2 the surface heat transfer is plotted for an automobile engine running at 840 rpm, a 7/1 compression ratio and an intake pressure of 43.1 psia as given by Overbye *et al* (Overbye, V. D., J. E. Bennethum, O. A. Uyehara and P. S. Myers, 1961: Unsteady heat transfer in engines. *SAE Trans., 61*, 461-494.) Equation (6.12.4) was then used to compute the first 36 harmonics. The results are presented in terms of amplitudes and phases of a cosine series Eq. (2.8.7).

Both of these studies neglect the radiation heat away from the exterior wall of the cylinder. For a discussion of the role of radiation in this problem, see Annand, W. J. D., 1963: Heat transfer in the cylinder of reciprocating internal combustion engines. *Proc. Inst. Mech. Eng., 177*, 973-990.

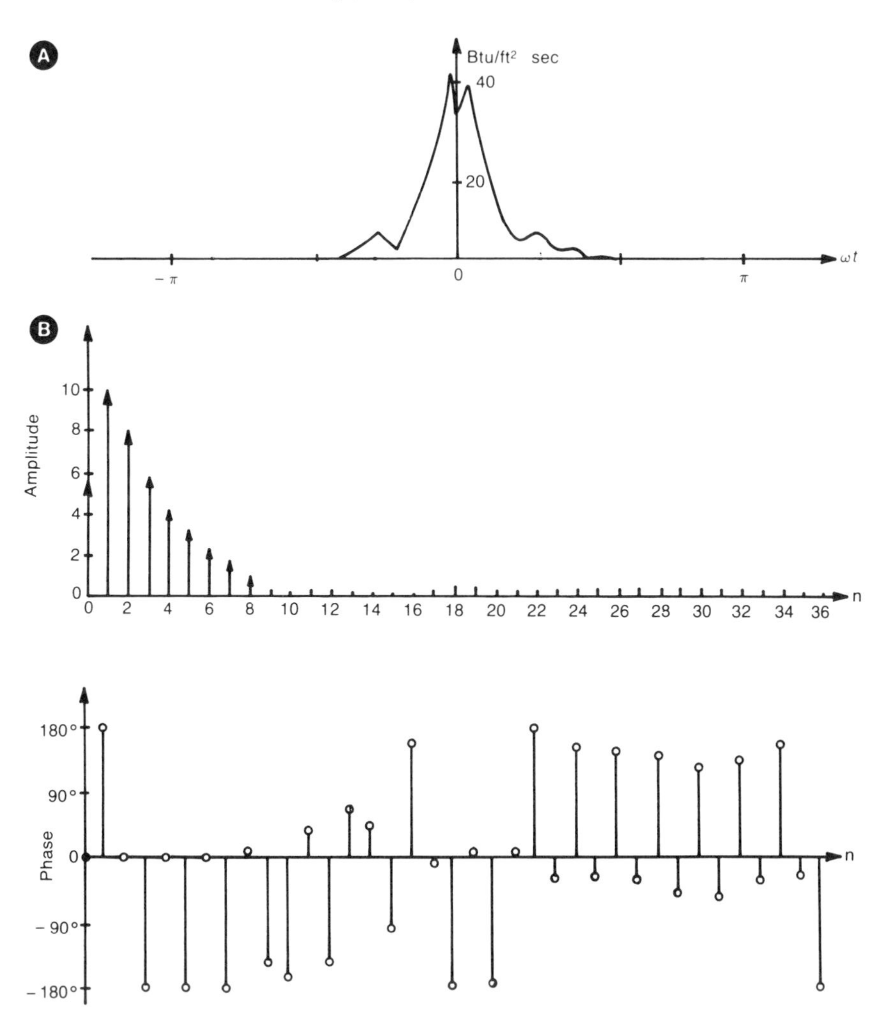

Fig. 6.12.2: The surface heat flux from a supercharged internal combustion engine as given by (A) experiment and (B) Eq. (6.9.4) written as a cosine series with amplitudes and phases.

6.13 THE SOLUTION OF THE HEAT EQUATION ON A SEMI-INFINITE HALF-SPACE USING SIMILARITY VARIABLES: KELVIN'S ESTIMATE OF THE AGE OF THE EARTH

A number of important heat conduction problems can be approximated by heat flow in a semi-infinite plane. Because the domain is infinite, we can no longer use Fourier series but must turn to Fourier integrals. We shall do so in Section 6.17. For the present we solve the semi-infinite problem through the use of the similarity variable. The problem consists of solving

$$u_t = a^2 u_{xx} \qquad \textbf{(6.13.1)}$$

with the initial-boundary conditions

$$u(x,0) = T_0,\ u(0,t) = T_S \text{ and } u(\infty,t) = T_0 \qquad \textbf{(6.13.2)}$$

To begin with, it is convenient to introduce the *dimensionless* temperature

$$\theta = \frac{u - T_0}{T_S - T_0} \qquad \textbf{(6.13.3)}$$

as a new dependent variable. The equation for θ is identical with the one for u

$$\theta_t = a^2 \theta_{xx} \qquad \textbf{(6.13.4)}$$

but the initial-boundary conditions on θ are simpler

$$\theta(x,0) = 0,\ \theta(0,t) = 1 \text{ and } \theta(\infty,t) = 0. \qquad \textbf{(6.13.5)}$$

The method of solving Eq. (6.13.4) has its origin in an approach known as *similarity*. This method is based on the idea that the only length scale in the problem (that is, the quantity that has the dimensions of length other than x itself) is $at^{1/2}$, the characteristic *thermal diffusion distance*. It is assumed that, in this circumstance, θ is not a function of x and t separately, but rather it is a function of the dimensionless ratio

$$\eta = \frac{x}{2at^{1/2}} \cdot \qquad \textbf{(6.13.6)}$$

The factor of 2 is introduced to simplify future results.

The dimensionless parameter η is known as the *similarity variable*. The solutions at different times are "similar" to each other in the sense that the spatial dependence at one time can be obtained from the spatial dependence at a different time by stretching the coordinate x by the square root of the ratio of the time. The characteristic thermal diffusion length is the distance over which the effects of a sudden, localized change in temperature can be felt after the time t has elapsed from the onset of the change.

To rewrite the equation in terms of η, we use the chain rule

$$\theta_t = \frac{d\theta}{d\eta}\eta_t = \frac{d\theta}{d\eta}\left(-\frac{x}{4\,at^{1/2}}\frac{1}{t}\right) = \frac{d\theta}{d\eta}\left(-\frac{\eta}{2t}\right) \quad \textbf{(6.13.7)}$$

$$\theta_x = \frac{d\theta}{d\eta}\eta_x = \frac{d\theta}{d\eta}\left(\frac{1}{2at^{1/2}}\right) \quad \textbf{(6.13.8)}$$

$$\theta_{xx} = \frac{1}{2at^{1/2}}\frac{d^2\theta}{d\eta^2}\eta_x = \frac{1}{4a^2t}\frac{d^2\theta}{d\eta^2}. \quad \textbf{(6.13.9)}$$

Therefore

$$-\eta\frac{d\theta}{d\eta} = 1/2\frac{d^2\theta}{d\eta^2}. \quad \textbf{(6.13.10)}$$

The boundary conditions become

$$\theta(0) = 1 \text{ and } \theta(\infty) = 0. \quad \textbf{(6.13.11)}$$

Our ability to transform the partial differential equation into an ordinary differential equation in η confirms the correction of our definition of η in terms of x and t.

Integrating Eq. (6.13.10) in conjunction with the boundary conditions Eq. (6.13.11), we have

$$\theta(x,t) = 1 - \frac{2}{\sqrt{\pi}}\int_0^{x/(2at^{1/2})} e^{-y^2}\,dy \quad \textbf{(6.13.12)}$$

The integral in Eq. (6.13.12) is called the error function and tables of this function are available. In Fig. 6.13.1 the solution (6.13.12) has graphed for various positions and times.

The heat flux at the surface $x = 0$ is given by

$$-\varkappa\, u_x(0,t) \quad \textbf{(6.13.13)}$$

or

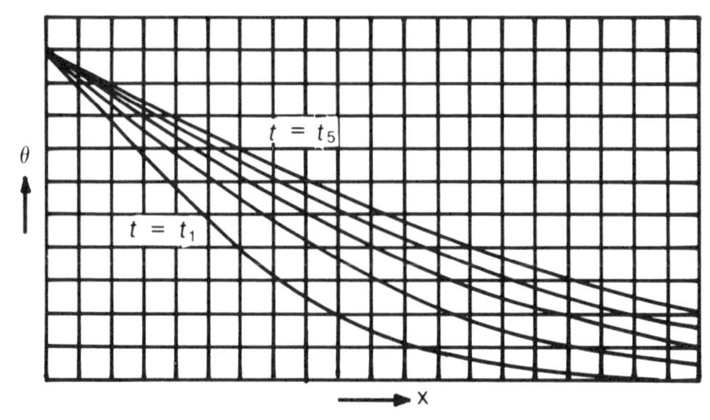

Fig. 6.13.1: The thermal diffusion of heat into a semi-infinite slab, initially at the temperature zero, at the times t_1, t_2, ...

$$\frac{(T_S - T_0)}{a\sqrt{\pi t}} . \tag{6.13.14}$$

The surface heat flux is infinite at $t = 0$ because of the sudden application of the temperature T_S at $t = 0$.

In the mid-1800s William Thompson, later Lord Kelvin, used this result to explain the conductive cooling of the earth and to estimate the age of the earth. He hypothesized that the earth was formed at a uniform high temperature T_0 and that its surface was subsequently maintained at the lower temperature T_S. He assumed that a thin, crustal boundary layer developed as the earth cooled. Because this layer would be thin compared with the radius of the earth, he reasoned that our one-dimensional model could be applied. Using Eq. (6.13.14) he calculated that the age of the earth t_0 was given by

$$t_0 = \frac{T_0 - T_S}{\pi a^2 T_x^2 (0,t)} \tag{6.13.15}$$

where $T_x (0,t)$ is the present crustal thermal gradient. With $T_x (0,t) = 25$ K/km, $T_0 - T_S = 2000$ K, and $a^2 = 1$ mm²/sec, the predicted age of the earth is 65 million years. It was not until radioactivity was discovered (circa 1900) that this youthful estimate could be reconciled with observations.

6.14 THE APPLICATION OF LAPLACE TRANSFORMS TO THE SOLUTION OF PROBLEMS IN HEAT CONDUCTION

In the previous chapter we showed that many problems in wave motion are most conveniently solved by the method of Laplace transforms. This is also true for heat conduction or diffusion. Once again we take the Laplace transform with respect to time. From the definition of Laplace transforms, we have

$$\mathcal{L}[u(x,t)] = \bar{u}(x,s) \tag{6.14.1}$$

$$\mathcal{L}[u_t(x,t)] = s\,\bar{u}(x,s) - u(x,0) \tag{6.14.2}$$

and

$$\mathcal{L}\,[u_{xx}(x,t)] = \bar{u}_{xx}(x,s). \tag{6.14.3}$$

The resulting ordinary differential equation, known as the *auxiliary equation*, is then solved using the boundary conditions. The function $u(x,0)$ is defined by the initial condition. The final step is the inversion of the Laplace transform $\bar{u}(x,s)$.

Quite often $\bar{u}(x,s)$ is equal to $f(x,s)/(sF(x,s))$ where x is treated as a parameter and both $f(x,s)$ and $F(x,s)$ are polynomials (the degree of $f(x,s)$ is equal or less than that of $F(x,s)$) where $F(x,s)$ has N simple zeros, s_n, which cannot equal zero. In those instances partial fractions give us

$$u(x,t) = \frac{f(x,0)}{F(x,0)} + \sum_{n=1}^{\infty} \frac{f(x,s_n)}{s_n F'(x,s_n)} \exp(s_n t) \tag{6.14.4}$$

The prime denotes partial differentiation with respect to s. In those cases when the transform $\bar{u}(x,s)$ cannot be written as the ratio of two polynomials, recourse to Bromwich's integral and integration on the complex plane is necessary. See Section 3.13.

To illustrate these concepts, we solve the heat conduction problem in a plane slab of thickness $2L$ with radiation at the surface. At the time $t = 0$, the slab at the constant temperature of unity is placed in a medium at the lower constant temperature of zero. If $u(x,t)$ denotes the temperature, a^2 the thermal diffusivity, h the relative emissivity, t time, and x the distance perpendicular to the face of the slab measured from the middle of the slab, then the governing equation is

$$u_t = a^2\, u_{xx} \tag{6.14.5}$$

with

$$u(x,0) = 1, \tag{6.14.6}$$

$$u_x(L,t) + h\, u(L,t) = 0 \tag{6.14.7}$$

and

$$u_x(-L,t) - h\, u(-L,t) = 0. \tag{6.14.8}$$

The transformed heat equation is

$$a^2 \frac{d^2\overline{u}(x,s)}{dx^2} - s\,\overline{u}(x,s) = -1. \tag{6.14.9}$$

If we write $s = a^2q^2$, Eq. (6.14.9) becomes

$$\frac{d^2\overline{u}(x,q)}{dx^2} - q^2\,\overline{u}(x,q) = -\frac{1}{a^2}. \tag{6.14.10}$$

From Eqs. (6.14.7) and (6.14.8) it is clear that $\overline{u}(x,q)$ is an even function of x and the solution of Eq. (6.14.10) is

$$\overline{u}(x,q) = \frac{1}{s} + A\cosh(qx) \tag{6.14.11}$$

where

$$\cosh(qx) = (e^{qx} + e^{-qx})/2. \tag{6.14.12}$$

From Eq. (6.14.7) or (6.14.8) it follows that

$$qA\sinh(qL) + \frac{h}{s} + hA\cosh(qL) = 0 \tag{6.14.13}$$

and

$$\overline{u}(x,q) = \frac{1}{s} - \frac{h\cosh(qx)}{s(q\sinh(qL) + h\cosh(qL))}. \tag{6.14.14}$$

The two terms on the right-hand side of Eq. (6.14.14) may be inverted separately. The first term is simply unity. The second term may be inverted with help from Eq. (6.14.4) if we take $f(x,s) = h\cosh(qx)$ and $F(x,s) = q\sinh(qL) + h\cosh(qL)$. Because $f(x,s)/F(x,s)$ is an even function of q, it is a single-valued function of s. The numerator has no poles; the denominator has zeros when

$$-\lambda\sin(\lambda L) + h\cos(\lambda L) = 0 \tag{6.14.15}$$

or

$$\lambda L\tan(\lambda L) = hL \tag{6.14.16}$$

where $q = i\lambda$. The roots of Eq. (6.14.16) are all real and nonzero if h is positive. Consequently we have

$$\frac{f(x,0)}{F(x,0)} = 1 \qquad \textbf{(6.14.17)}$$

$$f(x,s) = h \cos(\lambda x). \qquad \textbf{(6.14.18)}$$

Because

$$s \frac{dF(x,s)}{ds} = q/2 \frac{dF(x,q)}{dq}$$

$$= q[(1+hL) \sinh(qL) + qL \cosh(qL)]/2, \qquad \textbf{(6.14.19)}$$

we have for $q = i\lambda$ that

$$s \frac{dF(x,s)}{ds} = -\lambda/2 \, [(1+hL) \sin(\lambda L) + \lambda L \cos(\lambda L)]. \qquad \textbf{(6.14.20)}$$

The inversion of Eq. (6.14.14) is

$$u(x,t) = 1 - \left[1 - 2 \sum_{n=1}^{\infty} \frac{h \cos(\lambda_n x)}{\lambda_n \, [(1+hL) \sin(\lambda_n L) + \lambda_n L \cos(\lambda_n L)]} \exp(-a^2 \lambda_n^{\,2} t)\right]$$

(6.14.21)

or

$$u(x,t) = 2 \sum_{n=1}^{\infty} \frac{h \cos(\lambda_n x)}{\lambda_n \, [(1+hL) \sin(\lambda_n L) + \lambda_n L \cos(\lambda_n L)]} \exp(-a^2 \lambda_n^{\,2} t)$$

(6.14.22)

Now, because $h/\lambda_n = \tan(\lambda_n L)$ and $hL = \lambda_n L \tan(\lambda_n L)$, **(6.14.23)**

we have

$$① = \frac{h}{\lambda_n \, [(1+hL) \sin(\lambda_n L) + \lambda_n L \cos(\lambda_n L)]}$$

$$= \frac{\tan(\lambda_n L)}{[1+\lambda_n L \tan(\lambda_n L)] \sin(\lambda_n L) + \lambda_n L \cos(\lambda_n L)} \qquad \textbf{(6.14.24)}$$

$$① = \frac{\tan(\lambda_n L)}{\sin(\lambda_n L) + \lambda_n L/\cos(\lambda_n L)} \quad (6.14.25)$$

$$= \frac{\sin(\lambda_n L)}{\lambda_n L + \sin(\lambda_n L)\cos(\lambda_n L)} \quad (6.14.26)$$

so that

$$u(x,t) = 2\sum_{n=1}^{\infty} \frac{\sin(\lambda_n L)}{\lambda_n L + \sin(\lambda_n L)\cos(\lambda_n L)} \cos(\lambda_n x)\, e^{-a^2\lambda_n^2 t} \quad (6.14.27)$$

which is the usual form of the solution.

For large values of a^2t/L^2, one or two terms of Eq. (6.14.26) are all that are necessary to provide a satisfactory approximation to the value of $u(x,t)$. But the solution (6.14.26) is unsuitable for computation or for exhibiting the nature of the solution when a^2t/L^2 is small. To find a form suitable under such circumstances, we expand Eq. (6.14.14) as follows:

$$\bar{u}(x,q) = \frac{1}{s} - \frac{h\,(e^{qx} + e^{-qx})}{sq(e^{qL} - e^{-qL}) + hs(e^{qL} + e^{-qL})} \quad (6.14.28)$$

$$= \frac{1}{s} - \frac{he^{qx}}{s(h+q)e^{qL}}(1 + e^{-2qx})\left(1 + \frac{h-q}{h+q}e^{-2qL}\right)^{-1} \quad (6.14.29)$$

$$\bar{u}(x,q) = \frac{1}{s} - \frac{h}{s(h+q)}e^{-q(L-x)}(1 + e^{-2qx}) + \left(1 - \frac{h-q}{h+q}e^{-2qL} + \ldots\right) \quad (6.14.30)$$

$$= \frac{1}{s} - \frac{h}{s(h+q)}e^{-q(L-x)} - \frac{h}{s(h+q)}e^{-q(L+x)} + \frac{h(h-q)}{s(h+q)^2}e^{-q(3L-x)} - \ldots \quad (6.14.31)$$

Now

$$\mathcal{L}^{-1}\left(\frac{he^{-qx}}{s(h+q)}\right) = 1 - \operatorname{erf}\left(\frac{x}{2\sqrt{a^2t}}\right) - \exp(h^2a^2t + hx)\left[1 - \operatorname{erf}\left(\frac{x}{2\sqrt{a^2t}} + h\sqrt{a^2t}\right)\right] \quad (6.14.32)$$

where

$$\operatorname{erf}(x) = \frac{2}{\sqrt{\pi}} \int_0^x e^{-u^2}\, du \qquad \textbf{(6.14.33)}$$

and is known as the error function. For large values of x,

$$1 - \operatorname{erf}(x) \simeq \frac{e^{-x^2}}{\sqrt{\pi}} \left\{ \frac{1}{x} - \frac{1}{2x^3} + \frac{3}{4x^5} - \frac{1 \bullet 3 \bullet 5}{2^3 x^7} + \cdots \right\} \qquad \textbf{(6.14.34)}$$

If we neglect all but the first three terms of (6.14.30), we find

$$u(x,t) = 1 - \left\{ 1 - \operatorname{erf}\left(\frac{L-x}{2\sqrt{a^2t}}\right) - \exp(h^2a^2t + h(L-x)) \right.$$

$$\left. \times \left[1 - \operatorname{erf}\left(\frac{L-x}{2\sqrt{a^2t}} + h\sqrt{a^2t}\right)\right]\right\} \qquad \textbf{(6.14.35)}$$

$$- \left\{ 1 - \operatorname{erf}\left(\frac{L+x}{2\sqrt{a^2t}}\right) - \exp(h^2a^2t + h(L+x)) \right.$$

$$\left. \times \left[1 + \operatorname{erf} \frac{L+x}{2\sqrt{a^2t}} + h\sqrt{\alpha^2t} \right]\right\} \qquad \textbf{(6.14.36)}$$

When $L-x$ or $L+x$ is not small, we may use Eq. (6.14.33) to approximate these expressions. However, Eq. (6.14.34) applies whether $L-x$ or $L+x$ is small or not.

6.15 THE TEMPERATURE REACHED IN THE INTERFACE BETWEEN A BRAKE DRUM AND LINING

In Section 6.10 we applied the method of separation of variables to find the temperature reached in disc brakes during a single braking. In this section we shall essentially rework this problem using Laplace transforms. Once again we ignore the errors introduced by replacing the cylindrical portion of the drum by a flat rectangular plate. For constant deceleration the rate of heat generation is taken to diminish linearly with time t as given by

$$N(1 - Mt).$$

Unlike our earlier calculation we will ignore the radiation of heat away from the surface.

Consider the linear flow of heat in a solid initially at temperature zero and bounded by a pair of infinite parallel planes at $x = 0$ and $x = L$. At $x = 0$ there is no flow of heat perpendicular to the plane. At $x = L$ there is a flux $N(1 - Mt)$ into the solid. This differs from the case of disc brakes where heat is introduced through both surfaces. If $u(x,t)$, $\varkappa$ and a^2 denote the temperature, thermal conductivity and diffusivity, respectively, then the heat equation is

$$u_t = a^2 u_{xx} \qquad (0 < x < L) \tag{6.15.1}$$

with the boundary conditions that

$$u_x(0,t) = 0 \tag{6.15.2}$$

and

$$\varkappa u_x(L,t) = N(1 - Mt). \tag{6.15.3}$$

Introducing the Laplace transform of $u(x,t)$, defined as

$$\overline{u}(x,s) = \int_0^\infty u(x,t)\, e^{-st}\, dt, \tag{6.15.4}$$

the equation to be solved becomes

$$\frac{d^2\overline{u}}{dx^2} - \frac{s}{a^2}\overline{u} = 0 \tag{6.15.5}$$

subject to

$$\frac{d\overline{u}(0,s)}{dx} = 0 \tag{6.15.6}$$

and

$$\frac{d\overline{u}(L,s)}{dx} = \frac{N}{\varkappa}\left(\frac{1}{s} - \frac{M}{s^2}\right). \tag{6.15.7}$$

The solution to Eq. (6.15.5) is

$$\overline{u}(x,s) = A\cosh(qx) + B\sinh(qx) \tag{6.15.8}$$

where $q = s^{1/2}/a$. Using the boundary conditions, the solution becomes

$$\overline{u}(x,s) = \frac{N}{\varkappa}\left(\frac{1}{s} - \frac{M}{s^2}\right)\frac{\cosh(qx)}{q\sinh(qL)} \tag{6.15.9}$$

Replacing the hyperbolic functions in terms of exponentials and expanding in a series by the binomial theorem, we obtain

$$\overline{u}(x,s) = \frac{N}{\varkappa}\left(\frac{1}{s} - \frac{M}{s^2}\right)\frac{1}{q}\left(e^{-q(L-x)} + e^{-q(L+x)}\right)\sum_{n=0}^{\infty} e^{-2nqL} \tag{6.15.10}$$

$$\overline{u}(x,s) = \frac{N}{\varkappa}\left(\frac{1}{sq} - \frac{M}{s^2 q}\right)$$

$$\times \sum_{n=0}^{\infty} \exp\{-q[(2n+1)L-x]\} + \exp\{-q[(2n+1)L+x]\} \tag{6.15.11}$$

Inverting the transform (see Section 3.7), we have

$$u(x,t) = 2t^{1/2}\sum_{n=0}^{\infty} i\mathrm{erfc}\left[\frac{(2n+1)L-x}{2at^{1/2}}\right] + i\mathrm{erfc}\left[\frac{(2n+1)L+x}{2at^{1/2}}\right]$$

$$- 8Nt^{3/2}\sum_{n=0}^{\infty} i^3\mathrm{erfc}\left[\frac{(2n+1)L-x}{2at^{1/2}}\right] + i^3\mathrm{erfc}\left[\frac{(2n+1)L+x}{2at^{1/2}}\right] \tag{6.15.12}$$

where

$$6\, i^3\mathrm{erfc}\,(x) = i\mathrm{erfc}\,(x) - 2x\, i^2\mathrm{erfc}\,(x) \tag{6.15.13}$$

$$i^2\mathrm{erfc}\,(x) = ((1+2x^2)\,\mathrm{erfc}\,(x) - \frac{2}{\sqrt{\pi}}\,(x)\,e^{-x^2})/2 \tag{6.15.14}$$

$$i\mathrm{erfc}\,x = \frac{1}{\sqrt{\pi}}\,e^{-x^2} - x\,\mathrm{erfc}\,(x) \tag{6.15.15}$$

and

$$\mathrm{erfc}\,(x) = 1 - \mathrm{erf}(x) \tag{6.15.16}$$

where erf(x) is the error function defined in the Section 6.13.

Newcomb (Newcomb, T. P., 1959: The flow of heat in a parallel-faced infinite solid. *Brit. J. Appl. Phys., 9*, 370-372.) has used Eq. (6.15.12) to determine the transient temperature at the interface between a brake drum and lining. In Fig. 6.15.1 the temperature in the brake lining is presented at various places within the lining ($s = x/L$) if $a^2 = 3.3 \times 10^{-3}$ cm²/sec, $\varkappa = 1.8 \times 10^{-3}$ cal/cm sec °C, $L = 0.48$ cm, and $N = 1.96$ cal/cm² sec.

Exercises

1. Solve $$u_{tt} - a^2(u - T) = u_t$$

subject to the conditions that $u_x(0,t) = u_x(1,t) = 0$ and $u(x,0) = T_0$.

2. Solve $$u_t = u_{xx}$$

subject to the conditions that $u_x(0,t) = 0$, $u(1,t) = t$, and $u(x,0) = 0$.

3. Solve $$u_t = u_{xx}$$

subject to the conditions $u(0,t) = 0$, $u(1,t) = 1$, and $u(x,0) = 0$.

4. Solve $$u_t = u_{xx}$$

subject to the conditions that $u(x,0) = x$, $u_x(0,t) = 1$ and $u(x,t)$ is bounded as x goes to infinity.

5. Solve $$u_t = u_{xx}$$

subject to the conditions that $u(x,0) = e^{-x}$, $u(0,t) = 1$, and $u(x,t)$ is bounded as x tends to infinity.

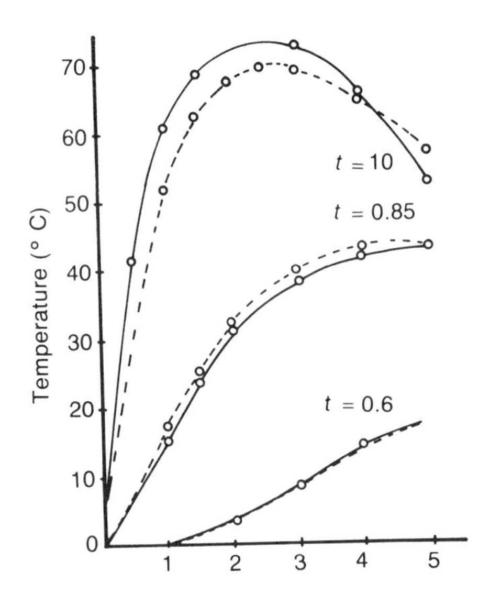

Fig. 6.15.1: Typical curves of transient temperature at different locations in a brake lining. Circles denote experimental measurements by Newcomb and crosses are calculated temperatures. Taken with permission from Newcomb, T. P., 1958: The flow of heat in a parallel-faced infinite solid. *Brit. J. Appl. Phys., 9*, 377.

6. Solve $$u_t - u_{xx} = 1$$

subject to the conditions $u(x,0) = u(1,t) = 0$ and $u(0,t) = 0$.

7. Given that $u_t = u_{rr} + \frac{2}{r} u_r$ for $0<r<1$ and $t>0$

with $u(r,0) = 0$, $u_r(1,t) = 1$ and $u(r \to 0,t)$ is finite, show that

$$u(r,t) = 3t - r^2/2 - 0.3 - \frac{2}{r} \sum_{n=1}^{\infty} \frac{\sin(\lambda_n r)\, e^{-\lambda_n^2 t}}{\lambda_n \sin(\lambda_n)}$$

where λ_n is the nth positive root of the equation $\tan(\lambda) = \lambda$. (Hint: Use the new dependent variable $v(r,t) = r\, u(r,t)$.)

8. In shallow aquifers, the height of the water table $u(x,t)$ is given by the Boussinesq equation

$$u_t = a^2 u_{xx}$$

where a^2 equals the hydraulic conductivity of the aquifer times the depth of the aquifer divided by the storage coefficient. If the height of the water table is raised to the height of u_0 at $x = L$ while it is maintained at zero at $x = 0$, show that the water table is given by

$$u(x,t) = u_0 \sum_{n=0}^{\infty} \operatorname{erfc}\left[\frac{(2n+1)L - x}{2at^{1/2}}\right] - \operatorname{erfc}\left[\frac{(2n+1)L + x}{2at^{1/2}}\right]$$

if the initial height is zero. Erfc is the complementary error function. Use Laplace transforms.

Luthin and Holmes (Luthin, J. N. and J. W. Holmes, 1960: An analysis of the flow of water in a shallow, linear aquifer, and of the approach to a new equilibrium after intake. *J. Geophys. Res.*, *65*, 1573-1576.) have used these results in an investigation of an aquifer in upper South-East district of South Australia.

9. Show that the solution of the heat equation

$$u_t = a^2 u_{xx}$$

subject to the initial condition $u(x,0) = 0$ and the boundary condition $u(ct,t) = T_0$ is

$$u(x,t) = T_0/2 \left\{ \operatorname{erfc}\left(\frac{x}{2a\sqrt{t}}\right) + \operatorname{erfc}\left(\frac{x}{2a\sqrt{t}} - \frac{c\sqrt{t}}{a}\right) \exp\left[-\frac{c\,(x-ct)}{a^2}\right] \right\} H(x-ct)$$

where $H(x)$ is the Heaviside step function and erfc is the complementary error function.

6.16 SEEPAGE OF A HOMOGENEOUS FLUID IN FISSURED ROCKS

In their study of the seepage of a homogeneous fluid in fissured rocks, Barenblatt *et al* (Barenblatt, G. I., Iu. P. Zheltov and I. N. Kochina, 1960: Basic concepts in the theory of seepage of homogeneous liquids in fissured rocks (strata). *Applied Mathematics and Mechanics (PMM), 24*, 1286-1303.) solved an equation similar to

$$u_t - u_{xxt} = u_{xx} \tag{6.16.1}$$

with the boundary conditions

$$u(0,t) = 1 - e^{-t} \text{ and } u(\infty,t) = 0 \tag{6.16.2}$$

and the initial condition

$$u(x,0) = 0. \tag{6.16.3}$$

We have included this problem because its solution by Laplace transform illustrates the inversion of a transform that has branch points.

Applying the Laplace transform of Eq. (6.16.1) and the boundary conditions and initial condition, we have

$$\frac{d^2\overline{u}}{dx^2} - \frac{s}{1+s}\overline{u} = 0 \tag{6.16.4}$$

with

$$\overline{u}(0,s) = \frac{1}{s(1+s)} \tag{6.16.5}$$

and

$$\overline{u}(\infty,s) = 0. \tag{6.16.6}$$

The general solution of Eq. (6.16.4) which satisfies the boundary condition is

$$\overline{u}(x,s) = \frac{1}{s(1+s)} \exp(-(\frac{s}{1+s})^{1/2} x). \tag{6.16.7}$$

The inversion of Eq. (6.16.7) is found by the Bromwich integral:

$$u(x,t) = \frac{1}{2\pi i} \int_{c-\infty i}^{c+\infty i} \frac{e^{st}}{s(1+s)} \exp(-\,(\frac{s}{1+s})^{1/2}\, x)\, ds$$

(6.16.8)

where c is to the right of the imaginary axis in the complex s-plane.

As will be seen later, the evaluation of the integral (6.16.8) is facilitated by the change of variables so that $s = z-1$ and

$$u(x,t) = \frac{e^{-t}}{2\pi i} \int_{c-\infty i}^{c+\infty i} \frac{e^{zt}}{z(z-1)} \exp(-\,(\frac{z-1}{z})^{1/2}\, x)\, dz$$

(6.16.9)

where c now runs parallel to the imaginary axis and is greater than one.

Because of the square root in the exponential in Eq. (6.16.9), the integrand of Eq. (6.16.9) is a multivalued function. A complete

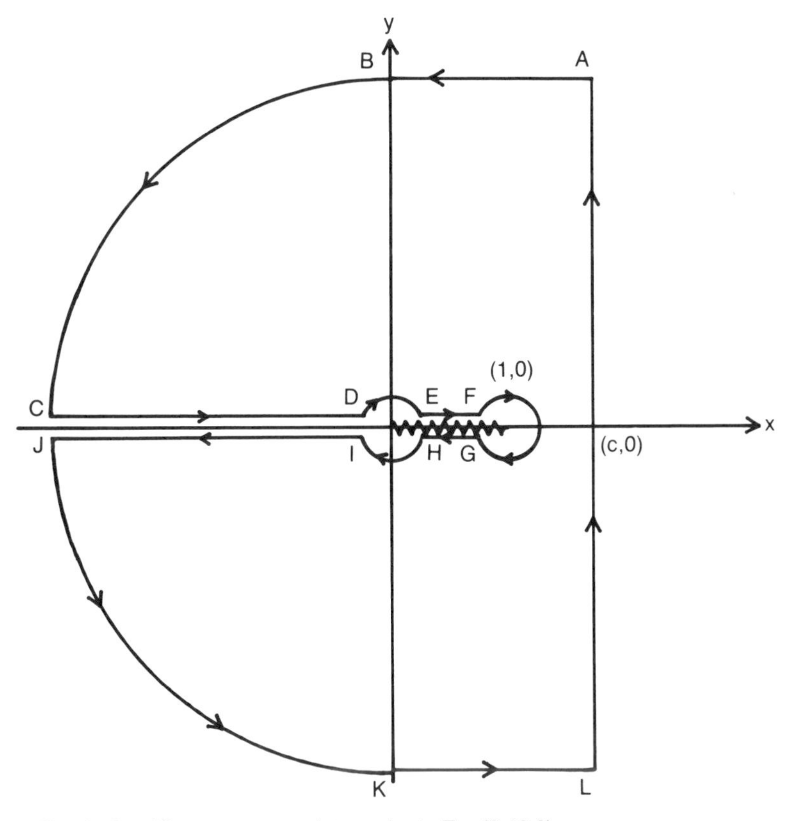

Fig. 6.16.1: The contour used to evaluate Eq. (6.16.9).

analysis shows that taking any closed path which includes $z = 0$ and $z = 1$ inside the contour does not result in any change in the argument of the integrand while any closed path that only includes $z = 0$ or $z = 1$ in the interior does. Consequently $z = 0$ and $z = 1$ are the branch points of the integrand. A convenient choice for the branch cut is the line segment lying along the real axis running between $z = 0$ and $z = 1$.

At this point we shall convert the contour integral running from $c - \infty i$ to $c + \infty i$ into a closed contour so that we can apply the residue theorem. Figure 6.16.1 shows the contour that will be used. The contribution from the arcs ABC and JKL may be shown to be negligibly small in a manner similar to our proof of *Jordan's lemma* in Section 9 of Appendix B. The contribution from the line segment CD cancels the contribution from IJ. Consequently, the dumbbell-shaped contour shown in Fig. 6.16.2 is equivalent to the contour $ABCDEFGHIJKL$ shown in Fig. 6.16.1. Because there are no singularities inside the closed contour, the value given by the contour shown in Fig. 6.16.2 must equal the negative of the integral from $c - \infty i$ and $c + \infty i$.

Along C_1, $z = \epsilon e^{i\theta}$ and $dz = i\epsilon e^{i\theta} d\theta$ so that

$$① = \frac{e^{-t}}{2\pi i} \int_{C_1} \frac{e^{zt}}{z(z-1)} \exp\left[-\left(\frac{z-1}{z}\right)^{1/2} x\right] dz$$

$$= \frac{e^{-t}}{2\pi} \lim_{\epsilon \to 0} \int_{-\pi}^{\pi} \frac{\exp(\epsilon t e^{i\theta})}{(-1 + \epsilon e^{i\theta})} \exp\left[-\left(1 - \frac{e^{-i\theta}}{\epsilon}\right)^{1/2} x\right] d\theta$$

(6.16.10)

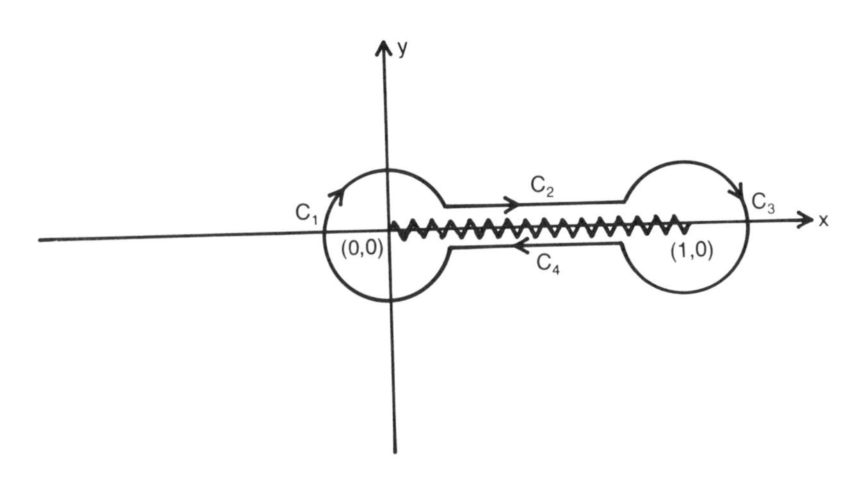

Fig. 6.16.2: Another contour used in the evaluation of Eq. (6.16.9).

$$① = -\frac{e^{-t}}{2\pi} \lim_{\epsilon \to 0} \int_{-\pi}^{\pi} \exp(-\frac{e^{i\theta/2 - i\pi/2}}{\sqrt{\epsilon}} x) d\theta = 0$$

(6.16.11)

Along C_3, we have $z = 1 + \epsilon e^{i\theta}$, $dz = i\epsilon e^{i\theta}\, d\theta$ and

$$\frac{e^{-t}}{2\pi i}\int_{C_3} \frac{e^{zt}}{z(z-1)} \exp\left[-(\frac{z-1}{z})^{1/2} x\right] dz$$

$$= \frac{1}{2\pi} \lim_{\epsilon \to 0} \int_{\pi}^{-\pi} \frac{\exp(\epsilon t\, e^{i\theta})}{1 + \epsilon e^{i\theta}} \exp\left[-(\frac{\epsilon e^{i\theta}}{1 + \epsilon e^{i\theta}})^{1/2} x\right] d\theta$$

(6.16.12)

$$= -1 \qquad \textbf{(6.16.13)}$$

Along C_2, $z = \sigma e^{0i}$, $dz = d\sigma$ and

$$\frac{e^{-t}}{2\pi i}\int_{C_2} \frac{e^{zt}}{z(z-1)} \exp\left[-(\frac{z-1}{z})^{1/2} x\right] dz$$

$$= \frac{e^{-t}}{2\pi i}\int_0^1 \frac{e^{\sigma t}}{\sigma(\sigma-1)} \exp\left[-i\ (\frac{1-\sigma}{\sigma})^{1/2} x\right] d\sigma \qquad \textbf{(6.16.14)}$$

Along C_4, $z = \sigma e^{2\pi i}$, $dz = d\sigma$ and

$$\frac{e^{-t}}{2\pi i}\int_{C_4} \frac{e^{zt}}{z(z-1)} \exp\left[-(\frac{z-1}{z})^{1/2} x\right] dz$$

$$= \frac{e^{-t}}{2\pi i}\int_1^0 \frac{e^{\sigma t}}{\sigma(\sigma-1)} \exp\left[i(\frac{1-\sigma}{\sigma})^{1/2} x\right] d\sigma \qquad \textbf{(6.16.15)}$$

Therefore,

$$u(x,t) = 1 + \frac{1}{\pi}\int_0^1 \frac{e^{(\sigma-1)t}}{\sigma(\sigma-1)} \sin\left[(\frac{1-\sigma}{\sigma})^{1/2} x\right] d\sigma \qquad \textbf{(6.16.16)}$$

$$u(x,t) = 1 - \frac{1}{\pi} \int_0^1 \frac{e^{-\eta t}}{(1-\eta)} \sin\left[\left(\frac{\eta}{1-\eta}\right)^{1/2} x \right] d\eta \qquad \textbf{(6.16.17)}$$

$$= 1 - \frac{2}{\pi} \int_0^\infty \frac{1}{v} \sin(vx) \exp\left(-\frac{v^2 t}{1+v^2}\right) dv \qquad \textbf{(6.16.18)}$$

where $\eta = 1 - \sigma$ and $v^2 = \eta/(1-\eta)$.

6.17 THE SOLUTION OF THE ONE-DIMENSIONAL HEAT EQUATION IN AN INFINITE SLAB USING THE FOURIER INTEGRAL

We now reconsider the problem of one-dimensional heat flow in a rod of infinite length with insulated sides. In Section 6.13 we solved this problem through the introduction of a similarity variable. In general this substitution does not work and we must use Fourier transforms.

We assume that an initial distribution is prescribed along the slab, $u(x,0) = f(x)$. Although there are no boundary conditions because the slab is infinite, we do require that the solution remains bounded as we go to either positive or negative infinity. Employing the product technique as before, we have $u(x,t) = X(x)\ T(t)$

with

$$T' + a^2k^2\ T = 0 \qquad \textbf{(6.17.1)}$$

and

$$X'' + k^2X = 0 \qquad \textbf{(6.17.2)}$$

The solution to Eq. (6.15.1) and Eq. (6.15.2) which satisifies the boundness condition is

$$X(x) = A \cos(kx) + B \sin(kx) \qquad \textbf{(6.17.3)}$$

and

$$T(t) = \exp(-k^2a^2t). \qquad \textbf{(6.17.4)}$$

In this particular problem we do not have any boundary condition which limits our choice of k and we must take *all* possible values. From the principle of superposition, this summation of all the product solutions gives

$$u(x,t) = \int_0^\infty [A(k)\cos(kx) + B(k)\sin(kx)] \quad \exp(-k^2a^2t)\,dk. \tag{6.17.5}$$

The initial conditions are satisfied by choosing

$$\begin{matrix} A(k) \\ B(k) \end{matrix} = \frac{1}{\pi}\int_{-\infty}^{\infty} f(x) \begin{matrix} \cos(kx) \\ \sin(kx) \end{matrix}\,dx \tag{6.17.6}$$

because the initial conditions have the form of a Fourier integral

$$f(x) = \int_0^\infty [A(k)\cos(kx) + B(k)\sin(kx)]\,dk \tag{6.17.7}$$

when $t = 0$.

We can obtain several important results by rewriting Eq. (6.17.7) as

$$u(x,t) = \frac{1}{\pi}\int_0^\infty \left[\int_{-\infty}^{\infty} f(\xi)\cos(k\xi)\,d\xi\,\cos(kx) + \int_{-\infty}^{\infty} f(\xi)\sin(k\xi)\,d\xi\,\sin(kx)\right] \exp(-k^2a^2t)\,dk. \tag{6.17.8}$$

Combining terms,

$$u(x,t) = \frac{1}{\pi}\int_0^\infty \left\{\int_{-\infty}^{\infty} f(\xi)\cos(k\xi)\cos(kx) + \sin(k\xi)\sin(kx)\,d\xi\right\} e^{-k^2a^2t}\,dk \tag{6.17.9}$$

$$= \frac{1}{\pi}\int_0^\infty \left\{\int_{-\infty}^{\infty} f(\xi)\cos[k(\xi - x)]\,d\xi\right\} \times \exp(-k^2a^2t)\,dk \tag{6.17.10}$$

If the order of integration is now reversed, we may write

$$u(x,t) = \frac{1}{\pi}\int_{-\infty}^{\infty} f(\xi)\left\{\int_0^\infty \cos[k(\xi - x)]\exp(-k^2a^2t)\,dk\right\} d\xi. \tag{6.17.11}$$

The inner integral is called the source function. Its value may be found through an integration in the complex plane and is equal to

$$\left(\frac{\pi}{4a^2t}\right)^{1/2} \exp\left(-\frac{(\xi - x)^2}{4a^2t}\right) \text{ with } t>0. \tag{6.17.12}$$

This gives us, finally, a new form for the temperature distribution:

$$u(x,t) = \frac{1}{\sqrt{4a^2\pi t}} \int_{-\infty}^{\infty} f(\xi) \exp\left(-\frac{(\xi-x)^2}{4a^2t}\right) d\xi. \quad \textbf{(6.17.13)}$$

In particular, if the initial temperature is given by

$$u(x,0) = \begin{matrix} T_0 & x>0 \\ -T_0 & x<0 \end{matrix} \quad \textbf{(6.17.14)}$$

then

$$u(x,t) = \frac{T_0}{\sqrt{4a^2\pi t}}\left[\int_{-\infty}^{0} - \exp\left(-\frac{(\xi-x)^2}{4a^2t}\right) d\xi \right.$$

$$\left. + \int_{0}^{\infty} \exp\left(-\frac{(\xi-x)^2}{4a^2t}\right) d\xi \right] \quad \textbf{(6.17.15)}$$

$$= \frac{T_0}{\sqrt{\pi}} \int_{-x/2at^{1/2}}^{\infty} e^{-u^2}\, du - \int_{x/2at^{1/2}}^{\infty} e^{-u^2}\, du$$

$$\textbf{(6.17.16)}$$

$$= \frac{T_0}{\sqrt{\pi}} \int_{-x/2at^{1/2}}^{x/2at^{1/2}} e^{-u^2}\, du$$

$$= \frac{2T_0}{\sqrt{\pi}} \int_{0}^{x/2at^{1/2}} e^{-u^2}\, du = T_0 \operatorname{erf}\left(\frac{x}{2a\sqrt{t}}\right). \quad \textbf{(6.17.17)}$$

where erf(x) is the error function introduced in Section 6.13.

Exercises

1. Find $u(x,t)$ if

(a)

$$f(x) = \begin{matrix} 1 & |x|<a \\ 0 & \text{otherwise} \end{matrix}$$

(b)

$$f(x) = \exp(-\alpha|x|)$$

(c)

$$f(x) = \begin{matrix} 0 & -\infty<x<0 \\ T_0 & 0<x<a \\ 0 & a<x<\infty \end{matrix}$$

2. Solve the heat equation on the half-plane where

$u_x(0,t) = 0$, $t>0$ and

$$u(x,0) = \begin{matrix} T_0 & 0<x<a \\ 0 & x>a \end{matrix}.$$

3. Redo problem 2 with $u(0,t) = 0,\ t>0$.
4. Redo problem 2 with $u(0,t) = T_0,\ t>0$ and

$$u(x,0) = T_0(1 - e^{-\alpha x}),\ x>0.$$

5. If $f(x)$ is the unit singularity function $\delta(x)$ (so that a heat source is present at $x = 0$ at the instant $t = 0$), show that the solution (5.17.13) is of the form

$$u(x,t) = \frac{1}{2a^2\sqrt{\pi t}} \exp\left[-\frac{x^2}{4a^2t}\right].$$

6.18 DUHAMEL'S THEOREM OR THE SUPERPOSITION INTEGRAL: TEMPERATURE OSCILLATIONS IN A WALL HEATED BY AN ALTERNATING CURRENT

In Section 6.10 we found the temperature distribution within a disc brake during the stopping of a car. Because the boundary conditions were time dependent, we were forced to find a particular solution that removed this time dependence before we could apply separation of variables. Often finding this particular solution is impossible.

In Section 6.15 we solved a similar problem by Laplace transforms. This technique allowed us to eliminate the time dependence in the boundary conditions through the use of transforms. The principal drawback in using this technique is the difficulty in obtaining the inverse transform.

In this section we solve a problem with time-dependent boundary conditions by first solving a simpler problem with the time-dependent boundary condition replaced by unity. We will then build the more complicated solution from this simpler solution.

Let us solve the heat conduction problem

$$u_t = a^2 u_{xx} \tag{6.18.1}$$

with

$$u(0,t) = 0 \tag{6.18.2}$$

$$u(L,t) = f(t) \tag{6.18.3}$$

and

$$u(x,0) = 0 \tag{6.18.4}$$

As a first step, we solve

$$A_t = a^2 A_{xx} \tag{6.18.5}$$

with

$$A(0,t) = 0 \tag{6.18.6}$$

$$A(L,t) = 1 \qquad (6.18.7)$$

and

$$A(x,0) = 0 \, . \qquad (6.18.8)$$

Separation of variables yields the solution

$$A(x,t) = \frac{x}{L} + \frac{2}{\pi} \sum_{n=1}^{\infty} \frac{(-1)^n}{n} \sin(\frac{n\pi x}{L}) \exp(-\frac{a^2n^2\pi^2t}{L^2}) \qquad (6.18.9)$$

Consider the following case. Suppose that the temperature at the end $x = L$ is maintained at zero until $t = \tau_1$ and then raised to the temperature of unity. The temperature at $x = L$ is then maintained at unity thereafter. The resulting temperature distribution will be zero everywhere when $t < \tau_1$ and be given by $A(x,t-\tau_1)$ for $t > \tau_1$. We have merely shifted our time axis so that the initial condition occurs at $t = \tau_1$.

Consider an analogous, but more complicated, situation of the temperature at the end position $x = L$ being held at $f(0)$ from $t = 0$ to $t = \tau_1$ at which time it is abruptly raised by an amount $f(\tau_1) - f(0)$ to the value $f(\tau_1)$. This temperature is then held until $t = \tau_2$ when it is again abruptly raised by an amount $f(\tau_2) - f(\tau_1)$. We can imagine this process continuing up to the instant $t = \tau_n$. Because of linear superposition, the temperature distribution at any given time is equal to the sum of these temperature increments:

$$\begin{aligned} u(x,t) &= f(0)\, A(x,t) \\ &+ [f(\tau_1) - f(0)]\, A(x,t - \tau_1) \\ &+ [f(\tau_2) - f(\tau_1)]\, A(x,t-\tau_2) + \dots \\ &+ [f(\tau_n) - f(\tau_{n-1})]\, A(x,t-\tau_n) \end{aligned} \qquad (6.18.10)$$

where τ_n is the time of the most recent temperature change. If we write

$$\Delta f_k = f(\tau_k) - f(\tau_{k-1}) \qquad (6.18.11)$$
$$\Delta \tau_k = \tau_k - \tau_{k-1}, \qquad (6.18.12)$$

Eq. (6.18.10) becomes

$$u(x,t) = f(0)A(x,t) + \sum_{k=1}^{\infty} A(x,t-\tau_k) \frac{\Delta f}{\Delta \tau}_k \Delta \tau_k \qquad (6.18.13)$$

Consequently, in the limit of $\Delta\tau_k \rightarrow 0$, Eq. (6.18.13) becomes

$$u(x,t) = f(0)A(x,t) + \int_0^t A(x,t-\tau)\, f'(\tau)\, d\tau \qquad \textbf{(6.18.14)}$$

assuming that $f(t)$ is differentiable. Eq. (6.18.14) is known as the *superposition integral* or *Duhamel's theorem*. An alternative form is obtained by integration by parts:

$$u(x,t) = f(t)\, A(x,0) - \int_0^t f(\tau)\, \frac{\partial A(x,t-\tau)}{\partial \tau}\, d\tau \qquad \textbf{(6.18.15)}$$

$$u(x,t) = f(t)\, A(x,0) + \int_0^t f(\tau)\, \frac{\partial A(x,t-\tau)}{\partial t}\, d\tau \qquad \textbf{(6.18.16)}$$

because

$$\frac{\partial A(x,t-\tau)}{\partial \tau} = -\frac{\partial A(x,t-\tau)}{\partial t}\, . \qquad \textbf{(6.18.17)}$$

To illustrate Eq. (6.18.14), suppose $f(t) = t$. Then by Eq. (6.18.14)

$$u(x,t) = \int_0^t \frac{x}{L} + \frac{2}{\pi} \sum_{n=1}^{\infty} \frac{(-1)^n}{n} \times \sin(\frac{n\pi x}{L}) \exp\left[-\frac{a^2n^2\pi^2}{L^2}(t-\tau)\right] d\tau \qquad \textbf{(6.18.18)}$$

$$= \frac{xt}{L} - \frac{2}{\pi}\frac{L^2}{a^2\pi^2} \sum_{n=1}^{\infty} \frac{(-1)^n}{n^3} \times \sin(\frac{n\pi x}{L}) \left[1 - \exp(-\frac{a^2n^2\pi^2 t}{L^2})\right] \qquad \textbf{(6.18.19)}$$

In Fig. 6.18.1 the solution (6.18.19) has been graphed for various values of x/L and a^2t/L^2, the Fourier number.

In addition to finding solutions to heat conduction problems with time-dependent problems with time-dependent boundary conditions, Duhamel's theorem may also be applied to the nonhomogeneous heat equation when the source term is time dependent. This technique was used by Jeglic (Jeglic, F., 1962: An analytical determination of temperature oscillations in a wall heated by alternating current. *NASA Technical Note No. D-1286.*) in obtaining the tem-

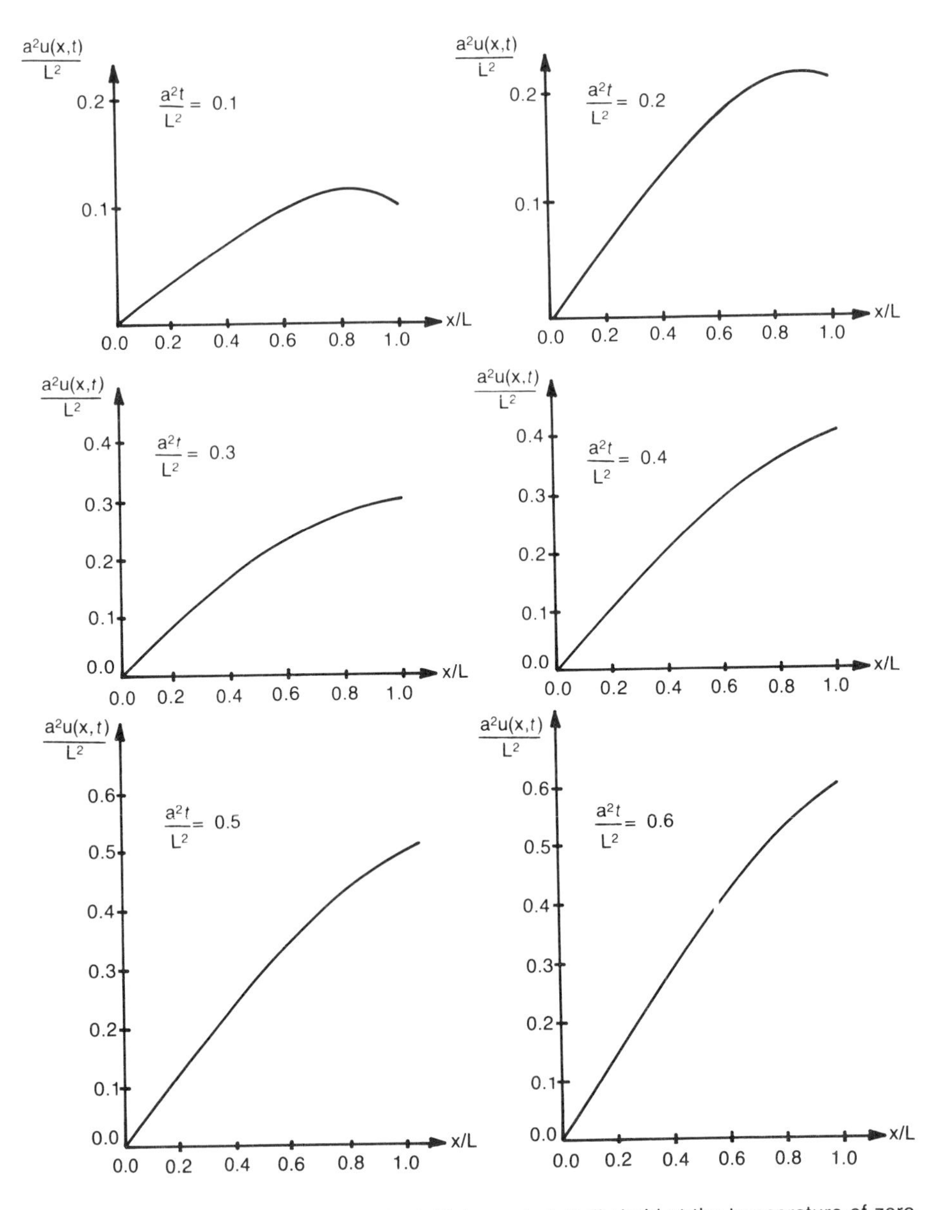

Fig. 6.18.1: Temperature within a slab of thickness L, initially held at the temperature of zero, when u(0,t) = 0 and u(L,t) = t.

perature distribution within a slab which is heated by a sinusoidal heat source. If we assume that the flat plate has a surface area A and depth L, then the heat equation for the plate when electrically heated by an alternating current of frequency ω is

$$u_t - a^2 u_{xx} = \frac{2q}{\varrho C_p LA} \sin^2(\omega t) \qquad \textbf{(6.18.20)}$$

where q is the average heating rate caused by the current, ϱ the density, C_p the specific heat at constant pressure and a^2 the diffusivity. We assume that the inner wall is insulated

$$u_x(0,t) = 0 \tag{6.18.21}$$

while the outer wall is allowed to radiate to free space at the temperature of zero

$$\varkappa\, u_x(L,t) + h\, u(L,t) = 0 \tag{6.18.22}$$

where $\varkappa$ is the thermal conductivity and h the heat transfer coefficient. The slab is initially at the temperature of zero:

$$u(x,0) = 0. \tag{6.18.23}$$

To solve the heat equation, we first solve the simpler problems of

$$A_t - a^2 A_{xx} = 1 \tag{6.18.24}$$

with

$$A_x(0,t) = 0 \tag{6.18.25}$$

$$\varkappa\, A_x(L,t) + h\, A(L,t) = 0 \tag{6.18.26}$$

and

$$A(x,0) = 0 \tag{6.18.27}$$

The solution $A(x,t)$ is often called the *indicial admittance* because it is the response of the system to forcing by the step function $H(t)$.

In our study of the nonhomogeneous heat equation we solved a heat conduction problem very similar to Eq. (6.18.24)-(6.18.27). Here we assume that the general solution $A(x,t)$ comprises a steady-state solution $w(x)$ plus the transient solution $v(x,t)$ where

$$a^2\, w''(x) = -1 \tag{6.18.28}$$

$$w'(0) = 0 \tag{6.18.29}$$

$$\varkappa\, w'(L) + h\, w(L) = 0 \tag{6.18.30}$$

$$v_t = a^2\, v_{xx} \tag{6.18.31}$$

$$v_x(0,t) = 0 \tag{6.18.32}$$

$$\varkappa\, v_x(L,t) + h\, v(L,t) = 0 \tag{6.18.33}$$

and

$$v(x,0) = -\, w(x) \tag{6.18.34}$$

From the solution found in Section 6.5, we find that $w(x)$ must be

$$w(x) = \frac{L^2 - x^2}{2a^2} + \frac{\varkappa L}{ha^2}\,. \tag{6.18.35}$$

Turning to the transient solution $v(x,t)$, we use separation of variables and find

$$v(x,t) = \sum_{n=1}^{\infty} C_n \cos(\frac{k_n x}{L}) \exp(-\frac{a^2 k_n^2 t}{L^2}) \qquad \textbf{(6.18.36)}$$

where k_n is the nth root of the transcendental equation

$$k_n \tan(k_n) = \frac{hL}{\varkappa} = h^*. \qquad \textbf{(6.18.37)}$$

In Table 6.18.1 the first six roots of Eq. (6.18.37) are given for various values of $hL/\varkappa$.

The Fourier coefficient C_n of the orthogonal expansion in $\cos(k_n x/L)$ which occurs at the time $t = 0$ is given by

$$C_n = \frac{\int_0^L -w(x) \cos(\frac{k_n x}{L})\, dx}{\int_0^L \cos^2(\frac{k_n x}{L})\, dx} \qquad \textbf{(6.18.38)}$$

$$= \frac{-\frac{L^3}{a^2 k_n^3} \sin(k_n)}{\frac{L}{2k_n} [k_n + \sin(2k_n)/2]} \qquad \textbf{(6.18.39)}$$

Table 6.18.1: The First Six Roots of the Equation $k_n \tan(k_n) = h^*$.

h^*	k_1	k_2	k_3	k_4	k_5	k_6
0.001	0.3162	3.14191	6.28334	9.42488	12.56645	15.70803
0.002	0.04471	3.14223	6.28350	9.42499	12.56653	15.70809
0.005	0.07065	3.14318	6.28398	9.42531	12.56677	15.70828
0.010	0.09830	3.14477	6.28478	9.42584	12.56717	15.70860
0.020	0.14095	3.14795	6.28637	9.42690	12.56796	15.70924
0.050	0.22176	3.15743	6.29113	9.43008	12.57035	15.71115
0.100	0.31105	3.17310	6.29906	9.43538	12.57432	15.71433
0.200	0.43284	3.20393	6.31485	9.44595	12.58226	15.72068
0.500	0.65327	3.29231	6.36162	9.47748	12.60601	15.73972
1.000	0.86033	3.42562	6.43730	9.52933	12.64529	15.77128
2.000	1.07687	3.64360	6.57833	9.62956	12.72230	15.83361
5.000	1.31384	4.03357	6.90960	9.89275	12.93522	16.01066
10.000	1.42887	4.30580	7.22811	10.20026	13.21418	16.25936
20.000	1.49613	4.49148	7.49541	10.51167	13.54198	16.58640
∞	1.57080	4.71239	7.85399	10.99557	14.13717	17.27876

$$C_n = -\frac{2L^2 \sin(k_n)}{a^2 k_n^2 [k_n + \sin(2k_n)/2]} \quad \textbf{(6.18.40)}$$

Combining Eq. (6.18.36) and (6.18.40) gives

$$v(x,t) = -\frac{2L^2}{a^2} \sum_{n=1}^{\infty} \frac{\sin(k_n)}{k_n^2 \; [k_n + \sin(2k_n)/2]}$$

$$\times \cos(\frac{k_n x}{L}) \exp(-\frac{a^2 k_n^2 t}{L^2}) \quad \textbf{(6.18.41)}$$

Consequently, $A(x,t)$ is

$$A(x,t) = \frac{\varkappa L}{ha^2} + \frac{L^2 - x^2}{2a^2}$$

$$-\frac{2L^2}{a^2} \sum_{n=1}^{\infty} \frac{\sin(k_n)}{k_n^2 \, [k_n + \sin(2k_n)/2]} \cos(\frac{k_n x}{L}) \exp\,(-a^2 kn^2 t/L^2) \quad \textbf{(6.18.42)}$$

We now wish to use the solution (6.18.42) to find the temperature distribution within the slab when heated by a time-dependent source $f(t)$. As in the case of a time-dependent boundary condition, we imagine that we can break the process into an infinite number of small changes to the heating which occur at the times $t = \tau_1$, $t = \tau_2$, etc. Consequently the temperature distribution at the time t following the change at $t = \tau_n$ and before the change at $t = \tau_{n+1}$ is

$$u(x,t) = f(0)\, A(x,t) + \sum_{k=1}^{\infty} A(x,t-\tau_k) \frac{\Delta f}{\Delta \tau_k} \Delta t_k \quad \textbf{(6.18.43)}$$

where

$$\Delta f_k = f(\tau_k) - f(\tau_{k-1}) \quad \textbf{(6.18.44)}$$

$$\Delta \tau_k = \tau_k - \tau_{k-1} \quad \textbf{(6.18.45)}$$

If we take the limit of $\Delta \tau_k \rightarrow 0$, we have

$$u(x,t) = f(0)\, A(x,t) + \int_0^t A(x,t-\tau)\, f'(\tau)\, d\tau \quad \textbf{(6.18.46)}$$

$$= f(t)\, A(x,0) + \int_0^t f(\tau) \frac{\partial A(x,t-\tau)}{\partial t} d\tau \quad \textbf{(6.18.47)}$$

In our present problem,

$$f(t) = \frac{2q}{\varrho C_p LA} \sin^2(\omega t) \qquad \textbf{(6.18.48)}$$

and

$$f'(t) = \frac{2q\omega}{\varrho C_p LA} \sin(2\omega t) \qquad \textbf{(6.18.49)}$$

Therefore,

$$u(x,t) = \frac{2q\omega}{\varrho C_p LA} \int_0^t \sin(2\omega\tau) \left\{ \frac{\varkappa L}{ha^2} + \frac{L^2 - x^2}{2a^2} \right\}$$

$$- \frac{2L^2}{a^2} \sum_{n=1}^{\infty} \frac{\sin(k_n)}{k_n^2 \, [k_n + \sin(2k_n)/2]} \cos\left(\frac{k_n x}{L} \right) \exp\left[- \frac{a^2 k_n^2 (t-\tau)}{L^2} \right]$$

$$\textbf{(6.18.50)}$$

$$u(x,t) = \frac{q}{\varrho C_p LA} \left(\frac{\varkappa L}{ha^2} + \frac{L^2 - x^2}{2a^2} \right) [- \cos(2\omega\tau)] \Big|_0^t$$

$$- \frac{2q\omega}{\varrho C_p LA} \frac{2L^2}{a^2} \sum_{n=1}^{\infty} \frac{\sin(k_n) \exp(-a^2 k_n^2 t/L^2)}{k_n^2 \, [k_n + \sin(2k_n)/2]} \cos\left(\frac{k_n x}{\mathrm{L}} \right)$$

$$\times \int_0^t \sin(2\omega\tau) \exp\left(- \frac{a^2 k_n^2 \, \tau}{L^2}\right) d\tau \qquad \textbf{(6.18.51)}$$

$$= \frac{qL}{\varrho C_p A a^2} \left\{ 2 \sin^2 (\omega t) \left[\frac{\varkappa}{hL} + \frac{1 - (x/L)^2}{2} \right] \right.$$

$$- \sum_{n=1}^{\infty} \frac{4 \sin(k_n) \cos(k_n x/L)}{k_n^2 \, [\, k_n^2 + \sin (2k_n)/2] \, (4 + a^4 k_n^4 / L^4 \omega^2)}$$

$$\left. \times \left[\frac{a^2 k_n^2}{L^2 \omega} \sin(2\omega t) - 2 \cos (2\omega t) + 2 \exp \left(- \frac{a^2 k_n^2 t}{L^2} \right) \right] \right\}$$

$$\textbf{(6.18.52)}$$

In Fig. 6.18.2 Eq. (6.18.52) has been plotted for $hL/\varkappa = 0.1$ and $a^2/(L^2\omega) = 1.0$ for the case when $x = 0$ and $x = L$.

Exercises

1. Solve the heat equation when $u(0,t) = 0$, $u_x(L,t) = t$ and $u(x,0) = 0$.
2. Solve the heat equation when $u(0,t) = 0$, $u(L,t) = T_0 \sin(\omega t)$ and $u(x,0) = 0$.
3. Solve the heat equation when $u(0,t)$ - 0, $u(L,t) = T_0(1-e^{-bt})$ and $u(x,0) = 0$.
4. Consider the partial differential equation

$$\frac{\partial u}{\partial t} = f(t) + \frac{1}{R}\frac{\partial^2 u}{\partial x^2} + S\frac{\partial^2}{\partial x^2}\left(\frac{\partial u}{\partial t}\right)$$

with R and S greater than zero. If $u(x,0) = 0$ and $u(1,t)$ and $u(-1,t)$ equal 0, use Laplace transforms to show that the solution which decays with time is

$$u(x,t) = \int_0^t f(\tau)\, A(x,t-\tau)\, d\tau$$

where

$$A(x,t) = \sum_{n=0}^{\infty} \frac{4(-1)^n(1+\lambda_n RS)}{\pi(2n+1)} \cos\left[\frac{(2n+1)\pi x}{2}\right] \exp(\lambda_n t)$$

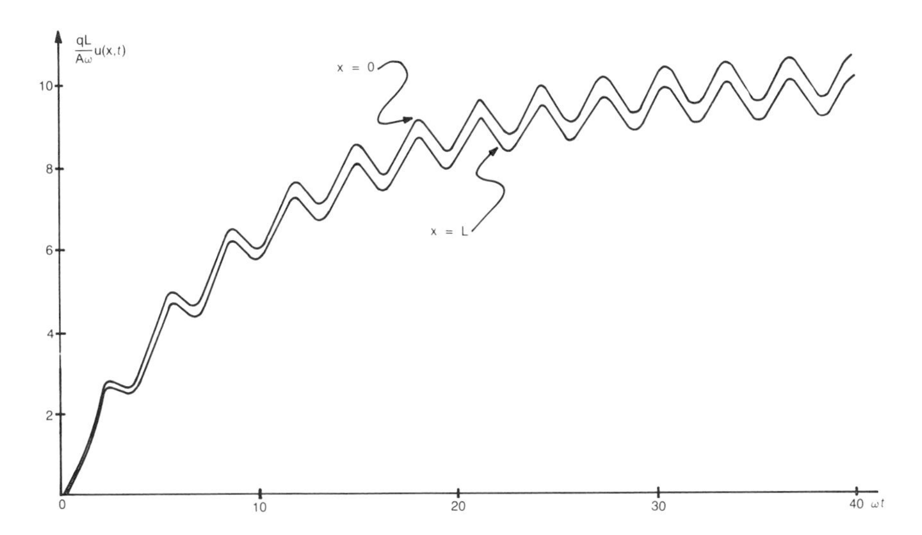

Fig. 6.18.2: The temporal evolution of the temperature field within a slab of thickness L as it is heated by an alternating current of frequency ω. The end x = 0 is perfectly insulated while the end x = L radiates to free space.

and

$$\lambda_n = \frac{-\pi^2(2n+1)^2}{4R + \pi^2(2n+1)^2 RS.}$$

Answers

P. 327

1. $u(x,t) = \frac{4A}{\pi} \sum_{n=1}^{\infty} \frac{1}{(2n-1)} \sin[(2n-1)x] \exp[-a^2(2n-1)^2 t]$

2. $u(x,t) = \frac{3}{4} \sin(x) \exp(-a^2 t) - \frac{1}{4} \sin(3x) \exp(-9a^2 t)$

3. $u(x,t) = \frac{1}{2} + \frac{1}{2} \cos(2x) \exp(-4a^2 t)$

4. $u(x,t) = \frac{8}{\pi} \sum_{n=1}^{\infty} \frac{1}{(2n-1)^3} \sin[(2n-1)x] \exp[-(2n-1)^2 a^2 t]$

5. $u(x,t) = \frac{4}{\pi} \sum_{n=1}^{\infty} \frac{(-1)^n}{(2n-1)^2} \sin[(2n-1)x] \exp[-(2n-1)^2 a^2 t]$

6. $u(x,t) = -16 \sum_{n=1}^{\infty} \frac{(-1)^n}{(2n-1)^3} \cos\left(\frac{(2n-1)x}{2}\right) \exp[-a^2 (2n-1)^2 t/4]$

7. $u(x,t) = \frac{\pi}{4} - \frac{2}{\pi} \sum_{n=1}^{\infty} \frac{1}{(2n-1)^2} \cos[(2n-1)x] \exp[-a^2(2n-1)^2 t]$

8. $u(x,t) = \sin(x) \exp(-a^2 t)$

9. $u(x,t) = T_0 + \frac{4}{\pi}(T_1 - T_0) \sum_{n=1}^{\infty} \frac{1}{(2n-1)} \sin\left[\frac{(2n-1)\pi x}{L}\right] \exp\left[-\frac{a^2(2n-1)^2\pi^2 t}{L^2}\right]$

10. $u(x,t) = T_0 + \frac{4}{\pi} \sum_{n=1}^{\infty} \left[\frac{2T_1(-1)^n}{(2n-1)^2\pi} - \frac{T_0}{(2n-1)}\right] \cos\left[\frac{(2n-1)\pi x}{2L}\right]$

$$\times \exp\left[-\frac{a^2(2n-1)^2\pi^2 t}{4L^2}\right]$$

11. $u(x,t) = (T_0 + \Delta T/2) + \frac{\Delta T x}{L}$

$$+ \frac{4\Delta T}{\pi^2} \sum_{n=1}^{\infty} \frac{1}{(2n-1)^2} \cos\left[\frac{(2n-1)\pi x}{L}\right] \exp\left[-\frac{a^2 (2n-1)^2\pi^2 t}{L^2}\right]$$

12. $$u(x,t) = T_0 + \frac{4(T_1 - T_0)}{\pi} \sum_{n=1}^{\infty} \frac{1}{(2n-1)}$$

$$\times \sin\left[\frac{(2n-1)\pi x}{2L}\right] \exp\left[-\frac{(2n-1)^2\pi^2 a^2 t}{4L^2}\right]$$

13. $$u(x,t) = (T_0 + T_1)/2 + \frac{2}{\pi}(T_0 - T_1) \sum_{n=1}^{\infty} \frac{(-1)^{n+1}}{(2n-1)}$$

$$\times \cos\left(\frac{(2n-1)\pi x}{2L}\right) \exp\left[-\frac{(2n-1)^2\pi^2 a^2 t}{4L^2}\right]$$

P. 340

1. $$u(x,t) = \frac{4}{\pi} \sum_{n=1}^{\infty} \frac{(-1)^n}{(2n-1)^4} \sin[(2n-1)x]$$

$$\times \{1 - \exp[-(2n-1)^2 t]\}$$

2. $$u(x,t) = -\frac{e^{-x}}{a^2} + \sum_{n=1}^{\infty} A_n \sin [(2n-1)x/2]$$

$$\times \exp[-(2n-1)^2 a^2 t/4]$$

where

$$A_n = \frac{2}{\pi} \int_0^{\pi} \left[f(x) + \frac{e^{-x}}{a^2}\right] \sin[(2n-1)x/2]\, dx$$

3. $$u(x,t) = \frac{A}{\omega} \sin(\omega t) + A_0/2 + \sum_{n=1}^{\infty} A_n \cos(nx)\, e^{-a^2 n^2 t}$$

$$A_0 = \frac{2}{\pi} \int_0^{\pi} f(x)\, dx$$

$$A_n = \frac{2}{\pi} \int_0^{\pi} f(x) \cos(nx)\, dx$$

4. $$u(x,t) = \frac{P}{a^2} x(x-L) + \frac{2PL^2}{a^2\pi^3} \sum_{n=1}^{\infty} \frac{(n^2\pi^2 - 2)(-1)^n + 2}{n^3}$$

$$\times \sin\left(\frac{n\pi x}{L}\right) \exp\left(-\frac{a^2 n^2 \pi^2 t}{L^2}\right)$$

6. $u(r,t) = \sum_{n=1}^{\infty} \frac{2(-1)^n}{n\pi} \frac{\sin(n\pi r)}{r} \exp(-a^2 n^2 \pi^2 t)$

P. 357

1. $u(x,t) = T_1 + \frac{T_0 - T_1}{1 + h} \frac{x}{L} + \sum_{n=1}^{\infty} A_n \sin(k_n \frac{x}{L}) \exp(-\frac{a^2 k^2_n t}{L^2})$

where

$$\tan(k_n) + hk_n = 0$$

$$A_n = \frac{\frac{2}{L}\int_0^L \left[f(x) - T_1 - \frac{T_0 - T_1}{1 + h} \frac{x}{L}\right] \sin(k_n \frac{x}{L})\, dx}{1 + h\cos^2(k_n)}$$

2. $u(x,t) = 200 \sum_{n=1}^{\infty} \frac{\sin(k_n)}{k_n [1 + \sin^2(k_n)/hL]} \cos(k_n \frac{x}{L})$

$$\times \exp(-\frac{a^2 k_n^2 t}{L^2})$$

where

$$k_n \tan(k_n) = hL$$

3. $u(x,t) = 70 +$ the solution to problem 2.

4. $u(x,t) = \sum_{n=1}^{\infty} \frac{200L \sin(k_n)}{k_n^2 (1 + \cos^2(k_n)/hL)} \sin(k_n \frac{x}{L})$

$$\times \exp(-\frac{a^2 k^2_n t}{L^2})$$

where

$$k_n \cot(k_n) = -hL.$$

P. 376

1. $u(x,t) = T_0 e^{-a^2 t} + T(1 - e^{-a^2 t})$

2. $u(x,t) = t - (1 - x^2)/2$

$$-\sum_{n=1}^{\infty} \frac{4\cos[(2n-1)\pi x/2] \exp[-(2n-1)^2 \pi^2 t/4]}{(2n-1)\pi/2 \quad \sin[(2n-1)\pi/2]}$$

3. $u(x,t) = x + 2 \sum_{n=1}^{\infty} \frac{\sin(n\pi x) \exp(-n^2\pi^2 t)}{n\pi \cos(n\pi)}$

4. $u(x,t) = x + 2t^{1/2}\, i\mathrm{erfc}[x/(2t^{1/2})]$

5. $u(x,t) = e^{t-x} + \mathrm{erf}\,(\frac{x}{2\sqrt{t}}) - e^{t+x}\,\mathrm{erfc}\,(\frac{x}{2\sqrt{t}} + t^{1/2})/2$

$- e^{t-x}\,\mathrm{erfc}\,(\frac{x}{2\sqrt{t}} - t^{1/2})/2$

6. $u(x,t) = \frac{x(1-x)}{2} - \sum_{n=1}^{\infty} \frac{4\cos[(2n-1)\pi(x-1/2)]}{(2n-1)\pi \sin[(2n-1)\pi/2]}$

$\times \exp[-(2n-1)^2\pi^2 t]$

P. 390

1. $u(x,t) = xt - \frac{32L^3}{a^2\pi^4} \sum_{n=1}^{\infty} \frac{(-1)^n}{(2n-1)^4} \sin\left[\frac{(2n-1)\pi x}{2L}\right]$

$\times [1 - \exp(-\frac{a^2(2n-1)^2\pi^2 t}{4L^2})]$

2. $u(x,t) = T_0 \left\{ \frac{x}{L} \sin(\omega t) + \frac{2\lambda\omega}{\pi} \sum_{n=1}^{\infty} \frac{(-1)^n n}{n^4 + \lambda^2\omega^2} \right.$

$\left. \times \left[\cos(\omega t) + \frac{\lambda\omega}{n^2}\sin(\omega t) - e^{-n^2 t}\right] \sin(\frac{n\pi x}{L}) \right\}$

where

$$\lambda = L^2/(\pi^2 a^2)$$

3. $u(x,t) = \frac{T_0 x}{L}(1 - e^{-bt}) - \frac{2T_0}{\pi} e^{-bt} \sum_{n=1}^{\infty} \left\{ \frac{(-1)^n}{n} - \frac{(-1)^n n}{n^2 - b\lambda} \right\}$

$\times \sin(\frac{n\pi x}{L})$

$- \frac{2b\lambda T_0}{\pi} \sum_{n=1}^{\infty} \frac{(-1)^n}{n(n^2 - b\lambda)} \sin(\frac{n\pi x}{L}) \exp(-\frac{n^2 t}{\lambda})$

$$u(x,t) = \frac{T_0 x}{L} - T_0\, e^{-bt}\, \frac{\sin\left(\frac{(b\lambda)^{1/2}\pi x}{L}\right)}{\sin\,[(b\lambda)^{1/2}\pi]}$$

$$-\frac{2b\lambda T_0}{\pi} \sum_{n=1}^{\infty} \frac{(-1)^n}{n(n^2 - b\lambda)} \sin\left(\frac{n\pi x}{L}\right) \exp\left(-\frac{n^2 t}{\lambda}\right)$$

because

$$x = -\frac{2L}{\pi} \sum_{n=1}^{\infty} \frac{(-1)^n}{n} \sin\left(\frac{n\pi x}{L}\right) \qquad 0<x<L$$

$$\sin(ax) = -\frac{2}{\pi} \sin(aL) \sum_{n=1}^{\infty} \frac{(-1)^n\, n}{n^2 - a^2L^2/\pi^2} \sin\left(\frac{n\pi x}{L}\right) \quad 0<x<L$$

Chapter 7

The Potential or Laplace's Equation

THE POTENTIAL EQUATION, OR LAPLACE'S EQUATION, IN TWO dimensions is

$$u_{xx} + u_{yy} = 0.$$

This equation describes such diverse fields as the equilibrium (time independent) displacements of a two-dimensional membrane to a steady-state temperature field. Many other physical phenomena—gravitational and electrostatic potential, certain fluid flows—are described by this equation, thus making it fundamental in the fields of mathematics, physics, and engineering.

7.1 THE DERIVATION OF LAPLACE'S EQUATION

In the previous chapter, we solved the one-dimensional heat equation. We often saw that after the transients died away we were left with a steady-state solution. Frequently it is important to find this steady-state solution of the heat equation when the temperature varies in both the x and y directions. To solve this problem we must rederive the heat equation.

Let us imagine a thin flat plate of heat-conducting material between sheets of insulation. We assume that a sufficient time has passed so that the temperature depends only on the position in the x-y plane. We now apply the law of conservation of energy (in rate form) to a small rectangle with sides of length Δx and Δy.

Let $q_x(x,y)$ and $q_y(x,y)$ denote the heat flow rates in the x- and y-direction, respectively. Conservation of energy requires that

$$\text{rate in} - \text{rate out} = 0. \quad \textbf{(7.1.1)}$$

if there is no storage or generation. Because

$$\text{rate in} = q_x(x,\ y+\Delta y/2)\Delta y + q_y\ (x+\Delta x/2,\ y)\Delta x \quad \textbf{(7.1.2)}$$

and

$$\text{rate out} = q_x\ (x+\Delta x, y+\Delta y/2)\Delta y + q_y(x+\Delta x/2, y+\Delta y)\Delta x. \quad \textbf{(7.1.3)}$$

If we assume that the plate has unit thickness, we find

$$\begin{aligned}&(q_x(x,y+\Delta y/2) - q_x(x+\Delta x, y+\Delta y/2)\ )\ \Delta y\\ &+\ (q_y(x+\Delta x/2, y) - q_y(x+\Delta x/2, y+\Delta y)\ )\Delta x = 0.\end{aligned} \quad \textbf{(7.1.4)}$$

On dividing through by $\Delta x \Delta y$, we see two differences quotients on the left-hand side of the equation. In the limit as Δx and Δy tend to zero, they become partial derivatives, leaving

$$\frac{\partial q_x}{\partial x} + \frac{\partial q_y}{\partial y} = 0. \quad \textbf{(7.1.5)}$$

All functions are evaluated at the point (x,y).

We now employ Fourier's law to eliminate the rates q_x and q_y. After doing so, we have

$$\frac{\partial}{\partial x}\ (a_x^{\,2}\ \frac{\partial u}{\partial x}) + \frac{\partial}{\partial y}\ (\ a_y^{\,2}\ \frac{\partial u}{\partial y}) = 0. \quad \textbf{(7.1.6)}$$

In the case of isotropic material $a_x^{\,2} = a_y^{\,2} =$ constant, our equation reduces to

$$u_{xx} + u_{yy} = 0 \quad \textbf{(7.1.7)}$$

which is the two-dimensional, steady-state heat equation.

The solutions of Laplace's equation (called harmonic functions) are fundamentally different from those encountered with the heat and wave equation. The heat and wave equations describe the evolution of some phenomena. Laplace's equation, on the other hand, describes things at equilibrium. Consequently, any change in the boundary conditions will affect to some degree the entire domain because a change to any one point will cause its neighbors to change in order to reestablish the equilibrium. That will, in turn, affect others. Because all these points are in equilibrium, this modification must occur instantaneously.

Another important principle is the *maximum principle*: If

$$u_{xx} + u_{yy} = 0$$

in a region, then $u(x,y)$ cannot have a relative maximum or minimum *inside* the region unless $u(x,y)$ is constant. (See Courant, R. and D. Hilbert, 1962: *Methods of Mathematical Physics. Vol. II.* New York: Interscience Publishers, 326-331 for the proof.) If we think of $u(x,y)$ as a steady-state temperature distribution, this principle is clearly true because at any one point the temperature cannot be greater than at all other nearby points. If that were so, heat would flow away from the hot point to cooler points nearby, thus eliminating the hot spot when equilibrium was once again restored.

It is often useful to consider the two-dimensional Laplace's equation in other coordinate systems. In polar coordinates, where

$$x = r\cos(\theta),\ y = r\sin(\theta) \text{ and } z = z, \tag{7.1.8}$$

Laplace's equation becomes

$$\frac{\partial^2 u}{\partial r^2} + \frac{1}{r}\frac{\partial u}{\partial r} + \frac{\partial^2 u}{\partial z^2} = 0 \tag{7.1.9}$$

if the problem possesses axisymmetry. Several solutions to this equation are found in Section 7.4. On the other hand, if the solution is independent of height, Laplace's equation becomes

$$\frac{\partial^2 u}{\partial r^2} + \frac{1}{r}\frac{\partial u}{\partial r} + \frac{1}{r^2}\frac{\partial^2 u}{\partial \theta^2} = 0 \tag{7.1.10}$$

and an example of its solution is worked out in Section 7.11, Eqs. (7.11.3)-(7.11.21).

In spherical coordinates, we have

$$x = r\cos(\phi)\sin(\theta),\ y = r\sin(\phi)\sin(\theta), \text{ and } z = r\cos(\theta) \tag{7.1.11}$$

where

$$r^2 = x^2 + y^2 + z^2, \tag{7.1.12}$$

θ is the angle measured *down* to the point from the z-axis (colatitude) and ϕ the angle made between the x-axis and the projection of the point on the x-y plane. In the case of axisymmetry (no ϕ dependence), Laplace's equation becomes

$$\frac{\partial}{\partial r}\left(r^2 \frac{\partial u}{\partial r}\right) + \frac{1}{\sin(\theta)}\frac{\partial}{\partial \theta}\left(\sin(\theta)\frac{\partial u}{\partial \theta}\right) = 0. \tag{7.1.13}$$

Its solution is given in Section 7.5.

7.2 BOUNDARY CONDITIONS

Because Laplace's equation involves time-independent phenomena, we need only specify boundary condition. As we discussed in Section 6.2, these boundary conditions may be classified as

1) Dirichlet condition : u given
2) Neumann condition : u_n given
3) Robin's condition : $u + \alpha u_n$ given

along any section of the boundary. (By u_n we mean the directional derivative in the direction normal or perpendicular to the boundary.) If all the boundaries have Neumann conditions, the solution is not unique, for if u is a solution, so is u plus a constant.

Finally we note that the boundary conditions must be specified along each side of the boundary. These sides may be at infinity as in problems with semi-infinite domains. We must specify values along the entire closed boundary because we could not have an equilibrium solution if a portion of the domain were undetermined.

7.3 THE SOLUTION OF LAPLACE'S EQUATION IN A RECTANGULAR REGION: THE MOTION OF GROUNDWATER IN A SMALL DRAINAGE BASIN

Over a century ago, a French hydraulic engineer named Henry Darcy published the results of a laboratory experiment on the flow of water through sand. He showed that the amount of discharge is directly proportional to the gradient of the sum of the elevation of the point of measurement plus the pressure head $(p/\varrho g)$. This sum is called the hydraulic head. In combination with conservation of mass, Darcy's law describes the flow of groundwater in an isotropic, homogeneous aquifer under steady-state conditions and may be written as

$$u_{xx} + u_{yy} = 0 \qquad \textbf{(7.3.1)}$$

where u is the hydraulic head or groundwater potential.

To illustrate the application of separation of variables to Laplace's equation, we shall determine the hydraulic head within a basin that lies beneath a shallow valley. See Fig. 7.3.1. Following the work of Toth (Toth, J., 1962: A theory of groundwater motion in a small drainage basin in central Alberta, Canada. *J. Geophys. Res.*, *67*, 4375-4387.), the boundary conditions are

$$u(x,z_0) = gz_0 + gcx \qquad \textbf{(7.3.2)}$$

$$u_x(0,y) = u_x(L,y) = 0 \qquad \textbf{(7.3.3)}$$

$$u_y(x,0) = 0 \qquad \textbf{(7.3.4)}$$

where g is the gravitational constant and c gives the slope of the topography. Eqs. (7.3.3) and (7.3.4) state that there is no flow

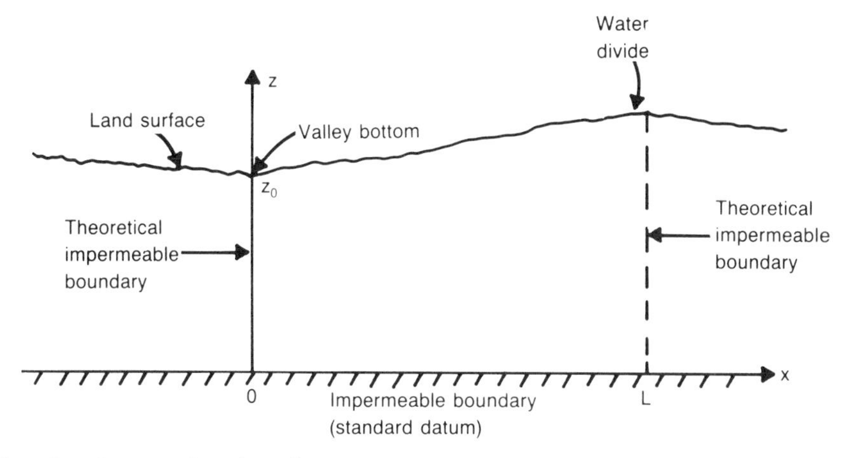

Fig. 7.3.1: Cross section of a valley.

through the bottom and sides of the aquifier. Equation (7.3.2) gives the fluid potential at the water table where z_0 is the elevation of the water table above the standard datum. The term *gcx* in Eq. (7.3.2) expresses the increase of the potential from the valley bottom toward the water divide. On average it closely follows the topography.

If we assume that $u(x,y)$ has a product form of $u(x,y) = X(x)Y(y)$, then Eq. (7.3.1) becomes

$$X''\,Y + X\,Y'' = 0. \tag{7.3.5}$$

This equation can be separated by dividing through by XY, yielding

$$-\frac{X''}{X} = \frac{Y''}{Y}\,. \tag{7.3.6}$$

Both sides of Eq. (7.3.6) must be constant, but the sign of that constant is not obvious. If we try a positive constant (say μ^2), Eq. (7.3.6) becomes two ordinary differential equations

$$X'' - \mu^2 X = 0 \tag{7.3.7}$$

and

$$Y'' + \mu^2 Y = 0 \tag{7.3.8}$$

which have the solutions

$$X(x) = A\cosh(\mu x) + B\sinh(\mu x) \tag{7.3.9}$$

and

$$Y(y) = C\cos(\mu x) + D\sin(\mu x). \tag{7.3.10}$$

Because the boundary conditions Eq. (7.3.3) imply

$$X'(0) = X'(L) = 0,$$

both A and B must be zero, leading to a solution $u(x,y) = 0$.

When a separation constant of zero is tried, we find

$$X_0(x) = 1 \text{ and } Y_0(y) = A_0/2 + B_0y.$$

However, because $Y'_0(0) = 0$ from Eq. (7.3.4), $B_0 = 0$. Thus the particular solution for $k = 0$ is $u_0(x,y) = A_0/2$.

Finally, taking both members of Eq. (7.3.6) equal to $-k^2$, we have

$$X'' + k^2X = 0 \tag{7.3.11}$$

and

$$Y'' - k^2Y = 0. \tag{7.3.12}$$

The first of these equations, along with the boundary conditions $X'(0) = X'(L) = 0$, gives

$$X_n(x) = \cos(k_nx) \tag{7.3.13}$$

with

$$k_n = \frac{n\pi}{L}, \qquad n = 1,2,3, \ldots \tag{7.3.14}$$

The function $Y_n(y)$ which accompanies the $X_n(x)$ is

$$Y_n(y) = A_n \cosh(k_ny) + B_n \sinh(k_ny) \tag{7.3.15}$$

We must take $B_n = 0$ because $Y'_n(0) = 0$.

We now have a product X_nY_n, which satisfies Laplace's equation and most of the boundary conditions. By the principle of linear superposition

$$\mathrm{u(x,y)} = A_0/2 + \sum_{n=1}^{\infty} A_n \cos\left(\frac{n\pi x}{L}\right)\cosh\left(\frac{n\pi y}{L}\right) \tag{7.3.16}$$

The nonhomogeneous boundary condition (7.3.2) is yet to be satisfied. Applying that condition we have

$$u(x,z_0) = gz_0 + gcx = A_0/2 + \sum_{n=1}^{\infty} A_n \cos\left(\frac{n\pi x}{L}\right) \cosh\left(\frac{n\pi z_0}{L}\right) \tag{7.3.17}$$

We recognize Eq. (7.3.17) as a Fourier half-range cosine series and can write down immediately

$$A_0 = \frac{2}{L}\int_0^L (gz_0 + gcx)dx \tag{7.3.18}$$

$$\cosh\left(\frac{n\pi z_0}{L}\right) A_n = \frac{2}{L}\int_0^L (gz_0 + gcx) \cos\left(\frac{n\pi x}{L}\right) dx. \quad \textbf{(7.3.19)}$$

Performing the integration,

$$A_0 = 2gz_0 + gcL \quad \textbf{(7.3.20)}$$

and

$$A_n = -\frac{2gcL}{\pi^2}\frac{[1 - (-1)^n]}{n^2 \cosh(n\pi z_0/L)} \quad \textbf{(7.3.21)}$$

Finally, we can write everything as:

$$u(x,y) = gz_0 + gcL/2 - \frac{4gcL}{\pi^2}\sum_{m=0}^{\infty}\frac{\cos\left[\frac{(2m+1)\pi x}{L}\right]\cosh\left[\frac{(2m+1)\pi y}{L}\right]}{(2m+1)^2\cosh\left[\frac{(2m+1)\pi z_0}{L}\right]}$$

(7.3.22)

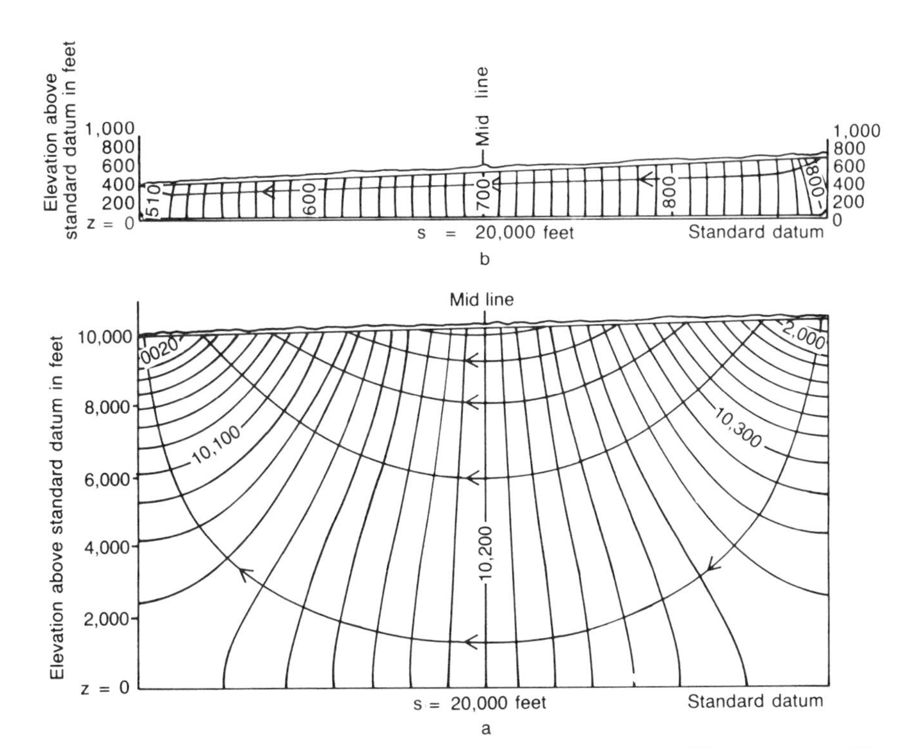

Fig. 7.3.2: Two-dimensional potential distribution and flow patterns from different depths of the horizontal impermeable boundary. Taken from Toth. J., 1962: A theory of groundwater motion in a small drainage basin in central Alberta, Canada. *J. Geophys. Res., 67,*, 4375-4387.

In Fig. 7.3.2 two graphs given by Toth are presented for two different aquifiers. Water flows from the elevated land (on the right) into the valley (on the left).

Exercises

1. Solve $u_{xx} + u_{yy} = 0$

with the following boundary conditions:

(a) $u_x(0,y) = 0$ and $u(a,y) = u(x,0) = u(x,b) = 1$
(b) $u_x(0,y) = u(a,y) = u(x,0) = 0$ and $u(x,b) = 1$
(c) $u_y(x,0) = u(x,b) = u(a,y) = 0$ and $u(0,y) = 1$
(d) $u(a,y) = u(x,0) = u(x,b) = 0$ and $u(0,y) = 1$
(e) $u_y(x,0) = u_y(x,b) = 0$ and $u(0,y) = u(a,y) = 1$
(f) $u(0,y) = u(x,0) = 1$ and $u_x(a,y) = u_y(x,b) = 0$
(g) $u(0,y) = u(a,y) = 1$ and $u_y(x,0) = u(x,b) = 0$
(h) $u(0,y) = u(x,0) = 1$ and $u(a,y) = u(x,b) = 0$
(i) $u(0,y) = u(a,y) = u(x,0) = 0$ and

$$= \begin{cases} 2x/a & 0<x<a/2 \\ 2(a-x)/a & a/2<x<a \end{cases}$$

2. The equation describing the hydraulic head $h(x,y)$ of a rectangular island on which a recharging well of a rate R is located at the center is given by Poisson's equation:

$$h_{xx} + h_{yy} = -\frac{R}{T}$$

where T is the product of the hydraulic conductivity and aquifier thickness. If the elevation of the water level in the well is 20 units above sea level and the water table is at sea level around the island $h(-a,y) = h(a,y) = h(x,b) = h(x,-b) = 0$, find the water table anywhere in the island.

3. *Subsurface temperature due to periodic surface temperature and topography.* Variations in the earth's surface temperature can arise as a result of topographic undulations and the altitude dependence of temperature in the earth's atmosphere. To find the temperature within the solid earth, solve the two-dimensional Laplace's equation with the boundary conditions:

$$T(x,0) = T_0 + \Delta T \cos\left(\frac{2\pi x}{\lambda}\right)$$

where λ is the wavelength of the spatial temperature variation and $T(x,y)$ is bounded as y approaches infinity.

4. *Groundwater flow under undulating topography.* Toth (Toth, J., 1963: A theoretical analysis of groundwater flow in small drainage basins. *J. Geophys. Res., 88,* 4795-4812.) generalized his analysis of groundwater in an aquifer where the water table follows the topography. Find the groundwater potential if the groundwater potential along the top is $u(x,z_0) = g(z_0 + cx + a \sin(bx))$ and

$$u_x(0,y) = u_x(s,y),\ u_y(x,0) = 0$$

where g is the gravitational constant.

5. Find the solution to Laplace's equation in rectangular coordinates if $u_x(0,y) = u_x(L,y) = 0$, $u(x,b) = u_1$ and

$$u(x,0) = \begin{cases} f(x) & 0 \leq x < \alpha \\ 0 & \alpha < x \leq L. \end{cases}$$

7.4 THE ELECTROSTATIC POTENTIAL WITHIN A CYLINDER

In this section we find the electrostatic potential $u(r,z)$ inside a closed cylinder of length L and radius a. The base and lateral surfaces are held at the potential 0 while the upper surface is held at potential V.

Because the potential varies in only r and z, the two-dimensional potential equation in cylindrical coordinates reduces to

$$\frac{1}{r}\frac{\partial}{\partial r}\left(r\frac{\partial u}{\partial r}\right) + \frac{\partial^2 u}{\partial z^2} = 0 \qquad \textbf{(7.4.1)}$$

with the boundary conditions that

$$u(a,z) = u(r,0) = 0 \text{ and } u(r,L) = V. \qquad \textbf{(7.4.2)}$$

If we can solve this problem by separation of variables, then $u(r,z) = R(r)Z(z)$ and

$$\frac{1}{rR}\frac{d}{dr}\left(r\frac{dR}{dr}\right) = -\frac{1}{Z}\frac{d^2Z}{dz^2} = -\frac{k^2}{a^2} \qquad \textbf{(7.4.3)}$$

We have assumed a negative separation constant because a positive or zero separation constant yields trival solutions in the radial direction. On the other hand, a negative separation constant results in

$$\frac{1}{r}\frac{d}{dr}\left(r\frac{dR}{dr}\right) + \frac{k^2}{a^2}R = 0. \qquad \textbf{(7.4.4)}$$

The solution to Eq. (7.4.4) are the Bessel functions $J_0(kr/a)$ and $Y_0(kr/a)$. Because $Y_0\ (kr/a)$ becomes infinite at $r = 0$, the only permissible solution is $J_0(kr/a)$. The condition that

$$u(r,z) = R(a)Z(z) = 0$$

forces us to choose k's such that $J_0(k) = 0$. Therefore, the solution in the radial direction is $J_0(k_n r/a)$ where $J_0(k_n) = 0$ for $n = 1,2,3,$ . . .

In the z direction, we have

$$\frac{d^2Z}{dz^2} + \frac{k_n{}^2}{a^2} Z = 0. \qquad \textbf{(7.4.5)}$$

The general solution to Eq. (7.4.5) is

$$Z_n(z) = A_n \sinh\left(\frac{k_n z}{a}\right) + B_n \cosh\left(\frac{k_n z}{a}\right). \qquad \textbf{(7.4.6)}$$

Because $u(r,0) = R(r)\,Z(0) = 0$ and $\cosh(0) = 1$, B_n must equal zero. Therefore, the general product solution is

$$u(r,z) = \sum_{n=1}^{\infty} A_n J_0\left(\frac{k_n r}{a}\right) \sinh\left(\frac{k_n z}{a}\right) \qquad \textbf{(7.4.7)}$$

The arbitrary constant A_n is given by the condition that $u(r,L) = V$. Consequently,

$$u(r,L) = V = \sum_{n=1}^{\infty} A_n J_0\left(\frac{k_n r}{a}\right) \sinh\left(\frac{k_n L}{a}\right) \qquad \textbf{(7.4.8)}$$

where

$$\sinh\left(\frac{k_n L}{a}\right) A_n = \frac{2V}{a^2 J_1(k_n)^2} \int_0^L r\, J_0\left(\frac{k_n r}{a}\right) dr \qquad \textbf{(7.4.9)}$$

from Eq. (1.7.58) and (1.7.66).

$$\sinh\left(\frac{k_n L}{a}\right) A_n = \frac{2\,V}{k_n{}^2 J_1(k_n)} \left(\frac{k_n r}{a}\right) J_1\left(\frac{k_n r}{a}\right) \Bigg|_0^a \qquad \textbf{(7.4.10)}$$

$$= \frac{2\,V}{k_n J_1(k_n)}. \qquad \textbf{(7.4.11)}$$

The final answer is

$$u(r,z) = 2V \sum_{n=1}^{\infty} \frac{J_0(k_n r/a)}{k_n J_1(k_n)} \frac{\sinh(k_n z/a)}{\sinh(k_n L/a)}. \qquad \textbf{(7.4.12)}$$

In Fig. 7.4.1 the solution (7.4.12) is graphed for the case when $L = a$.

Let us now consider a similar, but slightly different, problem where the ends are held at zero while the lateral side has the value V. Once again, the governing equation is Eq. (7.4.1) with the boundary conditions that

$$u(r,0) = u(r,L) = 0 \text{ and } u(a,z) = V. \qquad \textbf{(7.4.13)}$$

Separation of variables yields

$$\frac{1}{rR}\frac{d}{dr}\left(r\frac{dR}{dr}\right) = -\frac{1}{Z}\frac{d^2Z}{dz^2} = \frac{k^2}{L^2} \qquad \textbf{(7.4.14)}$$

with $Z(0) = Z(L) = 0$. We have chosen a positive separation constant because a negative constant would give hyperbolic functions in z which cannot satisfy the boundary conditions. A separation constant of zero would give a straight line for $Z(z)$. Applying the boundary conditions gives a trival solution. Consequently the only solution in the z direction which satisfies the boundary conditions is

$$Z_n(z) = \sin(n\pi z/L).$$

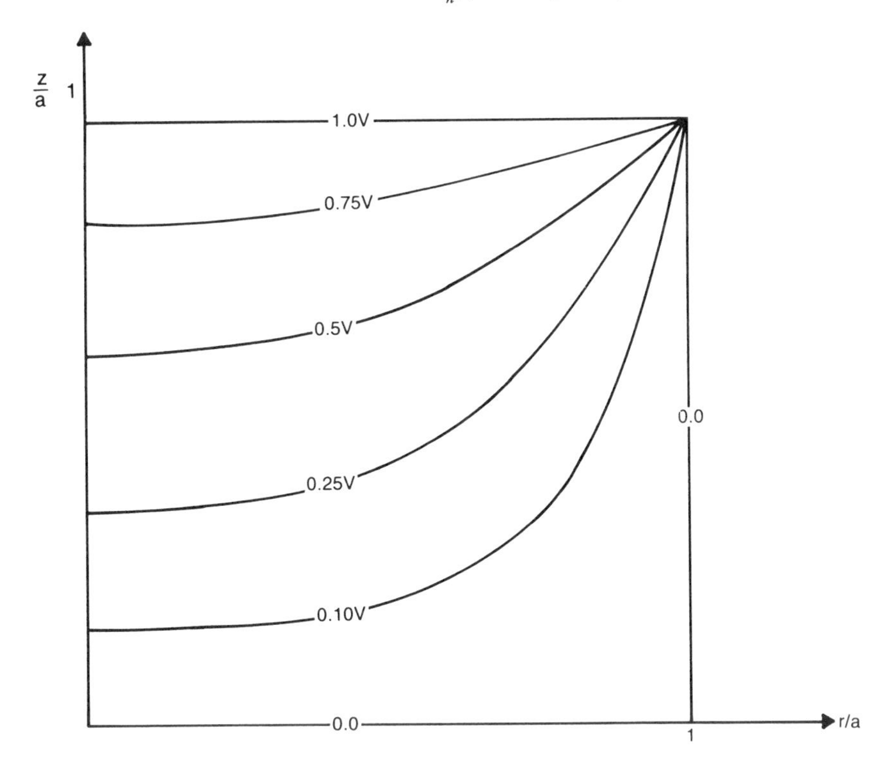

Fig. 7.4.1: The steady-state potential within a cylinder of equal radius and height a when the top is maintained at V while the lateral side and bottom are at potential zero.

In the radial direction, the differential equation is

$$\frac{1}{r}\frac{d}{dr}\left(r\frac{dR}{dr}\right) - \frac{k_n^2}{L^2} R = 0 \tag{7.4.15}$$

which is very similar to that of Bessel. Indeed, the general solution is

$$R(r) = c_1 I_0\left(\frac{k_n r}{L}\right) + c_2 K_0\left(\frac{k_n r}{L}\right) \tag{7.4.16}$$

where I_0 and K_0 are given by the power series:

$$I_0(x) = \sum_{k=0}^{\infty} \frac{1}{(k!)^2}\left(\frac{x}{2}\right)^{2k} = J_0(ix) \tag{7.4.17}$$

and

$$K_0(x) = \frac{i\pi}{2}\,[J_0(ix) + i\,Y_0(ix)]. \tag{7.4.18}$$

These solutions are referred to as modified Bessel functions of the first and second kind, respectively, of order zero. Because $K_0(x)$ behaves as $-\ell n(x)$ as $x \to 0$, we must discard this solution and our solution in the radial direction becomes $R(r) = I_0(k_n r/L)$. Hence, the product solution is

$$u_n(r,z) = A_n I_0\left(\frac{n\pi r}{L}\right)\sin\left(\frac{n\pi z}{L}\right) \tag{7.4.19}$$

and the general solution is a sum of the particular solutions:

$$u(r,z) = \sum_{n=1}^{\infty} A_n I_0\left(\frac{n\pi r}{L}\right)\sin\left(\frac{n\pi z}{L}\right). \tag{7.4.20}$$

Finally, we use the boundary condition that $u(a,z) = V$ to compute A_n. This condition gives

$$u(a,z) = V = \sum_{n=1}^{\infty} A_n I_0\left(\frac{n\pi a}{L}\right)\sin\left(\frac{n\pi z}{L}\right) \tag{7.4.21}$$

so that

$$I_0\left(\frac{n\pi a}{L}\right) A_n = \frac{2}{L}\int_0^L V \sin\left(\frac{n\pi z}{L}\right) dz \tag{7.4.22}$$

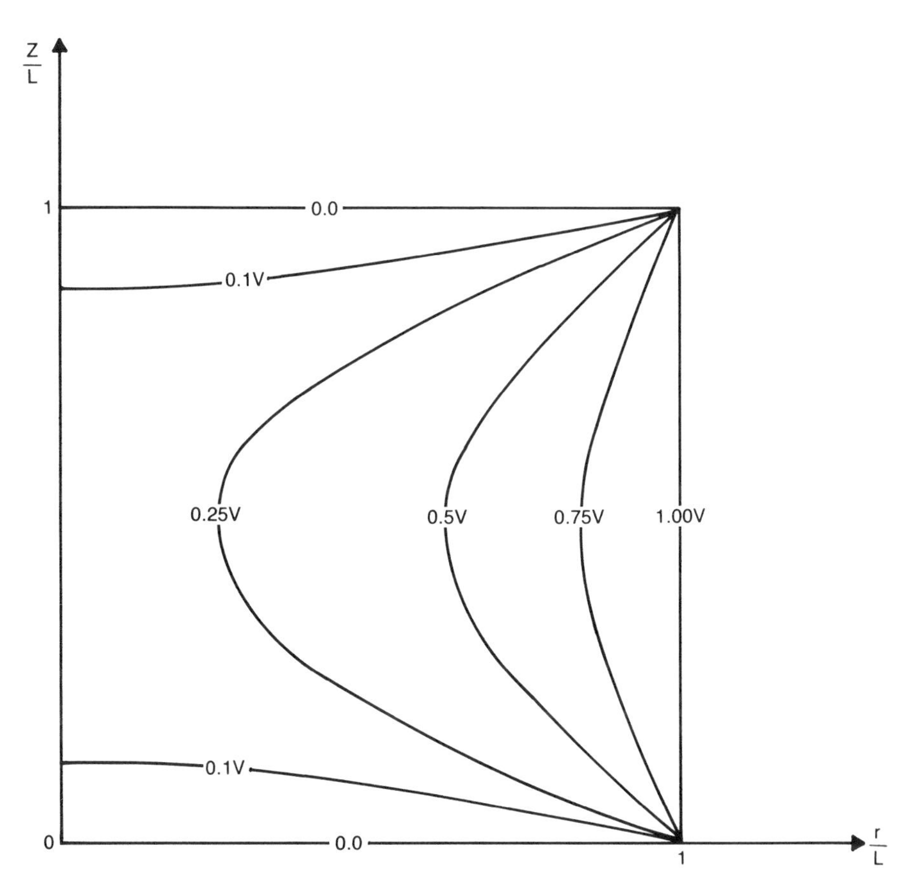

Fig. 7.4.2: Potential within a conducting sphere when the top and bottom are held at the potential zero while the lateral sides are held at V.

$$I_0\left(\frac{n\pi a}{L}\right)A_n = \frac{2V}{n\pi}\left(1 - (-1)^n\right). \quad \textbf{(7.4.23)}$$

Therefore, the final answer is

$$u(r,z) = \frac{4V}{\pi}\sum_{n=1}^{\infty}\frac{I_0\left(\frac{(2n-1)\pi r}{L}\right)\sin\left(\frac{(2n-1)\pi z}{L}\right)}{I_0\left(\frac{(2n-1)\pi a}{L}\right)(2n-1)} \quad \textbf{(7.4.24)}$$

In Fig. 4.7.2 the solution (7.4.24) is graphed for the case where $L = a$.

Exercises

1. Find the solution to

$$\frac{\partial^2 u}{\partial r^2} + \frac{1}{r}\frac{\partial u}{\partial r} + \frac{\partial^2 u}{\partial z^2} = 0$$

with

$$u_r(a,z) = 0 \text{ and } u(r,H) = 0$$

and

$$u_z(r,0) = \begin{cases} 1 & 0<r<r_0 \\ 0 & r_0<r<a \end{cases}.$$

2. Solve

$$\frac{\partial^2 u}{\partial r^2} + \frac{1}{r}\frac{\partial u}{\partial r} - \frac{u}{r^2} + \frac{\partial^2 u}{\partial z^2} = 0$$

with $u_z(r,h) = Ar$ and $u(r,0) = 0$ for $0 \le r \le R$, $u(R,z) = 0$ for $0 \le z \le h$.

3. Solve

$$\frac{\partial^2 u}{\partial r^2} + \frac{1}{r}\frac{\partial u}{\partial r} - \frac{u}{r^2} + \frac{\partial^2 u}{\partial z^2} = 0$$

with $u(r,0) = u(r,1) = 0$ for $0 \le r \le a$ and $u(a,z) = z$ for $0<z<1$.

4. Resolve problem 3 with

$$u_r(a,z) = u(r,0) = 0 \text{ and } u_z(r,h) = 2r.$$

7.5 THE ELECTROSTATIC POTENTIAL WITHIN A CONDUCTING SPHERE

Solutions to Laplace's equation not only describe two-dimensional, steady-state groundwater flows but also give the potential in two-dimensional electrostatic problems. To illustrate this, we will find the potential at any point P within a conducting sphere of radius a. At the surface the potential is held at V_0 in the hemisphere $0<\theta<\pi/2$ and $-V_0$ for $\pi/2<\theta<\pi$. This problem is obviously best solved in spherical coordinates.

The potential $u(r,\theta)$ is given by Laplace's equation in spherical coordinates

$$\frac{\partial}{\partial r}\left(r^2 \frac{\partial u}{\partial r}\right) + \frac{1}{\sin(\theta)}\frac{\partial}{\partial \theta}\left(\sin(\theta) \frac{\partial u}{\partial \theta}\right) = 0. \qquad \textbf{(7.5.1)}$$

To solve Eq. (7.5.1) we use separation of variables

$$u(r,\theta) = R(r)\Theta(\theta). \quad (7.5.2)$$

Substituting into Eq. (7.5.1) yields

$$\frac{1}{R}\frac{d}{dr}\left(r^2\frac{dR}{dr}\right) = -\frac{1}{\Theta}\frac{1}{\sin(\theta)}\frac{d}{d\theta}\left[\sin(\theta)\frac{d\Theta}{d\theta}\right] = k^2 \quad (7.5.3)$$

or

$$r^2 R'' + 2r R' - k^2 R = 0 \quad (7.5.4)$$

and

$$\frac{1}{\sin(\theta)}\frac{d}{d\theta}\left[\sin(\theta)\frac{d\Theta}{d\theta}\right] + k^2\Theta = 0 \quad (7.5.5)$$

A common substitution is to replace θ with $\mu = \cos(\theta)$. Then as θ varies from 0 to π, μ goes from 1 to -1. Equation (7.5.5) becomes

$$\frac{d}{d\mu}\left((1-\mu^2)\frac{d\Theta}{d\mu}\right) + k^2\Theta = 0 \quad (7.5.6)$$

Eq. (7.5.6) is Legendre's equation which was discussed in Section 1.6. Consequently if the solution is going to be finite at the poles, then

$$k^2 = n(n+1) \quad (7.5.7)$$

and

$$\Theta_n(\theta) = P_n(\mu) = P_n(\cos(\theta)) \quad (7.5.8)$$

where $n = 0, 1, 2, 3, \ldots$

Turning to Eq. (7.5.4), this equation is known as the equidimensional equation or Euler-Cauchy linear differential equation. One method of solving this equation consists of introducing a new independent variable s so that

$$r = e^s \text{ or } s = \ell n(r). \quad (7.5.9)$$

There then follows

$$\frac{d}{dr} = \frac{ds}{dr}\frac{d}{ds} = e^{-s}\frac{d}{ds} \quad (7.5.10)$$

and

$$\frac{d^2}{dr^2} = \frac{d}{dr}(e^{-s}\frac{d}{ds}) = e^{-s}\frac{d}{ds}(e^{-s}\frac{d}{ds}) = e^{-2s}(\frac{d^2}{ds^2} - \frac{d}{ds}). \tag{7.5.11}$$

Substituting into Eq. (7.5.4), we find

$$\frac{d^2y}{ds^2} + \frac{dy}{ds} - n(n+1)y = 0. \tag{7.5.12}$$

Equation (7.5.12) is a second-order, constant-coefficient ordinary differential equation with the solution:

$$y(s) = A\, e^{ns} + B\, e^{-(1+n)s} \tag{7.5.13}$$

$$y(r) = A \exp\,[n\, \ell n(r)] + B \exp\,[-(1+n)\, \ell n(r)] \tag{7.5.14}$$

$$= A \exp[\ell n(r^n)] + B \exp[\ell n(r^{-1-n})] \tag{7.5.15}$$

$$= A\, r^n + B\, r^{-1-n} \tag{7.5.16}$$

In our particular case, $R(r)$ may be written as

$$R_n(r) = A_n\,(\frac{r}{a})^n + B_n\,(\frac{r}{a})^{-1-n} \tag{7.5.17}$$

The constant a, the radius of the sphere, is introduced to simplify future calculations.

Using the results from Eq. (7.5.8) and (7.5.17), the solution to Laplace's equation in axisymmetric problems is given by

$$u(r,\theta) = \sum_{n=0}^{\infty} \left[A_n\,(\frac{r}{a})^n + B_n\,(\frac{r}{a})^{-1-n} \right] P_n\,[\cos(\theta)] \tag{7.5.18}$$

In our particular problem we must take $B_n = 0$ because the solution becomes infinite at $r = 0$ otherwise. If we were solving the problem for the potential in the domain $a<r<\infty$, we would take $A_n = 0$ because the potential must approach zero as $r \to \infty$.

Finally we must evaluate A_n. Finding the potential at the surface of the sphere, we obtain

$$u(a,\mu) = \sum_{n=0}^{\infty} A_n\, P_n\,(\mu) = \begin{cases} V_0 & 0<\mu\leq 1 \\ -V_0 & -1\leq\mu<0 \end{cases}. \tag{7.5.19}$$

Upon examining Eq. (7.5.19), we see that it is merely an expansion in Legendre polynomials of the function

$$f(\mu) = \begin{cases} V_0 & 0<\mu\leq 1 \\ -V_0 & -1\leq\mu<0 \end{cases}. \quad \textbf{(7.5.20)}$$

Consequently, from Eq. (1.6.38),

$$A_n = \frac{2n+1}{2}\int_{-1}^{1} f(\mu)P_n(\mu)\, d\mu \quad \textbf{(7.5.21)}$$

Because $f(\mu)$ is an odd function, we know immediately that $A_n = 0$ if n is even. When n is odd, however,

$$A_n = 2n+1\int_{0}^{1} V_0\, P_n\,(\mu)\, d\mu. \quad \textbf{(7.5.22)}$$

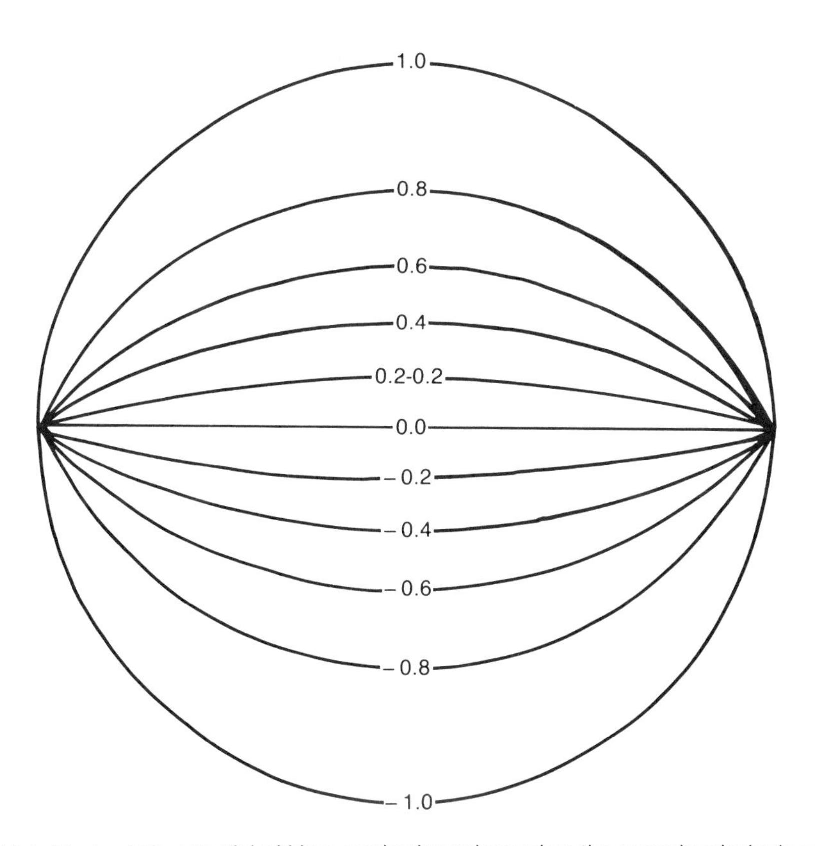

Fig. 7.5.1: Electrostatic potential within a conducting sphere when the upper hemispheric surface is held at $V_0 = 1$ and the lower surface at $V_0 = -1$.

This can be further simplified by using the relationship that

$$\int_x^1 P_n(t)\,dt = \frac{1}{2n+1}\left[P_{n-1}(x) - P_{n+1}(x)\right] \qquad \textbf{(7.5.23)}$$

where $n > 1$. In our problem we then have

$$A_n = \begin{cases} V_0 \quad (P_{n-1}(0) - P_{n+1}(0)) & \text{if } n \text{ is odd} \\ 0 & \text{if } n \text{ is even.} \end{cases} \qquad \textbf{(7.5.24)}$$

The first few terms are $A_1 = 3V_0/2$, $A_3 = -7V_0/8$ and $A_5 = 11V_0/16$. In Fig. 7.5.1 our solution is graphed.

7.6 THE HEATING OF A SPHERE IN THE SUNLIGHT

We now find the steady-state temperature field within a metallic sphere of radius a, which is placed in direct sunlight and allowed by radiatively cool. This classic problem, first solved by Rayleigh (Lord Rayleigh, 1870: On the value of the integral $\int_0^1 Q_n Q_{n}{}', d\mu$, $Q_{n'}$ $Q_n{}'$, being Laplace's coefficients of the orders n, n', with application to the theory of radiation. *Phil. Trans. (London), 160,* 579-590), requires the use of spherical coordinates with its origin at the center of the sphere and with the z-axis pointing toward the sun. With this choice for the coordinate system, the incident sunlight is given by

$$D(\theta) = \begin{cases} D(0)\cos(\theta) & 0 \le \theta < \pi/2 \\ 0 & \pi/2 < \theta \le \pi \end{cases} \qquad \textbf{(7.6.1)}$$

If we assume that heat dissipation takes place at the surface $r = a$ according to Newton's law of cooling and take the temperature of the surrounding medium to be zero, the solar heat absorbed by the surface dA must balance the Newtonian cooling at the surface plus the energy absorbed into the sphere's interior. This physical relationship can be written mathematically as

$$(1-\varrho)\,D(\theta)\,dA = \epsilon\,u(a,\theta)\,dA + \frac{\partial u(a,\theta)}{\partial r}\,dA, \qquad \textbf{(7.6.2)}$$

where ϱ is the reflectance of the surface (the albedo), ϵ the surface conductance or coefficient of surface heat transfer, and $\varkappa$ the thermal conductivity, or

$$\frac{\partial u}{\partial r} = \frac{1-\varrho}{\varkappa} D(\theta) - \frac{\epsilon}{\varkappa} u \text{ at } r = a. \tag{7.6.3}$$

If the sphere has reached thermal equilibrium, the temperature field within the sphere is described by Laplace's equation. In the previous section, we showed that solutions to Laplace's equation in axisymmetrical problems are made up of the solutions

$$u(r,\theta) = \sum_{n=0}^{\infty} \left[A_n \left(\frac{r}{a}\right)^n + B_n \left(\frac{r}{a}\right)^{-1-n} \right] P_n [\cos(\theta)] \tag{7.6.4}$$

In this particular problem we must take $B_n = 0$ because the solution would blow up at $r = 0$ otherwise. Therefore

$$u(r,\theta) = \sum_{n=0}^{\infty} A_n \left(\frac{r}{a}\right)^n P_n [\cos(\theta)]. \tag{7.6.5}$$

Differentiation gives

$$\frac{\partial u}{\partial r} = \sum_{n=0}^{\infty} A_n \frac{n\, r^{n-1}}{a^n} P_n [\cos(\theta)]. \tag{7.6.6}$$

Substituting into the boundary condition leads to

$$\sum_{n=0}^{\infty} A_n \left(\frac{n}{a} + \frac{\epsilon}{\varkappa}\right) P_n (\cos(\theta)) = \left(\frac{1-\varrho}{\varkappa}\right) D(\theta) \tag{7.6.7}$$

or

$$D(\theta) = \sum_{n=0}^{\infty} A_n \frac{n\varkappa + \epsilon a}{a(1-\varrho)} P_n [\cos(\theta)] \tag{7.6.8}$$

$$= \sum_{n=0}^{\infty} C_n P_n (\mu) \tag{7.6.9}$$

where

$$C_n = \frac{n\varkappa + \epsilon a}{a(1-\varrho)} A_n \text{ and } \mu = \cos(\theta). \tag{7.6.10}$$

The coefficients C_n are determined by

$$C_n = \frac{2n+1}{2} \int_{-1}^{1} D(\theta) P_n(\mu)\, d\mu \tag{7.6.11}$$

$$= \frac{2n+1}{2} D(0) \int_{0}^{1} \mu P_n(\mu)\, d\mu. \tag{7.6.12}$$

Evaluation of the first few coefficients gives

$$A_0 = \frac{1-\varrho}{4\epsilon} D(0), \qquad A_1 = \frac{a(1-\varrho)D(0)}{2(\varkappa + \epsilon a)}, \qquad A_2 = \frac{5a(1-\varrho)D(0)}{16(2\varkappa + \epsilon a)}$$

$$A_3 = 0 \qquad A_4 = -\frac{3a(1-\varrho)D(0)}{32\,(4\varkappa + \epsilon a)}$$

In Fig. 7.6.1 the temperature field within the interior of the sphere is graphed with $D(0) = 1200$ watts/m^2, $\varkappa = 45$ watts/m ° K, $\epsilon = 5$ watts/m^2 °K, $\varrho = 0$ and $a = 0.1\ m$. This corresponds to a cast iron sphere with blackened surface in sunlight.

7.7 THE POTENTIAL OUTSIDE OF A CONDUCTING SPHERE PLACED IN THE PRESENCE OF A POINT CHARGE

In this section we will find the potential at any point P which is due to a point charge $+q$ situated at $z = a$ on the z-axis when

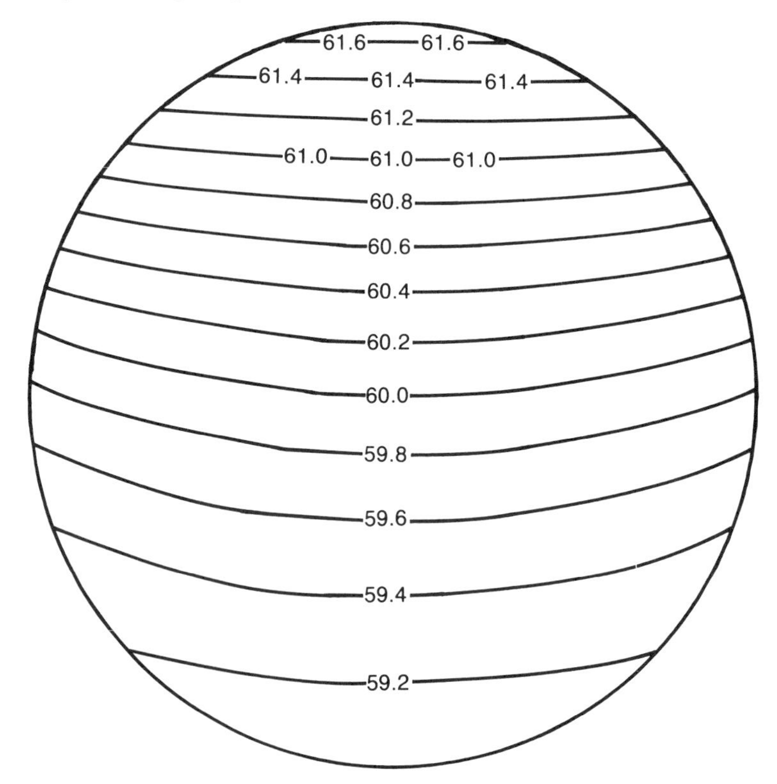

Fig. 7.6.1: The difference (in °C) between the temperature field within blackened, iron surface of radius 0.1 m and the surrounding medium when heated by sunlight and allowed to radiatively cool.

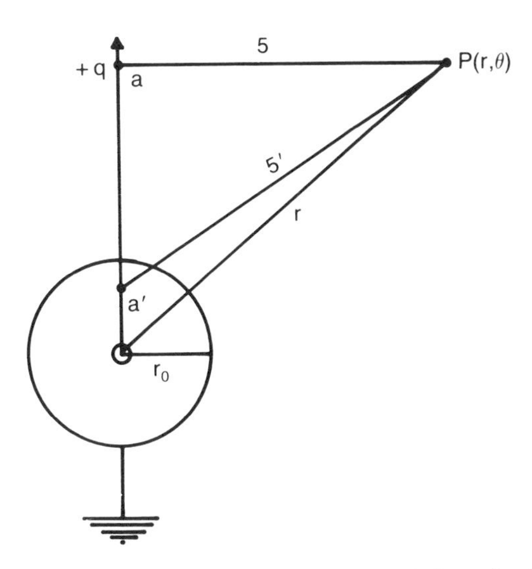

Fig. 7.7.1: Point charge +q in the presence of a grounded, conducting sphere.

a conducting, grounded sphere is placed with its origin at 0. See Fig. 7.7.1. Because of the principle of linear superposition, the total potential $v(r,\theta)$ can be written as a sum of the potential from the point charge and the potential $u(r,\theta)$ due to the induced charge on the sphere:

$$v(r,\theta) = \frac{q}{s} + u(r,\theta). \qquad \textbf{(7.7.1)}$$

In common with the first term q/s, $u(r,\theta)$ must be a solution of Laplace's equation. In Section 7.5 we showed that the general solution to Laplace's equation in axisymmetric problems is given by

$$u(r,\theta) = \sum_{n=0}^{\infty} \left[A_n \left(\frac{r}{r_0}\right)^{n} + B_n \left(\frac{r}{r_0}\right)^{-1-n} \right] P_n\,[\cos(\theta)]. \qquad \textbf{(7.7.2)}$$

Because we are interested in solutions valid *anywhere* outside of the sphere, we must take $A_n = 0$; otherwise, the solution will not die away as $r \to \infty$. Hence,

$$u(r,\theta) = \sum_{n=0}^{\infty} B_n \left(\frac{r}{r_0}\right)^{-1-n} P_n\,[\cos(\theta)]. \qquad \textbf{(7.7.3)}$$

The coefficients B_n are to be determined by the condition that $v(r_0,\theta) = 0$ or

$$\left.\frac{q}{s}\right|_{\text{on sphere}} + \sum_{n=0}^{\infty} B_n P_n\,[\cos(\theta)] = 0. \qquad \textbf{(7.7.4)}$$

We need to expand the first term on the left in terms of Legendre polynomials. From the law of cosines, we have

$$s = [r^2 + a^2 - 2ar\cos(\theta)]^{1/2}. \quad \textbf{(7.7.5)}$$

Consequently if $a > r$, then

$$\frac{1}{s} = \frac{1}{a}\left[1 - 2\cos(\theta)\left(\frac{r}{a}\right) + \left(\frac{r}{a}\right)^2\right]^{-1/2} \quad \textbf{(7.7.6)}$$

In Chapter 1, Section 1.6 we showed that

$$(1 - 2xz + z^2)^{-1/2} = \sum_{n=0}^{\infty} P_n(x)\, z^n \quad \textbf{(7.7.7)}$$

Consequently,

$$\frac{1}{s} = \frac{1}{a}\sum_{n=0}^{\infty} P_n[\cos(\theta)]\left(\frac{r}{a}\right)^n \quad \textbf{(7.7.8)}$$

At the surface of the sphere, we have

$$\sum_{n=0}^{\infty}\left[\frac{q}{a}\left(\frac{r_0}{a}\right)^n + B_n\right] P_n[\cos(\theta)] = 0. \quad \textbf{(7.7.9)}$$

This is only satisfied if the coefficient of every $P_n[\cos(\theta)]$ is zero, so that

$$B_n = -\frac{q}{a}\left(\frac{r_0}{a}\right)^n. \quad \textbf{(7.7.10)}$$

On substituting this back into Eq. (7.7.3), we find

$$u(r,\theta) = -\frac{q}{r}\frac{r_0}{a}\sum_{n=0}^{\infty}\left(\frac{r_0^2}{ar}\right)^n P_n[\cos(\theta)]. \quad \textbf{(7.7.11)}$$

The physical interpretation of Eq. (7.7.11) is as follows: Consider a point, such as a' (see Fig. 7.7.1) on the z-axis. If $r > a'$, the expression of $1/s'$ is

$$\frac{1}{s'} = \frac{1}{r}\sum_{n=0}^{\infty}\left(\frac{a'}{r}\right)^n P_n(\cos(\theta)) \qquad r > a' \quad \textbf{(7.7.12)}$$

Using Eq. (7.7.12), we can rewrite Eq. (7.7.11) as

$$u(r,\theta) = -q\frac{r_0}{a}\frac{1}{s'} \quad \textbf{(7.7.13)}$$

if we put $a' = r_0^2/a$. Our final result may now be written

$$v(r,\theta) = \frac{q}{s} - \frac{q'}{s'} \qquad \textbf{(7.7.14)}$$

provided q' is identified with $(r_0/a)q$. In other words: when a conducting, grounded sphere is placed near a point charge $+q$ it changes the potential in the same manner as would a point charge of opposite sign and magnitude $q' = (r_0/a)q$, placed at the point $a' = r^2_0/a$. The charge q' is said to be the *image* of q.

In Fig. 7.7.2 the solution (7.7.11) for $q = 1$, $r_0 = 1$ and $a = 2$ is graphed.

Exercises

1. The steady-state distribution of temperature within a sphere is governed by Laplace's equation. Using spherical coordinates, find that distribution if the upper half of the surface of radius b is maintained at the temperature $u = 100$ while the lower half is maintained at the temperature $u = 0$.

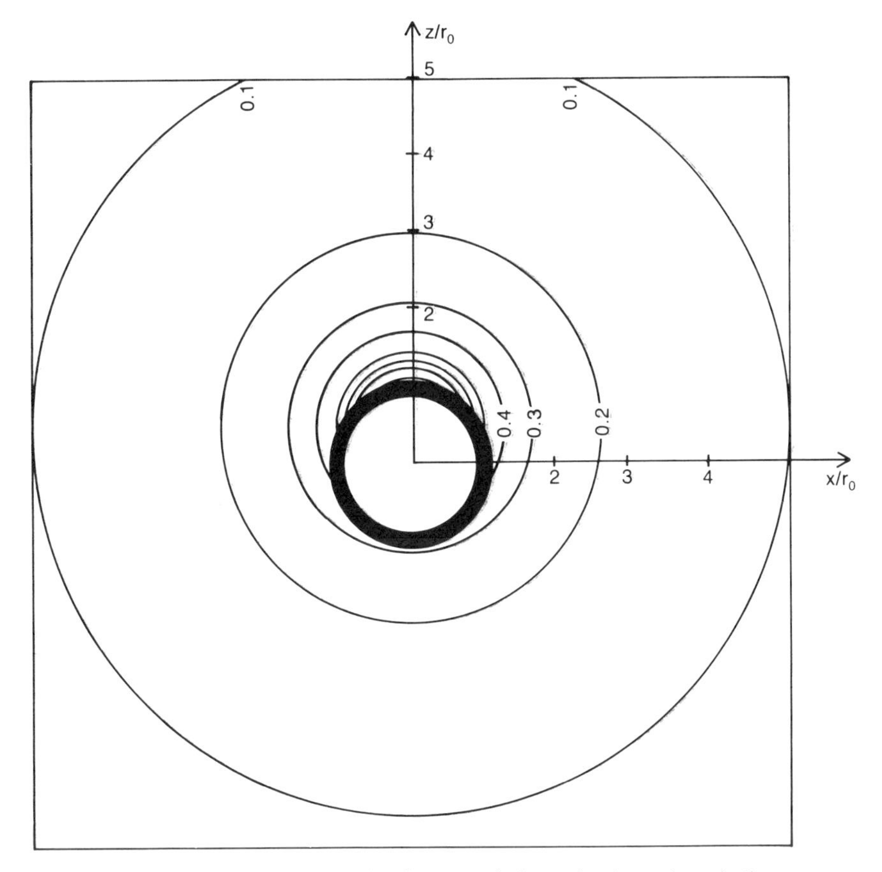

Fig. 7.7.2: Electrostatic potential outside of a grounded, conducting sphere in the presence of a point charge located at $z/r_0 = 2$. Contours are in units of 0.1 q/r_0.

2. Find the temperature within a sphere of radius a if the temperature along the surface is maintained at the temperature

$$u(a,\theta) = 100(\cos(\theta) - \cos^5(\theta)).$$

3. Find the steady-state temperature inside a sphere of radius a with the boundary condition

$$T_r(a,\theta) = \begin{array}{ll} q_0 & 0 \leq \theta < \theta_0 \\ 0 & \theta_0 < \theta < \pi - \theta \\ -q_0 & \pi - \theta_0 < \theta \leq \theta_0 \end{array}$$

4. Find the steady-state temperature within the northern hemisphere $(0 \leq \theta \leq \pi/2)$ of a sphere of radius a if the temperature on the equatorial plane is $T(r,\pi/2) = T_s$ and

$$T_r\,(a,\theta) = h\,[T_0 - T(a,\theta)]/k\ell$$

where T_0 is the temperature along the surface $r = a$ over the northern hemisphere. The parameters h and $k\ell$ are constants.

5. The surface of a sphere of radius a has a temperature of zero everywhere except in a spherical cap at the north pole (defined by the cone $\theta = \alpha$) where it is equal to T_0. Find the steady-state temperature within the sphere.

6. That portion of a sphere of radius a, lying between $\theta = \alpha$ and $\theta = \pi - \alpha$ has a temperature T_0 while the remaining portion has a temperature of zero. Find the temperature anywhere *outside* of the sphere.

7. Two concentric spheres have radii a and b $(b > a)$ and are divided into two hemispheres by the same horizontal plane. The upper hemisphere of the inner sphere and the lower hemisphere of the outer sphere are maintained at the temperature T_0. The other hemispheres are at a temperature of zero. Determine the steady-state temperature in the region $a \leq r \leq b$ as a series of Legendre polynomials.

8. A point charge q is placed at a distance b from the center of two concentric, grounded, conducting spheres of radii a and c, where $a < b < c$. Find the potential between $a < r < c$.

9. Find the temperature within the northern hemisphere of a sphere of radius a if the temperature on the equatorial plane is maintained at zero and the temperature on the surface is given by $k\,u_r(a,\theta) = h(T_a - T_c - u(a,\theta))$.

10. A dielectric sphere of radius R has a permanent polarization P that is uniform in direction and magnitude. The polarized sphere gives rise to an electric field inside and outside of this sphere.

(a) Show that the electrostatic potential inside and outside the sphere is

$$\phi_{\text{out}} = \sum_{n=0}^{\infty} A_n \left(\frac{R}{r}\right)^{n+1} P_n[\cos(\theta)]$$

and

$$\phi_{in} = \sum_{n=0}^{\infty} B_n \left(\frac{r}{R}\right)^n P_n [\cos(\theta)].$$

(b) Using the facts that ϕ must be continuous at $r = R$, $\hat{\mathbf{r}} \bullet (\epsilon_0 \mathbf{E}_{in} + \mathbf{P}) = \hat{\mathbf{r}} \bullet \epsilon_0 \mathbf{E}_{out}$ where $\hat{\mathbf{r}}$ is the unit vector pointing in the radial direction, ϵ_0 the dielectric constant, $\mathbf{P} = P\cos(\theta)$ and $\mathbf{E} = -\nabla\phi$, find ϕ_{in} and ϕ_{out}.

7.8 CONJUGATE HARMONIC FUNCTIONS: THE SEEPAGE OF STEADY RAINFALL THROUGH SOIL INTO DRAINS

In Section 7.3 the description of flow through an aquifer involved the concept of hydraulic head. However, in problems where the quantity of flow is specified, we do not know the level of the water table *a priori* and the value of $u(x,y)$ along the water table.

A method to circumvent this difficulty is to define a stream function $v(x,y)$ such that

$$q_x = (ku)_x = v_y \tag{7.8.1}$$

and

$$q_y = (ku)_y = -v_x \tag{7.8.2}$$

where q_x and q_y denote the pore velocity in the x and y direction and k is the hydraulic conductivity, respectively. The relationship between ku and v expressed in Eqs. (7.8.1) and (7.8.2) is called the *Cauchy-Riemann equations*. Cross differentiation shows that the stream function also satisfies Laplace's equation. Because both $ku(x,y)$ and $v(x,y)$ satisfy Laplace's equation and are related to each other through the Cauchy-Riemann equations, they are known collectively as *conjugate harmonic functions*. Lines of constant potential (equipotentials) are orthogonal to lines of constant stream function (streamlines) because the slope of an equipotential $ku(x,y) = a_1$ is

$$dy/dx = -(ku)_x/(ku)_y \tag{7.8.3}$$

while for a streamline $v(x,y) = b_1$ is

$$dy/dx = -v_x/v_y \tag{7.8.4}$$

so that the product of the slopes is, using the Cauchy-Riemann equations,

$$(ku)_x v_x/(ku)_y v_y = -v_y(ku)_y/(ku)_y v_y = -1 \quad \textbf{(7.8.5)}$$

and the curves are orthogonal.

The stream function gives the total discharge between two points A and B, because the difference between the values of the stream function corresponding to those points is

$$Q_{AB} = \int_A^B -q_x\, dy - q_y\, dx = -\int_A^B dv = v_A - v_B \quad \textbf{(7.8.6)}$$

We wish to use this information to solve for the groundwater flow that results from a steady rain (Kirkham, D., 1958: Seepage of steady rainfall through soil into drains. *Trans. Amer. Geophy. Union, 39*, 892-908.) If there were no way for the water to escape, then the water table would rise and the problem would not be steady-state. Consequently, we introduce drains, equally spaced, which will carry away all the water introduced into the system by the rainfall. See Fig. 7.8.1.

If R is the rate of rainfall (in cm/hr, for example), then the stream function along the boundaries are given by

$$v(x,0) = Rx \quad \textbf{(7.8.7)}$$

where we have integrated Eq. (7.8.2) along the top with $q_y = R$. Similarly along the left-hand side, the integration of Eq. (7.8.1) yields

$$v(0,y) = \begin{cases} 0 & 0<y<a \\ \dfrac{RL}{b-a}(y-a) & a<y<b \\ RL & b<y<h \end{cases} \quad \textbf{(7.8.8)}$$

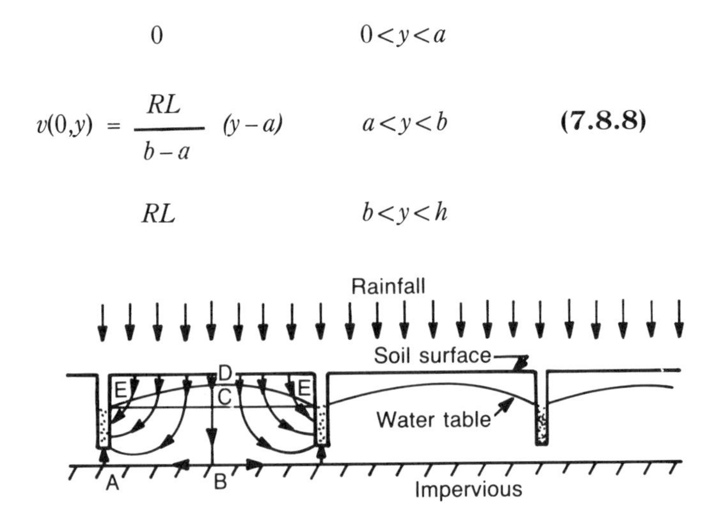

Fig. 7.8.1: Seepage of steady rainfall through soil into equally-spaced ditches. Taken from Kirkham, D., 1958: Seepage of steady rainfall through soil into drains. *Trans. Amer. Geophys. Union, 39*, 892-908.

because the drains are located between the depths of a and b, and

$$v(L,y) = RL \tag{7.8.9}$$

and

$$v(x,h) = RL \tag{7.8.10}$$

where L is half the distance between the drains and h is depth of the aquifer. We have assumed no flow through the right-hand side and bottom of the aquifer. Along these boundaries $v(x,y)$ is constant. Because the net flow through the entire aquifer must be zero, the value of this constant is chosen so that an integration around the entire aquifer is zero.

In Section 7.3 the value of the function along the vertical walls is zero. In the current problem, this is not true along any of the boundaries. We circumvent this difficulty by breaking the problem into two problems:

$$v(x,y) = r(x,y) + w(x,y) + RL \tag{7.8.11}$$

where

$$r_{xx} + r_{yy} = 0 \tag{7.8.12}$$

$$r(0,y) = r(L,y) = 0 \tag{7.8.13}$$

$$r(x,h) = 0 \tag{7.8.14}$$

$$r(x,0) = R(x-L) \tag{7.8.15}$$

and

$$w_{xx} + w_{yy} = 0 \tag{7.8.16}$$

$$w(x,0) = w(x,h) = 0 \tag{7.8.17}$$

$$w(L,y) = 0 \tag{7.8.18}$$

$$w(0,y) = \begin{cases} -RL & 0<y<a \\ \dfrac{RL}{b-a}(y-a) - RL & a<y<b \\ 0 & b<y<h \end{cases} \tag{7.8.19}$$

Turning our attention to the $r(x,y)$ problem first, we employ exactly the same method as in Section 7.3 and find

$$r(x,y) = \sum_{n=1}^{\infty} A_n \sin\left(\frac{n\pi x}{L}\right) \frac{\sinh\left[\frac{n\pi}{L}(h-y)\right]}{\sinh\left(\frac{n\pi h}{L}\right)} \qquad (7.8.20)$$

where

$$A_n = \frac{2}{L}\int_0^L R(x-L)\sin\left(\frac{n\pi x}{L}\right) dx \qquad (7.8.21)$$

$$= -\frac{2RL}{n\pi} \qquad (7.8.22)$$

On the other hand, the solution to $w(x,y)$ is

$$w(x,y) = \sum_{n=1}^{\infty} B_n \sin\left(\frac{n\pi y}{h}\right) \frac{\sinh\left[\frac{n\pi}{h}(L-x)\right]}{\sinh\left(\frac{n\pi L}{h}\right)} \qquad (7.8.23)$$

where

$$B_n = \frac{2}{h}\left\{-R\,L\int_0^a \sin\left(\frac{n\pi y}{h}\right) dy \right.$$

$$\left. + \mathrm{RL}\int_a^b \left(\frac{y-a}{b-a} - 1\right) \sin\left(\frac{n\pi y}{h}\right) dy \right\} \qquad (7.8.24)$$

$$= \frac{2RL}{\pi}\left\{-\frac{1}{n} + \frac{h}{(b-a)n^2\pi}\left[\sin\left(\frac{n\pi b}{h}\right) - \sin\left(\frac{n\pi a}{h}\right)\right]\right\} \qquad (7.8.25)$$

Once we have found $v(x,y)$, then we can integrate the Cauchy-Riemann equations to find the groundwater potential.

Kirkham worked out the geopotential for $L = 50$ ft and (A) $a = 0$ ft and $b = 1$ ft, (B) $a = 0$ ft and $b = 10$ ft and (C) $a = 0$ ft and $b = 0.5$ ft. Because we chose $y = 0$ to be located at the intersection of the water table and the ditch (see point E in Fig. 7.8.1), the equipotential defining the ditch is $u(x,y) = 0$. In Fig. 7.8.2 lines of equipotential (in arbitrary units) are plotted for these three cases. For these theoretical conditions, we see that conditions (A) and (C) could be realized in the field. However, condition (B) gives a ditch where the wall slope is in the wrong direction. Because this solution is not practical, it would not be physically realized.

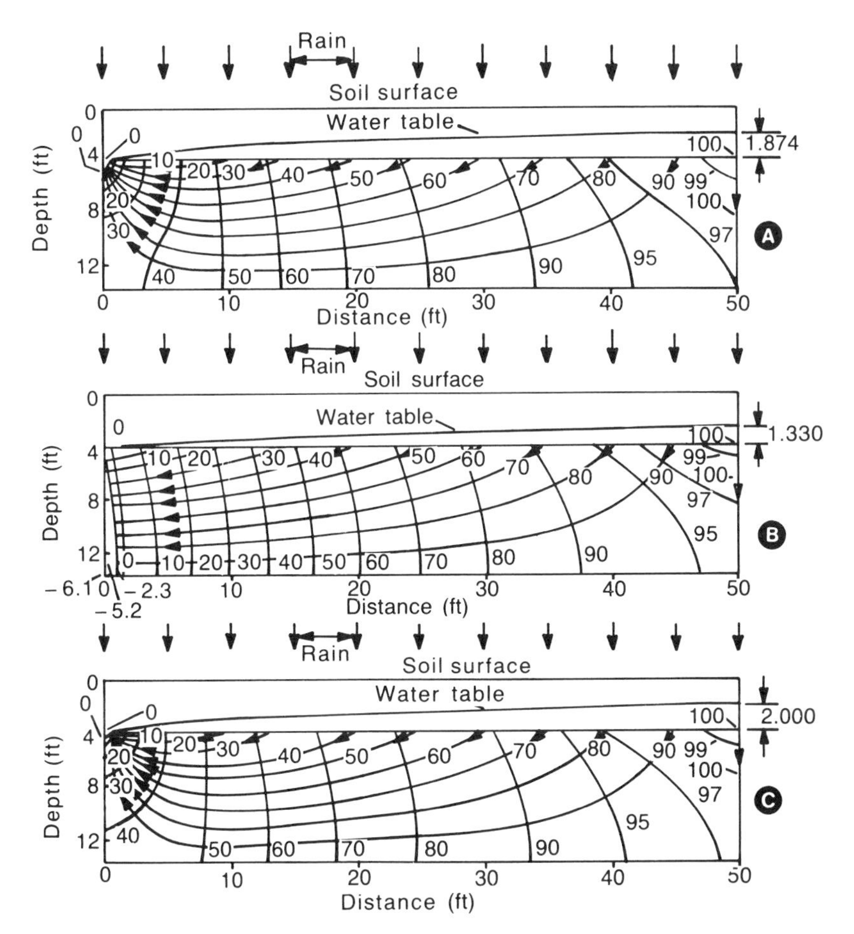

Fig. 7.8.2 Contours of equipotential and streamlines as rainfall flows through an aquifer to drainage ditches. The boundary conditions are given in the text. Taken from Kirkham, D., 1958: Seepage of steady rainfall through soil into drains. *Trans. Amer. Geophys. Union, 39,* 892-908.

7.9 CONFORMAL MAPPING

In the previous sections we have seen that for the simplest regions—the square, the cylinder, and sphere—Laplace's equation can be solved relatively simply even for some relatively complex boundary conditions. However, if the region has a complex structure, obtaining an analytic solution may be very difficult even if the boundary conditions are extremely simple. One possible solution to this difficulty is to find a transformation where the given region is transformed into a very simple geometry and our earlier techniques may be employed.

These transformations result not only in the regions being transformed but also the boundary conditions and the differential equation. Consequently, the transformation of greatest value is the one for which Laplace's equation itself remains invariant. To find such a transformation, let us introduce two new independent variables u and v. By the chain rule

$$\frac{\partial}{\partial x} = u_x \frac{\partial}{\partial u} + v_x \frac{\partial}{\partial v} = u_x \frac{\partial}{\partial u} - u_y \frac{\partial}{\partial v} \tag{7.9.1}$$

if we restrict ourselves to the condition that $v_x = -u_y$. Therefore

$$\frac{\partial \phi}{\partial x} = u_x \frac{\partial \phi}{\partial u} - u_y \frac{\partial \phi}{\partial v} \tag{7.9.2}$$

$$\frac{\partial^2 \phi}{\partial x^2} = u_{xx} \frac{\partial \phi}{\partial u} - u_{xy} \frac{\partial \phi}{\partial v} + u_x \frac{\partial^2 \phi}{\partial x \partial u} - u_y \frac{\partial^2 \phi}{\partial x \partial v} \tag{7.9.3}$$

$$= u_{xx} \frac{\partial \phi}{\partial u} - u_{xy} \frac{\partial \phi}{\partial v} + u_x^2 \frac{\partial^2 \phi}{\partial u^2} - 2u_x u_y \frac{\partial^2 \phi}{\partial u \partial v} + u_y^2 \frac{\partial^2 \phi}{\partial v^2} \tag{7.9.4}$$

Similarly we obtain

$$\frac{\partial^2 \phi}{\partial y^2} = u_{yy} \frac{\partial \phi}{\partial u} + u_{yx} \frac{\partial \phi}{\partial v} + u_y^2 \frac{\partial^2 \phi}{\partial u^2} + 2u_x u_y \frac{\partial^2 \phi}{\partial u \partial v} + u_x^2 \frac{\partial^2 \phi}{\partial v^2} \tag{7.9.5}$$

if $u_x = v_y$. Adding the last two results, we obtain

$$\nabla^2 \phi = \phi_{xx} + \phi_{yy} = (u_{xx} + u_{yy}) \frac{\partial \phi}{\partial u} + (u_x^2 + u_y^2)(\phi_{uu} + \phi_{vv}). \tag{7.9.6}$$

The coefficient of $\partial \phi / \partial u$ vanishes because $u_{xx} = v_{xy} = -u_{yy}$. Thus, unless $u_x^2 + u_y^2$ vanishes,

$$\phi_{uu} + \phi_{vv} = 0. \tag{7.9.7}$$

Our analysis has led us to the following conclusion. If we choose new independent variables u and v which satisfy the so-called Cauchy Riemann equations

$$u_x = v_y \text{ and } v_x = -u_y, \tag{7.9.8}$$

and $u_x^2 + u_y^2 \neq 0$, then we can transform the equation

$$\phi_{xx} + \phi_{yy} = 0 \tag{7.9.9}$$

into

$$\phi_{uu} + \phi_{vv} = 0. \tag{7.9.10}$$

The resulting solution in the u,v coordinates can then be back-transformed to give the solution to the original problem.

Because u and v satisfy the Cauchy-Riemann equations, the most direct method of obtaining u and v utilizes the properties of analytic functions from the theory of complex variables. A complex function is very similar to its counterpart in real variables. It provides a rule whereby, for each choice of a complex number

$$z = x + iy,$$

there corresponds another complex number $w(z) = u(x,y) + iv(x,y)$. Out of this large class of functions, there is a subset of functions which possess a well-defined first derivative. These functions are called *analytic* and $u(x,y)$ and $v(x,y)$ are referred to as *conjugate harmonic functions*. Conjugate harmonic functions satisfy the Cauchy-Riemann equations as well as Laplace's equation.

This method of transforming from x and y into the different coordinates u and v is called *conformal mapping*. The adjective *conformal* is used because any infinitismal configuration and its image conform, in the sense of being similar. This is not true, however, for large configuration which usually bears no resemblance to their image.

Conformal mappings are usually represented by two planes. In one of them, the original configuration is drawn on the complex z-plane where x is the absicca and iy the ordinate. In the other, the image is drawn on the complex w-plane where u is the absicca and iv the ordinate. In this way a correspondence is set up between points, curves, and regions and their images in the other place. In Fig. 7.9.1 the mapping for the transformation

$$w^2 = (u + iv)^2 = u^2 - v^2 + 2iuv = z = x + iy \tag{7.9.11}$$

or

$$x = u^2 - v^2 \text{ and } y = 2uv \tag{7.9.12}$$

is shown.

Out of the infinite number of possible conformal mappings, certain classics have emerged because of their usefulness in solving physical problems. We shall now examine a few of them.

(a) The transformation $w = z^m$, m a positive integer.

If we write both z and w in polar form, so that

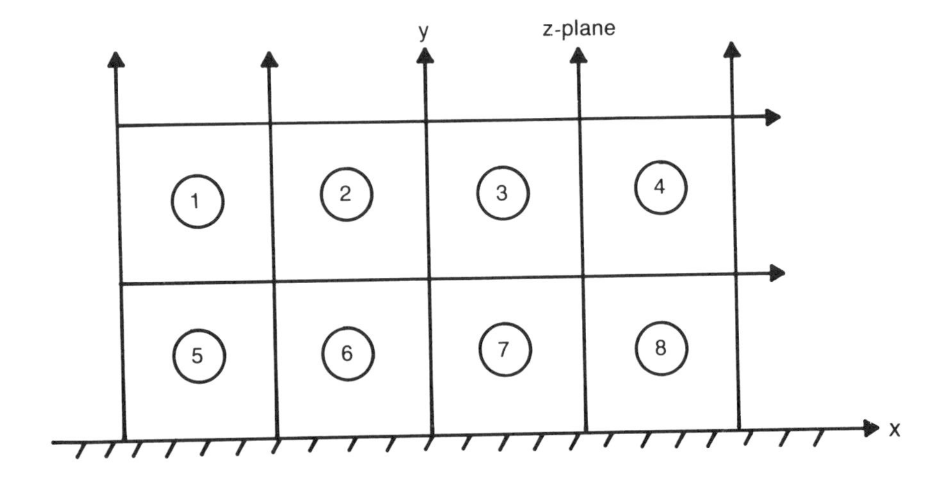

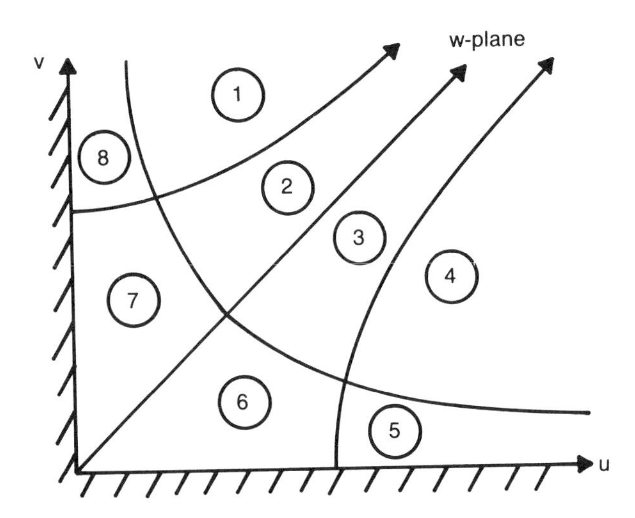

Fig. 7.9.1: The transformation $z = w^2$.

$$z = r\, e^{i\phi} \text{ and } w = R\, e^{i\Phi} \qquad \textbf{(7.9.13)}$$

then

$$w = R\, e^{i\Phi} = z^m = r^m\, e^{im\phi} \qquad \textbf{(7.9.14)}$$

or

$$R = r^m \text{ and } \Phi = m\phi. \qquad \textbf{(7.9.15)}$$

A sector in the z-plane of central angle $2\pi/m$ is "fanned out"

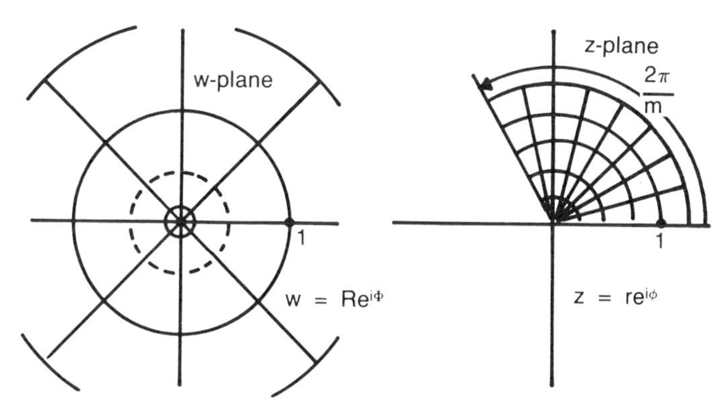

Fig. 7.9.2: The transformation $w = z^m$, m an integer. Reprinted from the *American Mathematical Monthly* (1932) with permission of the publishers.

to cover the entire w-plane, this sector also being stretched radially. (See Fig. 7.9.2, drawn for $m = 3$.) Note that a set of orthogonal curves in one plane transform into a set of orthogonal curves in the other plane.

(b) The transformation $w = e^z$.
If we set $w = R\, e^{i\Phi}$ and $z = x + iy$, then

$$R = e^x \text{ and } \Phi = y. \tag{7.9.16}$$

See Fig. 7.9.3 for a diagram.

(c) The transformation $w = \cosh(z)$
For this transformation

$$u + iv = \cosh(x + iy) = \cosh(x)\cosh(iy) + \sinh(x)\sinh(iy), \tag{7.9.17}$$

$$= \cosh(x)\cos(y) + i\sinh(x)\sin(y) \tag{7.9.18}$$

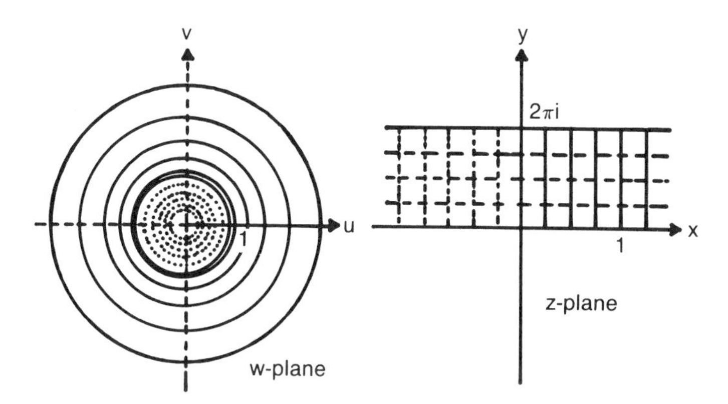

Fig. 7.9.3: The transformation $w = e^z$. Reprinted from the *American Mathematical Monthly* (1932) with permission of the publishers.

so that

$$u(x,y) = \cosh(x) \cos(y) \quad \textbf{(7.9.19)}$$

$$v(x,y) = \sinh(x) \sin(y) \quad \textbf{(7.9.20)}$$

or

$$\frac{u^2}{\cosh^2(x)} + \frac{v^2}{\sinh^2(x)} = 1 \quad \textbf{(7.9.21)}$$

$$\frac{u^2}{\cos^2(y)} + \frac{v^2}{\sin^2(y)} = 1 \quad \textbf{(7.9.22)}$$

This transformation is shown in Fig. 7.9.4, and it may be used to obtain the electrostatic field due to an elliptic cylinder, the electrostatic field due to a charged plane from which a strip has been removed, the circulation of liquid around an elliptical cylinder, the flow of liquid through a slit in a plane, etc.

(d) The transformation $w = z + e^z$.
In this transformation, we have

$$u + iv = x + iy + \exp(x + iy) \quad \textbf{(7.9.23)}$$

so that

$$u = x + e^x \cos(y) \quad \textbf{(7.9.24)}$$

$$v = y + e^x \sin(y). \quad \textbf{(7.9.25)}$$

This transformation is shown in Fig. 7.9.5 and gives the electrostatic field at the edge of a parallel plate condenser, the flow of liquid out of a channel into an open sea, etc.

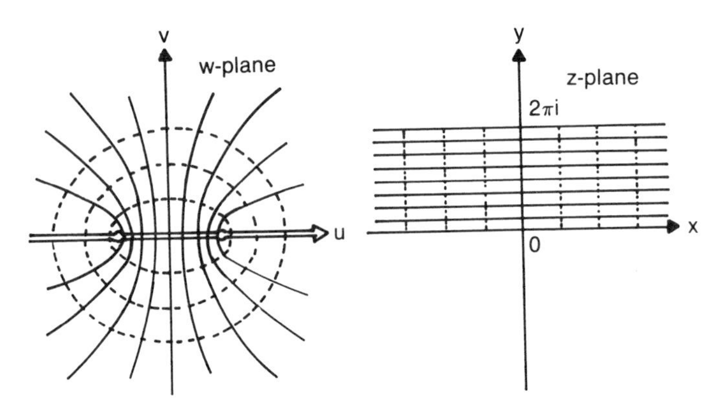

Fig. 7.9.4: The transformation w = cosh(z). Reprinted from the *American Mathematical Monthly* (1932) with permission of the publishers.

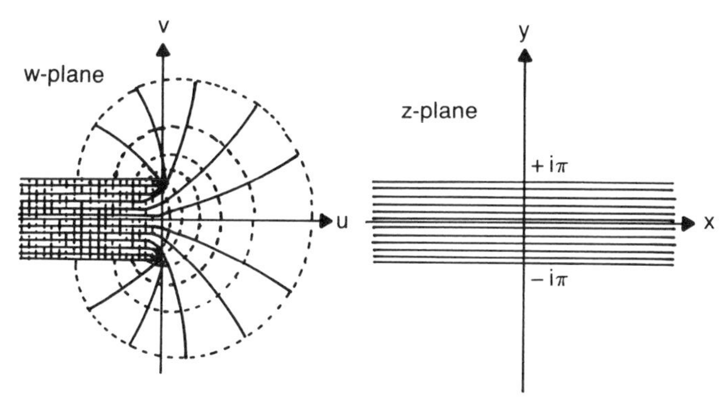

Fig. 7.9.5: The transformation w = z + e^z. Reprinted from the *American Mathematical Monthly* (1932) with permission of the publishers.

7.10 EXAMPLES OF CONFORMAL MAPPINGS IN SOLVING LAPLACE'S EQUATION

To illustrate the application of conformal mapping to the solution of the two-dimensional, steady-state heat equation, we find the solution $T(x,y)$ of Laplace's equation in the infinite sector $0 \leq \theta \leq \alpha$ where $T = 1$ when $0 < x < 1$ and $T = 0$ when $x > 1$, along the edge $\theta = 0$, and $T = 0$ everywhere along the edge $\theta = \alpha$ (see Fig. 7.10.1).

If we choose the mapping $w = z^{\pi/\alpha}$, then we can transform this problem to that of determining the solution of Laplace's equation in the halfplane $v \geq 0$. This reduces, when $v = 0$, to 1 when $0 < u < 1$ and to zero otherwise. To find $T(u,v)$, we utilize Schwarz integral formula (Section 2.17).

$$T(u,v) = \frac{1}{\pi} \int_0^1 \frac{v}{v^2 + (u - \lambda)^2} \, d\lambda$$

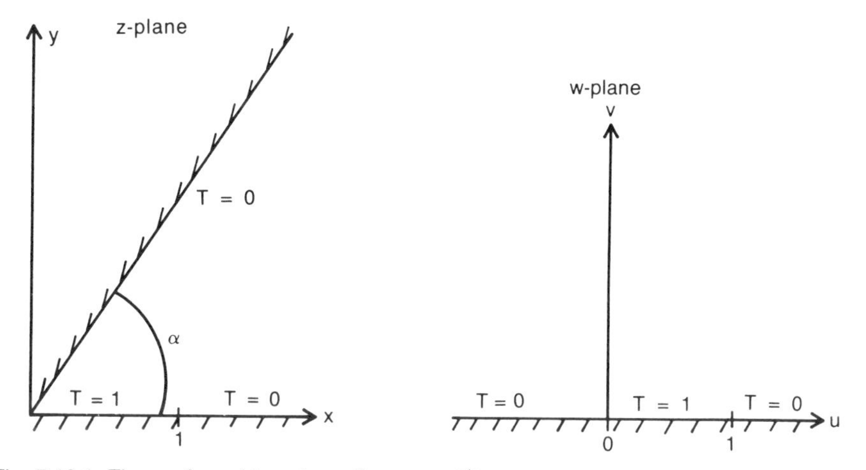

Fig. 7.10.1: The conformal transformation w = $z^{\pi/\alpha}$.

$$T(u,v) = \frac{1}{\pi}\left[\tan^{-1}\left(\frac{1-u}{v}\right) + \tan^{-1}\left(\frac{u}{v}\right)\right] \quad (7.10.1)$$

or

$$T(u,v) = \frac{1}{\pi}\cot^{-1}\left(\frac{u^2+v^2-u}{v}\right) \quad (7.10.2)$$

or, in polar coordinates (ϱ, ϕ),

$$T(\varrho,\phi) = \frac{1}{\pi}\cot^{-1}\left[\frac{\varrho - \cos(\phi)}{\sin(\phi)}\right] \quad (7.10.3)$$

From the conformal mapping there follows

$$\varrho = r^{\pi/\alpha} \text{ and } \phi = \frac{\pi\theta}{\alpha} \quad (7.10.4)$$

so that

$$T(r,\theta) = \frac{1}{\pi}\cot^{-1}\left[\frac{r^{\pi/\alpha} - \cos(\pi\theta/\alpha)}{\sin(\pi\theta/\alpha)}\right] \quad (7.10.5)$$

In Fig. 7.10.2 this solution is graphed when $\alpha = \pi/3$.

For our second example we shall determine the steady-state temperature within a semicircle bounded by the upper half of the circle $x^2 + y^2 = R^2$ and the x-axis. Along the curved boundary the temperature is equal to unity while it equals zero along the straight boundary. See Fig. 7.10.3.

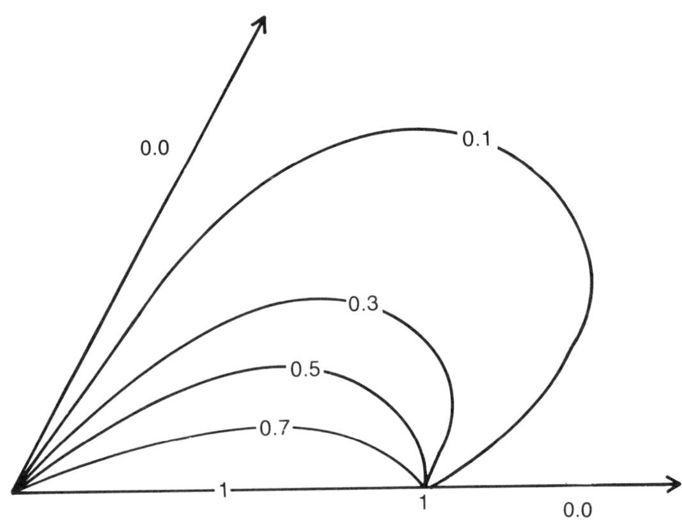

Fig. 7.10.2: The steady-state temperature witnin a wedge of width $\alpha = \pi/3$ where the portion from $x = 0$ to $x = 1$ is held at $T = 1$ and the remaining portion of the boundary is equal to zero.

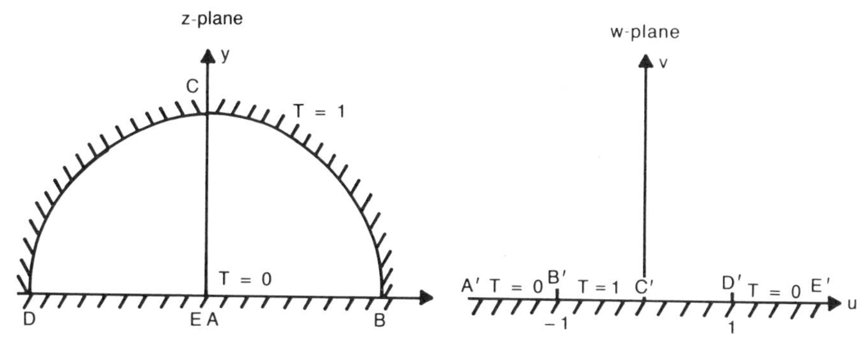

Fig. 7.10.3: The conformal transformation $w = -(\frac{z}{R} + \frac{R}{z})/2$.

To map this problem into the upper half-plane, we use the transformation

$$w = -\left(\frac{z}{R} + \frac{R}{z}\right)/2. \tag{7.10.6}$$

This transformation produces the following mappings:

$(0+,0) \rightarrow (-\infty,0)$, $(R,0) \rightarrow (-1,0)$, $(0,R) \rightarrow (0,0)$, $(-R,0) \rightarrow (1,0)$ and $(0-,0) \rightarrow (+\infty,0)$.

(See Fig. 7.10.3.) Consequently we have transformed the interior of a semicircle into the upper half-plane in the w-plane. In this new plane, $T = 1$ when $|u| < 1$ and is zero elsewhere along the u-axis.

We can immediately write down the solution in the w-plane as

$$\begin{aligned} T(u,v) &= \frac{1}{\pi}\int_{-\infty}^{\infty} \frac{f(\lambda)\, v}{v^2 + (u-\lambda)^2}\, d\lambda \\ &= \frac{1}{\pi}\int_{-1}^{1} \frac{v}{v^2 + (u-\lambda)^2}\, d\lambda \qquad (7.10.7) \\ &= \frac{1}{\pi}\left[\tan^{-1}\left(\frac{1-u}{v}\right) + \tan^{-1}\left(\frac{1+u}{v}\right)\right] \\ &= \frac{1}{\pi}\cot^{-1}\left(\frac{u^2 + v^2 - 1}{2v}\right) \qquad (7.10.8) \end{aligned}$$

Turning to the conformal mapping:

$$w = -\frac{1}{2}\left[\frac{x+iy}{R} + \frac{R}{x+iy}\right] = -\frac{1}{2}\left[\frac{x+iy}{R} + \frac{R(x-iy)}{x^2 + y^2}\right] \tag{7.10.9}$$

or

$$u(x,y) = -\frac{1}{2}\left[\frac{x}{R} + \frac{Rx}{x^2+y^2}\right],$$

$$v(x,y) = -\frac{1}{2}\left[\frac{y}{R} - \frac{Ry}{x^2+y^2}\right]. \qquad (7.10.10)$$

In polar coordinates, $x = r\cos(\theta)$ and $y = r\sin(\theta)$, we have

$$u(r,\theta) = -\frac{1}{2}\left(\frac{r}{R} + \frac{R}{r}\right)\cos(\theta) \qquad (7.10.11)$$

$$v(r,\theta) = \frac{1}{2}\left(\frac{R}{r} - \frac{r}{R}\right)\sin(\theta) \qquad (7.10.12)$$

and

$$u^2 + v^2 - 1 = \frac{1}{4}\left(\frac{R}{r} - \frac{r}{R}\right)^2 \sin^2(\theta). \qquad (7.10.13)$$

Therefore

$$T(r,\theta) = \frac{1}{\pi}\cot^{-1}\left[\frac{\frac{1}{4}\left(\frac{R}{r} - \frac{r}{R}\right)^2 - \sin^2(\theta)}{\left(\frac{R}{r} - \frac{r}{R}\right)\sin(\theta)}\right] \qquad (7.10.14)$$

$$= \frac{1}{\pi}\cot^{-1}\left[\frac{1 - \sin^2(\theta)/[(R/r - r/R)/2]^2}{2\sin(\theta)/((\frac{R}{r} - \frac{r}{R})/2)}\right] \qquad (7.10.15)$$

$$= \frac{2}{\pi}\tan^{-1}\left[\frac{2rR\sin(\theta)}{R^2 - r^2}\right] \qquad (7.10.16)$$

because $2\tan^{-1}(A) = \cot^{-1}[(1-A^2)/2A]$ if $A \geq 0$. The solution has been graphed in Fig. 7.10.4.

7.11 THE STEADY-STATE TEMPERATURE FIELD SURROUNDING A BURIED HEATING PIPE

In the previous section, we solved two heat conduction problems by mapping the domain into the upper half-plane and then using the Schwarz integral formula. Sometimes it is more convenient to map the domain into a square or annulus. To illustrate this we solve the heat conduction problem of a buried heating pipe. The pipe of radius a has its origin located at a distance h below

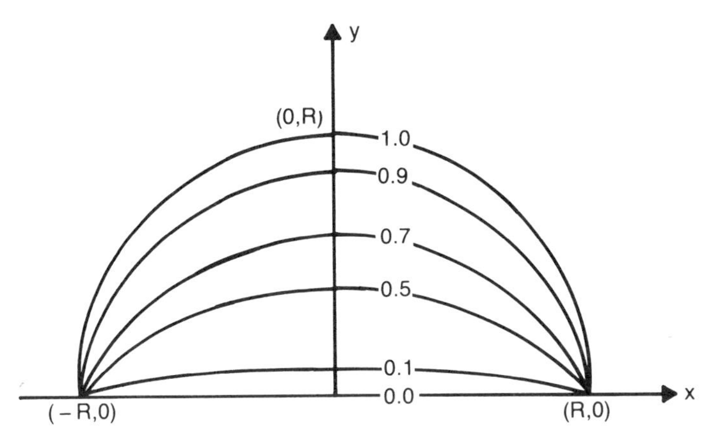

Fig. 7.10.4: Steady-state temperature distribution within a semicircle of radius R when the curved surface is held at T = 1 and the x-axis at T = 0.

the surface. At the rim of the pipe, the temperature is T_0 while the temperature at the earth's surface is taken to be zero (See Fig. 7.11.2).

Through the use of the transformation

$$w = \frac{z - c}{z + c} \tag{7.11.1}$$

where $c = (h^2 - a^2)^{1/2}$, the original domain is mapped into a circular annulus with an inner radius of $\dfrac{a}{h+c}$ and an outer radius of unity. (See Fig. 7.11.1.) This transformation (7.11.1) is a simple example of a more general transformation

$$w = \frac{\alpha z + \beta}{\gamma z + \delta} \tag{7.11.2}$$

where $\alpha\delta - \beta\gamma \neq 0$. This transformation is called the *bilinear* or *fractional transformation* and has the property that circles in the z-plane are mapped into circles in the w-plane. By circles we include cir-

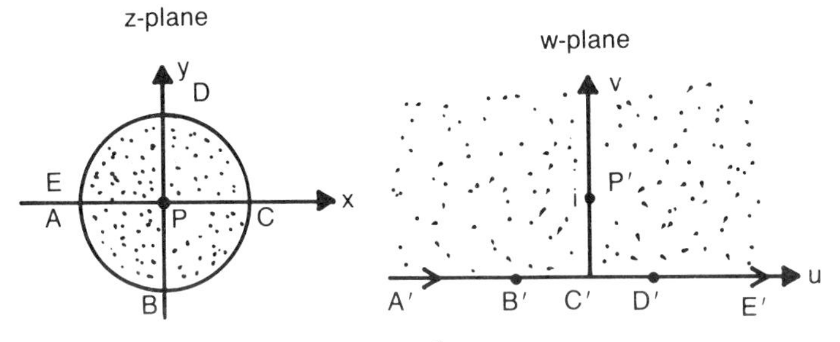

Fig. 7.11.1: The bilinear transformation $w = i\,\dfrac{1 - z}{1 + z}$.

cles of infinite radius (i.e., straight lines). The bilinear transformation is particularly useful in mapping circles into half planes and *vice versa*.

To illustrate this bilinear transformation, consider the transform

$$w(z) = i\,\frac{1-z}{1+z}\,.$$

We wish to find how this transformation maps the unit circle in the z-plane onto the w-plane.

The unit circle cuts the x-axis at $z = \pm 1$ and the y-axis at $z = \pm i$. At $z = 1$, $w(1) = 0$; at $z = i$, $w(i) = 1$; at $z = -i$, $w(-i) = -1$; and at $z = 0$, $w(0) = i$. (See Fig. 7.11.1.)

The point $z = -1$ poses some difficulty because the transformation becomes undefined. We can avoid this by considering

$$z = -1 + i\epsilon.$$

If ϵ is positive and infinitesimally small, then we are at a point just above the point $z = -1$; if ϵ is slightly negative, the point is just below $z = -1$. In either case $w(-1 + i\epsilon) = (2 - i\epsilon)/\epsilon$. Consequently a point just above $z = -1$ is mapped to $+\infty$ in the w-plane; a point just below $z = -1$, to $-\infty$. These results are summarized in Fig. 7.11.1.

As Fig. 7.11.2 shows, our use of conformal mapping has simplified the problem to solving Laplace's equation inside a circular annulus. The temperature along the inside ring $r_2 = \dfrac{a}{h+c}$ is $T = T_0$ while the temperature along the outside ring $r_1 = 1$ is $T = 0$.

Because of the geometry, we use polar coordinates and Laplace's equation in the plane becomes

$$\frac{\partial^2 T}{\partial r^2} + \frac{1}{r}\frac{\partial T}{\partial r} + \frac{1}{r^2}\frac{\partial^2 T}{\partial \theta^2} = 0 \qquad \textbf{(7.11.3)}$$

Assuming that we can solve the problem by the separation of variables, we look for the particular solution

$$T(r,\theta) = R(r)\Theta(\theta) \qquad \textbf{(7.11.4)}$$

Upon substituting this solution into Laplace's equation, we obtain

$$r^2 R''\,\Theta + r\,R'\,\Theta + R\Theta'' = 0 \qquad \textbf{(7.11.5)}$$

where a prime denotes differentiation with respect to the argument. By separating the variables, we obtain

$$\frac{1}{R}\,(r^2R'' + rR') = -\frac{\Theta''}{\Theta} = k^2 \qquad \textbf{(7.11.6)}$$

where k^2 is the separation constant. This condition implies the two

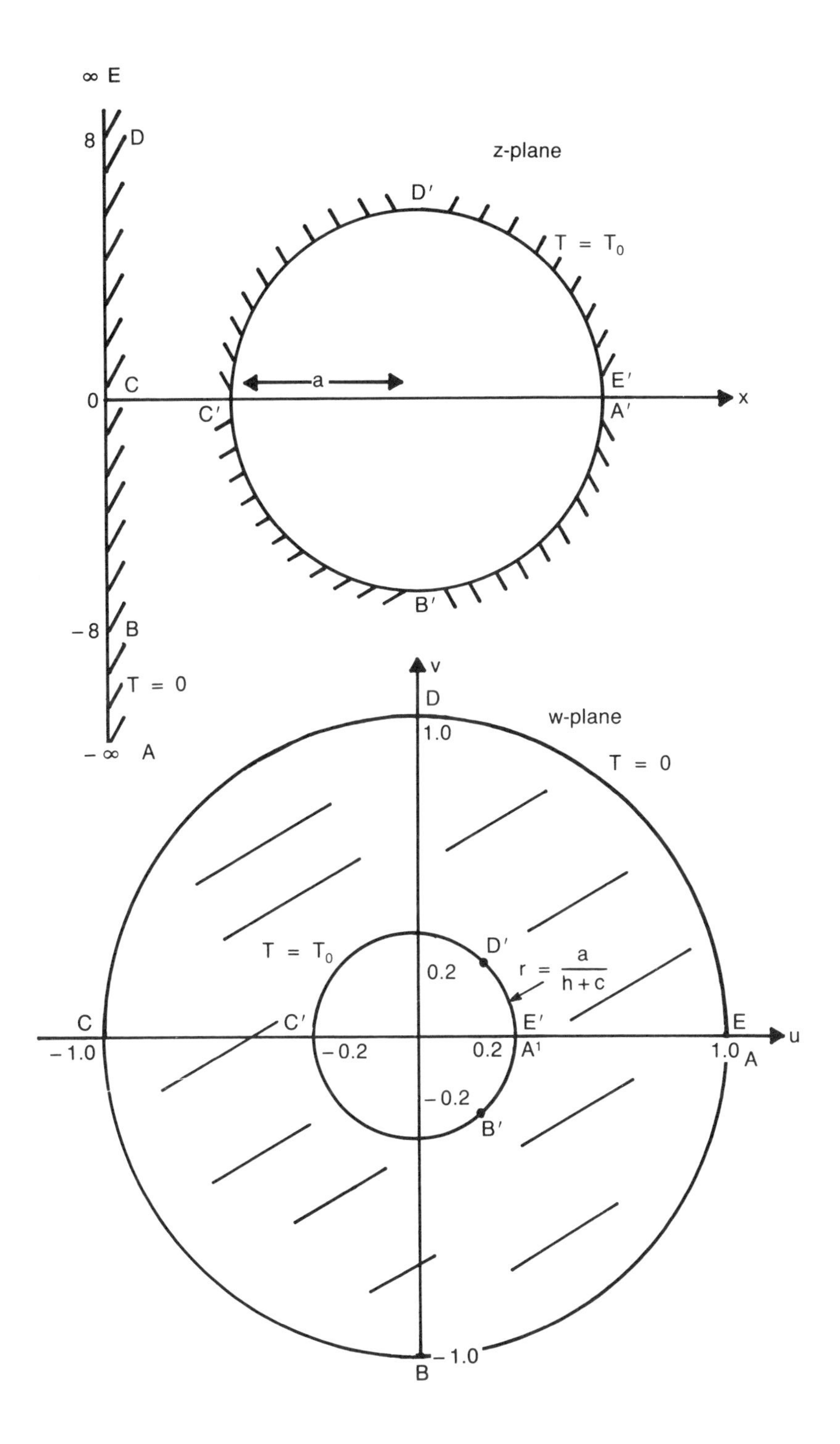

Fig. 7.11.2: The conformal transformation (7.11.1).

ordinary differential equations

$$r^2 R'' + r R' - k^2 R = 0 \qquad \textbf{(7.11.7)}$$

and

$$\Theta'' + k^2\Theta = 0. \qquad \textbf{(7.11.8)}$$

The sign of the separation constant is chosen in such a way that sines and cosines (rather than exponential functions) will occur in the θ direction. The solution presumably has to be periodic in that direction.

Equation (7.11.7) is the equidimensional equation with the general solution

$$R(r) = A_k r^k + B_k r^{-k} \text{ when } k \neq 0 \qquad \textbf{(7.11.9)}$$

and

$$R(r) = A_0 + B_0 \ln(r) \text{ when } k = 0 \qquad \textbf{(7.11.10)}$$

whereas the solution in the θ direction is

$$\Theta(\theta) = C_k \cos(k\theta) + D_k \sin(k\theta) \text{ if } k \neq 0 \qquad \textbf{(7.11.11)}$$

and

$$\Theta(\theta) = C_0 + D_0\theta \text{ if } k = 0 \qquad \textbf{(7.11.12)}$$

Thus any expression of the form

$$T(r,\theta) = a_0 + b_0 \ln(r) + \left[c_0 + d_0 \ln(r)\right] \theta$$

$$+ \sum_k (a_k r^k + b_k r^{-k}) \cos(k\theta) + (c_k r^k + d_k r^{-k}) \sin(k\theta), \qquad \textbf{(7.11.13)}$$

where k takes on arbitrary nonzero values, is a solution to Eq. (7.11.3). However, in order that T be periodic with a common period 2π, we must take

$$c_0 = d_0 = 0 \text{ and } k = n,\ n = 1,2,3,\ldots \qquad \textbf{(7.11.14)}$$

Hence we are lead to assume the solution to be

$$T(r,\theta) = a_0 + b_0 \ln(r)$$

$$+ \sum_{n=1}^{\infty} (a_n r^n + b_n r^{-n}) \cos(n\theta) + (c_n r^n + d_n r^{-n}) \sin(n\theta) \qquad \textbf{(7.11.15)}$$

In our problem we have the boundary conditions

$$T(1,\theta) = 0 \text{ and } T\left(\frac{a}{h+c}, \theta\right) = T_0 \qquad \textbf{(7.11.16)}$$

or

$$a_0 + b_0 \ln(r_1) + \sum_{n=1}^{\infty} (a_n r_1^n + b_n r_1^{-n}) \cos(n\theta)$$

$$+ (c_n r_1^n + d_n r_1^{-n}) \sin(n\theta) = 0 \qquad \textbf{(7.11.17)}$$

$$a_0 + b_0 \ln(r_2) + \sum_{n=1}^{\infty} (a_n r_2^n + b_n r_2^{-n}) \cos(n\theta)$$

$$+ (c_n r_2^n + d_n r_2^{-n}) \sin(n\theta) = T_0 \qquad \textbf{(7.11.18)}$$

where $r_2 = \dfrac{a}{h+c}$ and $r_1 = 1$. From the conditions at $r_1 = 1$ we find

$$a_0 = a_n = b_n = c_n = d_n = 0. \qquad \textbf{(7.11.19)}$$

From the condition at $r = r_2$, we have

$$b_0 = \frac{T_0}{\ln(r_2)} . \qquad \textbf{(7.11.20)}$$

Consequently the solution may be written as

$$T(r,\theta) = \frac{-T_0}{\ln\left(\frac{h+c}{a}\right)} \ln(r). \qquad \textbf{(7.11.21)}$$

Our final task is to transform back into the original domain. From the transformation, we have

$$w = u + iv = \frac{x+iy-c}{x+iy+c} \qquad \textbf{(7.11.22)}$$

or

$$u = \frac{(x-c)\ (x+c)+y^2}{(x+c)\ (x+c)+y^2} \qquad \textbf{(7.11.23)}$$

and

$$v = \frac{2yc}{(x+c)\ (x+c)+y^2} \qquad \textbf{(7.11.24)}$$

Because $r^2 = u^2 + v^2$, we finally obtain

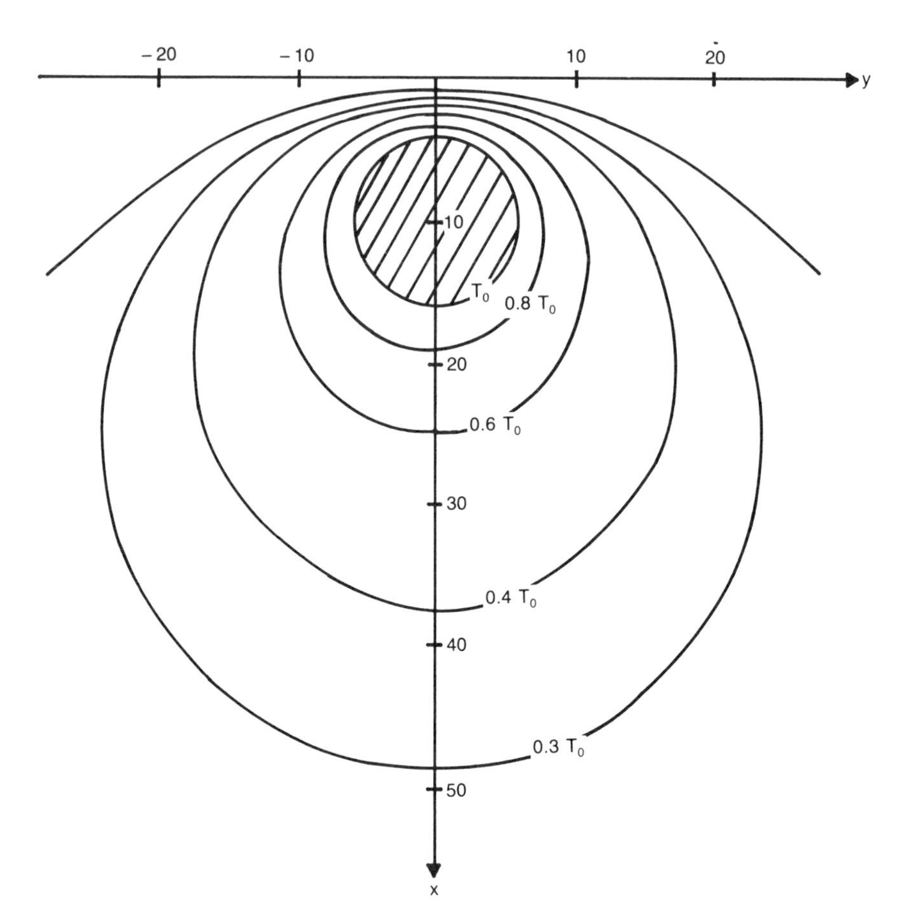

Fig. 7.11.3: Steady-state temperature distribution of a heated, buried pipe.

$$T(x,y) = \frac{-T_0}{\ell n((h+c)/a)} \, \ell n \left[\frac{\sqrt{(x^2 - c^2 + y^2)^2 + 4c^2y^2}}{(x+c)^2 + y^2} \right] \qquad \textbf{(7.11.25)}$$

In Fig. 7.11.3 is graphed the solution for the case when $h = 10\,m$ and $a = 6\,m$. Note that the highest temperature gradient lies directly above the buried pipe.

Exercises

1. Find the temperature distribution on a plate lying in the first quadrant with $T(x,0) = 0$ and $T(0,y) = 1$.

2. Find the steady-state temperature in a solid in the shape of an infinitely long cylindrical wedge if its boundary planes $\theta = 0$ and $\theta = \theta_0$ are kept at the constant temperatures of zero and T_0, respectively.

3. Derive a formula for the electrostatic potential V in the space interior to a long cylinder $r = 1$ if $V = 0$ on the quadrant

$0<\theta<\pi/2$ of the cylindrical surface and $V = 1$ in the rest $(\pi/2<\theta<2\pi)$ of that surface. Use $w = -i(z+1)/(z-1)$.

4. Find the potential V in the space between the plates $y=0$ and $y=\pi$ if $V = 0$ on the part $x>0$ of each of those plates and $V = 1$ on the parts $x < 0$. Use the transformation $w = e^z$.

5. Derive a formula for the electrostatic potential for the space bounded below by two half planes with $V = 0$ and a half cylinder with $V = 1$. Use $w = z + 1/z$.

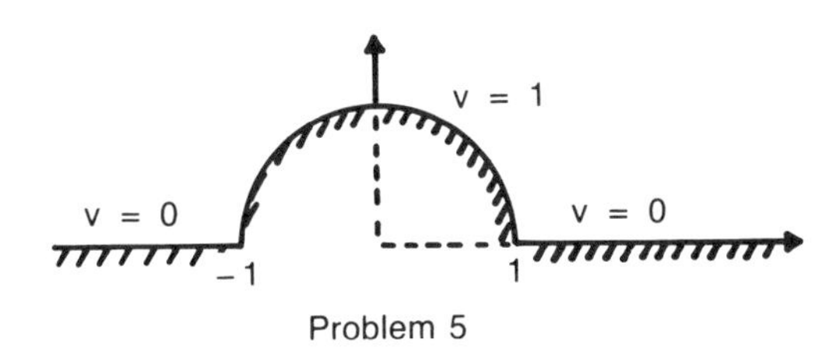

Problem 5

6. Find the steady-state temperature distribution in the semi-infinite strip $(0\leq x\leq a,\ 0<y\leq\infty)$, subject to the conditions that it equals unity along the end $y = 0$ and zero along the sides $x = 0$ and a. Use the conformal mapping $w = -\cos(\frac{\pi z}{a})$.

7. Two halves of a long hollow conducting cylinder of inner radius a are separated by small lengthwise gaps on each side, and are kept at different potentials V_1 and V_2. If θ is measured from the plane perpendicular to the plane through the gap, find the potential.

8. A long hollow conducting cylinder with inner radius a is divided into equal quadrants, alternate segments being at potential V and $-V$. Find the potential.

7.12 THE SLOSHING OF A FLUID WITHIN A PARTIALLY FILLED TANK

Nearly all of the liquid-filled containers with which we are familiar, ranging from the common teacup to the huge stabilization systems that employ anti-roll passive tanks in ocean-going ships, at one time or another posseses the common feature that the contained liquid has an unrestrained surface. The presence of this free surface allows for motions to occur within the fluid. These motions, in turn, may influence the container's stability if one of the natural oscillations should resonate with any forced motions. For these reasons, great technical interest exists in determining these oscillations.

In these problems the flow is assumed to be irrotational so that $\nabla\times\mathbf{u} = \mathbf{0}$ where $\mathbf{u}$ is the Eulerian velocity $(u,0,w)$. It follows from the vectorial identity $\nabla\times\nabla\phi = \mathbf{0}$ that $\mathbf{u}$ may be expressed as $-\nabla\phi$. Because the fluid is essentially incompressible, Laplace's equation

$$\nabla \bullet \mathbf{u} = \nabla^2 \phi = 0 \tag{7.12.1}$$

holds for ϕ throughout the fluid.

In addition to the potential equation in the interior, boundary conditions must be specified along the boundaries. At the solid walls, the velocity vanishes and $\phi_n = 0$ where n is the spatial variable normal to the wall.

The most intriguing aspect of this problem is the boundary condition along the free surface because the velocity is unknown there. From the momentum equations it is found that unless the pressure across a free surface is continuous, infinite accelerations will result. Consequently, along the free surface we require the dynamical boundary condition that

$$p = p_a \tag{7.12.2}$$

where p is the hydrodynamic pressure in the liquid and p_a the pressure of the air/vapor mixture.

This dynamical boundary condition must be expressed in terms of the potential. From the Eulerian momentum equations,

$$\frac{\partial u}{\partial t} + u \frac{\partial u}{\partial x} + w \frac{\partial u}{\partial z} = -\frac{1}{\varrho}\frac{\partial p}{\partial x} \tag{7.12.3}$$

and

$$\frac{\partial w}{\partial t} + u \frac{\partial w}{\partial x} + w \frac{\partial w}{\partial z} = -\frac{1}{\varrho}\frac{\partial p}{\partial z} - g. \tag{7.12.4}$$

where ϱ is the fluid's density and g the gravitational attraction. Because $u = -\phi_x$ and $w = -\phi_z$, Eq. (7.12.30) can be integrated with respect to x and Eq. (7.12.4) with respect to z to give

$$\frac{p}{\varrho} = \frac{\partial \phi}{\partial t} - gz - (u^2 + w^2)/2 + f(t) \tag{7.12.5}$$

where $f(t)$ is an arbitrary function of time. Because the velocities are unaffected by the value of $f(t)$, we set it equal to p_a/ϱ.

If it is assumed that the wave motions are of small amplitude, the quadratic terms u^2 and w^2 may be neglected. Consequently, the dynamic boundary condition becomes linear although it is along the surface $z = \eta(x,t)$ which is unknown. To circumvent this difficulty, the boundary condition is expanded about the equilibrium point located at $z = h$. Therefore, the boundary condition

$$\phi_t = g\eta \tag{7.12.6}$$

at $z = \eta$ becomes

$$\phi_t = g\eta \tag{7.12.7}$$

at $z = h$ where the quadratic and higher order terms are neglected. From the definition of w,

$$w = -\phi_z = \eta_t + u\eta_x \simeq \phi_t, \tag{7.12.8}$$

the dynamic boundary condition along the free surface is approximately given by

$$\phi_{tt} + g\phi_z = 0 \tag{7.12.9}$$

at $z = h$.

The simplest problem involves small-amplitude waves within a rectangular vessel. The governing equation and boundary conditions are then

$$\phi_{xx} + \phi_{zz} = 0 \tag{7.12.10}$$

$$\phi_x(0,z,t) = \phi_x(L,z,t) = \phi_z(x,0,t) = 0 \tag{7.12.11}$$

$$\phi_{tt}(x,h,t) + g\phi_z(x,h,t) = 0. \tag{7.12.12}$$

Assuming that the resulting motions are periodic in time, the solution to Eqs. (7.12.10)-(7.12.12) are taken to be

$$\phi(x,z,t) = X(x)Z(z)\cos(\omega t) \tag{7.12.13}$$

so that

$$\frac{1}{X}\frac{d^2X}{dx^2} = -\frac{1}{Z}\frac{d^2Z}{dz^2} = -k^2 \tag{7.12.14}$$

with

$$X'(0) = X'(L) = Z'(0) = 0 \tag{7.12.15}$$

and

$$g\,Z'(h) - \omega^2 Z(h) = 0. \tag{7.12.16}$$

where the separation constant has been taken to be negative. The boundary condition (7.12.15) results in trival solutions for a separation constant that is either positive or equal to zero.

A solution which satisfies Eqs. (7.12.14)-(7.12.15) is

$$\phi(x,z,t) = A_n \cosh\left(\frac{n\pi z}{L}\right) \cos\left(\frac{n\pi x}{L}\right) \cos(\omega t) \tag{7.12.17}$$

where $n = 1,2,3,4, \ldots$ Finally, we employ Eq. (7.12.16) and find the dispersion relationship that

$$\omega^2 = \frac{n\pi g}{L} \tanh\left(\frac{n\pi h}{L}\right). \tag{7.12.18}$$

Consequently, for a given h and L, Eq. (7.12.18) gives the corresponding frequency as n increases. The shape of normal modes is given by Eq. (7.12.17).

In the case of a cylindrical vessel, the equations are

$$\frac{1}{r}\frac{\partial}{\partial r}\left(r\frac{\partial \phi}{\partial r}\right) + \frac{\partial^2 u}{\partial z^2} = 0 \qquad \textbf{(7.12.19)}$$

with

$$\phi_z(r,0,t) = \phi_r(a,z,t) = 0 \qquad \textbf{(7.12.20)}$$

and

$$\phi_{tt}(r,h,t) + g\,\phi_z(r,h,t) = 0 \qquad \textbf{(7.12.21)}$$

where a is the radius of the cylinder. The solution is

$$\phi(r,z,t) = A\,J_0(kr)\cosh(kz)\cos(\omega t) \qquad \textbf{(7.12.22)}$$

where

$$\omega^2 = gk\tanh(kh) \qquad \textbf{(7.12.23)}$$

and k is given by the nth root of $J_1(ka) = 0$.

Except for these and a few other vessels, determining the normal modes of other vessels is quite difficult. Fox and Kuttler (Fox, D.W. and J.R. Kuttler, 1985: Sloshing frequencies. *ZAMP, 34,* 668-696.) have devised a technique that uses conformal mapping to transform the domain occupied by the fluid into a rectangular region. However, the boundary condition along the free surface is still too complicated to find analytic solutions. They then use approximate methods to compute the frequencies.

Exercises

1. Pshenichnov (Pshenichnov, G.I., 1972: Free oscillations of liquid in rigid vessels. *J. Appl. Math. Mech. (PMM), 36,* 229-235.) has shown that the oscillations with the highest frequencies within a partially filled vessel with a sloping bottom may be found approximately by solving the partial differential equation

$$\nabla_H^2\phi + \lambda^2\phi = 0$$

in the interior with

$$\frac{\partial \phi}{\partial n}\tan(\gamma) + \lambda\phi = 0$$

along the undisturbed surface where the angle between the outward pointing normal and the positive z axis is γ. ∇_H^2 is the Laplacian in the horizontal direction and $\lambda^2 = \omega^2/g$.

For a partially filled cone, the problem becomes

$$\frac{1}{r}\frac{\partial}{\partial r}\left(r\frac{\partial\phi}{\partial r}\right) + \frac{1}{r^2}\frac{\partial^2\phi}{\partial\theta^2} + \lambda^2\phi = 0$$

with

$$-\frac{\partial\phi}{\partial r}\tan(\theta_0) + \lambda\phi = 0 \text{ for } r = r_0.$$

Use separation of variables to find ϕ and the relationship for finding λ.

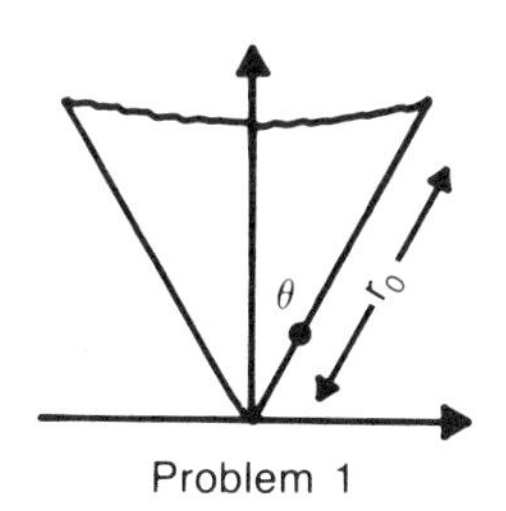

Problem 1

Answers

1. (a) $$u(x,y) = \frac{4}{\pi}\sum_{n=1}^{\infty}\frac{(-1)^n}{2n-1}\frac{\sinh\left[\frac{(2n-1)\pi y}{2a}\right] + \sinh\left[\frac{(2n-1)\pi(b-y)}{2a}\right]}{\sinh\left[\frac{(2n-1)\pi b}{2a}\right]} \times \cos\left[\frac{(2n-1)\pi x}{2a}\right]$$

$$+ \frac{4}{\pi}\sum_{n=1}\frac{1}{2n-1}\frac{\cosh\left[\frac{(2n-1)\pi x}{b}\right]}{\cosh\left[\frac{(2n-1)\pi a}{b}\right]}\sin\left[\frac{(2n-1)\pi y}{b}\right]$$

(b) $u(x,y) = y/b$

(c) $$u(x,y) = \frac{4}{\pi}\sum_{n=1}^{\infty}\frac{(-1)^{n+1}}{2n-1}\frac{\sinh\left[\frac{(2n-1)\pi(a-x)}{2b}\right]}{\sinh\left[\frac{(2n-1)\pi a}{2b}\right]}\cos\left[\frac{(2n-1)\pi y}{2b}\right]$$

(d) $$u(x,y) = \frac{4}{\pi}\sum_{n=1}^{\infty}\frac{1}{2n-1}\frac{\sinh\left[\frac{(2n-1)\pi}{b}(a-x)\right]}{\sinh\left[\frac{(2n-1)\pi a}{b}\right]}\sin\left[\frac{(2n-1)\pi y}{b}\right]$$

(e) $u(x,y) = 1$

(f) $$u(x,y) = \frac{4}{\pi} \sum_{n=1}^{\infty} \frac{1}{2n-1} \frac{\cosh\left[\frac{(2n-1)\pi(y-b)}{2a}\right]}{\cosh\left[\frac{(2n-1)\pi b}{2a}\right]} \sin\left[\frac{(2n-1)\pi x}{2a}\right]$$

$$+ \frac{4}{\pi} \sum_{n=1}^{\infty} \frac{1}{2n-1} \frac{\cosh\left[\frac{(2n-1)\pi(x-a)}{2b}\right]}{\cosh\left[\frac{(2n-1)\pi a}{2b}\right]} \sin\left[\frac{(2n-1)\pi y}{2b}\right]$$

(g) $$u(x,y) = \frac{4}{\pi} \sum_{n=1}^{\infty} \frac{(-1)^{n+1}}{2n-1} \frac{\sinh\left[\frac{(2n-1)\pi x}{2b}\right] + \sinh\left[\frac{(2n-1)\pi(a-x)}{2b}\right]}{\sinh\left[\frac{(2n-1)\pi a}{2b}\right]} \times \cos\left[\frac{(2n-1)\pi y}{2b}\right]$$

(h) $$u(x,y) = \frac{4}{\pi} \sum_{n=1}^{\infty} \frac{1}{2n-1} \frac{\sinh\left[\frac{n\pi}{a}(b-y)\right]}{\sin\left(\frac{n\pi b}{a}\right)} \sin\left(\frac{n\pi x}{a}\right)$$

$$+ \frac{4}{\pi} \sum_{n=1}^{\infty} \frac{1}{2n-1} \frac{\sinh\left[\frac{n\pi}{b}(a-x)\right]}{\sinh\left(\frac{n\pi a}{b}\right)} \sin\left(\frac{n\pi y}{b}\right)$$

(i) $$u(x,y) = \frac{8}{\pi^2} \sum_{n=1}^{\infty} \frac{(-1)^n}{(2n-1)^2} \frac{\sinh\left[\frac{(2n-1)\pi y}{a}\right]}{\sinh\left[\frac{(2n-1)\pi b}{a}\right]} \sin\left[\frac{(2n-1)\pi x}{a}\right]$$

2. $$h(x,y) = \frac{R(a^2-x^2)}{2T} - \frac{16Ra^2}{T\pi^3} \sum_{n=0}^{\infty} \frac{(-1)^n \cos\left[\frac{(2n+1)\pi x}{2a}\right] \cosh\left[\frac{(2n+1)\pi y}{2a}\right]}{(2n+1)^3 \cosh\left(\frac{(2n+1)\pi b}{2a}\right)}$$

3. $$T(x,y) = T_0 + \Delta T \cos\left(\frac{2\pi x}{\lambda}\right) e^{-2\pi y/\lambda}.$$

4. $u(x,y) = g\left[z_0 + cs/2 + \frac{a}{sb}(1 - \cos(bs)\right]$

$$+ 2\sum_{n=1}^{\infty}\left\{\frac{ab\,[1-\cos(bs)\cos(n\pi)]}{b^2 - n^2\pi^2/s^2} + \frac{cs^2}{n^2\pi^2}[(-1)^n - 1]\right\}$$

$$\times \frac{\cos(n\pi x/s)\cosh(n\pi y/s)}{s\cosh(n\pi z_0/s)}$$

5. $u(x,y) = u_1 + \frac{A_0}{L}(b-y)$

$$+ \sum_{n=1}^{\infty}\frac{A_n}{L}\left(\sinh\left(\frac{n\pi y}{L}\right) - \tanh\left(\frac{n\pi b}{L}\right)\cosh\left(\frac{n\pi y}{L}\right)\right)\cos\left(\frac{n\pi x}{L}\right)$$

where

$$2b\,A_0 = \int_0^{\alpha} f(x)\,dx - u_1 L$$

$$A_n = -\int_0^{\alpha} f(x)\left[\frac{\cos\left(\frac{n\pi x}{L}\right)}{\tanh\left(\frac{n\pi b}{L}\right)}\right]dx$$

P. 429

1. $u(r,z) = \frac{(z-H)r_0^2}{a^2} + 2r_0\sum_{n=1}^{\infty}\frac{J_1\left(\frac{\alpha_n r_0}{a}\right)}{\alpha_n^2 J_0(\alpha_n)^2}\frac{\sinh(\alpha_n(z-H)/a)}{\cosh(\alpha_n H/a)}J_0\left(\frac{\alpha_n r}{a}\right)$

where $J'_0(\alpha_n) = 0$ for $n = 1,2,3,\ldots$

2. $u(r,z) = -2A\sum_{n=1}^{\infty}\frac{\sinh(\alpha_n z)}{\alpha_n\cosh(\alpha_n h)}\frac{J_1(\alpha_n r)}{\alpha_n J_0(\alpha_n R)}$

where $J_1(\alpha_n R) = 0$.

3. $u(r,z) = -2\sum_{n=1}^{\infty}(-1)^n\frac{I_1(n\pi r)}{I_1(n\pi a)}\frac{\sin(n\pi z)}{n\pi}$

4. $u(r,z) = 4a^2\sum_{n=1}^{\infty}\frac{1}{\lambda_n a(\lambda_n^2 a^2 - 1)}\frac{J_1(\lambda_n r)}{J_1(\lambda_n a)}\frac{\sinh(\lambda_n z)}{\cosh(\lambda_n z)}$

where $J_0(\lambda_n a) = 0$.

P. 439

1. $T(r,\theta) = 50\left\{1 + \sum_{n=1}^{\infty}\left[P_{2n-2}(0) - P_{2n}(0)\right]\left(\frac{r}{b}\right)^{2n-1}P_{2n-1}[\cos(\theta)]\right\}$

2. $T(r,\theta) = 400 \left\{ \frac{1}{7} \frac{r}{a} P_1 [\cos(\theta)] \frac{1}{9} (\frac{r}{a})^3 P_3 [\cos(\theta)] - \frac{2}{63} (\frac{r}{a})^5 P_5 [\cos(\theta)] \right\}$

3. $T(r,\theta) = C_0 + \sum_{n=1,3,5,\ldots}^{\infty} C_n r^n P_n [\cos(\theta)]$

where

$$C_n = \frac{q0}{na^{n-1}} \{P_{n-1} [\cos(\theta_0)] - P_{n+1} [\cos(\theta_0)]\}$$

and C_0 is an arbitrary constant.

4. $\frac{T(r,\theta) - T_s}{T_0 - T_s} = \sum_{n=1,3,5,\ldots}^{\infty} \frac{2n+1}{1 + k_\ell n/ha} \times \left[\int_0^1 P_n(x)\, dx \right] (\frac{r}{a})^n P_n [\cos(\theta)]$

5. $T(r,\theta) = T_0 [1 - \cos(\alpha)]/2 + T_0/2 \sum_{n=1}^{\infty} \{P_{n-1} [\cos(\alpha)] - P_{n+1} [\cos(\alpha)]\} (\frac{r}{a})^n P_n [\cos(\theta)]$

6. $T(r,\theta) = T_0/2 \sum_{n=1}^{\infty} \{P_{2n-2} [\cos(\alpha)] - P_{2n} [\cos(\alpha)]\} \times (\frac{r}{a})^{-2n} P_{2n-1} [\cos(\theta)]$

7. $T(r,\theta) = T_0/2 \{1 - \sum_{n=1}^{\infty} \frac{b^{2n} + a^{2n}}{b^{4n-1} - a^{4n-1}} [P_{2n-2}(0) - P_{2n}(0)] r^{2n-1} P_{2n-1} [\cos(\theta)] + \sum_{n=1}^{\infty} \frac{a^{-2n-1} + b^{-2n-1}}{b^{-4n+1} - a^{-4n+1}} [P_{2n-2}(0) - P_{2n}(0)] r^{-2n} P_{2n-1} [\cos(\theta)]\}$

8. $u(r,\theta) = \frac{q}{4\pi} \sum_{n=0}^{\infty} \frac{b^{2n+1} - c^{2n+1}}{b^{n+1}(a^{2n+1} - c^{2n+1})} (r^n - \frac{a^{2n+1}}{r^{n+1}} P_n [\cos(\theta)]$

9. $u(r,\theta) = \sum_{n=0}^{\infty} \frac{(-1)^n (2n)! (T_a - T_\ell) (4n+3)}{2^{2n+1} (\frac{(2n+1)k}{ah} + 1) (n+1) (n!)^2} (\frac{r}{a})^{2n+1} P_{2n+1} [\cos(\theta)]$

10. $$\phi_{out} = \frac{R^3 P \cos(\theta)}{3\epsilon_0 r^2}$$

$$\phi_{in} = \frac{P}{3\epsilon_0} r \cos(\theta).$$

P. 460

1. $$T(x,y) = \frac{2}{\pi} \tan^{-1}\left(\frac{y}{x}\right)$$

2. $$T(x,y) = \frac{T_0}{\theta_0} \tan^{-1}\left(\frac{y}{x}\right)$$

3. $$V(x,y) = \frac{1}{2} + \frac{1}{\pi} \tan^{-1}\left\{\frac{(1+x)^2 + (1-y)^2 - 1}{1 - x^2 - y^2}\right\}$$

4. $$V(x,y) = \frac{1}{\pi} \tan^{-1}\left\{\frac{\sin(y)}{\sinh(x)}\right\} \quad \text{if } 0 \leq \tan^{-1}(t) \leq \pi$$

5. $$V(x,y) = \frac{1}{\pi} \cot^{-1}\left[\frac{(r^2-1)^2 - 4r^2\sin^2(\theta)}{4r(r^2-1)\sin(\theta)}\right]$$

$$= \frac{2}{\pi} \tan^{-1}\left[\frac{2r\sin(\theta)}{r^2 - 1}\right] = \frac{2}{\pi} \tan^{-1}\left(\frac{2y}{x^2+y^2-1}\right)$$

6. $$T(x,y) = \frac{2}{\pi} \tan^{-1}\left[\sin\left(\frac{\pi x}{a}\right) / \sinh\left(\frac{\pi y}{a}\right)\right]$$

7. $$V(r,\theta) = (V_1 + V_2)/2 + \frac{1}{\pi}(V_1 - V_2) \tan^{-1}\left[\frac{2ar}{a^2 - r^2} \cos(\theta)\right]$$

8. $$V(r,\theta) = \frac{2V}{\pi} \tan^{-1}\left[\frac{2r^2a^2\sin(\theta)}{a^4 - r^4}\right]$$

P. 464

1. $$\phi_n(r,\theta) = J_n(\lambda n) \cos(n\theta)$$

with $$-\left[J_{n-1}(\lambda r_0) - \frac{n}{\lambda r_0} J_n(\lambda r_0)\right] \tan(\theta_0) + J_n(\lambda r_0) = 0.$$

Appendix A

The Method of Frobenius

MANY OF THE ORDINARY DIFFERENTIAL EQUATIONS ENcountered in the solution of Sturm-Liouville problems do not have simple closed solutions. Consequently, we usually have two options, either solve it numerically or by a power series. Today, with the advent of high-speed digital computers, many choose the numerical approach. On the other hand, in order to understand the approach used by many before the age of modern computers as well as classical functions, some knowledge of power series methods is necessary. In this appendix, we present the most commonly used technique: the *method of Frobenius*.

A.1 THE SINGULAR POINTS OF A LINEAR SECOND-ORDER DIFFERENTIAL EQUATION

The method of Frobenius can be used to solve linear second-order differential equation

$$y'' + a_1(x)\, y' + a_2(x)\, y = 0 \qquad \textbf{(A.1.1)}$$

in terms of a power series about an *ordinary* or *regular singular* point $x = x_0$. The point $x = x_0$ is said to be an *ordinary* point of Eq. (A.1.1) if both the coefficients $a_1(x)$ *and* $a_2(x)$ remain finite at $x = x_0$. Otherwise, the point $x = x_0$ is said to be a *singular* point. There are two types of singular points. A *regular singular* point requires that *both* $(x-x_0)a_1(x)$ *and* $(x-x_0)^2a_2(x)$ remain finite at the point $x = x_0$. If this is not true, then the point is called an *irregular singular* point.

Consider the differential equation

$$x^3(1-x)^2y'' - 2x^2(1-x)y' + 3y = 0 \tag{A.1.2}$$

or

$$y'' - \frac{2}{x(1-x)}\, y' + \frac{3}{x^3(1-x)^2}\, y = 0 \tag{A.1.3}$$

so that

$$a_1(x) = -\frac{2}{x(1-x)} \tag{A.1.4}$$

and

$$a_2(x) = \frac{3}{x^3(1-x)^2} \tag{A.1.5}$$

Clearly the only singular points are at $x = 0$ and $x = 1$ because $a_1(x)$ and $a_2(x)$ become undefined there. The statement would still be true if only $a_1(x)$ or $a_2(x)$ became undefined at those points.

We check and see whether $x = 0$ is a regular or irregular singular point by computing

$$(x-x_0)a_1(x) = (x-0)\frac{-2}{x(1-x)} = -\frac{2}{1-x} \tag{A.1.6}$$

and

$$(x-x_0)^2a_2(x) = (x-0)^2\frac{3}{x^3(1-x)^2} = \frac{3}{x(1-x)^2}\,. \tag{A.1.7}$$

Although Eq. (A.1.6) remains finite at $x = 0$, Eq. (A.1.7) does not and the point $x = 0$ is an irregular singular point. We find for $x = 1$

$$(x-x_0)a_1(x) = (x-1)\frac{2}{x(x-1)} = \frac{2}{x} \tag{A.1.8}$$

and

$$(x-x_0)^2a_2(x) = (x-1)^2\frac{3}{x^3(x-1)^2} = \frac{3}{x^3}\,. \tag{A.1.9}$$

At the point $x = 1$ both Eq. (A.1.8) and (A.1.9) remain finite. Consequently $x = 1$ is a regular singular point.

A.2 EXAMPLES OF SOLVING A SECOND-ORDER ORDINARY DIFFERENTIAL EQUATION BY THE METHOD OF FROBENIUS

We illustrate the method of Frobenius by solving the differential equation

$$(2x^2 + x^3)y'' + (x + 3x^2)y' - (1 + 4x)y = 0 \tag{A.2.1}$$

about the regular singular point $x_0 = 0$. If the regular singular point has been at $x = x_0 \neq 0$, that point can be shifted to the origin by introducing the independent variable $t = x - x_0$ and proceeding as outlined below using t as the independent variable.

The method of Frobenius states that the most general solution is

$$y(x) = \sum_{k=0}^{\infty} A_k x^{k+s} \tag{A.2.2}$$

where both s and A_k are to be determined. Now

$$y'(x) = \sum_{k=0}^{\infty} (k+s)A_k x^{k+s-1} \tag{A.2.3}$$

and

$$y''(x) = \sum_{k=0}^{\infty} (k+s)(k+s-1)A_k x^{k+s-2} \tag{A.2.4}$$

so that

$$(2x^2 + x^3)y'' + (x + 3x^2)y' - (1 + 4x)y$$

$$= \sum_{k=0}^{\infty} 2(k+s)(k+s-1)A_k x^{k+s} + \sum_{k=0}^{\infty} (k+s)(k+s-1)A_k x^{k+s+1}$$

$$+ \sum_{k=0}^{\infty} (k+s)A_k x^{k+s} + \sum_{k=0}^{\infty} 3(k+s)A_k x^{k+s+1}$$

$$- \sum_{k=0}^{\infty} A_k x^{k+s} - \sum_{k=0}^{\infty} 4A_k x^{k+s+1} = 0 \tag{A.2.5}$$

Dividing out a common factor of x^s and grouping like terms together

$$\sum_{k=0}^{\infty} [2(k+s)^2 - k - s - 1] \; A_k x^k + \sum_{k=0}^{\infty} [(k+s)^2 + 2(k+s) - 4] \; A_k x^{k+1} = 0 \qquad \textbf{(A.2.6)}$$

Our ultimate goal is to collapse the two power series in Eq. (A.2.6) into one series. The difficulty arises from the first series beginning with the term $(2s^2 - s - 1)A_0 x^0$ while the second starts with $(s^2 + 2s - 4)A_0 x$. We need both of these series to begin with the same power of x. To get around this difficulty, suppose we write the first series as

$$(2s^2 - s - 1)A_0 + \sum_{n=1}^{\infty} [2(n+s)^2 - n - s - 1]A_n \, x^n \qquad \textbf{(A.2.7)}$$

(I have used n as the dummy integer, rather than k, to avoid any confusion of the first power series in Eq. (A.2.6) with the power series in Eq. (A.2.7).) Because the new power series in Eq. (A.2.7) now begins with $(2s^2 + 3)A_1 x$, the power series in Eq. (A.2.7) can be combined with the second power series in Eq. (A.2.6). This can be accomplished by defining $n = k + 1$ so that Eq. (A.2.7) becomes

$$(2s^2 - s - 1)A_0 + \sum_{k=0}^{\infty} [2(k+s+1)^2 - k - s - 2]A_{k+1} x^{k+1}. \qquad \textbf{(A.2.8)}$$

Consequently Eq. (A.2.6) can be written as

$$(2s^2 - s - 1)A_0 + \sum_{k=0}^{\infty} \left\{ [2(k+s+1)^2 - k - s - 2]A_{k+1} + [(k+s)^2 + 2(k+s) - 4]A_k \right\} x^{k+1} = 0 \qquad \textbf{(A.2.9)}$$

In summary, we can write two power series as one series by explicitly splitting up the power series with the *lowest* power in the leading term. One part consists of those powers of x that are *not* in common with the second power series (in our example, $(2s^2 - s - 1)A_0$). The second part consists of a power series representation of the remaining terms. Through a suitable change in the dummy index (in our case, $k = n + 1$), this new power series can be combined with the second power series.

The result stated in Eq. (A.2.9) must be true for all x. The only way that we can guarantee that this will occur is to require that each power of x to vanish separately. This is our entire motivation for grouping terms together by powers of x. Consequently

$$(2s^2 - s - 1)A_0 = 0 \qquad \textbf{(A.2.10)}$$

$$[2(k+s+1)^2 - k - s - 2]A_{k+1} + [(k+s)^2 + 2(k+s) - 4]A_k = 0 \tag{A.2.11}$$

where $k = 0, 1, 2, \ldots$

Because $A_0 \neq 0$ (or we would have a trival solution),

$$2s^2 - s - 1 = 0 \tag{A.2.12}$$

or

$$s_1 = 1 \text{ and } s_2 = -1/2. \tag{A.2.13}$$

Eq. (A.2.12) is called the *indicial equation*. In general there will be two unequal values for s that will give us our two independent solutions. However, if $s_1 = s_2$ we are clearly in trouble. What is less clear is the fact that when $|s_2 - s_1|$ differs by a positive integer, we can have a situation where we only have one solution of the form Eq. (A.2.2). The treatment of these exceptional cases is reserved for the next section. In any case, what follows will always work for the larger of the two indicial roots (See Hildebrand, F. B., 1962: *Advanced Calculus for Applications,* Englewood Cliffs, NJ: Prentice-Hall, Inc., p. 132.)

With these two indicial roots, we can use (A.2.11), commonly known as the *recurrence formula*, to find the A_k's in terms of A_0. For $s_1 = 1$, we have

$$[2(k+2)^2 - k - 3]A_{k+1} = -[(k+1)^2 + 2(k+1) - 4]A_k \tag{A.2.14}$$

or

$$A_{k+1} = -\frac{k^2 + 4k - 1}{2k^2 + 7k + 5} A_k. \tag{A.2.15}$$

Therefore

$$A_1 = \frac{1}{5} A_0 \tag{A.2.16}$$

$$A_2 = -\frac{2}{7} A_1 = -\frac{2}{35} A_0 \tag{A.2.17}$$

$$A_3 = -\frac{11}{27} A_2 = \frac{22}{945} A_0 \tag{A.2.18}$$

Because $y(x) = x^s(A_0 + A_1x + A_2x^2 + A_3x^3 + \ldots)$, a particular solution corresponding to $s_1 = 1$ is

$$y_1(x) = A_0x(1 + \frac{1}{5}x - \frac{2}{25}x^2 + \frac{22}{945}x^3 - \ldots). \quad \textbf{(A.2.19)}$$

For $s_2 = -1/2$ we find

$$A_{k+1} = -\frac{4k^2 + 4k - 19}{8k^2 + 4k - 4} A_k. \quad \textbf{(A.2.20)}$$

Therefore

$$A_1 = -\frac{19}{4} A_0 \quad \textbf{(A.2.21)}$$

$$A_2 = \frac{11}{8} A_1 = -\frac{209}{32} A_0 \quad \textbf{(A.2.22)}$$

$$A_3 = -\frac{5}{36} A_2 = \frac{1045}{1152} A_0 \quad \textbf{(A.2.23)}$$

Hence a second particular solution is

$$y_2(x) = A_0x^{-1/2}(1 - \frac{19}{4}x - \frac{209}{32}x^2 + \frac{1045}{1152}x^3 - \ldots) \quad \textbf{(A.2.24)}$$

The general solution is a linear combination of these two solution:

$$y(x) = c_1y_1(x) + c_2y_2(x). \quad \textbf{(A.2.25)}$$

For a second example, let us solve

$$y'' + y = 0 \quad \textbf{(A.2.26)}$$

which has the solution

$$y(x) = c_1 \cos(x) + c_2 \sin(x). \quad \textbf{(A.2.27)}$$

We substitute Eqs. (A.2.4) and (A.2.2) into Eq. (A.2.26) and find that

$$\sum_{k=0}^{\infty} (k+s)(k+s-1)A_k x^{k-2} + \sum_{k=0}^{\infty} A_kx^k = 0 \quad \textbf{(A.2.28)}$$

If we explicitly remove the first two terms in the first summation, we obtain

$$s(s-1)A_0x^{-2} + (s+1)sA_1x^{-1} + \sum_{n=2}^{\infty} A_n(n+s)(n+s-1)x^{n-2} + \sum_{k=0}^{\infty} A_kx^k = 0 \tag{A.2.29}$$

If we replace n by $k+2$ in the first summation,

$$s(s-1)A_0x^{-2} + s(s+1)A_1x^{-1} + \sum_{k=0}^{\infty} [A_{k+2}(k+s+2)(k+s+1) + A_k]\, x^k = 0 \tag{A.2.30}$$

Once again we require that each coefficient of a power of x vanish identically. Therefore

$$s(s-1)A_0 = 0 \tag{A.2.31}$$

$$s(s+1)A_1 = 0 \tag{A.2.32}$$

and

$$(k+s+2)(k+s+1)A_{k+2} + A_k = 0 \tag{A.2.33}$$

where $k = 0, 1, 2, \ldots$

In order not to have a trival solution, $A_0 \neq 0$ and

$$s = 0 \text{ and } 1. \tag{A.2.34}$$

The value of A_1 depends upon s. If $s = 1$, $A_1 = 0$ from Eq. (A.2.32). But, if $s = 0$, then A_1 remains a free parameter.

We first turn to the root $s = 1$. Then

$$A_{k+2} = -\frac{A_k}{(k+3)(k+2)} \tag{A.2.35}$$

so that

$$A_2 = -\frac{A_0}{6} = -\frac{A_0}{3!} \tag{A.2.36}$$

$$A_3 = -\frac{A_1}{12} = 0 \tag{A.2.37}$$

and

$$A_4 = -\frac{A_2}{20} = \frac{A_0}{120} = \frac{A_0}{5!} \tag{A.2.38}$$

Consequently the solution for $s = 1$ can be written

$$y_1(x) = A_0x(1 - \frac{x^2}{3!} + \frac{x^4}{5!} - \ldots) = A_0 \sin(x) \tag{A.2.39}$$

and we recover one of the known solutions. For $s = 0$ we have

$$A_{k+2} = -\frac{A_k}{(k+2)(k+1)} \tag{A.2.40}$$

so that

$$A_2 = -\frac{A_0}{2} = -\frac{A_0}{2!} \tag{A.2.41}$$

$$A_3 = -\frac{A_1}{6} = -\frac{A_1}{3!} \tag{A.2.42}$$

$$A_4 = -\frac{A_2}{12} = \frac{A_0}{4!} \tag{A.2.43}$$

$$A_5 = -\frac{A_3}{20} = \frac{A_1}{5!} \tag{A.2.44}$$

Consequently the second solution can be written

$$y_2(x) = A_0\left(1 - \frac{x^2}{2!} + \frac{x^4}{4!} - \ldots\right)$$

$$+ A_1\left(x - \frac{x^3}{3!} + \frac{x^5}{5!} - \ldots\right) \tag{A.2.45}$$

$$= A_0 \cos(x) + A_1 \sin(x). \tag{A.2.46}$$

Therefore, in the case $s = 0$ we have recovered *both* solutions. This does not mean that it is always best to work out the lower of the two roots in the hope of saving some work. It merely shows that having the extra arbitrary constant A_1 does not introduce any difficulties in the method.

Exercises

In problems 1-10, show that $x = 0$ is a regular singular point. Then find power series solutions for these ordinary differential equations.

1. $2xy'' + 5y' + xy = 0$
2. $3x(2+3x)y'' - 4y' + 4y = 0$
3. $x^2(4+x)y'' + 7xy' - y = 0$
4. $2x^2y'' + (x-x^2)y' - y = 0$
5. $2x^2y'' + 5xy' + (1+x)y = 0$
6. $9x^2y'' + (2+3x)y = 0$
7. $(2x^2+x^3)y'' - xy' + (1-x)y = 0$
8. $2x^2y'' - 3(x+x^2)y' + (2+3x)y = 0$

9. $3x^2y'' + (5x - x^2)y' + (2x^2 - 1)y = 0$
10. $4x^2y'' + x(x^2 - 4)y' + 3y = 0$

In problems 11-22, find the power series solutions to the following ordinary differential equations.

11. $2x^2y'' - 3(x + x^2)y' + 2y = 0$
12. $9x^2y'' + 9(x - x^2)y' + (x - 1)y = 0$
13. $4x^2(1 - x)y'' + 3x(1 + 2x)y' - 3y = 0$
14. $2x^2(1 - 3x)y'' + 5xy' - 2y = 0$
15. $4x^2(1 + x)y'' - 5xy' + 2y = 0$
16. $(4 + x)x^2y'' + x(x - 1)y' + y = 0$
17. $(8 - x)x^2y'' + 6xy' - y = 0$
18. $2x^2y'' + x(1 + x^2)y' - (1 + x)y = 0$
19. $2x^2y'' - xy' + (1 + x^2)y = 0$
20. $3x^2y'' + 2xy' + (x^2 - 2)y = 0$
21. $x^2(3 + x^2)y'' + 5xy' - (1 + x)y = 0$
22. $2xy'' - (1 + x^3)y' + y = 0.$

A.3 THE TREATMENT OF THE EXCEPTIONAL CASES

In certain cases the method outlined in the previous section fails and gives only one of the two possible solutions. One situation is when $s_1 = s_2$. Another occurs when $s_1 - s_2$ is a positive integer (where s_1 is the larger indicial root) and the recurrence formula for s_2 becomes undefined. In these instances when the differential equation, having $x = 0$ as a regular singular point, possesses only one solution

$$y_1(x) = \sum_{k=0}^{\infty} A_k x^{k+s_1} = A_0 u_1(x), \tag{A.3.1}$$

we state without proof that the second solution is given by

$$y_2(x) = C\, u_1(x) \ln(x) + \sum_{k=0}^{\infty} B_k x^{k+s_2} \tag{A.3.2}$$

where C is an arbitrary constant. (See Hildebrand, F. B., 1962: *Advanced Calculus for Application*. Englewood Cliffs (NJ): Prentice-Hall, Inc., Section 4.5 for the proof.)

For example, let us solve

$$x\, y'' - y = 0 \tag{A.3.3}$$

by the method of Frobenius. Substituting Eq. (A.2.4) and (A.2.2) into Eq. (A.3.3), we have

$$\sum_{n=0}^{\infty} (n+s)(n+s-1)A_n x^{n-1} + \sum_{k=0}^{\infty} A_k x^k = 0 \tag{A.3.4}$$

$$s(s-1)A_0x^{-1} + \sum_{n=1}^{\infty} (n+s)(n+s-1)A_nx^{n-1} + \sum_{k=0}^{\infty} A_kx^k = 0 \quad \textbf{(A.3.5)}$$

$$s(s-1)A_0x^{-1} + \sum_{k=0}^{\infty} [(k+s+1)(k+s)A_{k+1} + A_k]\, x^k = 0 \quad \textbf{(A.3.6)}$$

Consequently the indicial roots are $s = 0$ and $s = 1$ and the recurrence formula is

$$(k+s+1)(k+s)A_{k+1} + A_k = 0, \; k = 0,1,2,3, \ldots \quad \textbf{(A.3.7)}$$

For $s = 1$, we have

$$A_{k+1} = -\frac{A_k}{(k+2)(k+1)} \quad \textbf{(A.3.8)}$$

so that

$$A_1 = -\frac{A_0}{2} = -\frac{A_0}{2!1!} \quad \textbf{(A.3.9)}$$

$$A_2 = -\frac{A_1}{6} = \frac{A_0}{3!2!} \quad \textbf{(A.3.10)}$$

$$A_3 = -\frac{A_2}{12} = -\frac{A_0}{4!3!} \quad \textbf{(A.3.11)}$$

By inductive reasoning we find that

$$y_1(x) = A_0 \sum_{k=0}^{\infty} \frac{x^{k+1}}{k!(k+1)!} = A_0u_1(x). \quad \textbf{(A.3.12)}$$

On the other hand, for $s = 0$

$$(k+1)k\, A_{k+1} + A_k = 0 \quad \textbf{(A.3.13)}$$

which yields the absurd result that $A_0 = 0$ when $k = 0$. Consequently for our second solution we try

$$y_2(x) = C\, u_1(x)\, \ell n(x) + \sum_{k=0}^{\infty} B_kx^k \quad \textbf{(A.3.14)}$$

because $s_2 = 0$. By substituting directly into Eq. (A.3.3), we obtain

$$C[(xu_1'' - u_1)\ln(x) + 2u_1' - \frac{u_1}{x}]$$

$$+ \sum_{k=0}^{\infty} [(k+1)k\, B_{k+1} - B_k]\, x^k = 0. \qquad \textbf{(A.3.15)}$$

Because u_1 must satisfy Eq. (A.3.3), $xu_1'' - u_1 = 0$. The requirement that the coefficient of the power x^k vanish yields

$$(k+1)k\, B_{k+1} - B_k = -\frac{2k+1}{(k+1)!k!}\, C\ (k \geq 0) \qquad \textbf{(A.3.16)}$$

or

$$B_0 = C \qquad \textbf{(A.3.17)}$$

$$B_2 = B_1/2 - \frac{3}{4}\, C \qquad \textbf{(A.3.18)}$$

$$B_3 = \frac{1}{12}\, B_1 - \frac{7}{36}\, C. \qquad \textbf{(A.3.19)}$$

Hence the second solution becomes

$$y_2(x) = C[(x + x^2/2 + \frac{1}{12}\, x^3 + \ldots)\ln(x)$$

$$+1 - \frac{3}{4}\, x^2 - \frac{7}{36}\, x^3 + \ldots]$$

$$+ B_1(x + x^2/2 + \frac{1}{12}\, x^3 + \ldots) \qquad \textbf{(A.3.20)}$$

$$y_2(x) = B_1\, u_1(x) + C\, [(x + x^2/2 + \frac{1}{12}\, x^3 + \ldots)\, \ln(x)$$

$$+ 1 - \frac{3}{4}\, x^2 - \frac{7}{36}\, x^3 + \ldots] \qquad \textbf{(A.3.21)}$$

Once again, the solution for the smaller root actually contains *both* linearly independent solutions of Eq. (A.3.3).

Exercises

Find the general solution of each of the following differential equations.

1. $xy'' + y' + 2y = 0$
2. $xy'' + y' + 2xy = 0$
3. $x^2y'' - 3xy' + 4(1+x)y = 0$
4. $x^2y'' - x(1+x)y' + y = 0$
5. $x^2y'' - x(2x+3)y' + 4y = 0$
6. $x^2(1-x^2)y'' - 5xy' + 9y = 0$
7. $x^2y'' + x(x^2-1)y' + (1-x^2)y = 0$
8. $x^2y'' + x(2x-1)y' + x(x-1)y = 0$

15. $y_1 = x^2\left(1 - \frac{8x}{11} + \frac{32x^2}{55} - \ldots\right)$

$y_2 = x^{1/4}\left(1 - \frac{x}{4} + \frac{5x^2}{32} - \ldots\right)$

16. $y_1 = x\left(1 - \frac{x}{7} + \frac{2x^2}{77} - \ldots\right)$

$y_2 = x^{1/4}\left(1 - \frac{x}{16} + \frac{5x^2}{512} - \ldots\right)$

17. $y_1 = x^{1/2}\left(1 - \frac{x}{56} + \frac{3x^2}{9548} - \ldots\right)$

$y_2 = x^{-1/4}\left(1 + \frac{5x}{32} - \frac{15x^2}{10240} - \ldots\right)$

18. $y_1 = x\left(1 + \frac{x}{5} - \frac{2x^2}{35} - \frac{16x^3}{945} + \ldots\right)$

$y_2 = x^{-1/2}\left(1 - x - x^2/4 + \frac{x^3}{36} + \ldots\right)$

19. $y_1 = x\left(1 - \frac{x^2}{10} + \frac{x^4}{360} - \ldots\right)$

$y_2 = x^{1/2}\left(1 - \frac{x^2}{6} + \frac{x^4}{168} - \ldots\right)$

20. $y_1 = x\left(1 - \frac{x^2}{22} + \frac{x^4}{1496} - \ldots\right)$

$y_2 = x^{-2/3}\left(1 - x^2/2 + \frac{x^4}{56} - \ldots\right)$

21. $y_1 = x^{-1}\left(1 - x - \frac{3}{4}x^2 - \frac{x^3}{20} - \ldots\right)$

$y_2 = x^{1/3}\left(1 + \frac{x}{7} + \frac{23x^2}{1260} - \frac{179x^3}{16380} - \ldots\right)$

22. $y_1 = 1 + x - x^2/2 + \frac{x^3}{18} + \ldots$

$y_2 = x^{3/2}\left(1 - \frac{x}{5} + \frac{x^3}{70} + \frac{52x^3}{945} - \ldots\right)$

P. 480

1. $y_1 = 1 - 2x + x^2 + \ldots$

$$y_2 = y_1 \ell n(x) + 4x - 3x^2 + \frac{22x^3}{27} + \ldots$$

2. $y_1 = 1 - x^2/2 + \dfrac{x^4}{16} - \ldots$

$$y_2 = y_1 \ell n(x) + x^2/2 - \frac{3x^4}{32} + \frac{11x^6}{1728} - \ldots$$

3. $y_1 = x^2(1 - 4x + 16x^2 - \ldots)$

$$y_2 = y_1 \ell n(x) + x^2(8x - 12x^2 + \frac{176x^3}{27} - \ldots)$$

4. $y_1 = x\, e^x$

$$y_2 = x\, e^x \ell n(x) - x\,(x + \frac{3}{4} x^2 + \frac{11}{36} x^3 + \ldots)$$

5. $y_1 = x^2 (1 + 4x + 6x^2 + \ldots)$

$$y_2 = y_1 \ell n(x) - x^2 (6x + 13x^2 + \frac{124}{9} x^3 + \ldots)$$

6. $y_1 = x^3 (1 + \dfrac{3}{2} x^2 + \dfrac{15}{8} x^4 + \ldots)$

$$y_2 = y_1 \ell n(x) - x^5 (1/4 + \frac{13}{32} x^2 + \frac{101}{192} x^4 + \ldots)$$

7. $y_1 = x$

$$y_2 = x \ell n(x) - x^3(1/4 - \frac{x^2}{32} + \frac{x^4}{288} - \ldots)$$

8. $y_1 = 1 - x + \dfrac{x^3}{3} - \dfrac{5x^4}{24}$

$$y_2 = x^2 e^{-x}$$

9. $y_1 = x^{-1} (1 + x/2 + x^2/2 - \dfrac{x^4}{8} - \ldots)$

$$y_2 = x^2 (1 + x/2 + \frac{x^2}{20} - \frac{x^3}{60} - \ldots)$$

10. $y_1 = x^{-1}\left(1 - x - \frac{3x^2}{2} - \frac{9x^3}{8} - \ldots\right)$

$y_2 = x^2\left(1 - x + \frac{9x^2}{10} - \frac{17x^3}{30} + \ldots\right)$

11. $y_1 = x^3\left(1 + \frac{3x}{7} + \frac{3x^2}{14} + \frac{5x^3}{42} + \ldots\right)$

$y_2 = \frac{(1-x)^3}{x^3}$

12. $y_1 = 3 + 2x + x^2$

$y_2 = x^4(1 + 2x + 3x^2 + \ldots + (m+1)x^m + \ldots)$

13. $y_1 = x^6(30 - 42x + 28x^2 - \ldots)$

$y_2 = x^2\left(1 + x + x^2 + \frac{5x^3}{3} + \ldots\right) - \frac{1}{12} y_1 \ln(x)$

14. $y_1 = x^4(1 - 4x + 10x^2 - \ldots)$

$y_2 = x(1 + x/2 + x^2 - 10x^3 + \ldots) - 3y_1 \ln(x)$

15. $y_1 = x\left(3 + x^2/4 + \frac{3x^4}{2^7} + \frac{x^6}{2^9} + \ldots\right)$

$y_2 = x^{-3}\left(1 + \frac{3x^2}{4} + \frac{19x^4}{2^6} + \frac{5x^6}{2^9} + \ldots\right) - \frac{1}{16} y_1 \ln(x)$

Appendix B

Complex Variables

AS WE SHOWED IN SECTION 3.13 THE MOST POWERFUL METHod for inverting a Laplace transform is integration on the complex plane. This appendix is intended to assist those who are unfamiliar with this subject.

B.1 COMPLEX NUMBERS

A complex number is any number that can be written as

$$a + bi$$

where a and b are real and $i = \sqrt{-1}$. The symbol $z = x + iy$, which can stand for any of a set of complex numbers, is called a *complex variable*. The real part of z, usually denoted by $Re(z)$, is x while the imaginary part of z, $Im(z)$ is y. The *complex conjugate*, or briefly *conjugate*, of a complex number $a + bi$ is $a - bi$. The complex conjugate of a complex number z is often indicated by $\bar{z}$ or z^*.

Two complex numbers $a + bi$ and $c + di$ are equal if and only if $a = c$ and $b = d$.

Just as real numbers have the fundamental operations of addition, subtraction, multiplication, and division, so do complex numbers. These operations can be summarized as follows:

Addition $\quad (a + bi) + (c + di) = (a+c) + (b+d)i \qquad$ **(B.1.1)**

Subtraction $\quad (a + bi) - (c + di) = (a-c) + (b-d)i \qquad$ **(B.1.2)**

Multiplication

$$(a + bi)(c + di) = ac + bci + adi + i^2bd \quad \textbf{(B.1.3)}$$

$$= (ac - bd) + (ad + bd)i \quad \textbf{(B.1.4)}$$

Division

$$\frac{a + bi}{c + di} = \frac{a + bi}{c + di}\,\frac{c - di}{c - di} = \frac{ac - adi + bci - bdi^2}{c^2 + d^2} \quad \textbf{(B.1.5)}$$

$$= \frac{ac + bd + (bc - ad)i}{c^2 + d^2} \quad \textbf{(B.1.6)}$$

Another important property of complex variables is the *absolute value* or *modulus*. For a complex number $a + bi$, it is defined $|a + bi| = \sqrt{a^2 + b^2}$. It is readily proven that

$$|z_1 z_2 z_3 \ldots z_n| = |z_1||z_2||z_3| \ldots |z_n| \quad \textbf{(B.1.7)}$$

$$|z_1/z_2| = |z_1|/|z_2| \text{ if } z_2 \neq 0 \quad \textbf{(B.1.8)}$$

$$|z_1 + z_2 + z_3 + \ldots + z_n| \leq |z_1| + |z_2| + |z_3| + \ldots + |z_n| \quad \textbf{(B.1.9)}$$

$$|z_1 + z_2| \geq |z_1| - |z_2|. \quad \textbf{(B.1.10)}$$

It is important to remember that the use of inequalities with complex variables has only meaning when they involve absolute values.

Because a complex number $x + iy$ can be considered as an ordered pair of real numbers, we can represent such numbers by points in the xy plane, called the *complex plane*, by an *Argand diagram*. This is illustrated by Fig. B.1.1. This geometrical interpretation of a complex number leads to an alternative method of expressing a complex number: the polar form.

From the polar representation of x and y,

$$x = r\cos(\theta) \text{ and } y = r\sin(\theta), \quad \textbf{(B.1.11)}$$

where $r = \sqrt{x^2 + y^2}$ is called the *modulus, amplitude* or *absolute value* of z and θ is the *argument* or *phase*, we obtain

$$z = x + iy = r[\cos(\theta) + i\sin(\theta)]. \quad \textbf{(B.1.12)}$$

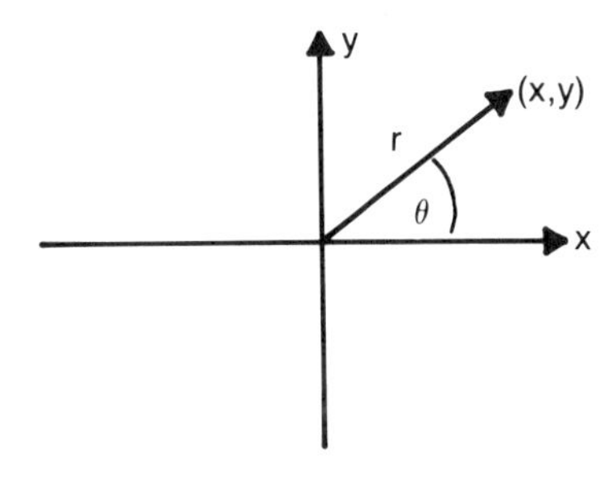

Fig. B.1.1: Argand diagram.

However, from the definition of the exponential,

$$\exp(i\theta) = \sum_{k=0}^{\infty} \frac{(i\theta)^k}{k!} . \qquad \textbf{(B.1.13)}$$

Expanding Eq. (B.1.13), we find that

$$e^{i\theta} = 1 - \frac{\theta^2}{2!} + \frac{\theta^4}{4!} - \frac{\theta^6}{6!} + \ldots + i(\theta - \frac{\theta^3}{3!} + \frac{\theta^5}{5!} - \frac{\theta^7}{7!} + \ldots) \qquad \textbf{(B.1.14)}$$

$$= \cos(\theta) + i \sin(\theta). \qquad \textbf{(B.1.15)}$$

Equation (B.1.15) is called *Euler's formula.* Consequently Eq. (B.1.12) can be expressed as

$$z = r\, e^{i\theta}. \qquad \textbf{(B.1.16)}$$

Furthermore, because

$$z^n = r^n\, e^{in\theta} \qquad \textbf{(B.1.17)}$$

by the law of exponents,

$$z^n = r^n\, [\cos(n\theta) + i \sin(n\theta)] \qquad \textbf{(B.1.18)}$$

Equation (B.1.18) is known as *deMoivre's theorem.*

To illustrate these concepts, let us simplify the following complex number:

$$\frac{3-2i}{-1+i} = \frac{3-2i}{-1+i}\,\frac{-1-i}{-1-i} = \frac{-3-3i+2i+2i^2}{1-i^2} = \frac{-5-i}{2}$$

$$= -\frac{5}{2} - \frac{i}{2} \qquad \textbf{(B.1.19)}$$

Let us now reexpress the complex number $-\sqrt{6} - \sqrt{2}\,i$ in polar form. Because $r = \sqrt{6+2}$

and $\theta = \tan^{-1}(y/x) = \tan^{-1}(2) = -5\pi/6,$

we have $-\sqrt{6} - \sqrt{2}i = 2\sqrt{2}\, e^{-5\pi i/6}.$ **(B.1.20)**

Finally, let us show that $e^{i\theta} = e^{i(\theta+2k\pi)}$ where $k = 0, \pm 1, \pm 2, \pm 3, \ldots$ From Euler's formula,

$$e^{i(\theta + 2k\pi)} = \cos(\theta + 2k\pi) + i \sin(\theta + 2k\pi) \qquad \textbf{(B.1.21)}$$

$$= \cos(\theta) + i \sin(\theta) = e^{i\theta}. \qquad \textbf{(B.1.22)}$$

Exercises

Simplify the following complex numbers.

1. $\dfrac{5i}{2+i}$

2. $\dfrac{5+5i}{3-4i} + \dfrac{20}{4+3i}$

3. $\dfrac{1+2i}{3-4i} + \dfrac{2-i}{5i}$

4. $(1-i)^4$

5. $i(1 - i\sqrt{3})(\sqrt{3} + i)$

Represent the following complex numbers in polar form:

6. $-i$

7. -4

8. $2 + 2\sqrt{3}\,i$

9. $-5 + 5i$

10. $2 - 2i$

11. $-1 + \sqrt{3}\,i$

B.2 FINDING ROOTS

The concept of finding roots of a number, which is rather straightforward in the case of real numbers, becomes more difficult in the case of complex numbers. By finding the *roots* of a complex number, we wish to find all the solutions w of the equation $w^n = z$ with $n>0$ for a given z.

If we put

$$z = r\,e^{i\phi} \tag{B.2.1}$$

for the given, and

$$w = R\,e^{i\Phi} \tag{B.2.2}$$

for the required solution, it follows that

$$w^n = R^n\,(e^{i\Phi})^n = R^n\,e^{in\Phi} = z = r\,e^{i\phi}. \tag{B.2.3}$$

The equality of the two vectors requires equality in their magnitude and conformity in their arguments up to multiples of 2π. Therefore

$$R^n = r \text{ and } n\Phi = \phi + 2k\pi,\ k = 0, \pm 1, \pm 2, \ldots \tag{B.2.4}$$

We thus have $R = r^{1/n}$, the uniquely determined real positive root being used for the absolute value of R, and

$$\Phi = \frac{\phi}{n} + k\,\frac{2\pi}{n} \quad (k = 0, \pm 1, \pm 2, \ldots). \tag{B.2.5}$$

Because the possible values of w_k repeat themselves as soon as k is increased or decreased by n, it is sufficient to take $k = 0,1,2, \ldots, (n-1)$. Therefore there are exactly n solutions:

$$w_k = R\exp(i\Phi_k) = r^{1/n}\exp\left(i\frac{\phi}{n} + k\,\frac{2\pi i}{n}\right),$$

with $\quad k = 0,1, \ldots, n-1.$ **(B.2.6)**

They are called the n roots of z. Geometrically the n corresponding points w_k are located on the circle with radius R with the origin at the point (0,0). These points are, however, the division points of the periphery when the circle is divided into n equal parts, the angle 2π being divided n-times, starting at the point of intersection with the solution of argument ϕ/n.

To illustrate these concepts, let us find all of the values of z for which $z^5 = -32$ and locate these values on the complex plane. Because

$$-32 = 32\,e^{i\pi} = 2^5\,e^{i\pi}, \tag{B.2.7}$$

we have

$$z_k = 2\exp\left(\frac{i\pi}{5} + \frac{2k\pi i}{5}\right), k = 0,1,2,3,4 \tag{B.2.8}$$

or

$$z_0 = 2\exp\left(\frac{i\pi}{5}\right) = 2\cos\left(\frac{\pi}{5}\right) + i\sin\left(\frac{\pi}{5}\right) \tag{B.2.9}$$

$$z_1 = 2\exp\left(\frac{3\pi i}{5}\right) = 2\cos\left(\frac{3\pi}{5}\right) + i\sin\left(\frac{3\pi}{5}\right) \tag{B.2.10}$$

$$z_2 = 2\exp(i\pi) = -2 \tag{B.2.11}$$

$$z_3 = 2\exp\left(\frac{7\pi i}{5}\right) = 2\cos\left(\frac{7\pi}{5}\right) + i\sin\left(\frac{7\pi}{5}\right) \tag{B.2.12}$$

$$z_4 = 2\exp\left(\frac{9\pi i}{5}\right) = 2\cos\left(\frac{9\pi}{5}\right) + i\sin\left(\frac{9\pi}{5}\right) \tag{B.2.13}$$

The location of these roots in the complex plane is shown in Fig. B.2.1.

As a second example, let us find the cube root of $-1 + i$ and locate them graphically. Because $-1 + i = \sqrt{2}\exp\frac{3\pi i}{4}$, we have

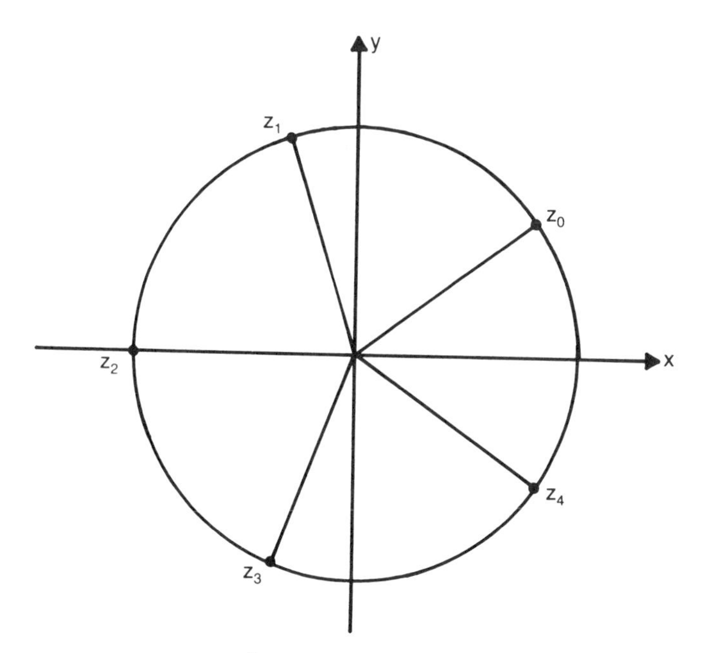

Fig. B.2.1: The zeros of $z^5 = -32$.

$$z_k = 2^{1/6} \exp\left(\frac{\pi i}{4} + \frac{2\pi k i}{3}\right), \; k = 0,1,2 \tag{B.2.14}$$

$$z_0 = 2^{1/6} \exp\left(\frac{\pi i}{4}\right) = 2^{1/6}\left[\cos\left(\frac{\pi}{4}\right) + i \sin\left(\frac{\pi}{4}\right)\right] \tag{B.2.15}$$

$$z_1 = 2^{1/6} \exp\left(\frac{11\pi i}{12}\right) = 2^{1/6}\left[\cos\left(\frac{11\pi}{12}\right) + i \sin\left(\frac{11\pi}{12}\right)\right] \tag{B.2.16}$$

$$z_2 = 2^{1/6} \exp\left(\frac{19\pi i}{12}\right) = 2^{1/6}\left[\cos\left(\frac{19\pi}{12}\right) + i \sin\left(\frac{19\pi}{12}\right)\right] \tag{B.2.17}$$

The location of these zeros on the complex plane are given in Fig. B.2.2.

Exercises

Extract all the possible roots of the following complex numbers.

1. $8^{1/6}$
2. $(-1)^{1/3}$
3. $(-i)^{1/3}$
4. $(5 - 12i)^{1/2}$
5. $(-27i)^{1/6}$

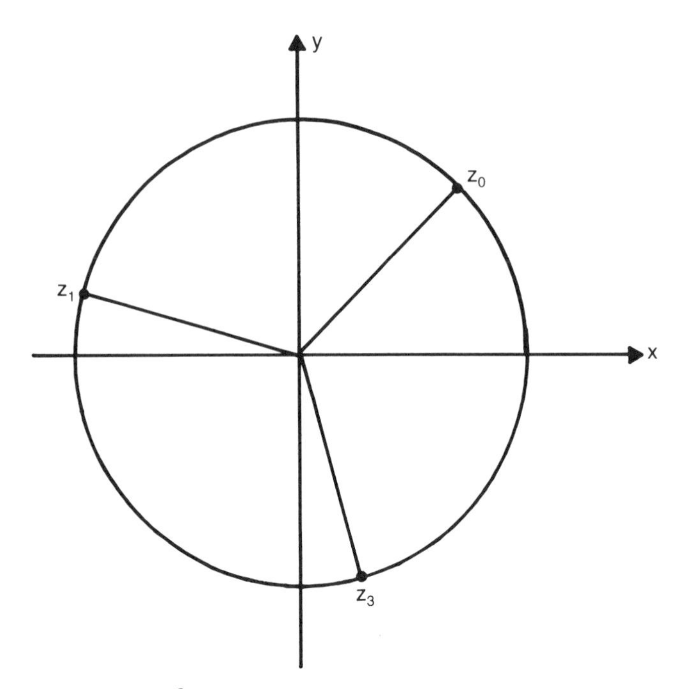

Fig. B.2.2: The zeros of $z^3 = -1 + i$.

B.3 ANALYTIC FUNCTIONS: THE CAUCHY-RIEMANN EQUATIONS

We have already introduced the concept of a complex variable $z = x + iy$ where x and y are variable. We now define another complex variable $w = u + iv$ so that for each value of z there corresponds a value of $w = f(z)$. Thus, as the variable z describes a curve in the z-plane, the corresponding values of w will trace out some curve in the w-plane, as shown in Fig. B.3.1.
that region. For example, if $w = 1/z^2$, the derivative

$$dw/dz = -2/z^3$$

exists at every z except $z = 0$. Thus $w = 1/z^2$ is analytic except at the origin. If the function is analytic everywhere in the complex plane, it is called *entire*.

The derivative of the function $w = f(z)$ is defined by

$$\frac{dw}{dz} = \lim_{\Delta z \to 0} \frac{\Delta w}{\Delta z} = \lim_{\Delta z \to 0} \frac{f(z + \Delta z) - f(z)}{\Delta z} \qquad \textbf{(B.3.1)}$$

Because $\Delta z = \Delta x + i\,\Delta y$, there are an infinite number of different ways of approaching this limit. For an analytic function, it is necessary to demand that the above limit be independent of the manner in which Δz approaches zero.

For example, if Δz is first taken in the x-direction and then in the y-direction, we obtain

$$\lim_{\Delta z \to 0} \frac{\Delta w}{\Delta z} = \lim_{\Delta x \to 0} \frac{\Delta u + i\,\Delta v}{\Delta x} = \frac{\partial u}{\partial x} + i\,\frac{\partial v}{\partial x} \qquad \textbf{(B.3.2)}$$

$$\lim_{\Delta z \to 0} \frac{\Delta w}{\Delta z} = \lim_{\Delta y \to 0} \frac{\Delta u + i\,\Delta v}{\Delta y} = \frac{\partial v}{\partial y} - i\,\frac{\partial u}{\partial y} \qquad \textbf{(B.3.3)}$$

For the limit to be unique, the above equations must be equal, or

$$\frac{\partial u}{\partial x} = \frac{\partial v}{\partial y} \qquad \textbf{(B.3.4)}$$

$$\frac{\partial u}{\partial y} = -\frac{\partial v}{\partial x} \qquad \textbf{(B.3.5)}$$

These equations which u and v must satisfy are known as the *Cauchy-Riemann equations*. They are the necessary conditions for the existence of the derivative. Thus an analytic function must satisfy the Cauchy-Riemann equations. Moreover, if $f(z)$ has a derivative at a given point z_0, it must also be continuous and single-valued at that point. The converse is not necessarily true, however.

To illustrate these results, let us show that $\sin(z)$ is analytic.

$$w = \sin(z) \qquad \textbf{(B.3.6)}$$

$$u + iv = \sin(x + iy) \qquad \textbf{(B.3.7)}$$

$$= \sin(x)\cos(iy) + \cos(x)\sin(iy) \qquad \textbf{(B.3.8)}$$

$$= \sin(x)\cosh(y) + i\cos(x)\sinh(y) \qquad \textbf{(B.3.9)}$$

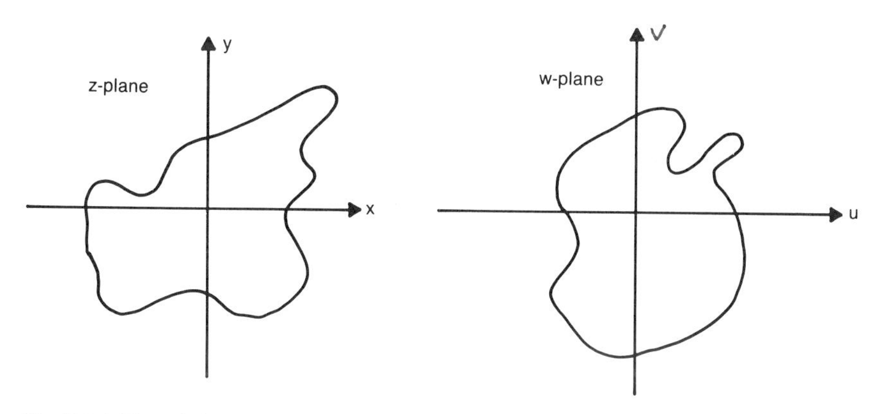

Fig. B.3.1: The relationship between the z-plane and w-plane given by $w = f(z)$.

so that

$$u = \sin(x)\cosh(y) \tag{B.3.10}$$

and

$$v = \cos(x)\sinh(y). \tag{B.3.11}$$

Differentiating each partially with respect to x and y in turn, we have

$$\frac{\partial u}{\partial x} = \cos(x)\cosh(y) \qquad \frac{\partial u}{\partial y} = \sin(x)\sinh(y) \tag{B.3.12}$$

$$\frac{\partial v}{\partial x} = -\sin(x)\sinh(y) \qquad \frac{\partial v}{\partial y} = \cos(x)\cosh(y) \tag{B.3.13}$$

and the Cauchy-Riemann equations are satisfied. The function $w = \sin(z)$ is analytic.

As a second example, consider

$$w = z^2 \tag{B.3.14}$$

Here

$$u + iv = (x + iy)^2 \tag{B.3.15}$$

$$= x^2 - y^2 + 2ixy \tag{B.3.16}$$

and $u(x,y) = x^2 - y^2$ and $v(x,y) = 2xy$. Therefore

$$\frac{\partial u}{\partial x} = 2x = \frac{\partial v}{\partial y} \tag{B.3.17}$$

and

$$\frac{\partial v}{\partial x} = 2y = -\frac{\partial u}{\partial y} \tag{B.3.18}$$

and the Cauchy-Riemann conditions are satisfied.

Finally, if

$$w = \frac{1}{z} \tag{B.3.19}$$

so that

$$u + iv = \frac{1}{x + iy} \tag{B.3.20}$$

$$= \frac{x}{x^2 + y^2} - \frac{iy}{x^2 + y^2}, \tag{B.3.21}$$

then

$$u = \frac{x}{x^2 + y^2} \text{ and } v = -\frac{y}{x^2 + y^2} \qquad \textbf{(B.3.22)}$$

$$\frac{\partial u}{\partial x} = \frac{(x^2 + y^2) - 2x^2}{(x^2 + y^2)^2} = \frac{y^2 - x^2}{(x^2 + y^2)^2} \qquad \textbf{(B.3.23)}$$

$$\frac{\partial v}{\partial y} = -\frac{(x^2 + y^2) - 2y^2}{(x^2 + y^2)^2} = \frac{y^2 - x^2}{(x^2 + y^2)^2} = \frac{\partial u}{\partial x} \qquad \textbf{(B.3.24)}$$

$$\frac{\partial v}{\partial x} = -\frac{0 - 2xy}{(x^2 + y^2)^2} = \frac{2xy}{(x^2 + y^2)^2} \qquad \textbf{(B.3.25)}$$

$$\frac{\partial u}{\partial y} = \frac{0 - 2xy}{(x^2 + y^2)^2} = -\frac{\partial v}{\partial x} \qquad \textbf{(B.3.26)}$$

Function is analytic at all points except the origin because partial derivatives cease to exist when both x and y are zero. At the origin the modulus of w becomes infinite. It has a singularity at the origin but is analytic everywhere else.

If a function is analytic, and therefore differentiable, we can obtain the derivative from any of the forms below:

$$\frac{dw}{dz} = \frac{\partial u}{\partial x} + i\frac{\partial v}{\partial x} \qquad \textbf{(B.3.27)}$$

$$= \frac{\partial v}{\partial y} - i\frac{\partial u}{\partial y} \qquad \textbf{(B.3.28)}$$

$$\frac{dw}{dz} = \frac{\partial v}{\partial y} + i\frac{\partial v}{\partial x} \qquad \textbf{(B.3.29)}$$

$$= \frac{\partial u}{\partial x} - i\frac{\partial u}{\partial y} \qquad \textbf{(B.3.30)}$$

For example, the derivative of $\sin(z)$ is

$$\frac{d}{dz}(\sin(z)) = \frac{\partial u}{\partial x} + i\frac{\partial v}{\partial x} \qquad \textbf{(B.3.31)}$$

$$= \cos(x)\cosh(y) - i\sin(x)\sinh(y) \qquad \textbf{(B.3.32)}$$

$$= \cos(x + iy) = \cos(z) \qquad \textbf{(B.3.33)}$$

$$\frac{d}{dz}(z^2) = 2x + 2yi \qquad \textbf{(B.3.34)}$$

$$= 2(x + iy) = 2z \qquad \textbf{(B.3.35)}$$

$$\frac{d}{dz}\left(\frac{1}{z}\right) = \frac{y^2 - x^2}{(x^2 + y^2)^2} + \frac{2ixy}{(x^2 + y^2)^2} \qquad \textbf{(B.3.36)}$$

$$= \frac{-1}{(x + iy)^2} = -\frac{1}{z^2} \qquad \textbf{(B.3.37)}$$

The results are the same as they would have been had z been real, and these examples are typical of an important general result.

An analytic function of a complex variable z behaves like a function of a real variable, as far as analytical results are concerned; and most of the results of real variable theory, including rules for differentiation, trigonometrical and hyperbolic identities, expansion in series, etc., may be carried over without alteration from real variable theory. Of particular importance is the applicability of l'Hospital rule with complex variables

$$\lim_{z \to z_0} \frac{f(z)}{g(z)} = \frac{f'(z_0)}{g'(z_0)} \qquad \textbf{(B.3.38)}$$

where $f(z)$ and $g(z)$ must be analytic at $z = z_0$. For example,

$$\lim_{z \to i} \frac{z^{10} + 1}{z^6 + 1} = \lim_{z \to i} \frac{10z^9}{6z^5} = \frac{5}{3} \lim_{z \to i} z^4 = \frac{5}{3} \qquad \textbf{(B.3.39)}$$

Although the Cauchy-Riemann equations provide us the means to determine whether a function is analytic, sometimes we only have either $u(x,y)$ or $v(x,y)$. However, by cross differentiation

$$u_{xx} = v_{xy} = -u_{yy} \qquad \textbf{(B.3.40)}$$

and

$$v_{xx} = -u_{xy} = v_{yy}. \qquad \textbf{(B.3.41)}$$

Therefore

$$u_{xx} + u_{yy} = 0 \qquad \textbf{(B.3.42)}$$

and

$$v_{xx} + v_{yy} = 0. \qquad \textbf{(B.3.43)}$$

That is, both $u(x,y)$ and $v(x,y)$ satisfy Laplace's equation if

$$f(z) = u + iv$$

is analytic. Consequently, if either u or v satisfies Laplace's equation, then the function is analytic. Because $u(x,y)$ and $v(x,y)$ satisfy Laplace's equation, they are each called *harmonic functions* and taken together are called *conjugate harmonic functions*.

To illustrate these concepts, let us show that

$$u(x,y) = e^{-x} [x \sin(y) - y \cos(y)]$$

is harmonic and find $v(x,y)$ such that $f(z)$ is analytic. Because

$$u_{xx} = -2 e^{-x} \sin(y) + x e^{-x} \sin(y) - y e^{-x} \cos(y) \qquad \textbf{(B.3.44)}$$

and

$$u_{yy} = -x e^{-x} \sin(y) + 2 e^{-x} \sin(y) + y e^{-x} \cos(y), \qquad \textbf{(B.3.45)}$$

then $u_{xx} + u_{yy} = 0$. From the Cauchy-Riemann equations,

$$v_y = u_x = e^{-x} \sin(y) - x e^{-x} \sin(y) + y e^{-x} \cos(y) \qquad \textbf{(B.3.46)}$$

$$v_x = -u_y = e^{-x} \cos(y) - x e^{-x} \cos(y) - y e^{-x} \sin(y) \qquad \textbf{(B.3.47)}$$

Integrating Eq. (B.3.46) with respect to y,

$$v(x,y) = y e^{-x} \sin(y) + x e^{-x} \cos(y) + g(x) \qquad \textbf{(B.3.48)}$$

Using Eq. (B.3.47),

$$-y e^{-x} \sin(y) - x e^{-x} \cos(y) + e^{-x} \cos(y) + g'(x)$$
$$= -y e^{-x} \sin(y) - x e^{-x} \cos(y) - y e^{-x} \sin(x). \qquad \textbf{(B.3.49)}$$

Therefore $g'(x) = 0$ or $g(x) =$ constant. Consequently

$$v(x,y) = e^{-x} [y \sin(y) + x \cos(y)] + \text{constant}. \qquad \textbf{(B.3.50)}$$

Exercises

1. Show that the following functions are analytic.
   (a) $f(z) = iz + 2$
   (b) $f(z) = e^{-x} [\cos(y) - i \sin(y)]$
   (c) $f(z) = z^3$
   (d) $f(z) = \cos(x) \cosh(y) - i \sin(x) \sinh(y)$
2. Find the derivative of the following functions.
   (a) $(1 + z^2)^{3/2}$
   (b) $(z + 2z^{1/2})^{1/3}$
   (c) $(1 + 4i)z^2 - 3z - 2$
   (d) $\dfrac{2z - i}{z + 2i}$
   (e) $(iz - 1)^{-3}$
3. Find the following limits.
   (a) $\lim_{z \to i} \dfrac{z^2 - 2iz - 1}{z^4 + 2z^2 + 1}$

(b) $\lim_{z \to 0} \dfrac{z - \sin(z)}{z^3}$

4. Show that the function z^* is nowhere differentiable.

B.4 ELEMENTARY ANALYTIC FUNCTIONS

(a) Exponential function $\exp(z)$.

We define the exponential function $\exp(z)$ in terms of the real-valued functions by the equation

$$\exp(z) = e^x [\cos(y) + i \sin(y)]. \quad \textbf{(B.4.1)}$$

Note that the $\exp(z)$ is single-valued for each z. The exponential function is an entire function, its derivative equals $\exp(z)$ and the law of exponents apply. Because

$$\exp(z + 2\pi i) = \exp(z) \exp(2\pi i) = \exp(z)$$

the exponential function is periodic with a period of $2\pi i$.

To illustrate the exponential function, let us find all of the roots of $\exp(z) = -1$. Because

$$e^x [\cos(y) + i \sin(y)] = -1, \quad \textbf{(B.4.2)}$$

$$e^x \cos(y) = -1 \quad \textbf{(B.4.3)}$$

and

$$e^x \sin(y) = 0. \quad \textbf{(B.4.4)}$$

Because e^x never equals zero, $\sin(y) = 0$ and

$$y = \pm n\pi, \ n = 0,1,2,3, \ldots$$

Because e^x is never negative, n must be an odd power and $x = 0$. Therefore the roots are $z = (\pi \pm 2n\pi)i, \ n = 0,1,2,3, \ldots$

(b) The trigonometric functions $\sin(z)$ and $\cos(z)$.

By definition

$$\cos(z) = \frac{1}{2}(e^{iz} + e^{-iz}) \quad \textbf{(B.4.5)}$$

and

$$\sin(z) = \frac{1}{2i}(e^{iz} - e^{-iz}). \quad \textbf{(B.4.6)}$$

Both $\sin(z)$ and $\cos(z)$ are entire functions. It is easily shown that

$$\frac{d}{dz}(\sin(z)) = \cos(z) \quad \textbf{(B.4.7)}$$

and

$$\frac{d}{dz}(\cos(z)) = -\sin(z). \quad \textbf{(B.4.8)}$$

Furthermore,

$$\cos(z) = \cos(x + iy) = \cos(x)\cosh(y) - i\sin(x)\sinh(y) \quad \textbf{(B.4.9)}$$

$$\sin(z) = \sin(x + iy) = \sin(x)\cosh(y) + i\cos(x)\sinh(y) \quad \textbf{(B.4.10)}$$

From these formulas, we have

$$\sin(iy) = i\sinh(y) \quad \textbf{(B.4.11)}$$

$$\cos(iy) = \cosh(y). \quad \textbf{(B.4.12)}$$

if we set $x = 0$.

These formulas may be used to find all of the roots of $\sin(z) = \cosh(4)$. Then

$$\sin(z) = \sin(x)\cosh(y) + i\cos(x)\sinh(y) = \cosh(4). \quad \textbf{(B.4.13)}$$

Taking the real and imaginary parts, we have

$$\sin(x)\cosh(y) = \cosh(4) \quad \textbf{(B.4.14)}$$

and

$$\cos(x)\sinh(y) = 0. \quad \textbf{(B.4.15)}$$

If $y = \pm 4$ and $x = (1 \pm 2n)\pi/2$, $n = 0,1,2,3, \ldots$, these equations are satisfied. Therefore $z = (1 \pm 4n)\pi/2 \pm 4i$ with $n = 0,1,2,3, \ldots$

(c) The hyperbolic functions $\sinh(z)$ and $\cosh(z)$.
By definition

$$\sinh(z) = (e^{z} - e^{-z})/2 \quad \textbf{(B.4.16)}$$

and

$$\cosh(z) = (e^{z} + e^{-z})/2. \quad \textbf{(B.4.17)}$$

Then

$$\frac{d}{dz}[\sinh(z)] = \cosh(z) \quad \textbf{(B.4.18)}$$

and

$$\frac{d}{dz}[\cosh(z)] = \sinh(z). \quad \textbf{(B.4.19)}$$

From the definition of $\sin(z)$, $\cos(z)$, $\sinh(z)$ and $\cosh(z)$, we have

$$\sinh(iz) = i\sin(z) \quad \textbf{(B.4.20)}$$

$$\cosh(iz) = \cos(z) \quad \textbf{(B.4.21)}$$

$$\sin(z) = i\sinh(z) \quad \textbf{(B.4.22)}$$

$$\cos(iz) = \cosh(z) \quad \textbf{(B.4.23)}$$

$$\sinh(z) = \sinh(x)\cos(y) + i\cosh(x)\sin(y) \quad \textbf{(B.4.24)}$$

$$\cosh(z) = \cosh(x)\cos(y) + i\sinh(x)\sin(y). \qquad \textbf{(B.4.25)}$$

There formuli can be used to find all of the roots of $\sinh(z) = 1$. Because

$$\sinh(z) = \sinh(x)\cos(y) + i\cosh(x)\sin(y) = i, \qquad \textbf{(B.4.26)}$$

the real and imaginary parts give

$$\sinh(x)\cos(y) = 0 \qquad \textbf{(B.4.27)}$$

and

$$\cosh(x)\sin(y) = 1. \qquad \textbf{(B.4.28)}$$

Because $\cosh(x) \neq 0$, $\sin(y) = 1$ or $y = (1 \pm 4n)\pi/2$ if $x = 0$. These values satisfy both Eqs. (B.4.26) and (B.4.27). Therefore

$$z = (1/2 \pm 2n)\pi i$$

if $n = 0,1,2,3, \ldots$

Exercises

Find all of the roots of the following equations.

1. $\exp(z) = -2$.
2. $\cosh(z) = 1/2$.
3. $\sin(z) = 2$.
4. $\cos(z) = i$.

B.5 LINE INTEGRALS: THE CAUCHY-GOURSAT THEOREM

So far, we have talked about complex numbers, complex functions and complex differentiation. We are now ready for integration.

Corresponding to the integral of a real variable, the integral of a complex variable between two (complex) limits can be defined. Because the z-plane is two-dimensional there is clearly greater freedom and hence ambiguity in what is meant by a complex integral. For example, we might ask whether the integral of some function between A and B depends upon the curve along which the integration is taken. What will be found is that in general it does. Consequently an important ingredient in any complex integration is the *contour* that is followed from point A to B.

We now ask how do we compute

$$\int_C f(z)\, dz? \qquad \textbf{(B.5.1)}$$

Let us deal with the definition now; the actual method will be illustrated by examples.

The key to treating complex line integrals is to break everything up into real and imaginary parts and so reduce to the familiar case of line integrals of real-valued functions which we know how to handle. Thus, we write

$$f(z) = u(x,y) + i\, v(x,y) \qquad \textbf{(B.5.2)}$$

as usual, and because $z = x + iy$, we have $dz = dx + idy$. Therefore,

$$\int_C f(z)\, dz = \int_C [u(x,y) + i\, v(x,y)]\, (dx + i\, dy) \qquad \textbf{(B.5.3)}$$

$$= \int_C u(x,y)\, dx - v(x,y)\, dy$$

$$+ i \int_C v(x,y)\, dx + u(x,y)\, dy. \qquad \textbf{(B.5.4)}$$

The exact method used to evaluate this integral depends upon the exact path specified. For example, let us evaluate $\int_C z^*\, dz$ from $z = 0$ to $z = 4 + 2i$ along the curve C given by (a) $z = t^2 + it$ and (b) the line from $z = 0$ to $z = 2i$ and then the line from $z = 2i$ to $z = 4 + 2i$.

For the first case, the points $z = 0$ and $z = 4 + 2i$ on C correspond to $t = 0$ and $t = 2$, respectively. Then the line integral equals

$$\int_0^2 (t^2 + it)^*\, d(t^2 + it) = \int_0^2 (t^2 - it)\, (2t + i)\, dt$$

$$= \int_0^2 (2t^3 - it^2 + t)\, dt \qquad \textbf{(B.5.5)}$$

$$= 10 - \frac{8}{3}\, i. \qquad \textbf{(B.5.6)}$$

The line integral for the second part equals

$$\int_C (x - iy)\, (dx + i\, dy) = \int_C x\, dx + y\, dy$$

$$+ i \int_C x\, dy - y\, dx \qquad \textbf{(B.5.7)}$$

The line from $z = 0$ to $z = 2i$ is the same as the line from (0,0) to (0,2) for which $x = 0$, $dx = 0$ and the line integral equals

$$\int_{y=0}^{y=2} [(0)\,(0) + y\, dy] + i \int_{y=0}^{y=2} [(0)\,(dy) - y\,(0)] = \int_{y=0}^{2} y\, dy = 2 \qquad \textbf{(B.5.8)}$$

The line from $z = 2i$ to $z = 4 + 2i$ is the same as the line from (0,2) to (4,2) for which $y = 2$, $dy = 0$ and the line integral equals

$$\int_{x=0}^{x=4} \left[x\,dx + (2)(0) \right] + i \int_{x=0}^{x=4} \left[(x)\,(0) - 2\,dx \right]$$

$$= \int_0^4 x\,dx + i \int_0^4 -2\,dx = 8 - 8i. \quad \textbf{(B.5.9)}$$

Then the required value is equal to $2 + 8 - 8i = 10 - 8i$. Note that $f(z) = z^*$ is *not* an analytic function. We found that two different line integrals between the points (0,0) to (4,2) yield different results.

Let us now do an integration with z^2 as the integrand along two paths from 0 to $2 + i$ shown in Fig. B.5.1. For the first integration, $x = 2t$ and $y = t$ where $0 \le t \le 1$. Because $z^2 = x^2 - y^2 + 2xy\,i$,

$$\int_{C_1} z^2\,dz = \int_{t=0}^{t=1} (4t^2 - t^2 + 4t^2 i)(2\,dt + i\,dt) \quad \textbf{(B.5.10)}$$

$$= \int_0^1 (3t^2 + 4t^2\,i)\,(2\,dt + i\,dt) \quad \textbf{(B.5.11)}$$

$$= \int_0^1 6t^2\,dt + 8t^2 i\,dt + 3t^2\,i\,dt - 4t^2\,dt \quad \textbf{(B.5.12)}$$

$$= \int_0^1 2t^2\,dt + 11t^2\,i\,dt$$

$$= \frac{2}{3}\,t^3 \Big|_0^1 + \frac{11}{3}\,i\,t^3 \Big|_0^1 = \frac{2}{3} + \frac{11}{3}\,i \quad \textbf{(B.5.13)}$$

For the second path the integral is taken along the straight line $y = 0$ and then the straight line $x = 2$. Then

$$\int_{C_2} z^2\,dz = \int_{x=0}^{x=2} \underset{y=0}{(x + iy)^2}\,\underset{dy=0}{(dx + idy)} + \int_{y=0}^{y=1} \underset{x=2}{(x + iy)^2}\,\underset{dx=0}{(dx + idy)}$$

(B.5.14)

$$= \int_0^2 x^2\,dx + i \int_0^1 (2 + iy)^2\,dy \quad \textbf{(B.5.15)}$$

$$= \frac{1}{3}\,x^3 \Big|_0^2 + i(4y + 4i\,\frac{y}{2} - \frac{y^3}{3}) \Big|_0^1 \quad \textbf{(B.5.16)}$$

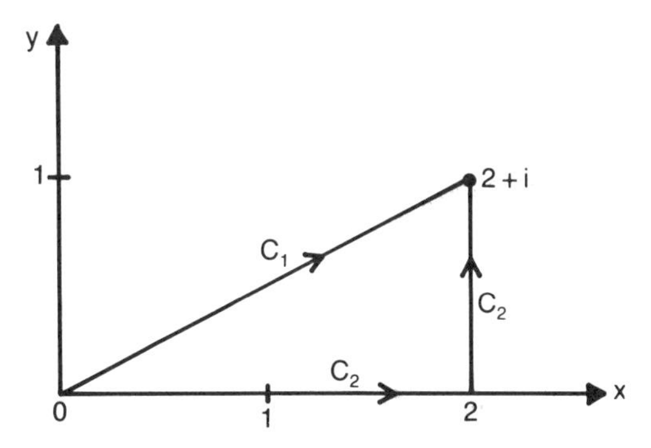

Fig. B.5.1: Contour used in a line integration given by Eqs. (B.5.9)-(B.5.16).

$$\int_{C_2} z^2\, dz = \frac{8}{3} + 4i - 2 - \frac{1}{3}\, i = \frac{2}{3} + \frac{11}{3}\, i \qquad \textbf{(B.5.17)}$$

We note that the results from two different paths are the same for this analytic function. The results from these two examples can be generalized into the following theorem:

The Cauchy-Goursat theorem: If a function $f(z)$ is analytic within and upon a closed contour, the line integral taken around the contour is zero.

$$\oint f(z)\, dz = 0 \qquad \textbf{(B.5.18)}$$

The notation $\oint$ denotes an integration around a closed curve where the integration is to proceed in the counterclockwise direction.

To prove this theorem, let the curve C be the specified contour around which $w = f(z)$ is to be integrated. We divide the region within C into a series of rectangles. If we consider the integration around the rectangle to be the product of the average value of w of each side and its length, we obtain

$$\left(w + \frac{\partial w}{\partial x}\frac{dx}{2}\right) dx + \left(w + \frac{\partial w}{\partial x}\, dx + \frac{\partial w}{\partial iy}\frac{d(iy)}{2}\right) d(iy)$$

$$+ \left(w + \frac{\partial w}{\partial x}\frac{dx}{2} + \frac{\partial w}{i\,\partial y}\, i\, dy\right)(-dx) + \left(w + \frac{\partial w}{i\partial y}\frac{i\, dy}{2}\right)(-idy)$$

$$= \left(\frac{\partial w}{\partial x} - \frac{\partial w}{i\partial y}\right)(idxdy) \qquad \textbf{(B.5.19)}$$

Substituting $w = u + iv$ into Eq. (B.5.18), we obtain the equation

$$\frac{\partial w}{\partial x} - \frac{\partial w}{i\,\partial y} = \left(\frac{\partial u}{\partial x} - \frac{\partial v}{\partial y}\right) + i\left(\frac{\partial v}{\partial x} + \frac{\partial u}{\partial y}\right). \qquad \textbf{(B.5.20)}$$

Because the function is analytic, the right-hand side of Eq. (B.5.19) is zero and the integration around each of these rectangles is zero.

We note next that in integrating around adjoining rectangles, each side is transversed in opposite directions, the net result being equivalent to integrating around the outer curve C. We therefore arrive at the result

$$\oint_C f(z)\,dz = 0 \qquad \textbf{(B.5.21)}$$

when the function $w = f(z)$ is analytic within and on the closed contour is zero. A corollary to the Cauchy-Goursat theorem states that if $f(z)$ is analytic in a given region, the line integral between two points in this region is independent of the path taken. This result was illustrated in the previous example.

To illustrate the Cauchy-Goursat theorem, let us integrate the function $w = z^2$ around the closed square contour shown in Fig. B.5.2. Along AB, $z = x$, $dz = dx$ and x varies from 0 to 1, so that

$$\int_{AB} w\,dz = \int_0^1 x^2\,dx = \frac{x^3}{3}\Bigg|_0^1 = \frac{1}{3} \qquad \textbf{(B.5.22)}$$

Along BC, $z = 1 + iy$, $dz = i\,dy$ and y varies from 0 to 1, therefore

$$\int_{BC} w\,dz = \int_0^1 (1 + iy)^2\,idy$$

$$= i\int_0^1 (1-y^2)\,dy - 2\int_0^1 y\,dy \qquad \textbf{(B.5.23)}$$

$$= i\left(y - \frac{y^3}{3}\right)\Bigg|_0^1 - 2\left(\frac{y^2}{2}\right)\Bigg|_0^1 = \frac{2}{3}i - 1 \qquad \textbf{(B.5.24)}$$

Along CD, $z = x + i$, $dz = dx$ and x varies from 1 to 0 (note the direction of the integration),

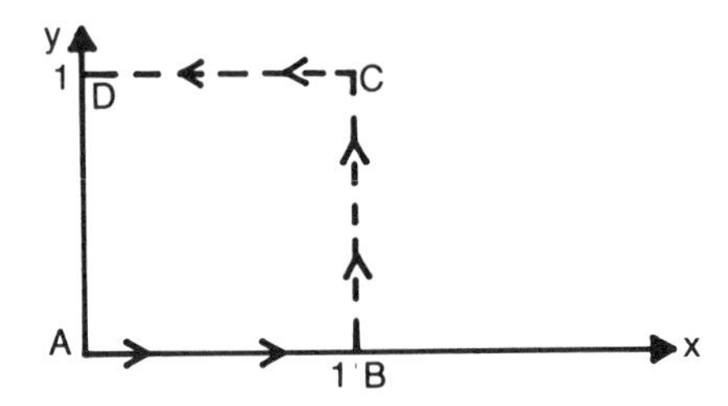

Fig. B.5.2: Contour used in the line integration (B.5.27).

$$\int_{CD} w\, dz = \int_1^0 (x+i)^2\, dx$$

$$= \int_1^0 (x^2 - 1)\, dx + 2i \int_1^0 x\, dx \qquad \textbf{(B.5.25)}$$

$$= \left(\frac{x^3}{3} - x\right)\Bigg|_1^0 + 2i \left(\frac{x^2}{2}\right)\Bigg|_1^0 = \frac{2}{3} - i. \qquad \textbf{(B.5.26)}$$

Finally, along DA, $z = iy$, $dz = i\, dy$ and y varies from 1 to 0,

$$\int_{DA} w\, dz = \int_1^0 (iy)^2\, i\, dy = -i \int_1^0 y^2\, dy$$

$$= -i\, \frac{y^3}{3}\Bigg|_1^0 = \frac{i}{3} \qquad \textbf{(B.5.27)}$$

Adding together these four separate results, the total integral round the complete square is

$$\oint_{ABCD} z^2\, dz = \frac{1}{3} + \left(\frac{2i}{3} - 1\right) + \left(\frac{2}{3} - i\right) + \frac{i}{3} = 0. \qquad \textbf{(B.5.28)}$$

Because $w = z^2$ is analytic at all finite points in the z-plane, Cauchy's theorem states that this integral must be zero.

For a second example, let us integrate the function $w = z$, which is obviously analytic, over two different paths between the two points $z = 0$ and $z = i$. Our first path is the direct straight path along the imaginary axis OA, where $z = iy$, $dz = i\, dy$ and y varies from 0 to 1.

$$\int_{OA} z\, dz = \int_0^1 iy\, i\, dy = -\int_0^1 y\, dy$$

$$= -y^2/2\,\Bigg|_0^1 = -1/2 \qquad \textbf{(B.5.29)}$$

As a second path let us take $OBCDA$ (Fig. B.5.3) where the contour goes left from 0 to the point $z = -1$ and to A via a circle of unit radius centered at the origin.

Along OB, $z = x$, $dz = dx$ and x varies from 0 to -1 and

$$\int_{OB} z\, dz = \int_0^{-1} x\, dx = x^2/2\,\Bigg|_0^{-1} = 1/2. \qquad \textbf{(B.5.30)}$$

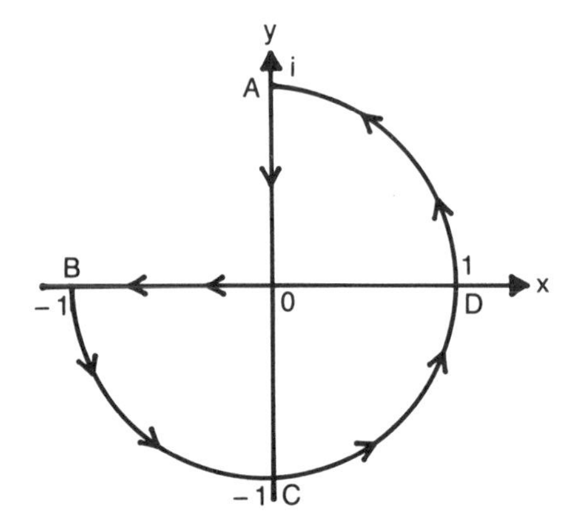

Fig. B.5.3: Contour used in the line integration (B.5.30).

On the unit circle, $z = e^{i\theta}$, $dz = i\,e^{i\theta}\,d\theta$ and varies from $-\pi$ to $\pi/2$.

$$\int_{BCDA} z\,dz = \int_{-\pi}^{\pi/2} e^{i\theta}\, i\, e^{i\theta}\,d\theta = i\int_{-\pi}^{\pi/2} e^{2i\theta}\,d\theta \quad \textbf{(B.5.31)}$$

$$= i\,\frac{e^{2i\theta}}{2i}\Bigg|_{-\pi}^{\pi/2} = e^{2i\theta}/2\Bigg|_{-\pi}^{\pi/2}$$

$$= [\cos(2\theta) + i\,\sin(2\theta)]/2\Bigg|_{-\pi}^{\pi/2} \quad \textbf{(B.5.32)}$$

$$= -\,1. \quad \textbf{(B.5.33)}$$

Thus, the integral for the path $OBCDA$ equals the sum of the integrals (B.5.29) and (B.5.32) or $-1/2$ as before.

Finally, let us integrate $w = z^{-1}$ around the unit circle centered at the origin. Here $z = e^{i\theta}$ on the circle, $dz = i\,e^{i\theta}\,d\theta$ and θ varies from 0 to 2π.

$$\oint z^{-1}\,dz = \int_{0}^{2\pi} e^{-i\theta}\, i\, e^{i\theta}\,d\theta = i\int_{0}^{2\pi} d\theta = 2\pi i \quad \textbf{(B.5.34)}$$

The result is not zero because z^{-1} has a singular point inside the contour; it ceases to be analytic at the origin. In the next sections we shall learn how to deal with these singular points.

Exercises

1. Evaluate $\oint (x-y)\,dz$ around the unit circle with its center at the origin.
2. Evaluate $\oint dz/z^2$ around a circle of radius r centered at the origin.

3. Show that $\oint \frac{dz}{(z-a)^n} = \begin{cases} 0 & \text{for } n \text{ an integer not equal to one} \\ 2\pi i & \text{for } n = 1. \end{cases}$

4. (a) $\oint \frac{dz}{z-3}$ (b) $\oint \frac{z\,dz}{\cos(z)}$ (c) $\oint \frac{e^z\,dz}{z-2}$ (d) $\oint e^{-z}\,dz$

where the integral is around the circle of radius 1 centered at $z = 0$.

B.6 SINGULARITIES

Points where a function ceases to be analytic are called singularities. For a single-valued function, there are two kinds of singularities.

(a) *Poles* (Unessential singularities). Poles are points where the function becomes infinite to a finite order. That is, the singularity can always be removed by multiplying the function by a suitable factor of finite index. For example, the function

$$f(z) = \frac{z}{(z - a)^n} \quad \textbf{(B.6.1)}$$

for n a finite integer has a pole of order n at $z = a$, which can be removed by multiplying the function by $(z - a)^n$. Thus

$$\lim_{z \to a} (z - a)^n f(z) \neq \infty . \quad \textbf{(B.6.2)}$$

(b) Essential singularity (Poles of infinite order). For an essential singularity, the limit

$$\lim_{z \to a} (z - a)^n f(z) \quad \textbf{(B.6.3)}$$

does not exist for a finite value of n. An example is

$$e^{1/z} = 1 + \frac{1}{z} + \frac{1}{2!z^2} + \frac{1}{3!z^3} + \ldots \quad \textbf{(B.6.4)}$$

which has an essential singularity at $z = 0$.

B.7 CAUCHY'S INTEGRAL FORMULA

We begin our study of how to integrate a function that contains a singularity by finding

$$\oint \frac{f(z)}{z-a}\,dz \quad \textbf{(B.7.1)}$$

where $f(z)$ is analytic within and on the contour C and the point a lies within the contour.

We begin our analysis by replacing the contour C by the con-

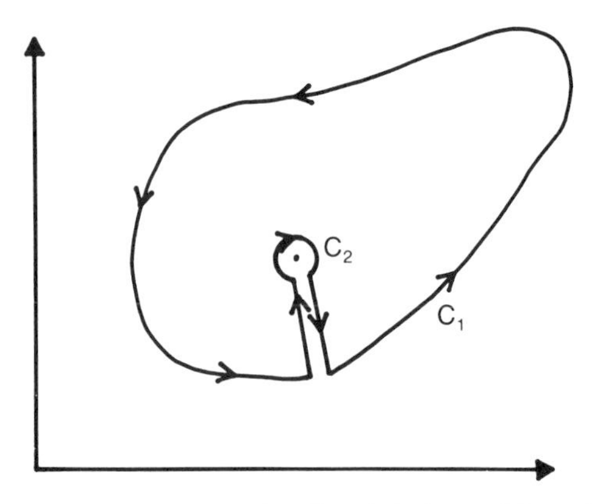

Fig. B.7.1: Contour used in proving Cauchy's integral formula.

tours C_1 and C_2 as shown in Fig. B.7.1. Because the integration as shown contains an analytic function, we have

$$\oint_{C_1} \frac{f(z)}{(z-a)}\, dz + \oint_{C_2} \frac{f(z)}{(z-a)}\, dz = 0 \tag{B.7.2}$$

Because the contribution to the integral from the path going into the singularity will cancel the contribution from the path going away from the singularity, the C_2 path can be replaced by an integration around a small circle of radius r centered at the point a. Furthermore, because $f(z)$ is analytic at a, we can approximate its value on the small circle by

$$f(z) \simeq f(a) + \epsilon \tag{B.7.3}$$

where ϵ is a small quantity. Substituting into Eq. (B.7.2), we obtain

$$\oint_{C_1} \frac{f(z)}{(z-a)}\, dz = -f(a) \int_{C_2} \frac{dz}{(z-a)} - \int_{C_2} \frac{\epsilon}{(z-a)}\, dz. \tag{B.7.4}$$

Because

$$z - a = r\, e^{i\theta} \tag{B.7.5}$$

$$dz = ir\, e^{i\theta}\, d\theta \tag{B.7.6}$$

Equation (B.7.4) becomes

$$\oint_{C_1} \frac{f(z)}{z-a} dz = f(a)\,(2\pi i) - i \int_{2\pi}^{0} \epsilon\, d\,\theta \tag{B.7.7}$$

We can now reduce the radius of the small circle so that $\epsilon \rightarrow 0$. Thus

$$f(a) = \frac{1}{2\pi i} \oint_{C} \frac{f(z)}{z-a}\, dz \tag{B.7.8}$$

where the contour C_1 approaches C. The function $f(z)$ at a is expressed by a contour integral around a path enclosing a.

This equation is known as Cauchy's integral formula. If we know the values of an analytic function at the boundary, we also know the value of the function at any point within the boundary. Because analytic functions are solutions to Laplace's equation, the equation suggests several physical interpretations. For example, the steady-state temperature at any point within a body is dependent on the temperature distribution on the boundary surface. The equilibrium deflection of a membrane stretched across a closed loop of wire depends on the shape of the wire and its elevation.

By taking n derivatives of Eq. (B.7.8), Cauchy's integral formula can be extended to

$$f^{(n)}(a) = \frac{n!}{2\pi i} \oint \frac{f(z)}{(z-a)^{n+1}}\, dz \qquad \textbf{(B.7.9)}$$

for $n = 1,2,3, \ldots$ (See Carrier, G. F., M. Krook and C. E. Pearson, 1966: *Functions of a complex variable*. New York: McGraw-Hill Book, Co., pp. 39-40 for the proof.)

To illustrate Cauchy's integral formula, let us find the value of the integral

$$\oint_C \frac{\sin(\pi z) + \cos(\pi z)}{(z-1)(z-2)}\, dz \qquad \textbf{(B.7.10)}$$

where C is the circle $|z| = 5$. Because

$$\frac{1}{(z-1)(z-2)} = \frac{1}{z-2} - \frac{1}{z-1}, \qquad \textbf{(B.7.11)}$$

we have

$$\oint_C \frac{\sin(\pi z) + \cos(\pi z)}{(z-1)(z-2)}\, dz = \oint_C \frac{\sin(\pi z) + \cos(\pi z)}{z-2}\, dz - \oint_C \frac{\sin(\pi z) + \cos(\pi z)}{z-1}\, dz \qquad \textbf{(B.7.12)}$$

By Cauchy's integral formula with $a = 2$ and $a = 1$, we have

$$\oint_C \frac{\sin(\pi z) + \cos(\pi z)}{z-2}\, dz = 2\pi i\,[\sin(2\pi) + \cos(2\pi)] = 2\pi i$$

$$\oint_C \frac{\sin(\pi z) + \cos(\pi z)}{z-1}\, dz = 2\pi i\,[\sin(1\pi) + \cos(1\pi)] = -2\pi i \qquad \textbf{(B.7.13)}$$

because $z = 1$ and $z = 2$ lie inside C and $\sin(\pi z) + \cos(\pi z)$ is analytic inside C. Then the required integral has the value

$$2\pi i - (-2\pi i) = 4\pi i.$$

Exercises

Evaluate

1. $\oint_C \frac{e^z}{z - 2}\, dz$ where C is the circle $|z| = 5$.

2. $\oint_C \frac{\sin(3z)}{z + \pi/2}\, dz$ where C is the circle $|z| = 3$.

3. $\oint_C \frac{\sin^6(z)}{z - \pi/6}\, dz$ where C is the circle $|z| = 2$.

4. $\oint_C \frac{e^{2z}}{(z + 1)^4}\, dz$ where C is the circle $|z| = 3$.

5. $\oint_C \frac{e^{iz}}{z}\, dz$ where C is the circle $|z| = 1$.

6. Show that $\frac{1}{2\pi i} \oint_C \frac{e^{zt}}{z^2 + 1}\, dz = \sin(t)$ if $t>0$ and C is the circle $|z| = 2$.

B.8 THEORY OF RESIDUES

Let $f(z)$ be analytic on and within the closed contour C except at $z = a$, where there is a pole of order n. Then

$$f(z) = \frac{f(z)\,(z - a)^n}{(z - a)^n} = \frac{F(z)}{(z - a)^n} \qquad \textbf{(B.8.1)}$$

where $F(z)$ is analytic on and within the closed contour, including the point a. Because $F(z)$ is analytic at a, we can expand $F(z)$ about a by the Taylor series:

$$F(z) = F(a) + F'(a)(z-a) + \frac{1}{2!}\, F''(a)(z-a)^2 + \ldots$$

$$+ \frac{F^{(n)}(a)\,(z-a)^n}{n!} + \ldots \qquad \textbf{(B.8.2)}$$

where all the derivatives $F^{(n)}(a)$ exist. Substituting Eq. (B.8.2) into Eq. (B.8.1), we obtain the series

$$f(z) = \frac{F(a)}{(z-a)^n} + \frac{F'(a)}{(z-a)^{n-1}} + \frac{F''(a)}{2!(z-a)^{n-2}} + \ldots$$

$$+ \frac{F^{(n-1)}(a)}{(n-1)!\,(z-a)} + \frac{F^{(n)}(a)}{n!} + \frac{F^{(n+1)}(a)}{(n+1)}(z-a) + \ldots$$

(B.8.3)

which is referred to as the *Laurent expansion*. We now consider the integral of (B.8.3)

$$\frac{1}{2\pi i}\oint f(z)\,dz = \frac{F(a)}{2\pi i}\oint \frac{dz}{(z-a)^n} + \frac{F'(a)}{2\pi i}\oint \frac{dz}{(z-a)^{n-1}} + \ldots$$

$$+ \frac{F^{(n-1)}(a)}{(n-1)!2\pi i}\oint \frac{dz}{z-a} + \frac{F^{(n)}(a)}{n!\,2\pi i}\oint dz$$

$$+ \frac{F^{(n+1)}(a)}{(n+1)!2\pi i}\oint (z-a)\,dz + \ldots$$

$$+ \frac{F^{(n+m)}(a)}{(n+m)!2\pi i}\oint (z-a)^m\,dz + \ldots \qquad \textbf{(B.8.4)}$$

In Eq. (B.8.4) all integrals on the right-hand side with the exception of one are zero. For example,

$$\oint (z-a)^n\,dz = 0 \text{ for } n = 0, 1, 2, \ldots \qquad \textbf{(B.8.5)}$$

because the function is analytic. From Cauchy's integral formula (B.7.9)

$$\oint \frac{dz}{(z-a)^n} = \begin{cases} 0 & \text{for } n \neq 1 \\ 2\pi i & \text{for } n = 1. \end{cases} \qquad \textbf{(B.8.6)}$$

Consequently, Eq. (B.8.4) reduces to

$$\frac{1}{2\pi i}\oint f(z)\,dz = \frac{F^{(n-1)}(a)}{(n-1)!} = R^{(n)}. \qquad \textbf{(B.8.7)}$$

The right-hand side of this equation is the coefficient of the $(z-a)^{-1}$ term of the Laurent expansion given by Eq. (B.8.7). It is the only coefficient of the Laurent expansion which affects the value of the integral of $f(z)$ around a closed contour C and it is called the *residue* $R^{(n)}$ of $f(z)$.

If there is more than one pole on or within the contour C, we can deform the path of integration as shown in Fig. B.8.1. The deformed contour C' now excludes all of the pole. Therefore

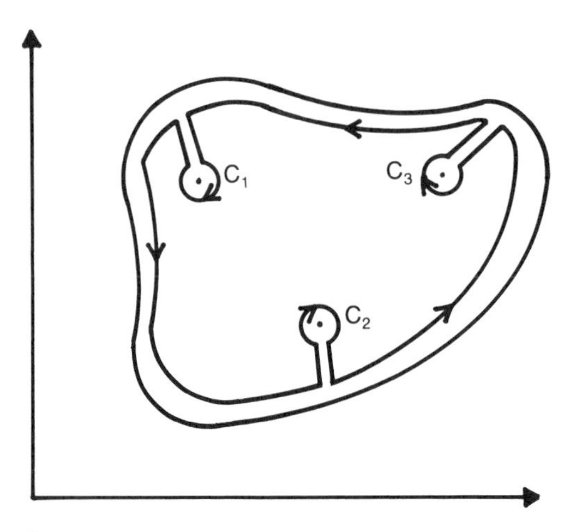

Fig. B.8.1: Contour used in proving the residue theorem.

$$\frac{1}{2\pi i}\oint_{C'} f(z)\,dz = 0 = \frac{1}{2\pi i}\oint_{C} f(z)\,dz - \frac{1}{2\pi i}\oint_{C_1} f(z)\,dz$$

$$-\frac{1}{2\pi i}\oint_{C_2} f(z)\,dz + \ldots \qquad \textbf{(B.8.8)}$$

Thus the integral around the outer contour C is

$$\frac{1}{2\pi i}\oint_{C} f(z)\,dz = R_1 + R_2 + \ldots \qquad \textbf{(B.8.9)}$$

Consequently, if $f(z)$ is analytic except for singularities at the poles, the integral around the closed path is equal to the sum of the residues of the poles. From Eqs. (B.8.1) and (B.8.7), the residue for the nth-order pole is

$$R^{(n)} = \frac{F^{(n-1)}(a)}{(n-1)!} = \lim_{z\to a} \frac{d^{n-1}}{dz^{n-1}}\left[\frac{f(z)\,(z-a)^n}{(n-1)!}\right] \qquad \textbf{(B.8.10)}$$

The function $f(z)$ is first multiplied by $(z-a)^n/(n-1)!$, differentiated with respect to z, $(n-1)$ times, after which we let $z = a$. For a simple pole, this reduces to

$$R^{(1)} = \lim_{z\to a} f(z)\,(z-a). \qquad \textbf{(B.8.11)}$$

In some cases the denominator of $f(z)$ cannot be readily factored. In those cases, let $f(z) = A(z)/B(z)$ contain only simple poles a_k. Then from Eq. (B.8.11) we can write

$$\Sigma R = \sum_{k=1}^{n} \lim_{z \to a_k} (z - a_k) \frac{A(z)}{B(z)} \qquad \textbf{(B.8.12)}$$

where a_k are the roots of $B(z) = 0$. Because the right-hand side of Eq. (B.8.12) is of the indeterminate form 0/0, l'Hospital's rule gives

$$\Sigma R = \sum_{k=1}^{n} \lim_{z \to a_k} \frac{(z - a_k)A'(z) + A(z)}{B'(z)} = \sum_{k=1}^{n} \frac{A(a_k)}{B'(a_k)}. \qquad \textbf{(B.8.13)}$$

Sometimes it is simpler to determine the residues by expansion of $f(z)$ into a Laurent series, particularly when $f(z)$ contains multiple-order poles.

To illustrate the residue theorem, let us compute

$$\int \frac{z^2}{z + 1} \, dz \qquad \textbf{(B.8.14)}$$

by the residue theorem. The function in this case is

$$f(z) = \frac{z^2}{z + 1} \qquad \textbf{(B.8.15)}$$

which has a simple pole at $z = -1$. From Eq. (B.8.11) the residue at $z = -1$ is

$$R = \lim_{z \to -1} (z + 1) \frac{z^2}{z + 1} = 1 \qquad \textbf{(B.8.16)}$$

and the integral around any closed path enclosing the point $z = -1$

$$\int \frac{z^2 \, dz}{z + 1} = 2\pi i. \qquad \textbf{(B.8.17)}$$

We can arrive at the same result by expanding $z^2/(z + 1)$ into a Laurent series about the pole $z = -1$. Letting $z + 1 = \lambda$, $z^2 = (\lambda - 1)^2$, and $dz = d\lambda$, we have

$$\frac{z^2}{z + 1} = \frac{(\lambda - l)^2}{\lambda} = \lambda - 2 + \frac{1}{\lambda} \cdot \qquad \textbf{(B.8.18)}$$

The residue is then the coefficient of $1/\lambda$, or 1, and we arrive at the value of $2\pi i$ for its integral.

For a second example, let us determine the residue of $f(z) = e^{iz}/(z^2 + a^2)$ where the contour encloses all the singularities. The poles are $z = \pm ia$, which are conjugate points on the imaginary axis. From Eq. (B.8.11), the residues are

$$R = \lim_{z \to -ia} \frac{e^{iz}}{(z - ia)} + \lim_{z \to ia} \frac{e^{iz}}{(z + ia)} \tag{B.8.19}$$

$$= -\frac{e^{a}}{2ia} + \frac{e^{-a}}{2ia} = \frac{i}{a} \sinh(a) \tag{B.8.20}$$

Consequently the integral around any closed contour which encloses the poles $z = \pm ia$ is

$$\oint \frac{e^{iz}\, dz}{z^2 + a^2} = \frac{2\pi i^2}{a} \sinh(a) = -\frac{2\pi}{a} \sinh(a). \tag{B.8.21}$$

Now, let us evaluate $\frac{1}{2\pi i} \oint_C \frac{e^{zt}}{z^2(z^2 + 2z + 2)} dz$ around the circle C with the equation $|z| = 5$. The integrand has a double pole at $z = 0$ and two simple poles at $z = -1 \pm i$ which are the roots of $z^2 + 2z + 2 = 0$. All of these poles lie within C. Therefore, the residue at $z = 0$ is

$$\lim_{z \to 0} \frac{1}{1!} \frac{d}{dz} \left\{ z^2 \left[\frac{e^{zt}}{z^2(z^2 + 2z + 2)} \right] \right\}$$

$$= \lim_{z \to 0} \frac{(z^2 + 2z + 2)\,(te^{zt}) - (e^{zt})(2z + 2)}{(z^2 + 2z + 2)^2} = \frac{t - 1}{2} \tag{B.8.22}$$

The residue at $z = -1 + i$ is

$$\lim_{z \to -1+i} \left\{ [z - (-1 + i)] \frac{e^{zt}}{z^2(z^2 + 2z + 2)} \right\}$$

$$= \lim_{z \to -1+i} \left(\frac{e^{zt}}{z^2} \right) \lim_{z \to -1+i} \left(\frac{z + 1 - i}{z^2 + 2z + 2} \right) \tag{B.8.23}$$

$$= \frac{e^{(-1+i)t}}{(-1+i)^2} \frac{1}{2i} = \frac{e^{(-1+i)t}}{4} \tag{B.8.24}$$

Similarly, for the residue at $z = -1 - i$, we have

$$\lim_{z \to -1-i} \left\{ [z - (-1 - i)] \frac{e^{zt}}{z^2(z^2 + 2z + 2)} \right\} = \frac{e^{(-1-i)t}}{4} \tag{B.8.25}$$

Then by the residue theorem,

$$\oint_C \frac{e^{zt}}{z^2(z^2+2z+2)}\, dz = 2\pi i \text{ (sum of the residues)} \quad \textbf{(B.8.26)}$$

$$= 2\pi i \left[\frac{t-1}{2} + \frac{e^{(-1+i)t}}{4} + \frac{e^{(-1-i)t}}{4}\right] \quad \textbf{(B.8.27)}$$

$$= 2\pi i \left[\frac{t-1}{2} + e^{-t}\cos(t)/2\right] \quad \textbf{(B.8.28)}$$

Therefore,

$$\frac{1}{2\pi i}\oint_C \frac{e^{zt}}{z^2(z^2+2z+2)}\, dz = \frac{t-1}{2} + e^{-t}\cos(t)/2 \quad \textbf{(B.8.29)}$$

Finally, let us evaluate the integral

$$\frac{1}{2\pi i}\oint_C f(z)\, dz = \frac{1}{2\pi i}\oint_C \frac{dz}{1 - a\, e^z} \quad 0<a<1 \quad \textbf{(B.8.30)}$$

where the contour is the circle $|z| < -\ell n(a)$. The integrand has an infinite number of simple poles at $z_m = -\ell n(a) \pm 2m\pi i$ with $m = 0, 1, 2, 3, \ldots$ (which lie outside the contour) and a $(n+1)$th order pole at $z = 0$. Because the straightforward evaluation of this integral by the residue theorem would require differentiating the denominator n times, we choose to evaluate Eq. (B.8.30) by expanding the contour so that it is a circle of infinite radius with a cut that excludes the simple poles at z_m. See Fig. B.8.2. Then, by the residue theorem,

$$\frac{1}{2\pi i}\oint_C f(z)\, dz = -\sum_{m=-\infty}^{\infty} [\text{Res of } f(z) \text{ at } z_m] + I \quad \textbf{(B.8.31)}$$

where I is the contribution from the circle at infinity. Because the residue of $f(z)$ at z_m is $-z_m^{-n-1}$, we have

$$\frac{1}{2\pi i}\oint_C f(z)\, dz = \sum_{m=-\infty}^{\infty} [2m\pi i - \ell n(a)]^{-1-n} + I. \quad \textbf{(B.8.32)}$$

Turning now to the evaluation of I, we have

$$I = \left|\frac{1}{2\pi}\int_0^{2\pi} \frac{(R\, e^{i\theta})^{-n}}{1 - a\exp\{R[\cos(\theta) + i\sin(\theta)]\}}\, d\theta\right| \quad \textbf{(B.8.33)}$$

$$\leq \frac{R^{-n}}{2\pi}\int_0^{2\pi} \frac{d\theta}{|1 - a\exp[R\cos(\theta)]|} \quad \textbf{(B.8.34)}$$

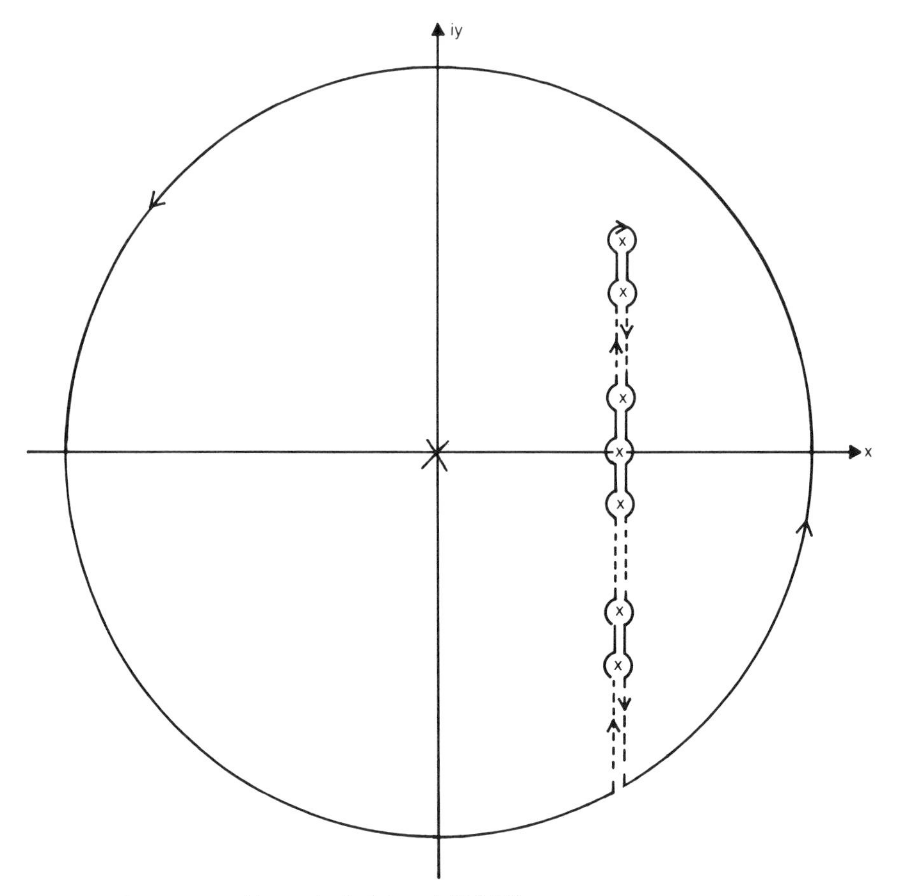

Fig. B.8.2: Contour used to evaluate integral (B.8.30).

Now as $R \to \infty$, we note that

$$| 1 - a \exp [R \cos(\theta)]|^{-1} \sim \begin{cases} \exp[-R \cos(\theta)] & 0 \leq \theta < \pi/2, 3\pi/2 \leq \theta \leq 2\pi \\ 1 & \pi/2 \leq \theta \leq 3\pi/2 \end{cases} \quad \textbf{(B.8.35)}$$

so that the integral in Eq. (B.8.34) is finite and equals to M, say. Consequently, for $n \leq 1$,

$$|I| \leq \lim_{R \to \infty} \frac{M}{2\pi} R^{-n} \to 0. \quad \textbf{(B.8.36)}$$

On the other hand, for $n = 0$, we have

$$2\pi I = \lim_{\epsilon \to 0} \left[\int_0^{\pi/2 - \epsilon} + \int_{\pi/2 + \epsilon}^{\frac{3\pi}{2} - \epsilon} + \int_{\frac{3\pi}{2} + \epsilon}^{2\pi} \right] \frac{d\theta}{1 - a \exp\{R[\cos(\theta) + i \sin(\theta)]\}} \tag{B.8.37}$$

Now, in the limit of $R \to \infty$,

$$\lim_{R \to \infty} \left(1 - a \exp\left\{R\left[\cos(\theta) + i \sin(\theta)\right]\right\}\right) = \begin{cases} 0 & 0 \leq \theta < \pi/2, \ \frac{3\pi}{2} < \theta \leq 2\pi \\ 1 & \pi/2 < \theta < \frac{3\pi}{2} \end{cases} \tag{B.8.38}$$

so that it immediately follows that $I = 1/2$ for $n = 0$.

Consequently, substituting all our results into Eq. (B.8.32),

$$\frac{1}{2\pi i} \oint_C f(z)\, dz = \begin{cases} s_0 (a) + 1/2 & \\ s_n (a) & n = 1,2,3, \ldots \end{cases} \tag{B.8.39}$$

where

$$s_n(a) = (-\ln(a))^{-1-n} + \sum_{m=1}^{\infty} [2m\pi i - \ln(a)]^{-n-1} + [-2m\pi i - \ln(a)]^{-n-1}, \tag{B.8.40}$$

$n \geq 1$ and

$$s_0(a) = 1/2 \ \frac{1+a}{1-a} \tag{B.8.41}$$

Exercises

1. Expand the function

$$f(z) = \frac{1}{z(z-1)}$$

into a Laurent series (a) about $z = 0$ for $|z| < 1$ and (b) about $z = 1$ for $|z-1| < 1$, and determine the residues at the poles.

2. Expand the function $f(z) = e^z/z^2$ into a Laurent series and determine its residue.

3. Evaluate the integral

$$\oint_C \frac{e^z\, dz}{\sin(nz)}$$

where (a) C is a circle with its center at $z = 0$ and radius $|z| < \pi/n$, (b) radius is specified as $\pi/n < |z| < 2\pi/n$.

4. Evaluate the integral

$$\oint_C \frac{dz}{z \cos(z)}$$

around the unit circle.

5. Evaluate

$$\oint_C \frac{\cosh(z)}{z^3}\, dz$$

where C is $|z| = 1$.

6. Evaluate

$$\oint_C \frac{2 + 3\sin(\pi z)}{z\,(z-1)^2}\, dz$$

where C is $|z| = 5$.

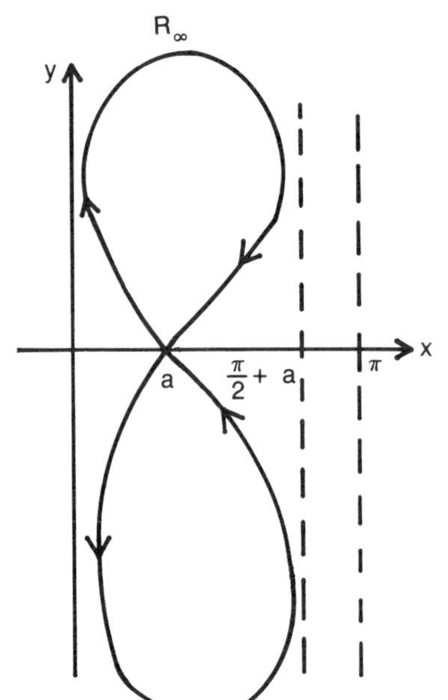

Problem 8

7. Evaluate

$$\frac{1}{2\pi i} \oint_C \frac{e^{zt}}{z(z^2+1)}\, dz, \; t>0$$

if C is $|z| = 5$.

8. Evaluate the integral

$$\frac{1}{2\pi i} \oint_C \frac{dz}{t - r\cos(z-a)}$$

where C is the infinite figure eight shown which crosses the x-axis at $x = a$. Consider the two cases of $t>r$ and $t<r$.

B.9 THE USE OF THE RESIDUE THEOREM IN THE EVALUATION OF DEFINITE INTEGRALS

One of the important applications of the theory of residues consists in the evaluation of certain types of real definite integrals. For example, let us evaluate the integral

$$\int_0^\infty \frac{dx}{1+x^2} = 1/2 \int_{-\infty}^\infty \frac{dx}{1+x^2} \cdot \qquad \textbf{(B.9.1)}$$

This integration occurs along the real axis. In terms of complex variables we can rewrite Eq. (B.9.1) as

$$\int_0^\infty \frac{dx}{1+x^2} = 1/2 \int_{-\infty}^\infty \frac{dz}{1+z^2} \qquad \textbf{(B.9.2)}$$

where the path is the line $Im(z) = 0$. The use of the residue theorem requires an integral along a closed contour. Let us choose the one pictured in Fig. B.9.1. Consequently

$$1/2 \oint_C \frac{dz}{1+z^2} = 1/2 \int_{-\infty}^\infty \frac{dz}{1+z^2} + 1/2 \int_R \frac{dz}{1+z^2} \qquad \textbf{(B.9.3)}$$

where R is the integration path along the semicircle at infinity. Clearly the integral at infinity must go to zero; otherwise, our choice of this contour is poor. Because $z = R\,e^{i\theta}$ and $dz = iR\,e^{i\theta}\,d\theta$,

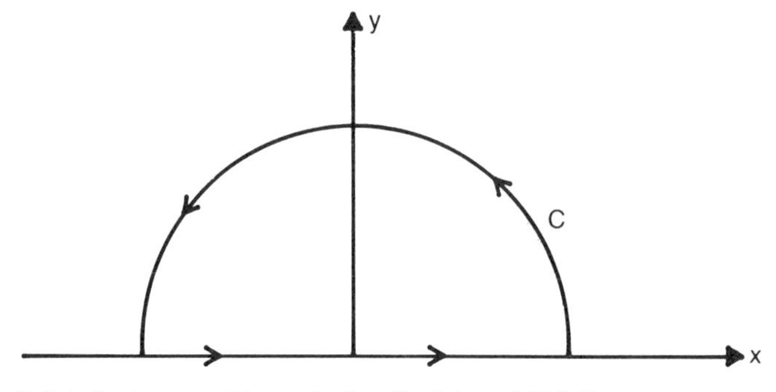

Fig. B.9.1: Contour used in evaluating the integral (B.9.1).

$$\int_R \frac{dz}{1+z^2} = \int_0^\pi \frac{iR\,e^{i\theta}\,d\theta}{1+R^2\exp(2i\theta)} \leq \int_0^\pi \frac{R\,d\theta}{R^2-1} \tag{B.9.4}$$

which tends to zero as $R \to \infty$. On the other hand, the residue theorem gives

$$\oint \frac{dz}{1+z^2} = 2\pi i \,(\text{Residue at } z = i \text{ of } \frac{1}{1+z^2}) \tag{B.9.5}$$

$$= 2\pi i \lim_{z \to i} \frac{z-i}{z^2+1} = 2\pi i\,(\frac{1}{2i}) = \pi. \tag{B.9.6}$$

Therefore,

$$\int_0^\infty \frac{dx}{1+x^2} = \frac{\pi}{2}\,. \tag{B.9.7}$$

This example illustrates the basic concepts of computing definite integrals by the residue theorem. A closed contour is introduced that includes the real axis. In addition to the desired integral, additional integrals are introduced. If the closed contour is properly chosen, these additional integrals either vanish or else they can be evaluated. We shall now show that for certain classes of general integrals, the integral along a circular arc at infinity will always vanish.

Theorem 1. If, on a circular are C_R with a radius R and center at the origin, $z\,f(z) \to 0$ uniformly as $R \to \infty$, then

$$\lim_{R \to \infty} \int_{C_R} f(z)\,dz = 0. \tag{B.9.8}$$

This follows from the fact that if $|z\,f(z)| \leq M_R$, then $|f(z)| \leq M_R/R$.

Because the length of C_R is αR, where α is the subtended angle, we have

$$\left|\int_{C_R} f(z)\, dz\right| \leq \frac{M_R}{R}\alpha R = \alpha M_R \rightarrow 0 \qquad \textbf{(B.9.9)}$$

because $M_R \rightarrow 0$ as $R \rightarrow \infty$.

A simple example of this theorem is the integral

$$\int_{-\infty}^{\infty} \frac{dx}{x^2+x+1} = \int_{-\infty}^{\infty} \frac{dz}{z^2+z+1} \qquad \textbf{(B.9.10)}$$

A quick check shows that $z/(z^2+z+1)$ tends to zero as $R \rightarrow \infty$. Therefore, if we use the contour pictured in Fig. B.9.1, we have

$$\int_{-\infty}^{\infty} \frac{dx}{x^2+x+1} = \oint_C \frac{dz}{z^2+z+1} \qquad \textbf{(B.9.11)}$$

$$= 2\pi i\left(\text{Residue at } z = -1/2 + \frac{\sqrt{3}}{2}i \text{ of } \frac{1}{z^2+z+1}\right) \qquad \textbf{(B.9.12)}$$

$$= 2\pi i \lim_{z \rightarrow -1/2 + \frac{\sqrt{3}}{2}i} \left(\frac{1}{2z+1}\right) \qquad \textbf{(B.9.13)}$$

$$= \frac{2\pi}{\sqrt{3}}. \qquad \textbf{(B.9.14)}$$

Theorem 2: Suppose that, on a circular arc C_R, with radius R and center at the origin, $f(z) \rightarrow 0$ uniformly as $R \rightarrow \infty$. Then,

$$1) \quad \lim_{R \rightarrow \infty} \int_{C_R} e^{imz} f(z)\, dz = 0 \quad (m > 0) \qquad \textbf{(B.9.15)}$$

if C_R is in the first and/or second quadrant. (This result is known as *Jordan's lemma*);

$$2) \quad \lim_{R \rightarrow \infty} \int_{C_R} e^{-imz} f(z)\, dz = 0 \quad (m > 0) \qquad \textbf{(B.9.16)}$$

if C_R is the third and/or fourth quadrant;

$$3) \quad \lim_{R \rightarrow \infty} \int_{C_R} e^{-mz} f(z)\, dz = 0 \quad (m > 0) \qquad \textbf{(B.9.17)}$$

if C_R is the second and/or third quadrants; and

$$4) \quad \lim_{R\to\infty} \int_{C_R} e^{-mz}\, f(z)\, dz = 0 \qquad (m > 0) \qquad \textbf{(B.9.18)}$$

if C_R is in the first and/or fourth quadrants.

We shall prove the first part; the remaining portions may be proved in a similar manner.

$$\left| \int_{C_R} e^{imz}\, f(z)\, dz \right| \le \int_{C_R} |e^{imz}|\,|f(z)|\,|dz|. \qquad \textbf{(B.9.19)}$$

Now

$$|dz| = R\, d\theta \qquad \textbf{(B.9.20)}$$

$$|f(z)| \le M_R \qquad \textbf{(B.9.21)}$$

$$|e^{imz}| = |\exp(imR\, e^{i\theta})| = |\exp\{ imR\, [\cos(\theta) + i \sin(\theta)]\}|$$

$$= e^{-mR\sin(\theta)} \qquad \textbf{(B.9.22)}$$

Therefore

$$|I_R| = \left| \int_{C_R} e^{imz}\, f(z)\, dz \right|$$

$$\le R\, M_R \int_{\theta_0}^{\theta_1} \exp\, [-mR \sin(\theta)]\, d\theta \qquad \textbf{(B.9.23)}$$

where $0 \le \theta_0 < \theta_1 \le \pi$. Because the last integrand is positive, the right-hand side of Eq. (B.9.23) is not decreased if we take $\theta_0 = 0$ and $\theta_1 = \pi$.

$$|I_R| \le R\, M_R \int_0^{\pi} e^{-mR\sin(\theta)}\, d\theta = 2RM_R \int_0^{\pi/2} e^{-mR\sin(\theta)}\, d\theta.$$

$$\textbf{(B.9.24)}$$

We cannot evaluate the integrals in Eq. (B.9.24). However, because

$$\sin(\theta) \ge \frac{2\theta}{\pi}, \qquad \textbf{(B.9.25)}$$

we can bound the value of the integral by

$$|I_R| \le 2\, R\, M_R \int_0^{\pi/2} e^{-2mR\theta/\pi}\, d\theta = \frac{\pi}{m}\, M_R\, (1 - e^{-mR}). \qquad \textbf{(B.9.26)}$$

If $m > 0$, I_R tends to zero with M_R as $R \to 0$.

To illustrate Jordan's lemma, let us evaluate

$$\int_{-\infty}^{\infty} \frac{\cos(kx)}{x^2 + a^2}\, dx \text{ with } a,k>0. \tag{B.9.27}$$

To compute this integral, we note that

$$\int_{-\infty}^{\infty} \frac{\cos(kx)}{x^2 + a^2}\, dx = Re\{\int_{-\infty}^{\infty} \frac{e^{ikx}}{x^2 + a^2}\, dx\}. \tag{B.9.28}$$

A quick check will show that the integrand of the right-hand side of Eq. (B.9.28) satisfies Jordan's lemma. Therefore

$$\int_{-\infty}^{\infty} \frac{e^{ikz}}{z^2 + a^2}\, dz = \oint_C \frac{e^{ikz}}{z^2 + a^2}\, dx \tag{B.9.29}$$

$$= 2\pi i(\text{Residue at } z = \text{ia of } \frac{e^{ikz}}{z^2 + a^2}) \tag{B.9.30}$$

$$= 2\pi i \lim_{z \to ia} \frac{(z - ia)e^{ikz}}{z^2 + a^2} = \frac{\pi}{a} e^{-ka} \tag{B.9.31}$$

Therefore, taking the real and imaginary parts of Eq. (B.9.31), we have

$$\int_{-\infty}^{\infty} \frac{\cos(kx)}{x^2 + a^2}\, dx = \frac{\pi}{a} e^{-ka} \tag{B.9.32}$$

$$\int_{-\infty}^{\infty} \frac{\sin(kx)}{x^2 + a^2}\, dx = 0 \tag{B.9.33}$$

Finally, let us evaluate the integral

$$I = \int_0^{\infty} e^{-rx} \frac{k_2 \cosh(x) - k_1 \sinh(x)}{k_2 \sinh(rx)\cosh(x) + k_1 \cosh(rx)\sinh(x)} \sin(qx)\, dx \tag{B.9.34}$$

by the methods of residues where r is the ratio of two integers n/m. Our analysis begins by considering the contour integral

$$\oint f(z)\, dz = \oint e^{-rz} \frac{k_2 \cosh(z) - k_1 \sinh(z)}{k_2 \sinh(rz)\cosh(z) + k_1 \cosh(rz)\sinh(z)} e^{iqz}\, az \tag{B.9.35}$$

where the closed contour is a rectangle with its vertices at 0, R,

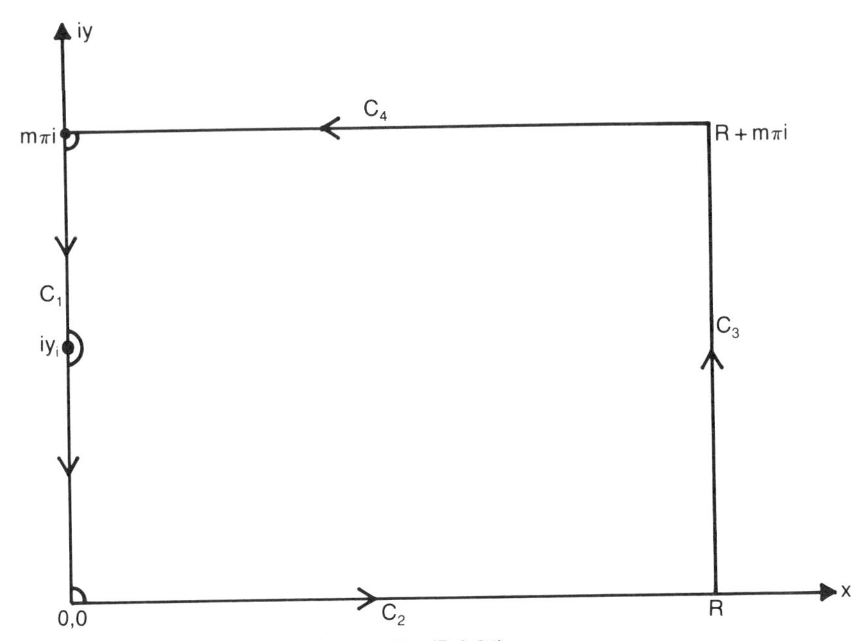

Fig. B.9.2: Contour used in evaluating Eq. (B.8.34).

$R + m\pi i$ and $m\pi i$. See Fig. B.9.2. Eventually we will take the limit $R \to \infty$. This use of a rectangle rather than a semicircle is very useful in evaluating integrals that contain hyperbolic functions.

A quick check of the integrand of (B.9.35) shows that it is analytic everywhere within the rectangle. However, along the imaginary axis, there are simple poles at $z = 0$, $m\pi i$ and the roots of $k_2 \sin(ry)\cos(y) + k_1 \cos(ry)\sin(y) = 0$ because $\cosh(iy) = \cos(y)$ and $\sinh(iy) = i \sin(y)$.

Turning to the evaluation of the integral along each leg of the rectangle, it is easily shown that

$$\int_{C_3} f(z)\, dz =$$

$$\lim_{R \to \infty} \int_R^{R + m\pi i} e^{-rz} \frac{k_2 \cosh(z) - k_1 \sinh(z)}{k_2 \sinh(rz)\cosh(z) + k_1 \cosh(rz)\sinh(z)} e^{iqz}\, dz = 0 \tag{B.9.36}$$

$$\int_{C_2} f(z)\, dz$$

$$= \int_0^{\infty} e^{-rz} \frac{k_2 \cosh(x) - k_1 \sinh(x)}{k_2 \sinh(rx)\cosh(x) + k_1 \cosh(rx)\sinh(x)} e^{iqx}\, dx \tag{B.9.37}$$

$$\int_{C_4} f(z)\, dz =$$

$$\int_{\infty}^{0} \frac{[e^{-rx-m\pi ri} k_2 \cosh(x+m\pi i) - k_1 \sinh(x+m\pi i)\, e^{-mq\pi+iqx}]}{k_2 \sin(rx+m\pi ri)\cosh(x+m\pi i)+k_1\cosh(rx+m\pi ri)\sinh(x+m\pi i)}\, dx \tag{B.9.38}$$

$$= -\int_0^{\infty} e^{-rx} \frac{(k_2 \cosh(x) - k_1 \sinh(x))\, e^{-mq\pi+iqx}}{k_2 \sinh(rx)\cosh(x)+k_1 \cosh(rx)\sinh(x)}\, dx \tag{B.9.39}$$

because $m\pi r = n\pi$.

The evaluation of the integral along the imaginary axis requires some thought because of the simple poles along that axis. This difficulty is usually circumvented by indenting the contour through the introduction of an arc of small radius ϱ so that integration through a singularity is avoided as shown in Fig. B.9.2. Clearly, the value of the integral is rather special and is given the unique name of *Cauchy principal value.*

The Cauchy principal value can be better understood by considering the integration of a real improper integral. For example, the integral $\int_{-1}^{2} dx/x$ does not exist in the strict sense because of the strong singularity at the origin. However, the integral would exist if the limit

$$\lim_{\substack{\epsilon\to 0\\ \delta\to 0}} \int_{-1}^{\epsilon} \frac{dx}{x} + \int_{\delta}^{2} \frac{dx}{x} \tag{B.9.40}$$

existed and had a unique value as ϵ and δ independently approach zero. Because this limit equals

$$\lim_{\epsilon,\delta\to 0} [\ell n(\epsilon) + \ell n(2) - \ell n(\delta)]$$

$$= \lim_{\epsilon,\delta\to 0} [\ell n(2) - \ell n(\delta/\epsilon)], \tag{B.9.41}$$

our integral would have the value of $\ell n(2)$ if we chose to let δ and ϵ equal each other. This particular limit is defined to be the Cauchy principal value of the improper integral and is usually written

$$P.V. \int_{-1}^{2} \frac{dx}{x} = \ell n(2). \qquad \textbf{(B.9.42)}$$

Let us apply this concept of principal value to our problem. For the contour C_1, we obtain

$$\int_{C_1} f(z)\, dz = P.V. \int_{m\pi}^{0} e^{-iry} \frac{k_2 \cos(y) - i\, k_1 \sin(y)}{k_2 \sin(ry) \cos(y) + k_1 \cos(ry) \sin(y)}\, dy$$

$$+ \sum_j \int_{\gamma_j} f(z)\, dz + \int_{\gamma_0} f(z)\, dz + \int_{\gamma_{m\pi}} f(z)\, dz. \qquad \textbf{(B.9.43)}$$

The first term in Eq. (B.9.43) gives the contribution from the line integration between the poles while the remaining terms give the contribution from the semicircles and quartercircles that were introduced to avoid integrating through the singularities.

To see how we can evaluate the second integral in terms of residues, consider the jth semicircle lying along the imaginary axis at $z_j = iy_j$. Because it is a simple pole, we know that the Laurent expansion for $f(z)$ about this point is

$$f(z) = c_{-1} \frac{1}{z - z_j} + c_0 + c_1 (z - z_j) + \ldots \qquad \textbf{(B.9.44)}$$

Letting $z - z_j = r\, e^{i\theta}$,

$$\int_{\gamma_j} f(z)\, dz = \lim_{r \to 0} c_{-1} \int_{\pi}^{0} \frac{ir\, e^{i\theta}}{r\, e^{i\theta}}\, d\theta$$

$$+ c_0 \int_{\pi}^{0} ir\, e^{i\theta} d\theta + \text{higher powers of } r \qquad \textbf{(B.9.45)}$$

$$= -\, i\pi c_{-1} = -\, 2\pi i\, (1/2 \text{ residue at } z = z_j). \qquad \textbf{(B.9.46)}$$

This clearly shows that we can employ the residue theorem provided we only take one-half of the value. Similarly for the quartercircles, we only take one quarter of the residue. Consequently,

$$\sum_j \int_{\gamma_j} f(z)\, dz = -\, i\pi \sum_j R_j \qquad \textbf{(B.9.47)}$$

while

$$\int_{\gamma_0} f(z)\, dz = -\, \pi i/2 \frac{k_2}{k_2 r + k_1} \qquad \textbf{(B.9.48)}$$

and

$$\int_{\gamma_{m\pi}} f(z)\,dz = -\pi i/2 \frac{k_2}{k_2 r + k_1} e^{-mq\pi} \qquad \textbf{(B.9.49)}$$

where R_j is the jth residue of $f(z)$ around the pole located at point $z_j = iy_j$.

Our final result is obtained by taking the imaginary part of the sum of the integrals:

$$(1 - e^{-mq\pi})\, I = \pi/2 \frac{k_2}{k_2 r + k_1} (1 + e^{-mq\pi}) + \pi\, Im(i \sum_j R_j)$$
$$- \frac{1}{q} (1 - e^{-mq\pi}) \qquad \textbf{(B.9.50)}$$

The last term in Eq. (B.9.50) follows from

$$Im\left[P.V. \int_{m\pi}^{0} e^{-iry} \frac{k_2 \cos(y) - i\, k_1 \sin(y)}{k_2 \sin(ry)\cos(y) + k_1 \cos(ry)\sin(y)} e^{-qy}\, dy \right]$$

$$= \int_{m\pi}^{0} \frac{-k_2 \cos(y) \sin(ry) - k_1 \sin(y) \cos(ry)}{k_2 \sin(ry) \cos(y) + k_1 \cos(ry) \sin(y)} e^{-qy}\, dy \qquad \textbf{(B.9.51)}$$

$$= -\int_{m\pi}^{0} e^{-qy}\, dy = \frac{1}{q}(1 - e^{-mq\pi}) \qquad \textbf{(B.9.52)}$$

Finally, we have that

$$Im(i\, R_j) = Re(R_j) = D_j\, e^{-qy_j} \qquad \textbf{(B.9.53)}$$

$$\text{where } D_j = \frac{k_2 \cos(y_j) \cos(ry_j) - k_1 \sin(y_j) \sin(ry_j)}{(k_2 r + k_1)\cos(ry_j)\cos(y_j) - (k_2 + rk_1)\sin(ry_j)\sin(y_j)} \qquad \textbf{(B.9.54)}$$

Exercises

Use the residue theorem to verify the following integrals.

1. $\displaystyle \int_0^\infty \frac{dx}{x^4+1} = \frac{\pi\sqrt{2}}{4}$

2. $\displaystyle \int_0^\infty \frac{x^2 dx}{x^6+1} = \frac{\pi}{6}$

3. $$\int_{0}^{\infty} \frac{dx}{(x^2+4x+5)^2} = \frac{\pi}{2}$$

4. $$\int_{0}^{\infty} \frac{dx}{(x^2+1)^2} = \frac{\pi}{4}$$

5. $$\int_{-\infty}^{\infty} \frac{x\,dx}{(x^2+1)(x^2+2x+2)} = -\frac{\pi}{5}$$

6. $$\int_{0}^{\infty} \frac{dx}{(x^2+1)(x^2+4)^2} = \frac{5\pi}{288}$$

7. $$\int_{-\infty}^{\infty} \frac{\sin(x)\,dx}{x^2+4x+5} = -\frac{\pi}{e}\sin(2)$$

8. $$\int_{0}^{\infty} \frac{\cos(x)\,dx}{(x^2+1)^2} = \frac{\pi}{2e}$$

9. $$\int_{-\infty}^{\infty} \frac{x\sin(ax)}{x^2+4}\,dx = \pi e^{-a}\sin(a)/2$$

10. $$\int_{0}^{\infty} \frac{8t^2\,dt}{(1+t^2)^2[t^2(a/h+1)+(a/h-1)]} = \frac{\pi}{4}\left(\frac{a}{h}\right)\left\{1-[1-(h/a)^2]^{1/2}\right\}$$

B.10 BRANCH POINTS: LOGARITHMS

If $w = f(z)$ is a single-valued function, each point of the z-plane corresponds to one point of the w-plane. In this section we examine multivalued functions where each point of the z-plane may correspond to more than one point of the w-plane.

Consider, for example, the function $w = z^{1/2}$ taken over a circular path. Changing to polar form,

$$z = r\,e^{i\theta} \tag{B.10.1}$$

$$w = r^{1/2}e^{i\theta/2}, \tag{B.10.2}$$

we find that the completion of one revolution of 0 to 2π radians in the z-plane corresponds to covering 0 to π radians in the w-plane. Continuing around the z-plane for the second time, 2π to 4π radians, w describes the angles π to 2π radians. Hence a point b in the z-plane corresponds to two points b_1 and b_2 in the w-plane. The two regions of the w-plane – 0 to π, π to 2π – are referred to as the two *branches* of the function $w = z^{1/2}$. The branch w_1 corresponds to the roots which lie above the real axis while the second branch contains roots that lie below the real axis. The value of z, which lies along the positive real axis gives rise to roots that lie along the entire real axis. Because both w_1 and w_2 (each of which are called a *Riemann surface*) have the real axis as a common boundary, we cannot have unique values for z being positive and real. This line of all of the points that do not give unique square roots is given the name of *branch cut.*

It is clear that there is nothing unique about our choice of the positive x-axis for the branch cut. Any rotation of the z-plane would merely result in a rotation of the results. Consequently, in this case, we can always choose the direction of the branch cut in any direction that we like. The point which is common to all of these branch cuts is given the name of *branch point.*

One of the most important functions that exhibits this behavior of branches is the logarithm. Because

$$z = r\, e^{i\theta + 2n\pi i} \qquad n = 0,1,2, \ldots \tag{B.10.3}$$

where we take

$$-\pi < \theta \leq \pi, \tag{B.10.4}$$

the logarithm is defined as

$$\log(z) = \log(r\, e^{i\theta + 2n\pi i}) = \ell n(r) + i\theta + 2n\pi i \tag{B.10.5}$$

where the $\ell n(r)$ is the *real* natural logarithm of a positive number. The function $\log(z)$ is therefore multiple-valued with infinitely many values. We shall call the *principal value* of $\log(z)$ the value given when $n = 0$.

It can be proven using these definitions that the derivative of the $\log(z)$ cannot exist along the line $\theta = \pm\pi$. Consequently the $\log(z)$ function is *not* analytic along this line. In other words, the function $\log(z)$ is analytic in the domain $r > 0$ and $-\pi < \theta < \pi$. At $r = 0$ $\log(z)$ becomes indefinite and the point $z = 0$ is a singular point. It is also the branch point for this particular function with the line $\theta = \pm\pi$ being the branch cut.

With this definition, we have

$$\frac{d}{dz}[\log(z)] = \frac{1}{z} \tag{B.10.6}$$

if $z \neq 0$ and $-\pi < \text{argument}(z) < \pi$, as well as

$$\log[\exp(z)] = z \tag{B.10.7}$$

$$\log(z_1) + \log(z_2) = \log(z_1 z_2) \tag{B.10.8}$$

and

$$z^{m/n} = \exp\left[\frac{m}{n}\log(z)\right]. \quad \textbf{(B.10.9)}$$

B.11 INTEGRATIONS THAT INVOLVE BRANCH POINTS

One of the most important applications of integration on the complex plane is its use in the inversion of Laplace transforms by Bromwich's integral. In physical problems not involving wave motion, the singularities associated with the transform are usually poles and the use of the residue theorem is straightforward. Where transmission lines, cables and acoustical problems are concerned, however, there is wave motion and multivalued functions often appear in the transform.

When branch points and cuts are present, integrations on the complex plane rapidly becomes more difficult. To avoid encroachment upon another branch, the branch cut must behave as a barrier and may not be crossed during the integration. This additional complication is somewhat mitigated by the fact that we can define the direction that the branch cut emanates from the branch point. However, because the path of the Bromwich integral runs parallel and to the right of the imaginary axis, we are essentially restricted to the angles $\pi/2$ to $3\pi/2$.

In the case of inversions with only poles, we chose a semi-infinite arc in the second and third quadrants as part of our contour. Because this contour would now cross the branch cut, we must modify it. This modification shall consist of distorting it so that it runs along the branch cut and around the branch point. See Fig. B.11.1, for example. Because the contribution from the contours along each side of the barrier must be computed, the branch cut is usually taken along the negative real axis.

To illustrate how the integration proceeds, let us evaluate

$$\frac{1}{2\pi i}\int_{c-i\infty}^{c+i\infty} \frac{e^{zt}}{(1+z)^{1/2}}\, dz \quad \textbf{(B.11.1)}$$

where c and t are any positive constants. The integrand has a branch point at $z = -1$. This follows from the fact that $e^{zt}/(1+z)^{1/2}$ does not return to its original value after any closed path is transversed about $z = -1$. This property of branch points was suggested in the previous section with the logarithm when a closed path around the branch point $z = 0$ resulted in the phase of the logarithm increasing by 2π. On the other hand the function $e^{zt}/(1+z)^{1/2}$ does not suffer any phase change if the closed path does *not* include $z = -1$.

The closed contour that we intend to use to evaluate Eq. (B.11.1) is shown in Fig. B.11.1. The contours C_3 and C_5 actually

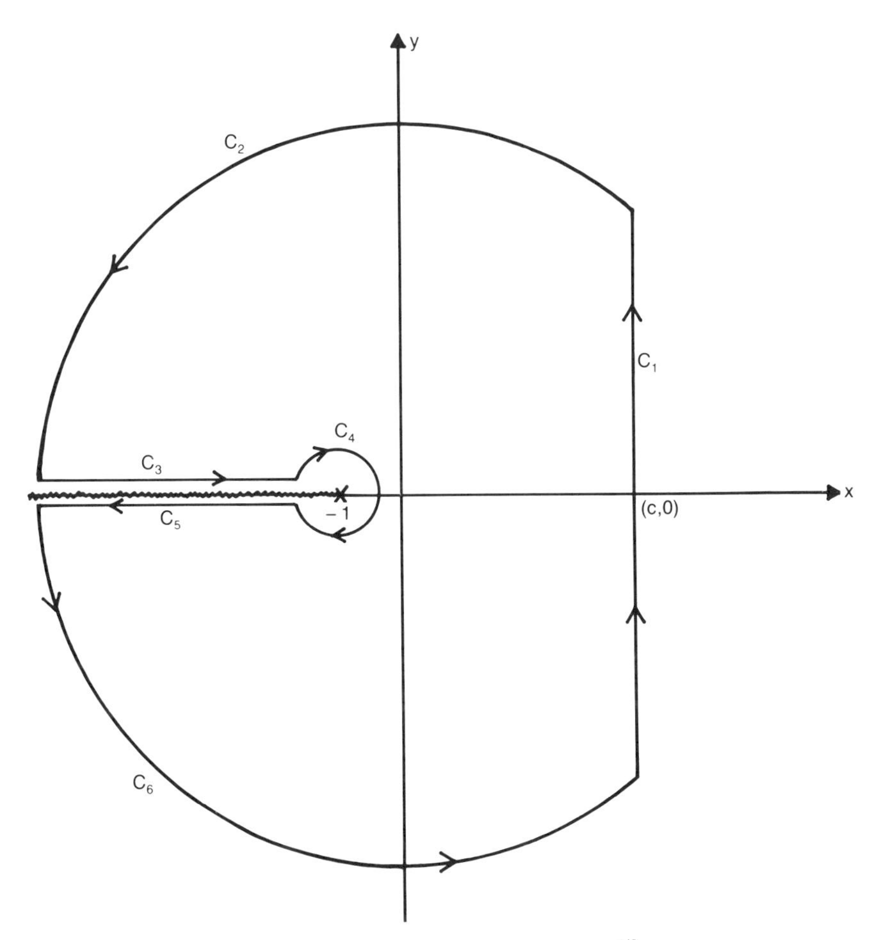

Fig. B.11.1: Contour used to invert the Laplace transform $(1+s)^{-1/2}$.

lie on the negative real axis but have been separated for visual purposes. The circular contour C_4 is a circle of radius ϵ while C_2 and C_6 represent arcs from a circle of radius R.

Because $e^{zt}/(z+1)^{1/2}$ is analytic within the closed curve

$$\oint \frac{e^{zt}}{(z+1)^{1/2}}\ dz = 0. \qquad \textbf{(B.11.2)}$$

We shall now evaluate each of the contours C_2 through C_6. The sum of their values equals the negative of the value for the contour C_1.

$$\int_{C_2} \frac{e^{tz}}{(z+1)^{1/2}}\, dz = \lim_{R\to\infty} \int_{\theta_0}^{\pi} \frac{\exp(Re^{i\theta}t)}{(1+R\exp(i\theta))} \ iR\, e^{i\theta}\, d\theta \qquad \textbf{(B.11.3)}$$

However,

$$\lim_{R \to \infty} \left| \int_{\theta_0}^{\pi} \frac{\exp(R\, e^{i\theta}\, t)}{(1 + R \exp(i\theta))} iR\, e^{i\theta} d\theta \right|$$

$$\leq \lim_{R \to \infty} \int_{\pi/2}^{\pi} \frac{\exp[Rt \cos(\theta)]}{R^{1/2}} R\, d\theta \qquad \textbf{(B.11.4)}$$

$$\leq \lim_{R \to \infty} \int_{0}^{\pi/2} \exp[-Rt \sin(\phi)]\, R^{1/2}\, d\phi \qquad \textbf{(B.11.5)}$$

$$\leq \lim_{R \to \infty} \int_{0}^{\pi/2} \exp(-2Rt\phi/\pi)\, R^{1/2}\, d\phi \qquad \textbf{(B.11.6)}$$

$$\leq \lim_{R \to \infty} \frac{\pi}{2R^{1/2}t}[1 - \exp(-Rt)] \to 0 \qquad \textbf{(B.11.7)}$$

where $\theta = \phi + \pi/2$ and $\sin(\phi) \leq 2\phi/\pi$ for $0 \leq \phi \leq \pi/2$. In Eq. (B.11.4) we have replaced θ_0 with $\pi/2$ because θ_0 tends to $\pi/2$ for finite c as $R \to \infty$. Similar considerations show that the contribution along C_6 is zero.

$$\int_{C_4} \frac{e^{zt}}{(z+1)^{1/2}} dz = \lim_{\epsilon \to 0} \int_{\pi}^{-\pi} \frac{\exp[(\epsilon e^{i\theta} - 1)t]}{[\epsilon \exp(i\theta)]^{1/2}} i\epsilon\, e^{i\theta}\, d\theta \qquad \textbf{(B.11.8)}$$

$$= e^{-t} \lim_{\epsilon \to 0} \int_{\pi}^{-\pi} i\epsilon^{1/2}\, e^{i\theta/2}(1 + \epsilon e^{i\theta} + \epsilon^2 e^{2i\theta} + \ldots)\, d\theta \qquad \textbf{(B.11.9)}$$

$$= e^{-t} \lim_{\epsilon \to 0} [-4i\, \epsilon^{1/2} + O(\epsilon^{3/2})] \to 0 \qquad \textbf{(B.11.10)}$$

$$\int_{C_3} \frac{e^{zt}}{(1+z)^{1/2}} dz = \lim_{\substack{R \to \infty \\ \epsilon \to 0}} \int_{-R}^{-\epsilon - 1} \frac{e^{zt}}{(1+z)^{1/2}} dz \qquad \textbf{(B.11.11)}$$

$$= - \lim_{\substack{R \to \infty \\ \epsilon \to 0}} \int_{R-\epsilon}^{\epsilon} \frac{e^{-(u+1)t}}{iu^{1/2}} du \qquad \textbf{(B.11.12)}$$

$$= i\, e^{-t} \int_{\infty}^{0} \frac{e^{-ut}}{u^{1/2}} du$$

$$\int_{C_3} \frac{e^{zt}}{(1+z)^{1/2}}\, dz = -i\, e^{-t} \int_0^{\infty} \frac{e^{-ut}}{u^{1/2}}\, du \quad \textbf{(B.11.13)}$$

$$= -i\, e^{-t} \frac{\pi^{1/2}}{t^{1/2}} \quad \textbf{(B.11.14)}$$

$$\int_{C_5} \frac{e^{zt}}{(1+z)^{1/2}} dz = \lim_{\substack{R\to\infty \\ \epsilon\to 0}} \int_{-1-\epsilon}^{-R} \frac{e^{zt}}{(1+z)^{1/2}}\, dz \quad \textbf{(B.11.15)}$$

$$= \lim_{\substack{R\to\infty \\ \epsilon\to 0}} \int_{\epsilon}^{R-1} \frac{e^{-(u+1)t}}{iu^{1/2}}\, du \quad \textbf{(B.11.16)}$$

$$= -i\, e^{-t} \int_0^{\infty} \frac{e^{-ut}}{u^{1/2}}\, du = -\frac{i\, e^{-t}\, \pi^{1/2}}{t^{1/2}} \quad \textbf{(B.11.17)}$$

where $1 + z = u\, e^{\pi i}$ and $(1+z)^{1/2} = u^{1/2}\, e^{-\pi i/2}$ on C_5. Consequently, adding all the contributions together, taking the negative of that quantity and dividing by $2\pi i$, we get

$$\frac{1}{2\pi i} \int_{c-i\infty}^{c+i\infty} \frac{e^{zt}}{(1+z)^{1/2}}\, dz = \frac{e^{-t}}{\sqrt{\pi t}} \quad \textbf{(B.11.18)}$$

When more than one branch point appears or the integrand is more complicated than our example, the integration becomes very difficult. Often an asymptotic expression for large or small t is sufficient. (The reader is referred to McLachlan, N.W., 1944: *Complex Variable and Operational Calculus with Technical Applications*. New York: The Macmillan Co., 355 pp. for further details.)

Once in a while, an integration will arise where the branch points and singularities lie within the contour. For this particular case, we establish the following useful theorem:

Theorem: If all the singularities and branch-points of a function $f(z)$ lie in a finite region of the z-plane and the value of $zf(z)$, as $|z|$ increases indefinitely, tends uniformly to M, then the value of

$$\int f(z)\, dz$$

around the simple curve is $2\pi iM$.

Proof: Because all the singularities and branch points lie within the contour, the contour may be deformed continuously to an infinite circle. Because $dz/z = i\, d\theta$, we have

$$\int f(z)\, dz = i \int_0^{2\pi} zf(z)\, d\theta = i \int_0^{2\pi} M\, d\theta = 2\pi iM \quad \textbf{(B.11.19)}$$

as $|z| \to \infty$.

Thus the value of $\int \frac{dz}{\sqrt{a^2 - z^2}}$ along any simple curve, which encloses the two points $z = a$ and $z = -a$, is 2π. The value of $\int \frac{dz}{\sqrt{(1 - z^2)(1 - k^2)}}$ around any simple curve enclosing the four points 1, -1, $1/k$ and $-1/k$ is zero, if k is any nonvanishing constant.

Exercises

1. Using the contour illustrated in the figure, evaluate

$$\oint_C \frac{e^{izt}\, dz}{(z^2 - a^2)\sqrt{z^2 + b^2}}$$

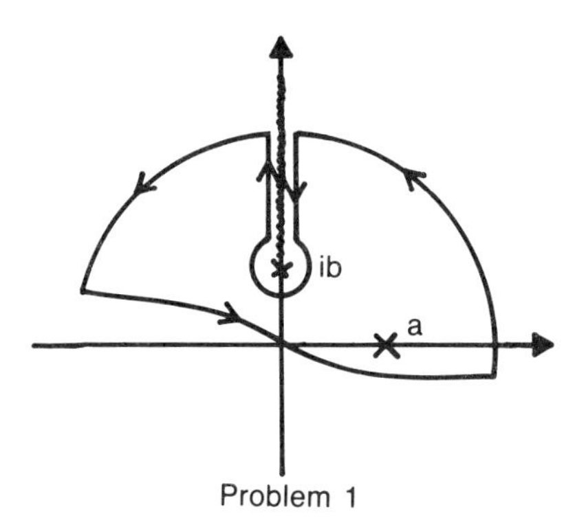

Problem 1

2. By evaluating the integral

$$\oint_C \frac{dz}{(z^2 - z^3)^{1/3}}$$

around the curve illustrated in the figure, evaluate

$$\int_0^1 \frac{dx}{(x^2 - x^3)^{1/3}}$$

Problem 2

3. For the contour illustrated below, replace the following integral with one involving only real variables:

$$\oint_C \left(\frac{1 + iz}{(1 + z^2)^{3/2}} + \frac{\sigma}{(1 + z^2)^{1/2}}\right) e^{i\sigma t}\, dz$$

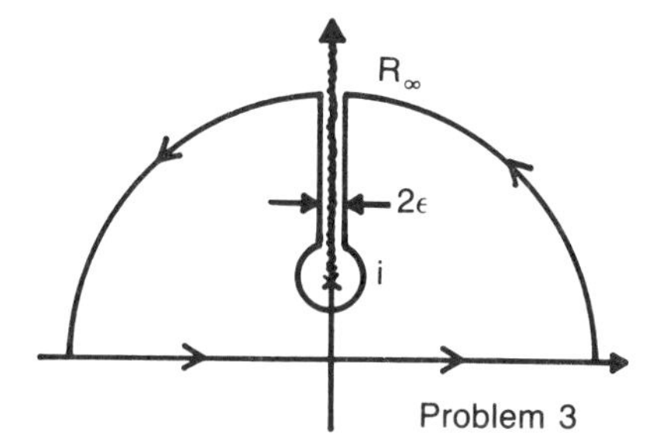

Problem 3

where the branch of $(z^2 + 1)^{1/2}$ is chosen so that it assumes the value of $i(y^2 - 1)^{1/2}$ on the segment $iy - \epsilon$ and the value of $i(y^2 - 1)$ on the segment $iy + \epsilon$.

4. Evaluate the following integral in terms of real parameters and an integral:

$$\frac{1}{2\pi i} \oint_C \frac{1 + z}{\sqrt{z}} \frac{dz}{z - a}$$

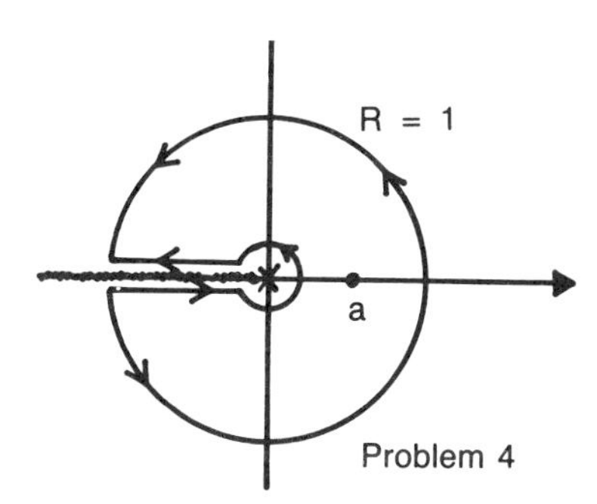

Problem 4

where C is the unit circle excluding the branch cut along the negative real axis. The point a lies within the circle.

5. Evaluate

$$\oint_C \frac{a - z}{z^2 \sqrt{z^2 - 2z + 2}}\, dz$$

where the branch is chosen so that the radical is positive for $Re(z) > 0$ and the contour below. (Hint: First show that a circular contour that includes both the branch cut and singularities is equal to zero. Then use the residue theorem to evaluate the left-hand side of the

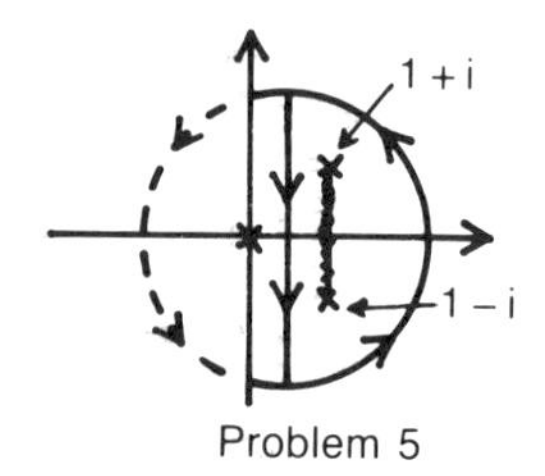

Problem 5

circle. The desire quantity must equal the negative of the value from the left-hand side.)

6. Evaluate

$$\oint_C \frac{z\,dz}{(z^2+a^2)\sqrt{z^2}-1(z-\sqrt{z^2}-1)^n}$$

where C is any circle excluding $z = \pm ia$ and including the branch cut. Choose the branch of $(z^2-1)^{1/2}$ so that $|z-(z^2-1)^{1/2}|>1$.

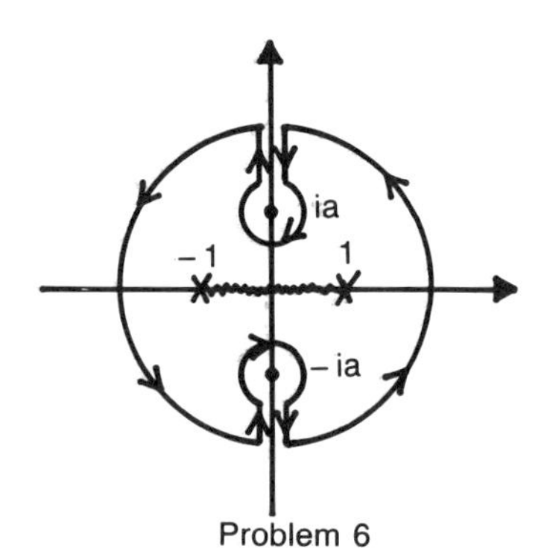

Problem 6

(Hint: Show that the circular contour which includes both the singularities and branch cut is zero. Then use the residue theorem to evaluate the contribution from inside the smaller circles. This must equal the negative of the answer.)

Answers

P. 488

1. $1+2i$
2. $3 - i$
3. $-2/5$
4. -4
5. $2 + 2i\ 3^{1/2}$
6. $\exp(3\pi i/2)$
7. $4\ \exp(i\pi)$
8. $4\ \exp(i\pi/3)$
9. $5\ 2^{1/2}\ \exp(3\pi i/4)$
10. $2\ 2^{1/2}\ \exp(7\pi i/4)$
11. $2\ \exp(2\pi i/3)$

P. 491

1. $\pm 2^{1/2}, \pm 1 \pm 3^{1/2}i$
2. $\cos(\pi/3) + i\sin(\pi/3), -1, \cos(5\pi/3) + i\sin(5\pi/3)$
3. $i, (\pm 3^{1/2} - i)/2$
4. $3 - 2i, -3 + 2i$
5. $3^{1/2}[\cos(\pi/4) + i\sin(\pi/4)]$
   $3^{1/2}[\cos(105°) + i\sin(105°)]$
   $3^{1/2}[\cos(165°) + i\sin(165°)]$
   $3^{1/2}[\cos(225°) + i\sin(225°)]$
   $3^{1/2}[\cos(285°) + i\sin(285°)]$
   $3^{1/2}[\cos(345°) + i\sin(345°)]$

P. ~~497~~ 500

2. (a) $3z(1+z^2)^{1/2}$
   (b) $\frac{1}{3}z^{-1/2}(z + 2z^{1/2})^{-2/3}(1 + z^{1/2})$
   (c) $(1 + 4i)z^2 - 3z - 2$
   (d) $\frac{2z - i}{z + 2i}$
   (e) $(iz - 1)^{-3}$

3. (a) $-1/4$
   (b) $1/6$

P. ~~500~~ 506

1. $\pi + \pi i$
2. 0
4. 0 in all four cases.

P. ~~506~~ 510

1. e^2
2. $2\pi i$
3. $\pi i/32$
4. $8\pi i e^{-2}/3$
5. $-\pi i$

P. ~~510~~ 517

1(a) $-\frac{1}{z} - 1 - z - z^2 - z^3 - z^4 - \ldots$, residue $= -1$

1(b) $\frac{1}{z-1} - 1 + (z-1) - (z-1)^2 + (z-1)^3 - \ldots$;
residue $= 1$

2. $\frac{1}{z^2} + \frac{1}{z} + \frac{1}{2!} + \frac{z}{3!} + \frac{z^2}{4!} + \ldots$

3. (a) $\dfrac{2\pi i}{n}$

(b) $\dfrac{2\pi i}{n}\left(1 - 2\cos\left(\dfrac{\pi}{n}\right)\right)$

4. $2\pi i$
5. πi
6. $-6\pi i$
7. $1 - \cos(t)$
8. $0 \quad t < r$

$-\dfrac{1}{\pi}\dfrac{1}{\sqrt{t^2 - r^2}} \quad t > r$

P. ~~517~~ 534

1. $\dfrac{ie^{iat}}{a\sqrt{b^2 + a^2}}$

2. $\dfrac{2\pi}{\sqrt{3}}$

3. $2\displaystyle\int_0^\infty \dfrac{1 + \sigma(y + 1)}{(1 + y)\sqrt{y^2 - 1}}\, e^{-\sigma y}\, dy$

4. $\dfrac{1 + a}{\sqrt{a}} - \dfrac{1}{\pi}\displaystyle\int_0^1 \dfrac{1 - t}{\sqrt{t}}\, \dfrac{dt}{t + a}$

5. $\pi i\left(\dfrac{2 - a}{\sqrt{2}}\right)$

6. $0 \quad n$ even

$\dfrac{2(-1)^{(n+1)/2}}{\sqrt{1 + a^2}\,(a + \sqrt{1 + a^2})^n} \quad n$ odd

Index

F

G

H

I

J

K

L

M

N

O

P

R

Edited by Roland Phelps

Notes

Notes

Notes

Notes

Notes

Notes

Notes

Notes